Preface

The first edition of this book evolved from lectures given at the University of Illinois during a one-semester course on light. The students were juniors, seniors, and first-year graduate students in physics, electrical engineering, mechanical engineering, and chemistry. The text was intended to serve in a one-semester to one-year course at the advanced undergraduate/first-year graduate level.

The second edition has been significantly revised throughout. Our main objective was to soften the approach and introduce consistency while maintaining a connection to rigorous concepts. The result is a book that is useful as a text as well as a general reference on the fundamentals of optics. Our revisions are based on five years' experience teaching a course in geometrical and physical optics at Rensselaer Polytechnic Institute. This course was taken by undergraduates— primarily sophomores—during the second semester of the academic year.

Optics presents an introduction to classical concepts of geometrical and physical optics. These are discussed with reference to fundamental theories of light: Fermat's principle, Huygen's principle, and Maxwell's equations. Everything from attenuated total reflection and geometrical aberration theory to spatial filtering, Gaussian beam optics, and statistical fluctuations is covered. In addition, we have presented the conventional groundwork for understanding practical optics: image formation, optical instruments, interference, diffraction, and polarization. The reader is given enough of the principles behind practical optical components and systems so that he or she can do effective laboratory work. With the background presented here, the student can take the next step: to enter into advanced treatments and the optics literature. The most significant omission of this book is the quantum theory of the interaction of light with matter. Consequently, there is no detailed discussion of laser action. However, this revised edition presents the most comprehensive elementary treatment of the optical processing of coherent light and Gaussian beams currently available in any textbook.

Some of the topics worked out in *Optics* will be found in few introductory textbooks. In Chapter 3, specific ray-tracing techniques are presented in the rigorous and paraxial limits. This information can serve as the basis for computerized ray-tracing techniques, and it leads to the use of the matrix technique for the presentation and exploitation of the concepts of lens action in the paraxial limit in the remainder of Chapter 3.

Our treatment of lens aberrations in Chapter 4 is the most straightforward and complete treatment available. Specific formulas are provided for the primary aberrations of a thin lens. Our approach to multiple-reflection interference in Chapter 5 is based on transfer matrices but has infrequently appeared in textbook form. It is a powerful formalism that can be computerized to deal with very complex problems such as the design of an interference filter (a problem presented at the end of the chapter). The treatment of diffraction in Chapters 6 and 7 is based on the transformation concept as found in advanced theory. However, it is presented here in its most simple and consistent form. This information can be directly applied in real problems that a practicing scientist or engineer might encounter.

Several specific modifications and additions in the revised edition are worthy of note.

- The SI system of units is used throughout.

- All introductory theory is condensed into Chapter 1, which is presented from an historical perspective.

- All the material relating to the interaction of light with matter is reorganized into Chapter 2.

- The matrix convention in Chapter 3 has been changed to conform to the most commonly found standard. More examples of optical imaging have been included, and a redundant treatment of image formation theory has been eliminated.

- The section on aberrations in Chapter 4 has been entirely rewritten.

- Chapter 5, which deals with interference, is new. The matrix method of multiple beam interference and many more examples of the application of interference are provided. Grating phenomena have been moved to this chapter as well.

- The details of Fresnel-Kirchhoff theory in Chapter 6 have been removed and placed in an appendix. Fourier mathematics is presented as a separate section within this chapter.

- More advanced topics in diffraction are found in Chapter 7. The notation has been simplified so as to bring out the easily understood transformation characteristics of the theory. This chapter contains new material on Gaussian beam optics.

- All material involving partial coherence is contained in Chapter 8, including incoherent image formation.
- Approximately half of the figures have been redrawn to emphasize clarity.

The revised version contains a significantly enlarged and broadened collection of problems at the ends of the chapters. There are enough different kinds of exercises to supplement instruction in a wide variety of courses.

The assumed prerequisites for a course taught from this text are introductory physics—including exposure to ideas of electricity, magnetism, and wave motion —and introductory calculus. Differential equations are discussed but only as a connection to wave theory. It is not expected that students be able to solve differential equations. Material from this text has been used with little difficulty in the course taught to sophomores at Rensselaer.

Instructors wishing to use this book for a one-semester course might follow these guidelines: Introductory theory (sections 1.1-1.5); light-matter interactions (sections 2.1.C, 2.2.B-2.2.E, 2.3); image formation and optical instruments (sections 3.1.A.1, 3.2.A, 3.3-3.5); stops (section 4.1.A); interference (5.1-5.6); far-field diffraction (sections 6.1, 6.2); near-field diffraction (sections 7.1, 7.2), and polarization (sections 9.1, 9.2).

We acknowledge the help of many colleagues and students for suggestions and remarks, especially R. D. Sard and H. Macksey. We are grateful to Nila Meredith, Nancy Fowler, Darcy Sorocco, and Geri Frank for their careful typing assistance, and to Marc de Peo for making some of the diffraction photographs. But most of all, we thank our families for their understanding and support during this project.

MILES V. KLEIN
Urbana, Illinois

THOMAS E. FURTAK
Troy, New York
August 1985

Contents

1 The Nature of Light

The study of optics covers those phenomena involving the production and propagation of light and its interaction with matter. Throughout history, philosophers and scientists have tried to explain what light is; in so doing, they have tested their evolving knowledge of our physical world. Although many early ideas have been proven false, others have been repeatedly verified by experimental tests. Among these are the concept of the finiteness of the speed of light; the principle of least time, which applies to the path of propagation; and the idea that light behaves like a wave.

1.1 Early Ideas and Observations

It is difficult for us to appreciate the mystery surrounding the nature of light and vision in the ancient world. Not only was the mechanism of the eye unknown, but fundamental optical principles that we take for granted were also obscure. In spite of this, motivated by interests in geometry, art, and deception (magic), the Greeks developed some relatively sophisticated notions.

A. Rectilinear Propagation

The earliest surviving optics record, Euclid's *Optics* (280 B.C.), recognized that *in homogeneous media, light travels in straight lines*. However, following the teaching of Plato, Euclid thought that "rays" of light originate in the eye and intercept those objects which end up being seen by the observer (Fig. 1.1). To the ancient philosopher, light was synonymous with vision. The speed with which the rays were thought to emerge from the eye was known to be very high, if not infinite. An observer with eyes closed could open them and immediately see the distance stars.

1

Fig. 1.1 The Greek impression of light was geometrical. All within the cone of vision was seen. Outside the cone, light had no meaning.

Hero of Alexandria in his *Catoptrics* (during the first century B.C.) also rationalized that, because light travels with infinite speed and therefore constant velocity, it must move in straight lines. This conclusion was based on analogy with mechanical events in which the concept that we now recognize as inertia plays an important role.

In the geometrical tradition of the Greeks, Hero identified the shortest point-to-point path that the light, by nature of its large velocity, was required to follow—the straight line. This shortest path concept is the earliest of the fundamental ideas concerning light that remain valid today. The underlying reason why light chooses the shortest path (or more rigorously the extremal path) was not understood, of course, until much later. (We will talk about that in due time.) This is a geometrical concept. Therefore, with regard to choosing the proper path, it makes no difference whether the light travels from the eye to the object or from the object to the eye or, for that matter, if the propagation is instantaneous. This is why Hero's idea worked, even though he too thought that the eye was the originating element and that the propagation speed was infinite.

We now understand that rectilinear propagation is not rigorously true because, through diffraction, light can bend around corners and, as explained by general relativity, light can be deflected by a strong gravitational field. This was not known to the ancient Greeks because the effects of diffraction are small and those described by relativity are observable only with sophisticated techniques and advanced instrumentation.

Today the *law of rectilinear propagation* has become one of the three principles of what we call "geometrical optics." The other two are the *law of reflection* and the *law of refraction*. The geometrical treatment is a phenomenological nonrelativistic approach wherein light is characterized by static rays and wherein the objects with which the light interacts are relatively large.

B. Reflection

At an interface between two different homogeneous optical media, incident light is, in general, partially transmitted and partially reflected. The interface might be a plane or a curved surface. In either case, the *surface normal* (the line perpendicular to the interface) at the point where an incident ray meets the interface is uniquely defined (see Fig. 1.2). The surface normal and the incident ray define a plane, the *plane of incidence*. The *law of reflection* states: *The reflected ray lies in the plane of incidence, and the angle of reflection,* θ'' *equals the angle of incidence,* θ.

The quantitative nature of the law of reflection was known in Aristotle's time and is documented in Euclid's book. Hero applied his "shortest path" principle to reflection and was able to geometrically prove the equality of the angles. Figure 1.3 reproduces the steps in Hero's proof. The initial conditions are illustrated in (*a*), where the interface AB and the surface normal OC are identified. The plane of incidence is the plane of the diagram. The incident ray forms angle θ with the normal. In (*b*), the reflected ray is constructed such that $\theta'' = \theta$ and two equal right triangles are formed. The path $\overline{POP''}$ must be demonstrated to be the shortest of all possible optical paths involving reflection at the interface. To achieve this, Hero drew the extension of $P''O$ to S as shown in (*c*), thus creating four congruent right triangles. Note that the marked lines PO and SO are equal in length. Thus $\overline{SOP''} = \overline{POP''}$.

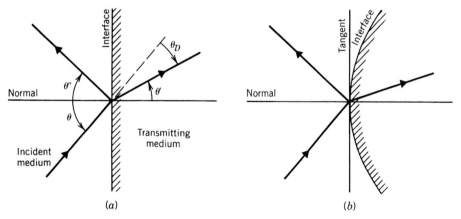

Fig. 1.2 Geometry of reflection and refraction at (*a*) a planar interface and (*b*) a curved interface.

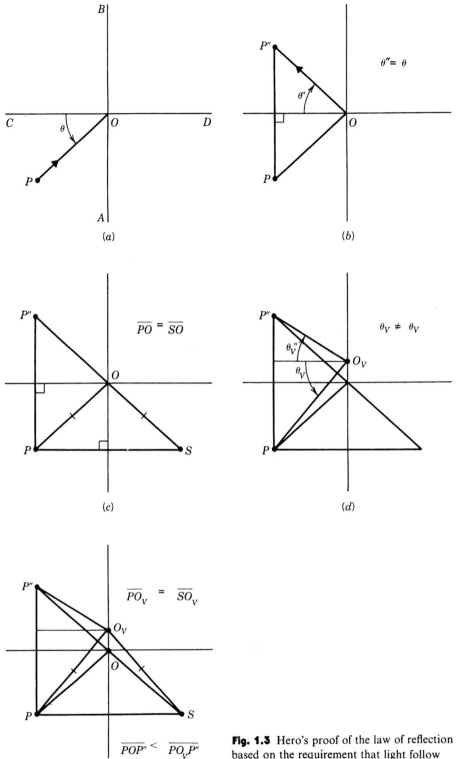

Fig. 1.3 Hero's proof of the law of reflection based on the requirement that light follow the shortest path.

If the law of reflection were not true, then other points, O_V (subscript V for "virtual" path), could be found and the resulting ray $PO_V P''$ would identify $\theta_V'' \neq \theta_V$, as shown in (d). In (e), $\overline{SO_V}$, which must be equal to $\overline{PO_V}$, is drawn. Thus $\overline{SO_V P''} = \overline{PO_V P''}$. However, it is clear that $\overline{SOP}$ (or $\overline{POP''}$) is less than $\overline{SO_V P}$ (or $\overline{PO_V P''}$), independent of the location of O_V on the interface, including points out of the plane of the diagram. Therefore, the ray that satisfies the law of reflection is also the shortest possible reflecting path.

The relative *distribution* of light *intensity* between the reflected component and the transmitted component was not properly explained until the 19th century. However, the more modern theory verifies that Hero's intuitive concept of light propagation along the "shortest path," and its consequences for geometrical optics, are correct.

C. Refraction

The geometry of refraction was experimentally studied by Claudius Ptolemy (100–170) and is reported in book V of his *Optics*. He recognized that the angle of deviation, θ_D in Fig. 1.2a, depended on the difference in density between the media forming the interface. He documented the quantitative relationship between θ' and θ, arriving at the empirical result that $\theta' = a\theta - b\theta^2$, where a and b are constants that depend on the two media. This expression approaches the correct result when θ is very small. However, for larger angles of incidence, this is, of course, not correct. In spite of its inaccuracy, Ptolemy's picture of refraction persisted for nearly 1500 years! People were simply not motivated to seek the correct answer.

Things began changing after A.D. 1280, when Italian artisans accidently discovered the spectacle lens. This development was still regarded as a curiosity by the educated community until Galileo Galilei (1564–1642), the famous Florentine mathematician, began experimenting with combinations of lenses that he ground for himself around 1609. Although the telescope was known before that time, Galileo was the first intellectual to take it seriously. Using his own instrument, which consisted of really good lenses, he discovered the moons of Jupiter and a host of other heavenly wonders.

Johannes Kepler (1571–1630), the great mathematician, optician, and astronomer at Cologne, Germany, for the first time summarized much of the known work with his *Dioptrice* (1609). This work was written after he had verified the discoveries of Galileo. It contains the theory of lenses and lens combinations. Kepler recognized that, provided the angles were small, the phenomenon of refraction followed the relation $\theta' = N\theta$ (where N is a constant depending on the two media). The resulting formalism is similar to that which we use today under the same limitation.

The true *law of refraction* ensures that: *The transmitted ray is in the plane of incidence, and the appropriate angles are related by* $\sin \theta' = N \sin \theta$.

Hero of Alexandria was able to derive the law of reflection using the shortest path principle in the first century B.C. The application of this principle in refraction is more complicated than in reflection. It requires knowledge about the finiteness of the speed of propagation of light, and how the speed depends on the propagation medium. This was one of the most confusing issues surrounding the theory of light. It was not satisfactorily worked out until 14 years after Kepler's death.

D. Theory of Light

Before the 17th century, knowledge about light was truly in the "Dark Ages." Although the Greeks had made significant progress with geometrical models in their time, this information, along with their other contributions, was suppressed in the years following the decline of Greek influence.

While the Western world struggled with barbarism, intellectual activity continued in the East along somewhat independent lines. Abu Ali Mohamed Ibn Al Hasan Ign Al Haytham (965–1039), or Alhazen for short, wrote a collection of seven books on optics in Baghdad around the year 1000. These are noted for their insightful comments concerning several key concepts.

Alhazen recognized that light sources illuminate objects, after which the light from the object is detected by the eye. He had a very good idea of how the optics of the eye worked. He described the operation of a "camera obscura." This was 500 years earlier than Leonardo da Vinci (1452–1519), who is usually credited with the discovery of the *pinhole camera* and its demonstration of the rectilinear propagation of light. In addition to these observations, Alhazen correctly hypothesized that light travels with a finite speed and that the speed is smaller in more dense media.

His physical picture was not correct, however, as it depended too much on

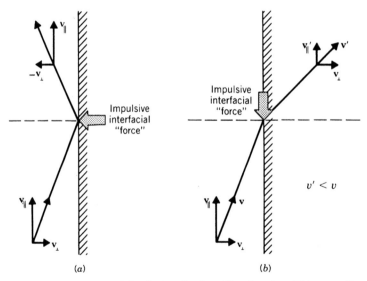

Fig. 1.4 Alhazen's models of (*a*) reflection, (*b*) refraction. This, as well as any other classical particle model, is conceptually incorrect, although Alhazen understood finite propagation and speed decrease in dense media.

mechanical analogies. Alhazen had the idea of light as a stream of particles that were subjected to surface forces on reflection and refraction. Refracted light particles were thought to be influenced by the forces that were only parallel to the surface, as shown in Fig. 1.4. He had many of the right answers, but rationalized them with incorrect reasons. The maturity of Alhazen's arguments influenced creative thought for more than 500 years, although the evolution of later ideas was not always direct.

The issues involving the nature of light were then and are now: (1) its speed of propagation, (2) the cause of the sensation we call color, (3) its tendency to travel in straight lines (rectilinear propagation), (4) the law of reflection, and (5) the phenomenon of refraction, whose "law" was not discovered until later. Any theory of light must deal with all of these characteristics. In addition, a comprehensive theory must deal with the phenomenon of interference and diffraction, which were not known at the time. It must also be able to explain the subtleties of relativistic effects and the details of light/matter interactions, which were not revealed until the 19th and 20th centuries.

1.2 The Particle Models

Further development of optical theory took place by virtue of the ideas of three prominent individuals: René du Perron Descartes (1596–1650, France), Pierre de Fermat (1601–1665, France), and Isaac Newton (1642–1727, England). There are fundamental differences in the philosophies of natural phenomena espoused by these early physicists. Most notable among the constrasts are the Cartesian (geometrically oriented) versus the Newtonian (force-oriented) points of view. These three early physicists are grouped together here because particle dynamics played an important role in their explanations of light. We have already mentioned that mechanical models are inadequate; however, Descartes and Newton had such dominating influence in their day that any study of optics is incomplete without some appreciation of their ideas. Among their lasting contributions—Descartes was the first to publish the correct form of the law of refraction, and Newton first explained refraction's chromatic character. Fermat developed the *principle of least time*, which was similar to Hero's shortest path principle. This, we have said, is a fundamental concept that is reinforced by modern theory.

A. Descartes

René Descartes' notions about light were consistent with his impressions of the physical world. To him, all things were related to geometry. Motion was the one fundamental "power" in nature. The only type of motion was that whereby a body passed from one geometrical state to another by successive steps. Motion could be communicated from one body to another only by impact. Cartesian matter was infinitely divisible and incompressible (because a void was thought to be impossible). These were, to Descartes, a priori truths.

With this background we can understand the Cartesian theory of light. According to Descartes light was a *tendency* toward movement that was transmitted through the all pervading medium, the "ether." This is similar to the way pressure is transmitted through a stick. To Descartes, this "tendency" followed the same laws that movement itself would follow. The "tendency," which Descartes equated with light, was propagated instantaneously, but mechanical movement analogies that required a finite time to evolve were used in discussing how the light would behave.

In 1637 Descartes published *La Dioptrique* in which the laws of optics were derived from his a priori truths. To discuss the action of light on encountering an interface, Descartes compared light to a mechanical particle. As an example he chose a tennis ball. In both reflection and refraction, the component of the velocity of his mechanical analog parallel to the interface was assumed to remain constant (see Fig. 1.5). The light, as a "tendency" toward motion, in reflection would have to follow the same path as that of a perfectly elastic rebounding ball.

In the mechanical analogy for refraction, the interface was a frail canvas. The speed after the encounter with the interface was assumed to be directly proportional to the initial speed $Nv' = v$. Conservation of the parallel component required $v' \sin \theta' = v \sin \theta$. Together these relations led Descartes to the law of refraction, which turns out to be experimentally correct: $\sin \theta' = N \sin \theta$. However, if the parallel component of the velocity were to remain constant, this required that the perpendicular component of the velocity be increased after refraction if the tennis

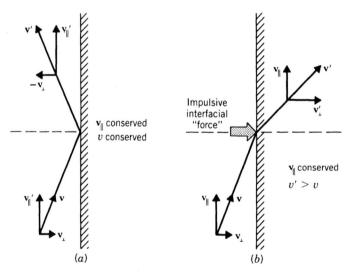

Fig. 1.5 Descartes' models of (*a*) reflection, (*b*) refraction. This shows the incorrect mechanical analogy that was required to explain refraction if $\mathbf{v}_\parallel$ was conserved. Light, as a "tendency" toward motion, was thought to follow the path of the analogy, but with infinite speed.

ball were to act as light does at an air-glass interface. For a tennis ball to behave like this, it would have to receive an impulse from the canvas. If Descartes had been more objective at this point, he would have recognized that for a real tennis ball this is impossible. He broke his own rules by perturbing the analogy to suit the phenomenon of refraction. Descartes' law was correct. Its theoretical foundation, however, was wrong.

The law of refraction had been independently discovered in 1621 by Willebrod Snell (1591-1626) in Leyden. Snell's work was not immediately published, and it was much more empirical than Descartes'. Although there is much controversy about plagiarism, Descartes was apparently unaware of Snell's work. Nevertheless, we in English-speaking countries still refer to the law of refraction as "Snell's law." The French refer to it as "Descartes' law."

B. Fermat

Pierre de Fermat had difficulty accepting Descartes' hypothesized relationship of light to tennis balls. He was most disturbed with the idea that light travels instantaneously.

[After Fermat, in 1676, Olaf Römer (1644-1710) reported on his observations, made in Denmark, of the innermost satellite of Jupiter. These demonstrated that Descartes' ideas were wrong and that light had a finite velocity. Römer's method was based on the fact that the period of the satellite was longer when the earth was receding from Jupiter. From this he concluded that light must take 11 minutes to travel a distance equal to the radius of the earth's orbit.

In 1664 Fermat wrote a letter in which he observed that nature always acts by the shortest path. He assumed that the reason the shortest path was preferred was that, for nature, this was the easiest course. This was an adaptation of Hero's principle. We have already seen how Hero was able to justify rectilinear propagation and the law of reflection. In the application of the same principle to refraction, Fermat introduced the concept of optical resistance as a characteristic of the medium. He presumed that light did possess a finite propagation speed and that this speed was inversely proportional to the optical resistance. He further supposed that the optical resistance was proportional to the density of the medium. Although the word "time" does not appear in Fermat's correspondence, his theory is frequently called the principle of least time. This elegant concept remains today as a fundamental starting point for geometrical optics. We will discuss *Fermat's principle* in more detail.

C. Fermat's Principle

Fermat did not have the advantage of knowing the calculus that was invented shortly after his time. His proof involved complicated algebraic expressions that he had difficulty reducing. Although Descartes had derived the expression for the law of refraction earlier, Fermat wished to find the true law from his own ideas. After several years of work, Fermat was astounded when he came up with the same law of refraction as had Descartes, whose philosophy he could not accept. Fermat's starting assumptions were correct whereas Descartes' were not. Descartes had to maintain an unphysical picture to arrive at the experimentally verified result. Calculus is ideally suited to handle Fermat's principle. Thus, although Fermat did not use it we will.

1. Optical Path Length. Fermat's optical resistance has become what we now call the *index of refraction*. We now also understand that a vacuum can exist and that in the vacuum light has a unique speed $c = 299{,}792{,}458$ m/s. If the speed of light in an optical medium (nonabsorbing) is v, then the index of refraction is defined by

$$n = \frac{c}{v} \tag{1.1}$$

The *optical path length* from point P to point P' is defined to be the line integral

$$\text{OPL } (PP') \equiv \int_P^{P'} n \, d\ell = \int_P^{P'} \frac{c}{v} \, d\ell \tag{1.2}$$

where $d\ell$ is a line element along the physical path of the light, the ray, from P to P'.

If the index of refraction is a function of position, $n = n(x, y, z)$, the OPL is still given by Eq. (1.2). Explicitly we would have

$$\text{OPL } (PP') = \int_P^{P'} n(x, y, z) \, d\ell$$

If the index of refraction is a constant, then $\text{OPL} = n\Delta\ell$ and the OPL is directly proportional to the geometrical distance $\Delta\ell$ along the ray.

The speed of light in the medium can also be expressed as $v = d\ell/dt$, so that along the ray one can write $c \, d\ell/v = c \, dt$ and hence,

$$\text{OPL } (PP') = \int_P^{P'} c \, dt = c\Delta t \tag{1.3}$$

Thus the OPL is proportional to the travel time for light along the path.

2. Mathematical Form for Fermat's Principle. Suppose that we have two points, P and P', and we wish to know the path of the light ray or rays that goes through both of them (see Fig. 1.6). Fermat's principle states: *Any deviation of the path from that taken by the true ray that is of first order in small quantities, will produce a deviation in optical path length that is at least second order in small quantities.*

The deviant path is known as a *virtual path*, since the light does not actually propagate along it. More precisely, the difference

$$I \equiv \text{OPL (virtual path)} - \text{OPL (true ray)} \tag{1.4}$$

must be at least of second order in the parameters that measure the deviation of the virtual path from the true ray (Fig. 1.6). This means, for instance, that if the

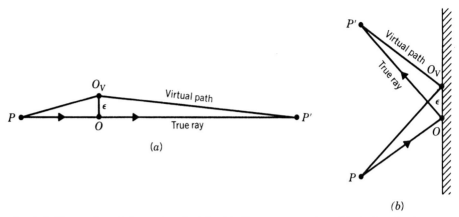

Fig. 1.6 Illustration of Fermat's principle: (a) direct propagation, (b) single reflection.

parameter ε is a measure of the maximum separation of the virtual path from the true path, then we must have, in the limit as $\varepsilon \to 0$,

$$I(\varepsilon) = \text{const} \cdot \varepsilon^2 \tag{1.5}$$

or a higher power of ε. The parameter ε must be a continuous "function" of the virtual path (such as "function" is called a *functional*) that equals zero when the virtual path coincides with the true ray.

A precise formulation of Fermat's principle requires the use of the calculus of variations and will not be presented here. The main point to be emphasized is that one can always define a quantitative measure of how much the virtual path deviates from the true ray.

Once the virtual rays have been defined in terms of ε, we can reexpress Eq. (1.4) in terms of a Taylor series.

$$I(\varepsilon) = \frac{d[\text{OPL}]}{d\varepsilon}\bigg|_{\varepsilon=0} \varepsilon + \frac{1}{2} \frac{d^2[\text{OPL}]}{d\varepsilon^2}\bigg|_{\varepsilon=0} \varepsilon^2 + \text{higher-order terms} \tag{1.6}$$

Comparing Eq. (1.6) with the statement of Fermat's principle in Eq. (1.5), we see that the principle is equivalent to the equation

$$\frac{d[\text{OPL}]}{d\varepsilon}\bigg|_{\varepsilon=0} = 0 \tag{1.7}$$

Thus the true ray is that for which the optical path is an *extremum* with respect to small deviations in optical path length. In Hero's words, we must seek the "shortest path." However, there may be more than one allowed path from P to P'. An example is shown in Figure 1.7. Each of the paths shown is an extremum. This is all that is required by Fermat's principle. It does not demand an absolute maximum or minimum.

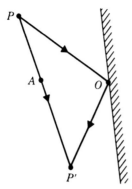

Fig. 1.7 The path $\overline{POP'}$ is a local minimum; $\overline{PAP'}$ is an absolute minimum.

When reflections or refractions are involved, the actual path followed by the ray may be either a local maximum or a minimum—the key point is that it must be *stationary* with respect to small changes in OPL.

3. Fermat's Principle and the Law of Refraction. We have already seen how Hero's shortest path principle led to the laws of rectilinear propagation and reflection, and how this is equivalent to Fermat's principles in uniform space. Fermat's contribution was to apply the principle to deduce the law of refraction.

In refraction, as in reflection, the optical laws are not restricted to planar interfaces. However, we can study these fundamental laws at planar surfaces without loss of generality because, at a point, real surfaces possess a well-defined tangent plane and surface normal (Fig. 1.2b). In the use of Fermat's principle, as we consider deviations from the true ray, we will investigate optical paths whose intercepts with the interface are infinitesimally removed from the true ray intercept. In the limit as the infinitesimal deviations go to zero, the surface formed by the locus of intercepts will always approach the tangent plane. We will therefore develop our proof using a planar interface.

In reflection, it is easy to see that the shortest path is that for which the incident ray, the reflected ray, and the surface normal all lie in the same plane, the plane of incidence. A similar restriction exists for the transmitted ray. As shown in Fig. 1.8, the shortest OPL lies in a plane perpendicular to the tangent plane.

Fig. 1.9 shows a source at P in the incident medium and an observation point at P' in the transmitting medium, both of which are fixed. The true ray passes through O at $(x, 0)$, while a virtual ray intercepts the interface at O_V, a distance ε away from O along the interface. With this geometry

$$\text{OPL}(PO_VP') = n\sqrt{(x + \varepsilon)^2 + z^2} + n'\sqrt{(x' - x - \varepsilon)^2 + z'} \qquad (1.8)$$

Fermat's principle requires that this be an extremum at $\varepsilon = 0$, as demonstrated by Eq. (1.7),

$$\frac{d[\text{OPL}]}{d\varepsilon}\bigg|_{\varepsilon=0} = \frac{nx}{\sqrt{x^2 + z^2}} - \frac{n'(x' - x)}{\sqrt{(x' - x)^2 + z'^2}} = 0 \qquad (1.9)$$

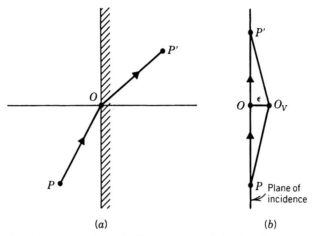

Fig. 1.8 In refraction, the light rays remain in the plane of incidence as required by Fermat's principle: (a) "side-on" view, (b) "top" view of the same refraction showing the longer virtual path $\overline{PO_V P'}$.

Figure 1.9 has been redrawn for clarity in Fig. 1.10, where the angular features are emphasized. There it can be seen that

$$\frac{x}{\sqrt{x^2 + z^2}} = \sin \theta$$

and

$$\frac{(x' - x)}{\sqrt{(x' - x)^2 + z'^2}} = \sin \theta'$$

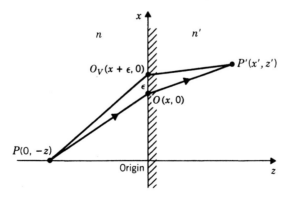

Fig. 1.9 Fermat's principle in refraction. PO is the true ray.

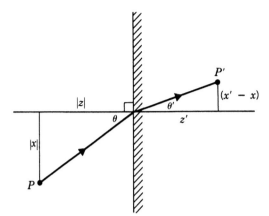

Fig. 1.10 Adaptation of Fig. 1.9.

Comparing this information with Eq. 1.9, we arrive at the law of refraction (Snell's law).

$$n' \sin \theta' = n \sin \theta \qquad (1.10)$$

Note that although this takes the same form of the law as derived by Descartes, $\sin \theta' = N \sin \theta$, the constant in Descartes' form is $N = v/v'$. The correct expression for the constant, from Eq. (1.10) is $N = n/n' = v'/v$.

4. Philosophical Basis for Fermat's Principle. We have not yet explained how the light "determines" the correct path. Light does not "know" ahead of time what the path of the true ray will be. It does not, as we have done following Fermat, "investigate" alternative paths. Propagation proceeds without hesitation along the path of the true ray. We must understand that Fermat's principle is a mathematical statement about the characteristics of nature. It is not the underlying cause but an empirical observation. The true mechanism behind Fermat's principle has to do with constructive interference of waves. In the classical theory of light, the waves are variations in electromagnetic fields. In the quantum theory of light, the waves are variations in photon probability amplitude. It turns out that extremal optical paths correspond to directions along which constructive interference is maximized. Fermat's principle can also be rationalized as a consequence of the structure of space and time. These details were not discovered until long after Fermat.

D. Newton

Isaac Newton was a giant of science. He was strikingly successful in developing a theory of mechanics that is still valid within its now recognized limits of applicability. It was natural that he attempt to apply his understanding of dynamics to the description of light. In doing so, he became the last of the strong proponents of the classical particle model. In 1704, he published *Opticks*, which became accepted as the most accurate description of light. Although there was considerable discussion in Newton's time about the suitability of a wave picture,

Newton's influence was so dramatic that further progress on this issue was delayed for nearly a century.

In *Opticks*, Newton described the results of careful experiments concerning refraction and what we now know as interference. One of the most famous of these is his two-prism experiment. Using light from the sun passing through a circular hole in a window shade, he showed that light of a given color underwent a deviation of the same angle at both prisms. This showed that the phenomenon of color was an intrinsic quality of light and that white light was a combination of rays of all the colors.

Newton thought that the rays were streams of particles that moved through an all pervading ethereal medium. The "size" of the particle was supposed to be related to the color of the light. He further hypothesized that the density of the ether was inversely proportional to the mass density of the matter through which the light was propagating. This idea was needed to rationalize the incorrect assumption that the velocity of the light particles was greater in more dense material. Newton believed that forces perpendicular to the interface acted on the light particles on reflection and refraction.

Newton is also famous for his observation of "Newton's rings" (see section 6.4D), an interference phenomenon that was actually reported by Robert Hooke 39 years earlier. Newton documented the details of the pattern of bright and dark rings that he observed in reflection from a spherical glass surface in contact with a flat glass surface (Fig. 6.22). He correctly concluded that the thickness of the air gap between the glass surfaces had an important role to play in the pattern. He showed that the dark rings appeared where the air gap was an even multiple of 1/89000th of an inch. In his interpretation, Newton really had to push his particle model. He hypothesized that on passing through the top interface into the air gap, the light was placed in alternating "fits" of easy transmission and reflection. When the light reached the bottom interface, depending on its "fit," it was either transmitted or reflected.

Newton's rings are caused by wave interference. Newton refused to believe in the wave picture primarily because of his thoughts concerning rectilinear propagation. According to Newton, if light were a wave, like sound, it would diffuse around obstacles and no shadows could be formed. We have already noted, however, that rectilinear propagation is not a "law" and that light does "bend" around corners in the phenomenon of diffraction. Early diffraction experiments were known to Newton, and he performed some of his own. However, his bias in favor of the particle model was too strong. He thought that the "bending" of a ray was caused by an attractive interaction between the light particles and the edges of a diffracting aperture.

1.3 The Wave Models

We must understand that the whole of geometrical optics can be correctly formulated without resorting to waves by starting with Fermat's principle.

However, interference, diffraction, and other physical optics phenomena require a wave picture.

A. The Mathematics of Wave Propagation

Before revealing the development of thought concerning the wave theory of light, we will review wave phenomena in general. We define a wave to be a traveling disturbance. Our first case has only one independent spatial variable, which represents the distance along the direction of propagation of the disturbance. The dependent variable, which represents the quantity doing the "waving," will be *transverse* and thus represent motion at right angles to the direction of propagation. We will eventually prove that light bears this transverse character.

It is appropriate here to use mechanical analogies to develop the wave formalism. The mathematics is the same as that for light, but the physical concepts are easier to understand. In transverse mechanical wave motion, the information is carried by a characteristic pattern of movement that is passed along in the medium from one location to the next (Fig. 1.11). We model this behavior using a one-dimensional chain of particles, which execute this transverse motion in turn according to Newton's laws.

1. Simple Oscillation. Each particle of the chain exerts forces on its neighbor particles. In this way, the movement is communicated from particle to particle to produce the wave motion. If we could somehow immobilize all the particles except one, then we could study its motion under the influence of the forces exerted by the other particles. Of particular importance is the special case where, in the thought

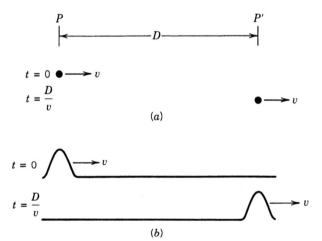

(a)

(b)

Fig. 1.11 Information propagation by (*a*) a moving particle and (*b*) a transverse wave. In (*b*), the "medium" passes the information from P to P' without itself moving along the direction from P to P'.

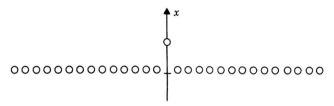

Fig. 1.12 A one-dimensional "medium" of particles, all of which are held immobilized except one.

experiment just described, the force that tends to restore the particle to its equilibrium position obeys Hooke's law, that is, is proportional to the particle's displacement from equilibrium, $F = -Cx$, where C is a constant (Fig. 1.12). If the particle's mass is M, Newton's second law of motion requires:

$$F = M \frac{d^2x}{dt^2} = -Cx$$

or

$$x + \frac{M}{C} \frac{d^2x}{dt^2} = 0 \qquad (1.11)$$

This second-order linear differential equation has a solution given by

$$x = x_0 \cos \phi$$

or

$$x = x_0 \, \mathrm{Re}[e^{i\phi}], \qquad (1.12)$$

where $\phi = 2\pi v t + \varphi$ is the *phase* of the motion, changing linearly with time, and φ is the *epoch angle*, or initial phase, which depends on the value of x at $t = 0$. Here v is the *frequency* of the oscillation with

$$v = \frac{1}{2\pi} \sqrt{\frac{C}{M}} \qquad (1.13)$$

The constant x_0 is the *amplitude*, which represents the maximum deviation from the equilibrium position.

   Although this might seem to be a special situation, the simple harmonic oscillator is actually the first-order approximation to many more complicated kinds of oscillations. This is true because a general restoring force can be expanded as a Taylor series about $x = 0$,

$$F = F \Big|_{x=0} + \frac{dF}{dx} \Big|_{x=0} x + \frac{1}{2} \frac{d^2F}{dx^2} \Big|_{x=0} x^2 + \cdots$$

If the displacement is small, then the dominating term is linearly proportional to x, just as we have assumed. The source of the force in a real medium would be the particle–particle interaction that is related to the cohesiveness of the medium. We need not know the details of this for our purposes.

2. Wave Equation. If we now relax the immobilization condition and allow all the particles of the chain to move, we arrive at the condition illustrated in Fig. 1.13. To simplify the complicated pairwise force law that correlates the motion of the particles with each other, we consider only near-neighbor interactions and assume that, in the first approximation, the restoring force is proportional to the combined difference in the x displacements between the nth particle and its two neighbors. Equation (1.11) becomes

$$(x_n - x_{n-1}) + (x_n - x_{n+1}) + \frac{M}{C} \frac{d^2 x_n}{dt^2} = 0 \qquad (1.14)$$

where we are specifying the motion of the particular particle labelled n in Fig. 1.13.

If all the particles are equally spaced, say by a distance b, then in going from $n - 1$ to n or from n to $n + 1$ we undergo a translation of $\Delta z = b$. We can reinterpret the difference terms in Eq. (1.14) as

$$(x_n - x_{n-1}) + (x_n - x_{n+1}) = -[(x_{n+1} - x_n) - (x_n - x_{n-1})] = -\Delta(\Delta x) \qquad (1.15a)$$

where Δ represents a change associated with integral changes in n. Rewrite the double difference as

$$\Delta(\Delta x) = \frac{\Delta(\Delta x)}{(\Delta z)^2} b^2$$

Now if b is very small the chain approximates a continuous one-dimensional medium and

$$\Delta(\Delta x) \rightarrow \frac{d^2 x}{dz^2} b^2 \qquad (1.15b)$$

Substituting Eqs. (1.15a) and (1.15b) into Eq. (1.14), we see that the equation of motion for a piece of the one-dimensional chain of particles is

$$\frac{\partial^2 x}{\partial z^2} - \frac{M}{b^2 C} \frac{\partial^2 x}{\partial t^2} = 0$$

We have used the partial derivative notation since x is a function of both z and t.

With some foresight into our final result we will set $b^2 C/M = v^2$, which for now can be thought of as just another constant. The new form for the equation of motion for a piece of the chain is now

$$\frac{\partial^2 x}{\partial z^2} - \frac{1}{v^2} \frac{\partial^2 x}{\partial t^2} = 0 \qquad (1.16)$$

Equation (1.16), a continuum equation, implies that it is possible to identify an arbitrary position z at which some mass will be found. Strictly speaking, there will always be gaps in a discrete particle model. However, as the gaps between particles are made smaller, the system becomes equivalent to a continuous medium.

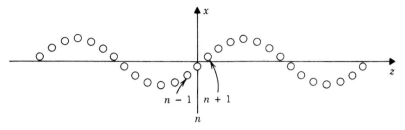

Fig. 1.13 A one-dimensional medium of particles wherein all particles are allowed to move parallel to the x-axis.

3. General Solution: Superposition Principle. The general solution of Eq. (1.16) takes the form

$$x(z, t) = f\left(t - \frac{z}{v}\right) + g\left(t + \frac{z}{v}\right) \tag{1.17}$$

where $f(u)$ and $g(u)$ are arbitrary functions of the variable u having well-defined second derivatives. By setting $u = t - z/v$ or $u = t + z/v$, we obtain a function of the two variables t and z. Before proving that Eq. (1.17) is a solution, we make a very important comment about the solutions of the wave equation (1.16) and almost all other wave equations, namely: *The sum of two solutions is also a solution.* This property of the solutions is often called the *superposition principle.* Using it, we can content ourselves with showing that the functions f and g in Eq. (1.17) separately satisfy the wave equation.

We give only the proof for f, which goes as follows. We differentiate $f(u)$ once with respect to z and t, where $u = t - z/v$.

$$\frac{\partial f}{\partial t} = \frac{df}{du} \quad ; \quad \frac{\partial f}{\partial z} = -\frac{1}{v}\frac{df}{du}$$

We differentiate again and obtain

$$\frac{\partial^2 f}{\partial t^2} = \frac{d^2 f}{du^2} \quad ; \quad \frac{\partial^2 f}{\partial z^2} = \frac{1}{v^2}\frac{d^2 f}{du^2}$$

These expressions can be directly combined to give Eq. (1.16). The proof for g proceeds in a similar way.

The two terms in Eq. (1.17) have the following simple interpretation. The term $f(t - z/v)$ describes a disturbance moving with velocity of magnitude v in the $+z$ direction with no change in size or shape. And $g(t + z/v)$ describes a similar undistorted disturbance moving with velocity v in the $-z$ direction.

To prove the statement about f, consider the displacment at time $t + \Delta t$ and compare it to the displacement at time t,

$$f\left(t + \Delta t - \frac{z}{v}\right) = f\left(t - \frac{z - v\Delta t}{v}\right) = f\left(t - \frac{z'}{v}\right)$$

where $z' = z - v\Delta t$. This means that the value of the function f at the point z and at

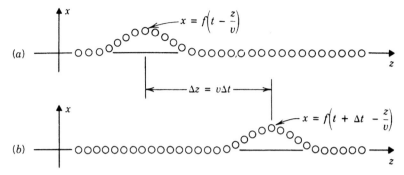

Fig. 1.14 Illustration of a disturbance moving in the $+z$ direction with velocity v. (a) The displacement at time t. (b) The displacement at time $t + \Delta t$. (b) is obtained from (a) by a rigid shift of the shape of the curve by an amount $\Delta z = v\,\Delta t$.

the later time $t + \Delta t$ is the same as it was at time t, not at the point z but at the point z', which lies to the left of z at a distance $v\Delta t$. Hence, the entire curve moves to the right with velocity v. This behavior is illustrated in Fig. 1.14. A similar argument applies to $g(t + z/v)$.

4. Harmonic Wave Solution. A particularly important special case among solutions to the wave equation (1.16) is the sinusoidal function of section 1.3A1. The form is the same as Eq. (1.12), only here the phase is given by

$$\Phi = 2\pi v\left(t \pm z/v \right) + \varphi \tag{1.18}$$

Each particle in the one-dimensional chain executes simple harmonic oscillation. If we took an instantaneous photograph of the chain, the displacement would be a sinusoidal function of z, as shown in Fig. 1.13. The solution is

$$x = x_0 \cos\left[2\pi v\left(t \pm \frac{z}{v} \right) + \varphi \right] \tag{1.19}$$

which can also be written as

$$x = x_0 \cos(\omega t \pm kz + \varphi) \tag{1.20}$$

where the *angular frequency* is $\omega = 2\pi v$ and the *wavenumber* is

$$k = 2\pi/\lambda = 2\pi v/v$$

The *wavelength* λ represents the crest-to-crest spatial distance between waves at a given time.

5. The Three-Dimensional Wave Equation. We have just discussed some of the important properties of wave motion in general that are already revealed in the one-dimensional case. Other important properties do not show up until we consider three spatial coordinates. We thus turn to an introductory discussion of waves and wavelike disturbances in three dimensions.

In three dimensions, the wave equation is generalized to

$$\frac{\partial^2 p}{\partial x^2} + \frac{\partial^2 p}{\partial y^2} + \frac{\partial^2 p}{\partial z^2} - \frac{1}{v^2}\frac{\partial^2 p}{\partial t^2} = 0 \quad \text{or} \quad \nabla^2 p - \frac{1}{v^2}\frac{\partial^2 p}{dt^2} = 0 \qquad (1.21)$$

Here $\mathbf{V}$ stands for the vector differential operator "del" or "nabla," whose coefficients in a rectangular coordinate system (mutually perpendicular axes x, y, and z specified by Descartes) are given by $(\partial/\partial x, \partial/\partial y, \partial/\partial z)$, and ∇^2 is an abbreviation for the scalar product or dot product of $\mathbf{V}$ with itself:

$$\mathbf{V} \cdot \mathbf{V} = \nabla^2 = \frac{\partial^2}{\partial x^2} + \frac{\partial^2}{\partial y^2} + \frac{\partial}{\partial z^2}$$

The function $p(x, y, z, t)$ could represent one of several physical quantities that can obey an equation such as (1.21). For instance, it could represent the pressure in an isotropic fluid, in which case the velocity v would be given by

$$v^2 = \frac{\beta}{\rho}$$

Here ρ is the fluid mass per unit volume and β is the bulk modulus (the reciprocal of the compressibility), defined as the ratio

$$\beta = \rho \frac{dp}{d\rho}$$

Light, as we will see, may be represented as electromagnetic waves. In free space each spatial component of the electric field $\mathbf{E}$ and the magnetic flux density $\mathbf{B}$ obeys an equation of the form of (1.21), with v replaced by the velocity of light, c. In addition to satisfying the wave equation, $\mathbf{E}$ and $\mathbf{B}$ must satisfy the equations of electromagnetic theory known as *Maxwell's equations*, which will be discussed in section 1.4. These extra equations imply that for the individual elementary sinusoidal components of the waves, the vectors $\mathbf{E}$ and $\mathbf{B}$ must be mutually perpendicular and also perpendicular to the direction of propagation. This *transverse* nature of light waves gives rise to various phenomena associated with the term *polarization*. Many other optical phenomena, however, do not depend on it. They can be described most simply by using a *scalar* wave theory such as that appropriate to sound waves in a fluid. Historically, the scalar model for light waves came first. In such a theory the vector nature of the optical disturbance (that is, the vector nature of $\mathbf{E}$ and $\mathbf{B}$) is overlooked (or is unrecognized as was the case in the 17th century) and the quantity that "waves" is treated as scalar. We will therefore use the scalar theory for much of our discussion and postpone the discussion of polarized light until later.

In the following few sections we will denote our dependent scalar variable by p. The reader may think of p as the pressure in a sound wave or as the electric field in a light wave. In the latter case, the results may not always be literally true because we have suppressed the vector nature of light. At any rate, our aim here is to discuss a few simple solutions of the three-dimensional wave equation (1.21).

We first remark that the superposition principle applies to the solutions of Eq. (1.21) for the same reason that it applies to the solutions of Eq. (1.16). Thus the sum of two solutions is also a solution, and complicated solutions can be obtained by the superposition of simple solutions.

6. Plane Disturbances and Waves. A simple situation is reached under the assumption that p is a function only of z and t. Then Eq. (1.21) reduces to Eq. (1.16) and has the same general solution, namely

$$p(z, t) = f\left(t - \frac{z}{v}\right) + g\left(t + \frac{z}{v}\right) \tag{1.22}$$

Here the disturbance is constant over each xy plane and consists of two parts, one propagating along the $+z$ axis with velocity of magnitude v (the f term) and one propagating along the $-z$ axis with velocity of magnitude v (the g term).

A *harmonic traveling plane wave* is a special case of Eq. (1.22) with, say, $g = 0$ and

$$p(z, t) = f\left(t - \frac{z}{v}\right) = A \cos\left[2\pi v\left(t - \frac{z}{v}\right) + \varphi\right] \tag{1.23}$$

Expressions exactly analogous to Eq. (1.22) could be written to describe uniform disturbances over the xz plane propagating along the $+y$ or $-y$ axis or unform disturbances over the yz plane propagating along the $+x$ or $-x$ axis.

What if the normal to the plane of the disturbance is not along a coordinate axis, but along an arbitrary direction having *unit vector* $\hat{s} = (\hat{s}_x, \hat{s}_y, \hat{s}_z)$ (where $\hat{s}_x^2 + \hat{s}_y^2 + \hat{s}_z^2 = 1$)? It is not difficult to see that a general plane disturbance propagating in the $+\hat{s}$ direction would be of the form

$$p(\mathbf{r}, t) = f\left(t - \frac{\mathbf{r} \cdot \hat{s}}{v}\right) = f\left(t - \frac{x\hat{s}_x + y\hat{s}_y + z\hat{s}_z}{v}\right) \tag{1.24}$$

First, we note that this expression is a solution of Eq. (1.21). Again let $u = t - \mathbf{r} \cdot \hat{s}/v$. We only have to notice that

$$\frac{\partial f}{\partial x} = -\frac{\hat{s}_x}{v} \frac{df}{du}$$

and that

$$\frac{\partial^2 f}{\partial x^2} = \frac{\hat{s}_x^2}{v^2} \frac{d^2 f}{du^2}$$

to obtain

$$\frac{\partial^2 f}{\partial x^2} + \frac{\partial^2 f}{\partial y^2} + \frac{\partial^2 f}{\partial z^2} = \frac{\hat{s}_x^2 + \hat{s}_y^2 + \hat{s}_z^2}{v^2} \frac{d^2 f}{du^2} = \frac{1}{v^2} \frac{d^2 f}{du^2} = \frac{1}{v^2} \frac{\partial^2 f}{\partial t^2}$$

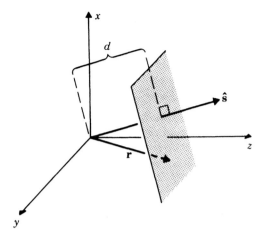

Fig. 1.15 The equation $\mathbf{r} \cdot \hat{\mathbf{s}} = d$ defines a plane perpendicular to the unit vector $\hat{\mathbf{s}}$ and a distance from the origin of d.

as required. As to the interpretation of Eq. (1.24), it gives a constant value for the disturbance p at a given value of time for all values of the position vector $\mathbf{r}$ obeying $\mathbf{r} \cdot \hat{\mathbf{s}} = \text{const}$. Such an equation defines a plane perpendicular to $\hat{\mathbf{s}}$. As the value of $\mathbf{r} \cdot \hat{\mathbf{s}}$ increases algebraically, the plane moves in the $+\hat{\mathbf{s}}$ direction (Fig. 1.15). If the system is harmonic, then an allowed plane wave solution is

$$p(\mathbf{r}, t) = A \cos\left[2\pi v\left(t - \frac{\mathbf{r} \cdot \hat{\mathbf{s}}}{v} \right) + \varphi \right]$$

or

$$p(\mathbf{r}, t) = A \, \text{Re}\{e^{i[2\pi v(t - \mathbf{r} \cdot \mathbf{s}/v) + \varphi]}\} \qquad (1.25)$$

We can simplify the notation, as was done for Eq. (1.20), by using the angular frequency $\omega = 2\pi v$ and by introducing the *wave vector* $\mathbf{k}$ where

$$\mathbf{k} = \frac{2\pi v}{v} \hat{\mathbf{s}} = \frac{\omega}{v} \hat{\mathbf{s}} = \frac{2\pi}{\lambda} \hat{\mathbf{s}} \qquad (1.26)$$

Therefore,

$$p(\mathbf{r}, t) = A \, \text{Re}\{e^{i(\omega t - \mathbf{k} \cdot \mathbf{r} + \varphi)}\} \qquad (1.27)$$

In using the exponential form for Eq. (1.27) in Eq. (1.21) and in other differential equations, we can ignore the "Re" restriction during the calculation if we remember that it is the real value of the final result that has physical significance. The exponential notation is a convenient mechanism with which to keep track of the phase of the function.

In *rectangular coordinates* the differential operators $\partial/\partial t$ and ∇ have a particular simple interpretation *when applied to an exponential plane wave function*. Consider

$$\frac{\partial}{\partial t}[A \, e^{i(\omega t - \mathbf{k} \cdot \mathbf{r} + \varphi)}] = i\omega \, A \, e^{i(\omega t - \mathbf{k} \cdot \mathbf{r} + \varphi)}$$

If wherever $\partial/\partial t$ is found we replace it by $i\omega$, the differential operator is replaced by its result. Likewise, consider one component of ∇ and

$$\frac{\partial}{\partial x}[A\, e^{i(\omega t - \mathbf{k}\cdot\mathbf{r}+\varphi)}] = -ik_x A\, e^{i(\omega t - \mathbf{k}\cdot\mathbf{r}+\varphi)}$$

Because $\nabla = \hat{\mathbf{x}}(\partial/\partial x) + \hat{\mathbf{y}}(\partial/\partial y) + \hat{\mathbf{z}}(\partial/\partial z)$, we can replace ∇ in differential operations on exponential plane waves by $-i\mathbf{k}$. Whenever the entire space-time dependence of a wave is described by the factor $\exp i(\omega t - \mathbf{k}\cdot\mathbf{r})$, then the space-time derivatives may be replaced by

$$\frac{\partial}{\partial t} \to i\omega \tag{1.28}$$

$$\nabla \to -i\mathbf{k} \tag{1.29}$$

Using these techniques, Eq. (1.21) becomes

$$(-i\mathbf{k})\cdot(-i\mathbf{k})p - \frac{1}{v^2}(i\omega)^2 p = 0$$

which is consistent with the definition of the wave vector, Eq. (1.26).

7. Spherical Disturbances and Waves. Another important simple solution of the three-dimensional wave equation is obtained when we make the assumption that the function $p(\mathbf{r}, t)$ has spherical symmetry about the origin; that is, we assume that

$$p(\mathbf{r}, t) = p(r, t) \tag{1.30}$$

only, where

$$r = \sqrt{x^2 + y^2 + z^2}$$

To calculate $\nabla^2 p$ in this case, we begin with

$$\frac{\partial p}{\partial x} = \frac{\partial p(r, t)}{\partial r}\frac{\partial r}{\partial x} = \left(\frac{\partial p}{\partial r}\right)\frac{x}{r}$$

and differentiate once more to obtain

$$\frac{\partial^2 p}{\partial r^2} = \frac{\partial}{\partial x}\left(\frac{x}{r}\frac{\partial p}{\partial r}\right) = \frac{1}{r}\frac{\partial p}{\partial r} + x\frac{\partial}{\partial x}\left(\frac{1}{r}\frac{\partial p}{\partial r}\right)$$

$$= \frac{1}{r}\frac{\partial p}{\partial r} + x\frac{\partial}{\partial r}\left(\frac{1}{r}\frac{\partial p}{\partial r}\right)\frac{\partial r}{\partial x}$$

$$= \frac{1}{r}\frac{\partial p}{\partial r} + \frac{x^2}{r}\left(-\frac{1}{r^2}\frac{\partial p}{\partial r} + \frac{1}{r}\frac{\partial^2 p}{\partial r^2}\right)$$

$$\frac{\partial^2 p}{\partial x^2} = \frac{1}{r}\frac{\partial p}{\partial r} - \frac{x^2}{r^3}\frac{\partial p}{\partial r} + \frac{x^2 \partial^2 p}{r^2 \partial r^2}$$

In the same way, we obtain the other derivatives:

$$\frac{\partial^2 p}{\partial y^2} = \frac{1}{r}\frac{\partial p}{\partial r} - \frac{y^2}{r^3}\frac{\partial p}{\partial r} + \frac{y^2}{r^2}\frac{\partial^2 p}{\partial r^2}$$

$$\frac{\partial^2 p}{\partial z^2} = \frac{1}{r}\frac{\partial p}{\partial r} - \frac{z^2}{r^3}\frac{\partial p}{\partial r} + \frac{z^2}{r^2}\frac{\partial^2 p}{\partial r^2}$$

The sum of these three equations yields

$$\nabla^2 p = \frac{3}{r}\frac{\partial p}{\partial r} - \frac{(x^2 + y^2 + z^2)}{r^3}\frac{\partial p}{\partial r} + \frac{(x^2 + y^2 + z^2)}{r^2}\frac{\partial^2 p}{\partial r^2}$$

$$= \frac{2}{r}\frac{\partial p}{\partial r} + \frac{\partial^2 p}{\partial r^2} = \frac{1}{r}\frac{\partial^2}{\partial r^2}(rp)$$

The wave equation then becomes

$$\frac{1}{r}\frac{\partial^2}{\partial r^2}[rp(r, t)] - \frac{1}{v^2}\frac{\partial^2}{\partial t^2}p(r, t) = 0$$

or

$$\frac{\partial^2}{\partial r^2}[rp(r, t)] - \frac{1}{v^2}\frac{\partial^2}{\partial t^2}[rp(r, t)] = 0$$

This differential equation for the function $[rp(r, t)]$ has mathematically the same form as the one-dimensional wave equation discussed earlier for the function $x(z, t)$. Thus the general mathematical solution can be written in the form.

$$rp(r, t) = f\left(t - \frac{r}{v}\right) + g\left(t + \frac{r}{v}\right)$$

The actual disturbance p then takes the form

$$p(r, t) = \frac{1}{r}f\left(t - \frac{r}{v}\right) + \frac{1}{r}g\left(t + \frac{r}{v}\right) \tag{1.31}$$

The first term in Eq. (1.31) can be interpreted as a spherically symmetric disturbance that originates at the origin and propagates outward with velocity of magnitude v. The amplitude of the disturbance falls off as $1/r$. In the case of a sound wave, such a disturbance could be caused by a small round source at the origin. The second term in Eq. (1.31) represents an incoming spherical disturbance that would be hard to realize experimentally. An outgoing spherical sine wave is obtained by letting f equal an expression similar to Eq. (1.23).

$$p(r, t) = \frac{A}{r}\cos\left[2\pi v\left(t - \frac{r}{v}\right) + \varphi\right] \tag{1.32}$$

[Naturally, A in this equation must have different dimensions than A in Eq. (1.23)].

B. Early Wave Theories

In 1665 two documents were published that independently hypothesized that light was a wave. They were *Physico Mathesis de Luminie, Coloribus et Iride* by Francesco Maria Grimaldi (1618–1663) in Bologna, and *Micrographia* by Robert Hooke (1635–1703) in England.

Grimaldi was the first to observe the phenomenon of diffraction, whereby complex patterns are formed outside the geometrical edge of a shadow. He compared light to waves in water and hypothesized that there existed an ethereal medium in which the waves propagated.

Hooke was an opponent of Newton's and directly confronted Newton in the Royal Society of London. He discovered the interference that leads to multicolored patterns in thin transparent bodies. He also drew the analogy between light and waves in water and accepted the ethereal medium hypothesis. Unlike Grimaldi, Hooke believed that the velocity of light was greater in dense media than in rare media. This led to a faulty picture of wave propagation. In Fig. 1.16 we see that Hooke's model required that the wavefront could not remain perpendicular to the ray after refraction. We now understand that such phenomena can occur in materials whose optical properties are functions of direction. Hooke's model, however, was designed to apply in general for optically isotropic media.

C. Huygens

Christian Huygens (1629–1695), an early Dutch physicist, formulated the first semiquantitative wave model for light and presented it in lecture form before the Royal Academy of Sciences in Paris in 1679. He was greatly influenced by the philosophy of Descartes but recognized the shortcomings of Descartes' theory of

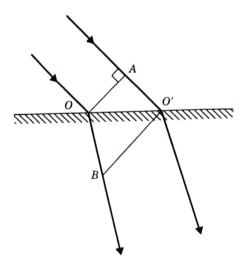

Fig. 1.16 Hooke's incorrect concept of refraction. The wavefront OA is initially perpendicular to the direction of the advancing rays. If the transmitted speed were greater than the incident speed, then the light would have to travel over OB in the same time required for another ray to travel over the distance AO'. Inside the medium the wavefront BO' would have to be inclined with respect to the rays.

light. He accepted the Cartesian view that geometry was the fundamental basis of physics. Therefore, light must be explicable with a mechanical analogy. Unlike Descartes, however, Huygens believed in the existence of the void. This enabled him to rationalize that the ethereal medium was elastic, consisting of particles whose separations could be varied. Given these characteristics, the ether would propagate a disturbance just as an elastic fluid does.

Huygens' wave theory of light is presented in *Traite de la Lumière*, which was printed in 1690. In his model, Huygens did not have in mind a train of transverse waves but rather an irregular series of single propagating pulses. Although the concept of interference was eventually invoked by others to place Huygens' ideas on a firm foundation, Huygens did not originally mention the concept of "phase" in his theory.

1. Huygens' Principle. We have seen in section 1.3A6 how a spherical wave or disturbance propagates in three dimensions, but we have not yet discussed the methods of exciting such disturbances. They clearly depend on the specific problem at hand and will vary with the nature of the waves. In a mechanical medium, such as a fluid, forces between one part of the medium and a nearby part transmit the disturbance. The physical *source* of the disturbance will act locally through similar forces on the medium near it. (Think of a loudspeaker emitting sound waves in air, for instance.) You might suspect that it would make little difference whether a given region of the medium was excited by an external source or by a nearby region of the medium. In fact, it is reasonable to consider each small excited region of the medium as a source of spherical waves of its own. The procedure suggested by Huygens' principle then is to imagine the disturbance at a given time as composed of the sum of many separate disturbances, each of which acts like a point source that radiates a spherically symmetric wavelike disturbance at a later time.

This scheme for thinking about the nature of wave propagation is called *Huygens' principle*. The method tells how to calculate the position in space of a disturbance at a later time if it is known at an earlier time. It does not give quantitative information about the resulting amplitude, although refinements of the procedure made by later mathematicians do.

Huygens' principle contains two important rules about the way secondary wavelets should be combined. We will state them as they apply to the situation shown in Fig. 1.17, where a source is sending a spherical disturbance onto a screen with an aperture in it. Suppose that at time t the disturbance can be localized to lie just inside the aperture surface σ and we wish to know where the disturbance is at time $t + \Delta t$. According to part 1 of Huygens' principle, we construct secondary spherical wavelets of radius of $v\Delta t$ originating at each point on σ.

We must now form the superposition of these secondary wavelets to produce the disturbance at time $t + \Delta t$. Note that the wavelets are most dense in the region included between the lines PA and PB in Fig. 1.17. This region is called the *region of geometrical brightness* because if light is assumed to travel outward from the source in straight lines or *rays*, this region will be illuminated, whereas the regions outside, the *regions of geometrical shadow*, will not be illuminated.

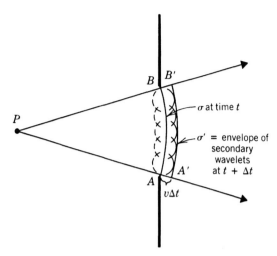

Fig. 1.17 Huygens' construction and the law of rectilinear propagation.

Note also that the regions in the geometrical shadow are reached by only a few of the secondary wavelets originating from σ. We assume in part 2 of Huygens' prescription that we can neglect these wavelets, and we determine the new disturbance σ' as the *envelope* of all the outgoing secondary wavelets. Each point on the envelope is tangent to one and only one Huygens' wavelet. The envelope ends abruptly at the edge of the geometrical shadow. The backward-moving wavelets shown dotted also have an envelope that would be propagating toward the source. They are the analog of the $g(t + z/v)$ terms in the one-dimensional case, whereas the forward-moving wavelets are the analog of the $f(t - z/v)$ terms. The backward moving wavelets are neglected in Huygens' prescription, without any justification beyond an appeal to the experimental fact that the back wave envelope does not occur in nature. Later versions of the theory are able to explain this nonoccurrence. The later versions also take account of the wavelets in the shadow, which give rise to diffraction effects.

A more detailed analysis will show that the entity that propagates in the way just described is not the optical disturbance itself, but rather those portions of it that are discontinuous as functions of time and space. These discontinuities are confined to surfaces and propagate according to Huygens' principle. The propagation of a continuous disturbance will be approximately described by this principle if the disturbance changes rapidly over distances that are small compared with the sizes of relevant physical objects.

Our ability to neglect those parts of the Huygens' wavelets that do not form the envelope, that is, those in the geometrical shadow, is closely connected to the localization of the changing part of the disturbance. This will be more clear in our discussion of diffraction (Chapters 6 and 7) where we explicitly consider these neglected parts of the wavelets.

To follow the propagation of the optical disturbance through space or through optical media, we have, in theory, only to make repeated applications of Huygens' principle. In the example of Fig. 1.17, to determine the location of the disturbance at time $t + 2\Delta t$ we would draw the envelope of all forward-moving wavelets of radius $v\Delta t$ originating on the surface σ'. This process can be repeated indefinitely.

The notion of *rays* follows quite simply from our idealized picture of a localized optical disturbance. In isotropic media: *Light rays are directed lines that are always perpendicular to the surface occupied by the disturbance at a given time and point along the direction of its motion.*

2. Proof of Laws of Geometrical Optics with Huygens' Principle. *a. Rectilinear Propagation.* A single ray has no meaning in the context of Huygens' principle because the wave surface, however small, will have a finite area, and infinitely many normals can be drawn through it. If the surface is limited—for example, by an aperture—there will be limited rays, such as *PB* in Fig. 1.17, that are normal to it at its edge. We now argue that these rays obey the law of rectilinear propagation in isotropic media.

The new surface σ' will also be sharply limited, as it is defined to be the outgoing envelope of all secondary Huygens' wavelets originating on σ. A wavelet centered at a point on the edge of σ, such as *B* in Fig. 1.17, will be tangent to σ' at its edge (point B' in the figure). There is no other wavelet beyond that centered at *B*; thus the envelope ends sharply at B'. Here σ and σ' are readily seen to be normal to the radius vector *PB*. If we repeat the Huygens' construction to determine yet another surface σ'', the extension of the line *PB* will just pass through its edge. Thus a line through *P* and *B* defines a limiting ray that obeys a law of rectilinear propagation.

It is the envelope prescription in Huygens' principle that gives well-defined edges to wave surfaces and allows us to speak of these limiting rays. When we give up the envelope prescription in favor of a quantitative treatment of the secondary wavelets, we should not be surprised to find that nongeometrical behavior results. In Huygens' day, this refinement was not appreciated because diffraction was very poorly understood. Newton's major objection to the wave theory was, in fact, that limiting rays could not be maintained. This was correct, however the size of the deviation is much smaller than what Newton expected.

b. Reflection and Refraction. Consider the plane interface represented in Fig. 1.18 by OO'''. A planar optical disturbance σ is originally found, at time t, just as the edge of the wavefront *A* touches the interface at *O*, as shown in Fig. 1.18a. At this instant the other edge of a wavefront A''' is a distance $\overline{A'''O'''}$ from the point O''' where it will meet the interface. This meeting will occur at time $t + \Delta t$, where $\Delta t = v/\overline{A'''O'''}$. According to Huygens' principle, any point on the original wavefront can be considered as a source of a secondary wavelet. We choose the points in sequential fashion as each point of the wavefront meets the interface. Thus starting at t in Fig. 1.18a, *A* will be the first part of the wavefront to act as a secondary wavelet source and this will be centered on the point *O*. Some time later, Fig. 1.18b, A' contacts the interface at O' leading to another secondary wavelet. In the mean time, the secondary wavelet that started at *O* has propagated into the incident

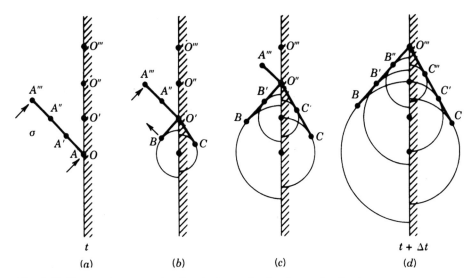

Fig. 1.18 Huygens' construction for the laws of refraction and reflection.

medium with speed v and into the transmitting medium with speed v'. Here we have assumed that $v' < v$ so that $\overline{OC} < \overline{OB}$. At a yet later time, Fig. 1.18c, a new secondary source is identified at O''', while the earlier sources have produced wavelets that have reached B and B' on the incident medium and C and C' in the transmitting medium. Finally, in Fig. 1.18d, time Δt has elapsed and the secondary wavelets have progressed even farther.

We have chosen two arbitrary times in Fig. 1.18b and c to illustrate the intermediate secondary wavelets. It should be clear that a continuum of opportunities exists and that secondary wavelets, each properly retarded, can be imagined along the entire interface from O to O'''.

The common tangents of the secondary wavelets identify the new location of the optical disturbance in the incident medium and in the transmitting medium.

In Fig. 1.18d, the secondary wavelet whose center is at O contributes to the reflected disturbance at B, which is a distance $v\Delta t$ from O. The same wavelet also contributes to the transmitted disturbance at C, which is a distance $v'\Delta t$ from O.

The situation in Fig. 1.18d has been redrawn in Fig. 1.19 for clarity. Also shown in Fig. 1.19a are the rays—the directions perpendicular to the wavefronts. In Fig. 1.19b the traditional angles are shown with all rays meeting the interface at a point.

In Fig. 1.19a line segment OO''' is a common hypotenuse for triangles $OA'''O'''$, OBO''', and OCO'''. Thus we can write

$$\overline{OO'''} = \frac{\overline{A'''O'''}}{\sin\theta} = \frac{\overline{OB}}{\sin\theta''} = \frac{\overline{OC}}{\sin\theta'} \tag{1.33}$$

From the previous discussion we can also conclude that

$$\Delta t = \frac{\overline{A'''O'''}}{v} = \frac{\overline{OB}}{v} = \frac{\overline{OC}}{v'} \tag{1.34}$$

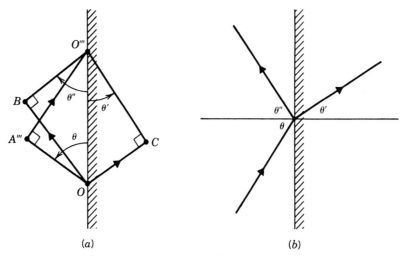

Fig. 1.19 Simplification of the angles involved in Fig. 1.18.

Combining Eqs. (1.33) and (1.34), we find

$$\frac{v\,\Delta t}{\sin\theta} = \frac{v\,\Delta t}{\sin\theta''} = \frac{v'\,\Delta t}{\sin\theta'} \tag{1.35}$$

From Eq. (1.35), which follows directly from the geometry of Fig. 1.19 and Huygens' principle, we can extract the law of reflection, $\theta'' = \theta$, and the law of refraction

$$\frac{\sin\theta'}{v'} = \frac{\sin\theta}{v}$$

or

$$n'\sin\theta' = n\sin\theta$$

where, as in Fermat's formalism, $n/n' = v'/v$.

3. Relationship to Other Theories. Huygens pointed out in his *Treatise* that the secondary wavelet construction gives rise to an optical disturbance that propagates according to the "principle of least time" presented by Fermat. The two approaches are equivalent methods for dealing with geometrical optics. However, Huygens' principle is more powerful for within it are contained the ideas necessary to explain diffraction. Huygens' book was published before Newton's *Opticks*. However, Newton was so highly respected that the scientific world accepted his ideas rather than those of Huygens'. As noted earlier, for nearly 100 years, further progress in optics was inhibited by this bias.

D. Definitive Experiments

The controversy surrounding the nature of light continued, in spite of Newton's dominance, throughout the 18th century. Active debate was renewed at the beginning of the 19th century following new experiments. As the implications of these data were gradually accepted, the classical particle models for light died in favor of the wave picture.

1. Young. In reporting the results of his observations of the patterns formed by thin films, Newton invoked the phenomenon of "fits" to explain the alternating regions of "easy reflection" and "easy transmission." This concept was totally unacceptable to Thomas Young (1773-1829), the London physician who, during the course of his studies, became familiar with optical theory through his interest in vision.

Young had the insight to combine the excellent observational work of Newton with the wave theory of Huygens, thus arriving at his "law of interference." He correctly surmised that light from a single source, after being split so as to follow optical paths of different lengths, would reinforce itself so as to produce a bright zone only if the difference of the routes was a multiple of a certain length. He thus explained "Newton's rings" by interference between light reflected from the top and the bottom surfaces of the air gap.

To further illustrate his point, he performed the now famous experiment in which "Young's interference fringes" are observed. We discuss the details of this experiment in Chapter 5. Figure 1.20 shows how he split sunlight into two identical sources using pinholes in shades. At each pinhole, secondary Huygens' wavelets diverge to recombine on the observation screen.

To understand Young's picture, we must assume that light behaves like a wave

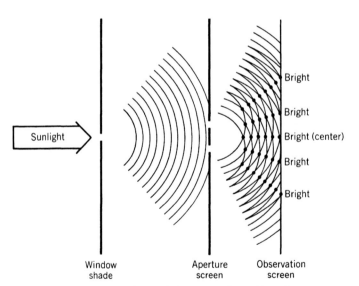

Fig. 1.20 Young's experimental demonstration of interference. Secondary Huygens' wavelets originate from the two holes in the aperture screen. Dots identify the locations for reinforcement as the wavelets combine.

and that the waves are periodic. The last idea was not originally part of Huygens' theory, but it is necessary to explain the regular pattern of fringes on either side of the central bright fringe. The circles in Fig. 1.20 represent the instantaneous locations of maxima in the periodic undulations of the optical disturbance. Young was able to deduce the spatial periodicity. For red light he found 1/36,000 of an inch, for violet light he found 1/60,000 of an inch. Here we have the first association of color with the wavelength of a periodic wave.

Young's presentation of his conclusions before the Royal Society in 1802 was met with ridicule. This was not only for the content of the paper but for his overconfident manner of presentation and for his direct attack on the ideas of Newton. Young was not mathematically inclined; thus he was unable to justify his "law of interference" in all cases for which he thought it applied. He tried unsuccessfully to explain the phenonemon of diffraction as resulting from the combination of just two beams of light from the edges of the diffracting object.

2. Fresnel. Augustin Fresnel (1788–1827) was a French government civil engineer who, although having been trained in mathematics, could not read English or Latin. Through his opposition to Napoleon, he lost his job, thus becoming free to devote time to his curiosity about the nature of light. Since he could not benefit from previous foreign publications, he set out to perform some very careful experiments of his own concerning diffraction. By using a small source (an image of the sun transmitted through a small hole covered with a drop of honey) and by directly observing the light with his eye rather than in projection on a screen, Fresnel was able to document the detailed patterns of fringes surrounding the shadows of small obstacles.

He discovered that as the observation point moved farther from the obstacle the locus of a given fringe did not follow a straight line but rather a hyperbola. This was completely inexplicable with the particle model in which the diffraction interaction was thought to occur as the light passed by the edges of the obstacle, after which it would have had to follow a straight line. Fresnel's observations showed that the fringe phenomenon in diffraction displays continuous deviation. The report of this observation, along with its explanation, was communicated to the Academy of Sciences in Paris during the years 1815 through 1818.

Fig. 1.21 illustrates Fresnel's idea for an aperture. The secondary wavelets combine to form the transmitted wavefront not just in the region of geometrical brightness but also in the shadow. Where constructive superposition occurs, there will be bright regions in the diffraction patterns. Because the wavelength of light is so small, the variations in illumination are restricted to lie close to the geometrical shadow. The proponents of Newtonian optics could not match Fresnel's elegant notions, but to help settle the issue they aided the establishment by the Academy of Sciencies of a competition to solicit the most convincing theoretical and experimental explanation of diffraction. Fresnel entered the competition with a report on his careful experiments that was backed by a mathematical model using the wave theory. He was selected as the winner in 1819. Thus formal and decisive recognition of the wave theory of light had finally come.

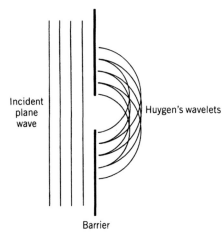

Fig. 1.21 Fresnel's method for explaining diffraction by an aperture. In the opening, all points on the incident wavefront are considered to be sources for a continuum of Huygen's wavelets.

3. Other Developments. *a. Polarization.* Since 1669, the phenomenon of "double refraction" had been one of the more puzzling problems of optics. In that year, in Copenhagen, Erasmus Bartholinus (1625-1698) showed that crystals of "Iceland spar" (which we now call calcite, $CaCO_3$) produced two refracted rays from a single incident beam. One ray, the "ordinary ray," followed Snell's law, while the other, the "extraordinary ray," was not always even in the plane of incidence.

Later it was determined by others that the two beams, on emerging from the crystal, possessed unique characteristics as if the rays had "sides." The two beams emerging from calcite behaved as if their sides were oriented at right angles with respect to each other. Etienne-Louis Malus (1775-1812) won the French Academy prize in 1810 with his discovery that reflected and scattered light also possessed this "sidedness," which he called "polarization" (see Fig. 1.22).

The existence of polarization was, in fact, hailed by the proponents of the particle theories since no explanation could be imagined by the wave theorists. This was because, in analogy to sound, the wave theories hypothesized *longitudinal* oscillations. That is, the motion of the particles of the medium were always thought to be parallel to the direction of propagation.

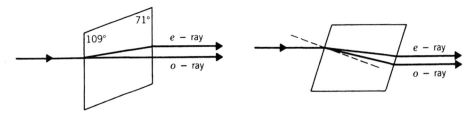

Fig. 1.22 Demonstration of double refraction characteristics of calcite. The *e*-ray, or extraordinary ray, is linearly polarized in the plane of the drawing. The *o*-ray, or ordinary ray, obeys Snell's law and is polarized perpendicular to the drawing.

In 1816 a collaborator of Fresnel's, Dominque-Francois Arago (1786–1853), relayed to Young in England the results of an experiment he had performed with Fresnel. Two oppositely polarized beams of light were observed *not* to interfere with each other, while in the same geometrical set-up, interference *was* observed with natural light or with light having one state of polarization. This stimulated Young to propose in 1817 that the oscillations in the optical disturbance were *transverse*, or perpendicular to the direction of propagation. The phenomenon of linear polarization would then be associated with the direction of the oscillation. Fresnel expanded on this idea by incorporating it into his mathematical theory of light. These works remain today as a useful parameterized theory of optical polarization and a testament to the success of the wave model of light.

b. Speed of Light in Water. Gradually, the entire scientific world accepted the wave theory of light. The continued success of the theory following Fresnel's contributions was impossible to ignore. The final blow to Newton's particle model came in 1850 when Jean Bernard Léon Foucault (1819–1868), a French physicist, completed his measurement of the speed of light in water. Figure 1.23 illustrates the principle behind his experiment. Light from a source at P passed through a beam splitter to a rotating mirror R then on to a stationary mirror at M. On its return the light again hit the rotating mirror, which had turned through an angle whose magnitude depended on the time of propagation from R to M and back to R. With air in the path (Fig. 1.23b) the reflected light was found at P'. In Fig. 1.23c, with water in the path the reflected light was at P'', indicating that the mirror had rotated through a larger angle and that the speed of light in water was less than in air.

Newton's classical particle model, as we have seen, required that the speed of light in optically dense media be greater than in air, whereas the wave theory, as initiated by Huygens, correctly predicted that the speed must be smaller in optically dense media. Throughout the last half of the 19th century the wave theory became universally accepted. The particle model was dead, for the time being.

1.4 The Electromagnetic Wave Model

Although it had been determined that light behaved like a wave by the middle of the 19th century, there was still a question about what it was that was "waving." Most physicists at that time believed in the existence of an ethereal medium in which the optical disturbance propagated. This medium had to possess some unusual characteristics. It had to pervade all space, thus being exceedingly rare. Yet Young had shown, following experiments with the phenomenon of polarization by Fresnel and Arago, that the light wave must be transverse. It was difficult to imagine a medium as rare as the ether that was stiff enough such that it could support the shear restoring force necessary in a tranverse mechanical wave. As we will see, the concept of the ether had been abandoned in the modern electromagnetic theory of light. As originally, proposed, however, the ether was imagined to be the vehicle of electromagnetic forces as well as for light.

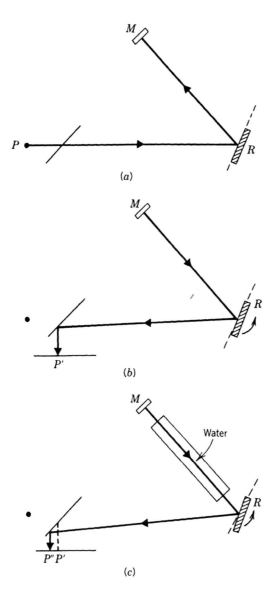

Fig. 1.23 Foucault's determina
of the speed of light in water.

The association between magnetism and light was first made in 1851
Michael Faraday (1791–1867), the gifted English experimentalist. In the course
works eventually published in *Experimental Researches in Electricity*, he observ
that the orientation of the linear polarization of a light beam could be rotated wh
the beam of light traveled through a piece of glass along the direction of the lines
force created by an electromagnet. This relationship caused Faraday to specula
that the ether that had been hypothesized in the theory of light also played :
important role in communicating the lines of force in magnetism.

In 1856, Wilhelm Weber (1804–1890) and Rudolph Kohlrausch (1809–1858) experimentally determined that the constant of proportionally between electric quantities and magnetic quantities was, within experimental error, the same as the speed of light. That this constant must bear the units of velocity is revealed from the fundamental definitions. The magnetic force per unit length between two long parallel current carrying wires in free space separated by distance r is given by

$$\frac{dF}{d\ell} = \mu_0 \frac{ii'}{2\pi r}$$

where μ_0 is a constant (Fig. 1.24a) known as the *permeability of free space*. The magnetic force has units of

$$\mu_0 \cdot \left(\frac{\text{Charge}}{\text{Time}}\right)^2$$

The electric force between two charges in free space separated by distance r is written

$$F = \frac{1}{4\pi\varepsilon_0} \frac{qq'}{r^2}$$

where ε_0 is a constant known as the *permitivity of free space* (Fig. 1.24b).

The electric force has units of

$$\frac{1}{\varepsilon_0} \cdot \left(\frac{\text{Charge}}{\text{Distance}}\right)^2$$

Because these are dimensionally equivalent, we must have $(\mu_0\varepsilon_0)^{-1/2}$ in the units of (distance/time). This was recognized a short time later as much more than a coincidence. But it was not until the Scots physicist James Clerk Maxwell (1831–1879) collected electromagnetic theory into a correlated set of mathematical relationships that the hypothesis that light behaved like an electromagnetic wave was justified.

Maxwell's theory, developed during 1861–1862, dealt with the ether as the medium in which electromagnetic phenomena existed. He proposed that the

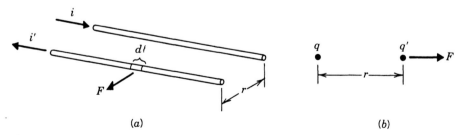

(a)   (b)

Fig. 1.24 (a) Magnetic force between long parallel wires carrying currents i and i'. (b) Electric force between like charges q and q'.

electric field was a "displacement" or distortion of charged ether particles from their equilibria, whereas the magnetic field was associated with rotating regions or "vortices" of charged ether particles. A change in the vortex structure of the ether (magnetic field) would produce displacement in the medium (electric field). This interrelationship is required in order that waves consisting of simultaneous fluctuations in the electric and magnetic fields be possible.

The mechanical character of Maxwell's original formalism was a natural product of the times. Even today students have a difficult time comprehending electric and magnetic fields without a mechanical analogy. Since Maxwell's proposal, we have learned that electric and magnetic fields are more properly thought of as *characteristics of space and relativity* (the relative motion of one reference frame with respect to another). Maxwell's equations for the fields, however, remain valid even in the absence of an ether, since they describe how these characteristics are related to each other and to static and dynamic material charges. We need not invent an ether whose condition of strain or rotation is connected with the action of electric and magnetic forces. To do so is only to further complicate the issue. Experimentally observed classical electromagnetic phenomena, including the propagation of light, are entirely explained in terms of the fields themselves. In addition, if the ether were to exist, then there was no reason to expect that the ether was at rest with respect to the earth, or, for that matter, with respect to the sun either. In the 19th-century model, the speed of light was linked to the ether reference frame. This required that the speed of light as measured on earth should depend on the direction of the measurement.

To detect this anisotropy was the object of the experiments by the American physicist Albert Abraham Michelson (1852-1931) in 1881 and with Edward Williams Morley (1838-1923) in 1887. Michelson constructed a device for this purpose, the *Michelson interferometer*. Details of its operation are described in section 5.4C. Here it is only important to recognize that the interferometer can be used to measure small differences in phase between two perpendicular beams. The propagation time for each of these two beams should have depended on the direction and speed of the interferometer's motion with respect to the proposed ether. Because the beams were perpendicular, their propagation times should have been different. The sensitivity of the instrument was sufficient to enable the phase difference resulting from the different propagation times to be detected. This experiment gave negative results when first performed and when repeated at other times of the year. This meant that light has a constant speed, independent of the motion of the measuring apparatus relative to the hypothesized ether. Because the ether exerts no influence on the light, it is best to adandon the concept altogether.

A. Maxwell's Equations

1. Integral Relationships. Most elementary physics courses deal with electromagnetism through integral equations (for example, Halliday and Resnick). The electromagnetic theory of light, as presented by Maxwell, is most directly deve-

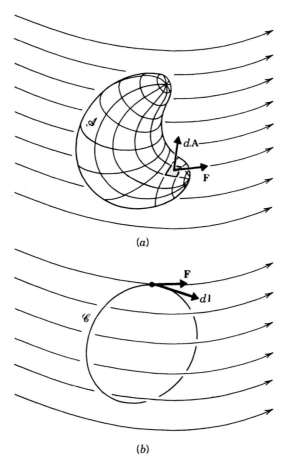

(a)

(b)

Fig. 1.25 (a) $\mathscr{A}$ is the integration surface over which Gauss' theorem is defined for the vector field. (b) $\mathscr{C}$ is the integration loop for Stoke's theorem.

loped if these equations can be recast in differential form. To do this, we require two general theorems from vector calculus.

a. Gauss' Theorem. Consider vector field **F**. That is, to each point in space (and at each time), a magnitude and direction for **F** is defined. Here **F** is a generalized quantity, but in our application we associate it with either the electric field **E** or the magnetic induction **B**. Consider also a closed surface $\mathscr{A}$, in the same space occupied by **F**. The net outflow or "flux" of **F** from $\mathscr{A}$ is

$$\oiint_{\mathscr{A}} \mathbf{F} \cdot d\mathbf{A}$$

where $d\mathbf{A}$ is directed along the outward normal to the surface and $|d\mathbf{A}| = d\mathscr{A}$ is the area of the surface element (Fig. 1.25). Gauss' theorem equates this flux to the volume integral (within the surface enclosed by $\mathscr{A}$) of another quantity that depends on the spatial variation of **F**.

$$\oint_{\mathscr{A}} \mathbf{F} \cdot d\mathbf{A} = \iiint_{\mathscr{V}} (\mathbf{V} \cdot \mathbf{F}) \, d\mathscr{V} \tag{1.36}$$

where $\mathscr{V}$ is the volume enclosed by $\mathscr{A}$. Here $\mathbf{V} \cdot \mathbf{F}$ is the "divergence" of $\mathbf{F}$ and is computed as a scalar or dot product of "del" and $\mathbf{F}$.

b. Stokes' Theorem. Within this same vector field consider now a closed curve $\mathscr{C}$. At each point along the curve identify the incremental segment $d\mathbf{l}$ whose direction is tangent to the curve. The total "circulation" of $\mathbf{F}$ around $\mathscr{C}$ is

$$\oint_{\mathscr{C}} \mathbf{F} \cdot d\mathbf{l}$$

We can also identify one of a number of open surfaces $\mathscr{A}$ whose edge is outlined by $\mathscr{C}$. Stokes' theorem relates the circulation of $\mathbf{F}$ to a different quantity, which is defined on $\mathscr{A}$.

$$\oint_{\mathscr{C}} \mathbf{F} \cdot d\mathbf{l} = \iint_{\mathscr{A}} (\mathbf{V} \times \mathbf{F}) \cdot d\mathbf{A} \tag{1.37}$$

Here $\mathbf{V} \times \mathbf{F}$ is the "curl" of $\mathbf{F}$ and is computed as a vector or cross product of "del" and $\mathbf{F}$.

2. Differential Form of Maxwell's Equations. *a. Macroscopic Fields.* When matter is present, part of the source terms—that is, the charge density and the current density—that appear in Maxwell's equations comes from "free" charges and currents and part comes from the response of matter to the fields (bound charges and currents). But the fields, in turn, depend on these same charges and currents, giving us a coupled system of equations.

The charges and currents associated with the atoms and molecules of matter have sharp discontinuities on an atomic scale, that is, over distances of about 0.1 nm. The resulting electric field $\mathbf{E}$ and magnetic induction $\mathbf{B}$ change very dramatically over similar distances. These then are *microscopic* fields. *Macroscopic* measuring devices will ordinarily not be able to observe the sharp discontinuities in the microscopic fields. It is very useful to introduce the macroscopic fields $\mathbf{E}$ and $\mathbf{B}$ obtained by averaging the microscopic fields over an element of volume that is macroscopically small but large enough to contain very many atoms or molecules. The sharp spatial fluctuations in the microscopic fields will be smoothed over, leaving well-behaved functions for the electric and magnetic quantities.

b. Maxwell's Version of Gauss' Law for $\mathbf{E}$. This is a statement of Coulomb's law in field terms. In words it says that the flux of the macroscopic electric field through a closed surface equals the total charge inside the surface divided by a constant. The constant is introduced in the SI system of units so that we can measure the charges in coulombs and the currents in amperes. The integral form of the law is

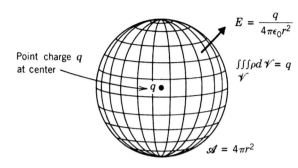

Point charge q at center

$E = \dfrac{q}{4\pi\epsilon_0 r^2}$

$\iiint_{\mathscr{V}} \rho\, d\mathscr{V} = q$

$\mathscr{A} = 4\pi r^2$

Fig. 1.26 Gauss' law for a point charge at the center of a spherical integration surface. $|E|$ is constant on the sphere.

$$\oiint_{\mathscr{A}} \mathbf{E}\cdot d\mathbf{A} = \frac{1}{\varepsilon_0}\iiint_{\mathscr{V}} \rho\, d\mathscr{V} \qquad (1.38)$$

where ρ is the spatially averaged total charge density and ε_0 is the permittivity of free space that is experimentally determined to be $\varepsilon_0 = 8.854187817 \times 10^{-12}$ $C^2/N - m^2$. Figure 1.26 illustrates the law for a point charge.

Using the mathematical relationship of Gauss' theorem, Eq. (1.36), Eq. (1.38) can be rewritten as

$$\iiint_{\mathscr{V}} (\mathbf{\nabla}\cdot \mathbf{E})\, d\mathscr{V} = \frac{1}{\varepsilon_0}\iiint_{\mathscr{V}} \rho\, d\mathscr{V}$$

which shows, because the limits of the integration are arbitrary, that an equivalent expression of Gauss' law for the electric field is

$$\mathbf{\nabla}\cdot\mathbf{E} = \frac{\rho}{\varepsilon_0} \qquad (1.39)$$

c. Maxwell's Version of Gauss' Law for **B**. In field terms this law results from the assumption that magnetic monopoles do not exist. (If they are found experimentally, this equation will require modification.) It says that the flux of the macroscopic magnetic induction through a closed surface is zero. In integral form, this statement is

$$\oiint \mathbf{B}\cdot d\mathbf{A} = 0 \qquad (1.40)$$

which can be transformed as in the preceding section into a differential equation.

$$\mathbf{\nabla}\cdot\mathbf{B} = 0 \qquad (1.41)$$

d. Maxwell's Version of Faraday's Law. This is the equation of magnetic induction. In words it says that the circulation of the macroscopic electric field around a closed curve (or the electromotive force) is equal to the negative of the

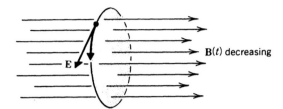

B(t) decreasing

E

Fig. 1.27 Faraday's law relates a change in magnetic flux to an electric field.

time rate of change of the macroscopic magnetic flux through the closed curve (Fig. 1.27). The integral form is

$$\oint_{\mathscr{C}} \mathbf{E} \cdot d\mathbf{l} = -\frac{d}{dt} \iint_{\mathscr{A}} \mathbf{B} \cdot d\mathbf{A} \tag{1.42}$$

Note that the surface $\mathscr{A}$ in this equation is not closed but is outlined by curve $\mathscr{C}$.

Using Stokes' theorem, Eq. (1.37), we can transform the left side of Eq. (1.42) and reverse the order of differentiation and integration on the right side to yield

$$\iint_{\mathscr{A}} (\nabla \times \mathbf{E}) \cdot d\mathbf{A} = - \iint_{\mathscr{A}} \frac{\partial}{\partial t} \mathbf{B} \cdot d\mathbf{A}$$

Partial derivative notation is used because **B** is a function of both time and space. This shows that an equivalent form for Faraday's law is

$$\nabla \times \mathbf{E} = -\frac{\partial \mathbf{B}}{\partial t} \tag{1.43}$$

e. Maxwell's Extension of Ampere's Law. Before Maxwell's work, Ampere's law connected a magnetic induction in the vicinity of a wire with only the current in the wire. In words it said that the circulation of the macroscopic magnetic induction around a closed curve was equal to a constant μ_0 times the total electric current through the curve.

$$\oint_{\mathscr{C}} \mathbf{B} \cdot d\mathbf{l} = \mu_0 \iint_{\mathscr{A}} \mathbf{J} \cdot d\mathbf{A} = \mu_0 I$$

Here I and **J** are spatially averaged total current and total current density through curve $\mathscr{C}$, respectively. Maxwell added a new term that acts as an additional source of the magnetic induction. He called this a "displacement current," which was related to the time rate of change of the electric flux through the closed curve. The constant μ_0 is required to ensure dimensional consistency. ($\mu_0 = 4\pi \times 10^{-7}$ N $\cdot$ sec^2/C^2 is the permeability of free space). That the new term was needed can be seen from Fig. 1.28, which illustrates the method of calculating the magnetic induction surrounding a wire that contains a time-varying current. A capacitor is also in the circuit. If supplied by a battery (not shown), the current will decrease

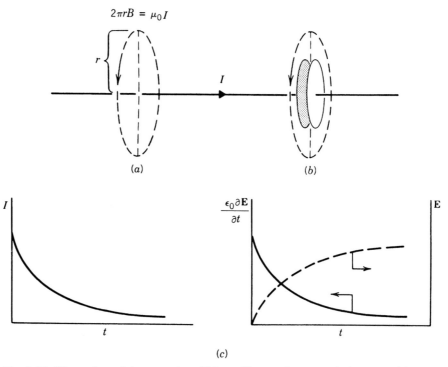

Fig. 1.28 Illustration of the necessity of Maxwell's term in Ampere's law. (a) This integration path encircles current I. (b) This integration path encircles a time-varying electric field but no current. (c) Representations of the current and the field showing the equivalence of the displacement current to the free current.

with time after connection of the battery as the charge on the capacitor plates increases. We use Ampere's law as originally formulated to find the magnetic induction surrounding the wire. To do this, we employ integration path (a). If we move the path into the gap between the plates, the magnetic induction would be zero by the original form of the law, as the free current within the gap is zero. Maxwell recognized that this was incorrect, because the magnetic induction would not drop to zero simply because there was a capacitor in the circuit. He proposed that the changing electric field in the gap would play the role of a current density, $\varepsilon_0 \, \partial \mathbf{E}/\partial t$, thus completing the circuit and providing a magnetic induction that was continuous.

The integral form of the new equation is

$$\oint_{\mathscr{C}} \mathbf{B} \cdot d\mathbf{l} = \mu_0 \iint_{\mathscr{A}} \left(\mathbf{J} + \varepsilon_0 \frac{\partial \mathbf{E}}{\partial t} \right) \cdot d\mathbf{A} \tag{1.44}$$

As with Faraday's law, this can be transformed to a differential relationship using Stokes' theorem, Eq. (1.37).

$$\iint\limits_{\mathscr{A}} (\mathbf{\nabla} \times \mathbf{B}) \cdot d\mathbf{A} = \mu_0 \iint\limits_{\mathscr{A}} \left(\mathbf{J} + \varepsilon_0 \frac{\partial \mathbf{E}}{\partial t} \right) \cdot d\mathbf{A}$$

which, since the limits on $\mathscr{A}$ are arbitrary, leads to

$$\mathbf{\nabla} \times \mathbf{B} = \mu_0 \left(\mathbf{J} + \varepsilon_0 \frac{\partial \mathbf{E}}{\partial t} \right) \tag{1.45}$$

In summary, we present the four equations of Maxwell in differential form with their original equation numbers in this section:

$$\text{I} \qquad \mathbf{\nabla} \cdot \mathbf{E} = \frac{\rho}{\varepsilon_0} \tag{1.39}$$

$$\text{II} \qquad \mathbf{\nabla} \cdot \mathbf{B} = 0 \tag{1.41}$$

$$\text{III} \qquad \mathbf{\nabla} \times \mathbf{E} = -\frac{\partial \mathbf{B}}{\partial t} \tag{1.43}$$

$$\text{IV} \qquad \mathbf{\nabla} \times \mathbf{B} = \mu_0 \left(\mathbf{J} + \varepsilon_0 \frac{\partial \mathbf{E}}{\partial t} \right) \tag{1.45}$$

The synthesis of electromagnetic phenomena represented in these four relatively simple equations remains one of the greatest achievements of physics. They are the starting point from which all classical electromagnetic effects can be explained—including the nature of light.

B. The Electromagnetic Wave Equation

Here we wish to demonstrate, following Maxwell, that light behaves like an electromagnetic wave, that is, a propagating disturbance involving time and space variations of coupled electric and magnetic fields. To do so, we will simplify the problem by considering light in a vacuum, such as exists in outer space. In this case the source terms in Eqs. (1.39) and (1.45) are zero and we need not perform spatial averages. The four Maxwell equations are, in this situation

$$\left. \begin{array}{lll} \text{I} & \mathbf{\nabla} \cdot \mathbf{E} & = 0 \\[2mm] \text{II} & \mathbf{\nabla} \cdot \mathbf{B} & = 0 \\[2mm] \text{III} & \mathbf{\nabla} \times \mathbf{E} & = -\dfrac{\partial \mathbf{B}}{\partial t} \\[4mm] \text{IV} & \mathbf{\nabla} \times \mathbf{B} & = \mu_0 \varepsilon_0 \dfrac{\partial \mathbf{E}}{\partial t} \end{array} \right\} \tag{1.46}$$

Starting with Eq. (1.46) III, we take the vector product of both sides of the equations with "del." (This is called "taking the curl" of the quantities.)

$$\mathbf{\nabla} \times (\mathbf{\nabla} \times \mathbf{E}) = \mathbf{\nabla} \times \left(-\frac{\partial \mathbf{B}}{\partial t} \right) = -\frac{\partial}{\partial t} (\mathbf{\nabla} \times \mathbf{B}) \tag{1.47}$$

where the order of the spatial and temporal derivatives have been interchanged on the right side. Now the left side of Eq. (1.47) can be simplified using the vector triple product identity,

$$\mathbf{A}_1 \times (\mathbf{A}_2 \times \mathbf{A}_3) = \mathbf{A}_2(\mathbf{A}_1 \cdot \mathbf{A}_3) - (\mathbf{A}_1 \cdot \mathbf{A}_2)\mathbf{A}_3.$$

Here

$$\mathbf{A}_1 = \mathbf{\nabla}, \mathbf{A}_2 = \mathbf{\nabla}, \text{ and } \mathbf{A}_3 = \mathbf{E}$$

Thus

$$\mathbf{\nabla} \times (\mathbf{\nabla} \times \mathbf{E}) = \mathbf{\nabla}(\mathbf{\nabla} \cdot \mathbf{E}) - \mathbf{\nabla} \cdot \mathbf{\nabla}\mathbf{E}$$

Because, by Eq. (1.46) I, $\mathbf{\nabla} \cdot \mathbf{E} = 0$, this simplifies to

$$\mathbf{\nabla} \times (\mathbf{\nabla} \times \mathbf{E}) = -\mathbf{\nabla} \cdot \mathbf{\nabla}\mathbf{E}$$

the right side of which we have written before as

$$-\mathbf{\nabla} \cdot \mathbf{\nabla}\mathbf{E} = -\nabla^2\mathbf{E}$$

Eq. (1.47) has become

$$-\nabla^2\mathbf{E} = -\frac{\partial}{\partial t}(\mathbf{\nabla} \times \mathbf{B})$$

Using Eq. (1.46) IV, this can be reexpressed in terms of $\mathbf{E}$ alone

$$\nabla^2\mathbf{E} = \mu_0\varepsilon_0\frac{\partial^2\mathbf{E}}{\partial t^2}$$

or

$$\nabla^2\mathbf{E} - \mu_0\varepsilon_0\frac{\partial^2\mathbf{E}}{\partial t^2} = 0 \tag{1.48}$$

In a similar way, take the curl of Eq. (1.46) IV and make use of Eq. (1.46) III to obtain

$$\nabla^2\mathbf{B} - \mu_0\varepsilon_0\frac{\partial^2\mathbf{B}}{\partial t^2} = 0 \tag{1.49}$$

Thus, each component of $\mathbf{E}$ and $\mathbf{B}$ obeys the usual three-dimensional wave equation, Eq. (1.21), for example,

$$\nabla^2 E_x - \mu_0\varepsilon_0\frac{\partial^2 E_x}{\partial t^2} = 0 \tag{1.50}$$

This is true provided that the speed of the electromagnetic wave is equal to $(\mu_0\varepsilon_0)^{-1/2}$. At the time of Maxwell's work this had already been experimentally demonstrated. Thus in free space,

$$v = c = (\mu_0\varepsilon_0)^{-1/2} \tag{1.51}$$

Although the acceptance of Maxwell's theory of light was slow at first, its close correspondence with experimental phenomena could not be ignored. By the end of the 19th century, the electromagnetic wave model of light was firmly established. The one remaining aspect that had not yet been resolved concerned the nature of the ether. The present view is that an etheral medium does not exist in any operational sense. The fields exist in free space, playing mutually supporting roles, in such a way that an energy carrying wave can be maintained. We must understand that the concept of oscillating electric and magnetic fields is the *most useful* mechanism with which to describe classical optical phenomena. This is the criterion against which any theory would be judged as successful. Until we encounter phenomena that do not obey the predictions of the model, we are justified in applying the model and treating light as an electromagnetic wave.

This is, for the most part, the approach taken in the rest of this book. Interference and diffraction can be adequately treated with the classical electro-magnetic wave theory of Maxwell. Many aspects of the interaction of light with matter can also be treated classically provided that the characteristics of the matter can be parameterized. Other aspects of light interactions with matter require a quantum theoretical approach. We will not touch on this, as it is more properly developed in a dedicated course in quantum optics. As we have mentioned before, the philosophical basis for geometrical optics is the simplest of all. Geometrical optics can be developed through an application of Fermat's principle alone.

C. Electromagnetic Wave Characteristics

The wave equations Eqs. (1.48) and (1.49) and their solutions are not independent of one another because Eqs. (1.46) must still apply. When wavelike solutions are obtained, the connections between **B** and **E** yield the transverse polarization properties predicted by the 19th-century experiments. As traveling waves, the solutions also show how energy can be carried through space by light.

1. Linearly Polarized, Harmonic Plane Waves. We already know that

$$\cos(\omega t - \mathbf{k} \cdot \mathbf{r} + \varphi) = \text{Re}\{e^{i(\omega t - \mathbf{k} \cdot \mathbf{r} + \varphi)}\}$$

obeys the free-space wave equation (1.50) if $\omega = c|k|$ and $\varphi = $ constant. Therefore, the following two equations will automatically be solutions of Eqs. (1.49) and (1.50):

$$\mathbf{E} = \text{Re}\{\mathbf{E}_0 e^{i(\omega t - \mathbf{k} \cdot \mathbf{r} + \varphi)}\} \tag{1.52a}$$

$$\mathbf{B} = \text{Re}\{\mathbf{B}_0 e^{i(\omega t - \mathbf{k} \cdot \mathbf{r} + \varphi)}\} \tag{1.52b}$$

where $\mathbf{E}_0$ and $\mathbf{B}_0$ are constant vectors associated with the maximum size of the oscillations. *Hereafter, when there is no chance of misunderstanding, we will drop the "Re" in equations such as (1.52).* It will be understood that the physical fields will be given by the real part of the complex fields appearing in our equations.

Because these are plane waves, we can use the operator shortcuts of Eqs. (1.28) and (1.29) in Maxwell equations (1.46) I and II to find

$$\nabla \cdot \mathbf{E} = -i\mathbf{k} \cdot \mathbf{E} = 0$$

and

$$\nabla \cdot \mathbf{B} = -i\mathbf{k} \cdot \mathbf{B} = 0$$

This shows that $\mathbf{E}$ and $\mathbf{B}$ must both be perpendicular to $\mathbf{k}$, which is along the direction of propagation. Thus $\mathbf{E}$ and $\mathbf{B}$ are both transverse oscillations.

To find the relationship between $\mathbf{E}$ and $\mathbf{B}$, apply the operator shortcuts to Maxwell equation (1.46) III. Thus,

$$\nabla \times \mathbf{E} = -\frac{\partial \mathbf{B}}{\partial t}$$

becomes

$$-i\mathbf{k} \times \mathbf{E} = -i\omega \mathbf{B}$$

or

$$\mathbf{B} = \frac{\mathbf{k} \times \mathbf{E}}{\omega} = \frac{1}{c}\frac{\mathbf{k} \times \mathbf{E}}{k}$$

Thus

$$\mathbf{B} = \frac{1}{c}\hat{\mathbf{s}} \times \mathbf{E} \tag{1.53}$$

where $\hat{\mathbf{s}} = \mathbf{k}/k$ is the unit vector in the propagation direction. Equation (1.53) says three important things: (1) $\mathbf{B}$ is perpendicular to $\mathbf{E}$, (2) $\mathbf{B}$ is in phase with $\mathbf{E}$, and (3) the magnitudes of $\mathbf{B}$ and $\mathbf{E}$ are related by $B = E/c$, for free space in the SI system of units.

The three vectors $\mathbf{E}$, $\mathbf{B}$, and $\mathbf{k}$ form a right-handed rectangular coordinate system. The resulting field pattern for a fixed value of time is sketched in Fig. 1.29. The lengths of the field vectors are proportional to the magnitudes of the fields

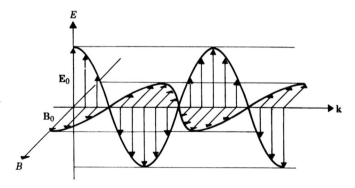

Fig. 1.29 Electromagnetic field patterns in a plane wave at fixed time.

along a line, coinciding with **k**, in real space. This pattern moves in the **k** direction with speed c as time evolves.

Note that $\mathbf{B}_0$ is uniquely determined once $\mathbf{E}_0$ and **k** are known, because $\mathbf{B}_0 = (\mathbf{k} \times \mathbf{E}_0)/ck$. Note also that $\mathbf{E}_0$ can take any direction in the transverse plane perpendicular to **k**. Two different waves of the type just discussed having the same ω and **k** may be superimposed, yielding, in general, elliptically polarized light. As discussed in Chapter 9, the nature of the ellipse will depend on the ratio of amplitudes and value of the phase difference for these two waves.

2. Energy Density and Energy Flux. Electromagnetic theory leads to the following expression for the energy density (in SI units) associated with electric and magnetic fields in free space:

$$U = \frac{1}{2}\left[\varepsilon_0 \mathbf{E} \cdot \mathbf{E} + \frac{1}{\mu_0}\mathbf{B} \cdot \mathbf{B}\right] \tag{1.54a}$$

But

$$\mathbf{B} \cdot \mathbf{B} = \frac{1}{c^2}\mathbf{E} \cdot \mathbf{E} = \varepsilon_0 \mu_0 \mathbf{E} \cdot \mathbf{E}$$

so

$$U = \varepsilon_0 \mathbf{E} \cdot \mathbf{E} \tag{1.54b}$$

The energy flow per unit time area, or energy flux density, in the direction of propagation is defined by the Poynting vector (John Henry Poynting, 1852–1914).

$$\mathbf{S} = \frac{1}{\mu_0}\mathbf{E} \times \mathbf{B} \tag{1.55}$$

This important vector gives the energy flux density or irradiance in an arbitrary direction (that which is identified by the unit vector $\hat{\eta}$) by means of the scalar product $\hat{\eta} \cdot \mathbf{S}$ (see Fig. 1.30).

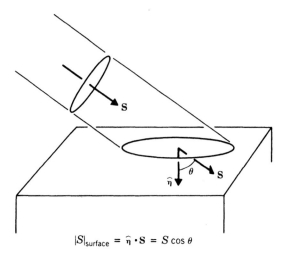

$$|S|_{\text{surface}} = \hat{\eta} \cdot S = S\cos\theta$$

Fig. 1.30 The energy flux density on a surface depends on the angle of incidence.

A quadratic expression in the fields such as S or U must be calculated with care when complex notation is used for the fields. For linearly polarized plane waves in vacuum, we have the following expressions for real physical fields:

$$\mathbf{E} = \mathbf{E}_0 \cos \phi, \quad \phi = \omega t - \mathbf{k} \cdot \mathbf{r} + \varphi$$

$$\mathbf{B} = \mathbf{B}_0 \cos \phi = \frac{\mathbf{k} \times \mathbf{E}_0}{ck} \cos \phi$$

Hence, S is given by

$$\mathbf{S} = \frac{1}{\mu_0} \mathbf{E}_0 \times \frac{(\mathbf{k} \times \mathbf{E}_0)}{ck} \cos^2 \phi = \varepsilon_0 c |\mathbf{E}_0|^2 \, \hat{\mathbf{s}} \cos^2 \phi \tag{1.56}$$

Because the time average over many cycles of $\cos^2(\omega t - \mathbf{k} \cdot \mathbf{r} + \varphi)$ is 1/2, the time average of **S** is given by

$$\langle \mathbf{S} \rangle = \frac{\varepsilon_0 c}{2} |\mathbf{E}_0|^2 \hat{\mathbf{s}} \tag{1.57}$$

This is the quantity to which our eyes or any detector would be sensitive. The time-averaged energy flow is along the wave vector **k**, (or $\hat{\mathbf{s}} = \mathbf{k}/k$) in the direction of propagation of the wave.

For a wave in vacuum, the energy density is given from Eq. (1.54) by

$$U = \varepsilon_0 |\mathbf{E}_0|^2 \cos^2 \phi$$

with a time average

$$\langle U \rangle = \frac{\varepsilon_0}{2} |\mathbf{E}_0|^2 \tag{1.58}$$

Note that magnitude of $\langle \mathbf{S} \rangle$ obeys

$$\langle S \rangle \equiv |\langle \mathbf{S} \rangle| = \langle U \rangle c \tag{1.59}$$

This is a general result

Energy flux density = (energy density) × (propagation speed)

Throughout this book, when there is no chance for confusion, we will eliminate the time-average brackets. In most cases, we are interested in the quantity that our eyes or a detector senses, so we will implicitly be dealing with the time-averaged values of S or U. To put things into perspective, suppose the E_0 has a magnitiude of 1 V/cm = 10^2 V/m. Then the flux density is

$$S = \frac{8.85 \times 10^{-12}}{2} \times 3 \times 10^8 \times (10^2)^2 = 13.3 \text{ W/m}^2$$

$$= 1.33 \text{ mW/cm}^2$$

3. Experimental Confirmation. The earliest documented proof that electromagnetic waves could be produced by manipulating electric and magnetic fields was provided by Heinrich Hertz (1857–1894) who in Kiel, Germany, in 1888 created

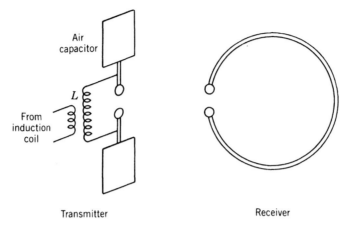

Fig. 1.31 Schematic representation of the device used by Hertz to prove that an electromagnetic wave could be generated from stored electromagnetic energy.

the first manufactured radio waves. This he accomplished with the apparatus shown schematically in Fig. 1.31. The transmitter was a resonant LC circuit, the capacitor of which consisted of large metal plates. The oscillation within the circuit could be detected when the voltage on the capacitor reached sufficient size so that a spark jumped across the gap. His apparatus had a resonant frequency of 5.5×10^7 cps (or Hz!). Electromagnetic waves were produced when the charges accelerated in the circuit. These were propagated to a detector circuit whose resonant frequency was the same as the transmitter. He was able to observe a spark in the receiver, indicating that what began as an LC circuit oscillation in one circuit was transmitted through space to the second LC circuit, thus supporting Maxwell's predictions.

Hertz went further, however, to show that these waves displayed interference characterized by a wavelength of 5.4 m. Thus the propagation speed had to be 2.97×10^8 m/s—within experimental error of being equal to the velocity of light. He also showed that these waves could be reflected by a wall, refracted by a hard pitch prism, and polarized by a wire grating. This proved that electromagnetic waves had all the characteristics associated with visible light.

1.5 Modern Developments

The electromagnetic wave model for light encompasses those aspects of geometrical and physical optics that are of concern to us in this text. Relativistic optics and quantum optics, developments of recent times, are required for a comprehensive picture of light. We will not develop these topics here, because to do so would require significant deviations from the classical optics areas—image formation, lens design, aberrations, radiometry, interference, diffraction, and polarization. We only mention the modern developments for the sake of completeness.

A. Relativistic Optics

Maxwell's equations, which specify the behavior of electric and magnetic fields, are mathematical statements of the physical laws that light obeys. In their original form these equations were closely connected with the mechanical properties of an ethereal medium. The Michelson–Morley experiment showed that the concept of the ether was unnecessary and that the speed of light was independent of the speed of the observer.

This point of view was inconsistent with the Newtonian concept of relative motion. In addition to this problem, it was also recognized near the turn of the century that Maxwell's equations changed if the observer were in motion. This also was a result of the application of Newtonian relativity.

The inconsistencies were resolved with the introduction of the theory of special relativity, which was eloquently stated in 1905 by Albert Einstein (1879–1955). Einstein started from the experimental fact that light has a speed that is independcent of the speed of the observer. He recognized that all nonaccelerated reference frames are totally equivalent for the performance of all physical experiments. From these facts he arrived at a new concept of relativity embodied in the *Lorentz transformations*. The Newtonian transformations from one reference frame to another are a special case of the Lorentz transformations in the limit that the relative speed between reference frames is very much less than the speed of light.

Maxwell's equations turn out to be invariant (preserved in form) when subjected to a Lorentz transformation. The new concept of relativity leads to a new formulation of the laws of motion. We see here the central role that light plays in the formation of physical theory.

B. Quantum Optics

The classical particle models for light were rejected through the last half of the 19th century. The success of the electromagnetic wave model was striking. We have seen how Maxwell's equations, on which the wave theory is based, are even consistent with the theory of special relativity. In 1900, some new ideas were born that brought back the "particle" (photon) concept of light. The classical framework was gone and in its place was a new way of describing, not only light, but also other natural phenomena. Once more, the search for a deeper understanding of light led to a reevaluation of physics. The new revolution gave rise to quantum physics. We call its application to light *quantum optics*. We will, for the most part, not discuss quantum optics in this text. The phenomena with which we are concerned are adequately explained with the electromagnetic model. In dealing with the interaction of light with matter, we parameterize the influence of the matter through quantities that can largely be regarded as experimentally determined.

REFERENCES

General

Born, Max, and Emil Wolf. *Principles of Optics*. Pergamon Press, Oxford, 1980.

Ditchburn, R. W. *Light*. Academic Press, London, 1976.

Driscoll, Walter G., and William Vaughan. *Handbook of Optics*. McGraw-Hill, New York, 1978.

Hecht, Eugene, and Alfred Zajac. *Optics*. Addison-Wesley, Reading, Mass., 1974.

Jenkins, Francis A., and Harvey E. White. *Fundamentals of Optics*. McGraw-Hill, New York, 1976.

Kingslake, Rudolf, Robert R. Shannon, and James C. Wyant, ed. *Applied Optics and Optical Engineering*, Vols. 1–9. Academic Press, New York, 1965–1967, 1981–1983.

Levi, Leo. *Applied Optics*. Wiley, New York, 1968.

Lipson, S. G., and H. Lipson. *Optical Physics*. Cambridge University Press, New York, 1969.

Longhurst, R. S. *Geometrical and Physical Optics*. Wiley, New York, 1967.

Mathiew, J. P. *Optics*. Pergamon Press, Oxford, 1975.

Meyer-Arendt, Jurgen R. *Introduction to Classical and Modern Optics*. Prentice-Hall, Englewood Cliffs, N.J., 1972.

Rossi, Bruno. *Optics*. Addison-Wesley, Reading, Mass., 1967.

Strong, J. *Concepts of Classical Optics*. Freeman, San Francisco, 1958.

Wolf, Emil, ed. *Progress in Optics*, Vols. 1–21. North Holland, Amsterdam, 1961–1984.

Young, M. *Optics and Lasers*. Springer-Verlag, Berlin, 1977.

Chapter 1

Arfken, G. *Mathematical Methods for Physicists*. Academic Press, New York, 1970.

Baker, B. B., and E. J. Copson. *The Mathematical Theory of Huygens' Principle*. Oxford University Press, London, 1969.

Bliss, Gilbert A. *Calculus of Variations*. Mathematical Association of America, 1944.

Born, Max, and Emil Wolf. *Principles of Optics*. Pergamon Press, Oxford, 1980.

Buchdahl, H. A. *An Introduction to Hamiltonian Optics*. Cambridge University Press, Cambridge, 1970.

Cook, David M. *The Theory of the Electromagnetic Field*. Prentice-Hall, Englewood Cliffs, N.J., 1975.

Ditchburn, R. W. *Light*. Academic Press, London, 1976.

Fowles, Grant R. *Introduction to Modern Optics*. Holt, Rinehart and Winston, New York, 1968.

Goldwin, Edwin. *Waves and Photons: An Introduction to Quantum Optics*. Wiley, New York, 1982.

Jackson, John D. *Classical Electrodynamics*. Wiley, New York, 1975.

Klauder, John R., and E. C. G. Sundarshan. *Fundamentals of Quantum Optics*. Benjamin, New York, 1968.

Kline, Morris, and Irwin W. Kay. *Electromagnetic Theory and Geometrical Optics*. Interscience, New York, 1965.

Loudon, R. *The Quantum Theory of Light*. Oxford Clarendon Press, London, 1973.

Misner, C. W., K. S. Thorne, and J. A. Wheeler. *Gravitation*. Freeman, San Francisco, 1973.

Newton, Isaac. *Optics*. Dover, New York, 1979, (originally published, 1704).

Ohanian, H. C. *Gravitation and Spacetime*. Norton, New York, 1976.

Pearson, J. M. *A Theory of Waves*. Allyn & Bacon, Boston, 1966.

Resnick, Robert. *Introduction to Special Relativity*. Wiley, New York, 1968.

Rindler, Wolfgang. *Essential Relativity*. Springer-Verlag, New York, 1977.

Ronchi, Vasco. *The Nature of Light*. Harvard University Press, Cambridge, Mass., 1971.

Rosser, W. G. V. *An Introduction to the Theory of Relativity*. Butterworths, London, 1964.

Rossi, Bruno. *Optics*. Addison-Wesley, Reading, Mass., 1957.

Sabra, A. I. *Theories of Light From Descartes to Newton*. Osbourne, London, 1967.

Shapiro, Alan E., ed. *The Optical Papers of Isaac Newton*. Cambridge University Press, Cambridge, 1984.

Stavroudis, O. N. *The Optics of Rays, Wavefronts and Caustics*. Academic Press, New York, 1972.

Weinstock, Robert. *Calculus of Variations*. Dover, New York, 1974, (originally published, 1952).

Whittaker, Edmund T. *A History of the Theories of Aether and Electricity*. Harper & Row, New York, 1960.

PROBLEMS

Section 1.1   Early Ideas and Observations

1. Identify several common optical phenomena that could have been observed by the ancient Greeks and that *cannot* be explained by the three laws of geometrical optics.

2. Reconstruct Hero's geometrical proof of the law of reflection in *three*-dimensions by considering virtual points O_V that are on the reflecting interface but not necessarily in the plane of incidence.

3. The following data were reported by Ptolemy after his studies of refraction:

θ	10°	20°	30°	40°	50°	60°	70°	80°
θ'	8°	15.5°	22.5°	29°	35°	40.5	45.5	50°

Using *his* law, determine the constants a and b. Using the *true* law of refraction, determine N, the ratio of the indices of refraction of the two media, to a precision compatible with the data.

4. Discuss the mechanism whereby the camera obscura is able to produce an image. For a device such as this, which has a front-to-back size of 10 cm, derive an expression for the magnification (defined as the image size divided by the object size) as a function of the object distance. How does this relation change if the interior of the camera is solid glass, a material whose index of refraction is 1.5 times that of air?

5. Using the law of rectilinear propagation, describe the shadow of the moon on the earth during an eclipse in quantitative terms. The mean diameters of the sun and moon are 1,390,600 km and 3,476 km, respectively, while their mean distances from the earth are 149×10^6 km and 38×10^4 km.

6. Using the power series expansion for $\sin \theta$ and $\sin \theta'$, determine the next two terms that act as corrections to Kepler's form of the law of refraction (that is, $\theta' = N\theta +$ correction terms). How does this compare with Ptolemy's relation? Find the value of θ beyond which the corrections are greater than 10 percent of θ'.

Section 1.2   The Particle Models

7. Newton thought that if light were a wave, then, similar to sound waves, it would diffract around the edges of obstacles (which it does). Estimate the relative sizes of sound diffraction patterns compared with optical diffraction patterns.

8. Obtain a copy of Newton's *Opticks* and examine his report on "Newton's rings." These data were used to support the particle model for light. Identify those aspects of his experiments that could also have been used to support the wave model. Use Newton's data to determine the wavelength of light in his experiment.

9. Determine how Römer measured the speed of light from his observations of Jupiter's moons. Using Römer's data, determine his value for the speed of light.

10. Use Fermat's principle as expressed by Eq. (1.7) to prove the law of reflection.

11. In the text we proved the law of refraction using Fermat's principle by considering virtual paths that were identified with the length of the deviation along the interface in the plane of incidence. Rederive the law of reflection by considering ε to be an *angular* deviation applied to θ, the angle of incidence.

12. Consider the path of light propagation from P to P', including reflection from a spherical mirror as shown in Fig. 1.32. Let P' be at the center of curvature of the mirror, of radius R, and P be on the z-axis. Use Fermat's principle to find the true path and show that the OPL is a maximum path for $z > R$ and a minimum path for $z < R$.

13. Point C is at the center of a reflecting sphere of radius R (see Fig. 1.33).

(**a**) Find values of α for which the line segments $PO + OP'$ will be actual paths of light rays from P to P' via reflection at O.

(**b**) Find an analytic expression for the square of the OPL as a function of α.

(**c**) Plot the square of the OPL as a function of α, showing quantitatively the extremal values of $(\text{OPL})^2$ and of the corresponding values of α.

14. At any point on an ideally spherical earth, you can identify an imaginary plane that is tangent to the earth's sphere. This defines the geometrical horizon. Using this concept, a true sunset time can be calculated. Explain in

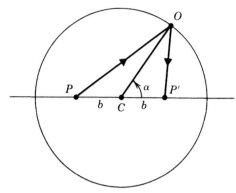

Fig. 1.33

qualitative terms, using Fermat's principle, why the observed sunset time is actually later than the true sunset time.

15. Consider a series of plane interfaces all parallel (see Fig. 1.34). At the first, the index changes from n_0 to n_1; at the second, from n_1 to n_2; at the mth, from n_{m-1} to n_m; and so on. Let θ_m be the angle of refraction at the mth interface and θ_{m-1} be the angle of incidence there.

Show that repeated application of Snell's law gives $n_0 \sin \theta_0 = n_m \sin \theta_m$.

16. Consider a case of light propagation in two dimensions where n is a function of z only; $n = n(z)$. Consider differential elements dz within which an increment of

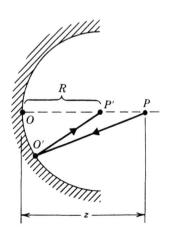

Fig. 1.32

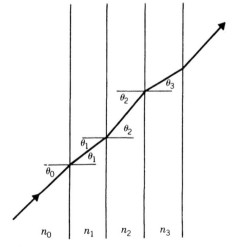

Fig. 1.34

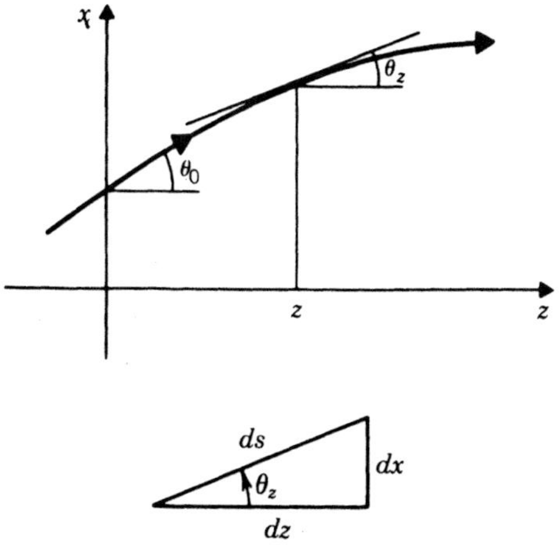

Fig. 1.35

the physical path is ds (see Fig. 1.35). In this element, θ_z is the angle that the ray makes with the z-axis and $\sin \theta_z = dx/ds$.

(a) Using Snell's law (which may be derived from Fermat's principle as we have shown), prove that

$$n(z) \sin \theta_z = n_0 \sin \theta_0$$

where n_0 and θ_0 are measured at $z = 0$.

(b) Show that

$$\left(\frac{\partial z}{\partial x}\right)^2 + 1 = \frac{n^2(z)}{n_0^2 \sin^2 \theta_0}$$

What happens when the right side of this equation equals unity?

This equation gives the slope of the ray in terms of $n(z)$ and the initial conditions at $z = 0$. It may be integrated to obtain $z = z(x)$ or $x = x(z)$ as the equation of the ray.

(c) Find the equation of the ray when $n(z) = a + bz$ if the ray passes through the origin at an angle of inclination θ_0 that is less than 90°. (*Caution:* x will not be a single-valued function of z when b is negative.)

Section 1.3   The Wave Models

17. For each of the following complex numbers z (where a, b, and c are real numbers), find the real part, the imaginary part, and reexpress the number in the form $|z|e^{i\phi}$: $z = (a + ib)^{-1}$; $z = (a + ib)^2$; $z = iae^{ic}$; $z = (1 + i)ae^{ic}$; $z = a + 1/ib + ic$; $z = (i/2)b$

18. The wave equation, Eq. (1.16), was derived from a consideration of discrete particles of mass m separated by distance b and subject to the restoring force $-Cx$. The velocity of the wave is $v = b\sqrt{C/m}$. Show that in the limit as $b \to 0$, with the mass per unit length (mass density) approaching μ, the velocity squared approaches T/μ, where T is the tension in the string of particles, now continuous.

19. Derive an expression for the kinetic energy of one of the particles in the chain of Fig. 1.13 as a function of time. Do the same for the potential energy as a function of time. Show that the solutions of the form of Eq. (1.20) lead to a propagation of the total mechanical energy.

20. Determine the form of the solutions to the difference equation 1.14. Try harmonic waves with $r = nb$. Here n is an integer because the wave only has meaning at the location of one of the masses. Find a relationship between ω and k for this wave.

21. A spherical disturbance propagating away from a source at the origin with velocity v is found to have the value $p_0(t)$ at a radius $r = a$. Find the disturbance $p(r, t)$ for $r \neq a$ and for all t. Repeat this calculation with the same source now moved to $\mathbf{r}' \neq 0$, so that on a sphere of radius a about $\mathbf{r}'$ the disturbance has the value $p_0(t)$. Find $p(\mathbf{r}, t)$ for all $\mathbf{r}$ and t.

22. Demonstrate that $Ar^{-1/2} \cos(\omega t - kr)$ is a solution to the wave equation for $r \gg 2\pi/k$, which corresponds to a cylindrical wave moving away from a line source at speed ω/k.

23. Identify which of the following functions would mathematically qualify as propagating disturbance solutions to the wave equation:

$$P = P_1(vt - z); P = P_2(vt + z);$$
$$P = P_3(vt - z)^3; P = P_4[(vt - z)^2 + z_0^2]^{-1};$$
$$P = P_5[vt - (z - z_0)]^{-1}; P = P_6 \exp[-(vt - z)^2];$$
$$P = P_0[\cos(\omega t - kz) + \cos(\omega t + kz)].$$

24. For the two cases that follow, express the phase of a spherical light wave in the form $at - br$, with t in seconds and r in nanometers.

(a) A wave whose wavelength is 500 nm traveling through a medium with an index of refraction equal to 1.5.

(b) A wave whose frequency is 10^{15} Hz in a vacuum.

25. A sinusoidal plane wave of angular frequency ω is propagating in the direction of the vector $3\hat{x} + 3\hat{y} + 4\hat{z}$. It has its maximum amplitude at the origin. Write down the expression for the phase as a function of time and position. Identify the equation of the constant phase surface that passes through the origin.

26. How far must we be from a point source that emits light at $\lambda = 600$ nm so that the resulting optical fields deviate from plane waves by less than one-eighth of a wavelength over an illuminated spot 1 cm in radius?

27. Use Huygens' construction to prove that there will be no transmitted, propagating wave if $\sin \theta > v/v'$.

Section 1.4 The Electromagnetic Wave Model

28. Prove by direct calculation that the time average of $\cos^2 \phi$ equals $\frac{1}{2}$ where ϕ is an arbitrary linear function of time.

29. Show that the time average of E^2 is equivalent to $(E^*E)/2$ if $E = E_0 \exp(i(\omega t - kr))$.

30. Prove the following identity: $\nabla \times (A_1 \cdot A_2) = A_2 \cdot (\nabla \times A_1) - A_1 \cdot (\nabla \times A_2)$.

31. Prove the vector triple product identity: $A_1 \times (A_2 \times A_3) = A_2(A_1 \cdot A_3) - (A_1 \cdot A_2)A_3$.

32. Given a vector F in rectangular coordinates, $F_x\hat{x} + F_y\hat{y} + F_z\hat{z}$. Write out the components in rectangular coordinates of $\nabla \cdot F$ and $\nabla \times F$.

33. Evaluate the gradient of r^{-1} where the gradient of $f(r)$ is defined as $\nabla f(r)$.

34. Let $f(r) = $ (constant) define a surface in three-dimensional space. Show that ∇f evaluated at $r = r_0$ is perpendicular to that surface at the point r_0.

35. At $y = 0$, $x = b$, find unit vectors perpendicular to
   (a) $x + y + z = 1$;
   (b) $x^2 + y^2 + z^2 = a^2$;
   (c) $x^2 + y^2 = az$.
In each case, make a sketch of your solution.

36. Use Gauss' theorem to calculate the net flux of r^2/r through a spherical surface of radius a about the origin.

37. Prove that a scalar quantity, the electrostatic potential V, may be defined such that $E = -\nabla V$. This ensures that E satisfies Faraday's law, provided the fields do not change with time. The electrostatic potential may be determined from an arbitrary charge distribution $\rho(r)$ through

$$V(r') = \frac{1}{4\pi\varepsilon_0} \iiint \frac{\rho(r)}{|r - r'|} d\mathcal{V}$$

Prove this result by an application of Gauss' law.

38. Determine the electrostatic potential that is associated with a dipole at the origin. A dipole consists of equal positive and negative charges ($+q$ at $z = +a/2$ and $-q$ at $z = -a/2$, where a is very much smaller than the distance r to the observation point). Express your answer in terms of the dipole moment $p = qa\hat{z}$ and the vector r. From this result, find the electric field at r.

39. A dynamic dipole is found at the origin with its time-dependent moment given by $p = qae^{i\omega t}\hat{z}$. At a position r the time dependence of this variation is recorded with a delay caused by the finite propagation time r/c from the origin to the observation point. For a single charge $qe^{i\omega t}$ a "retarded" potential can be defined

$$V = \frac{q}{4\pi\varepsilon_0 r'} \exp\left[i\omega\left(t - \frac{r'}{c}\right)\right]$$

where r' is the distance from the charge to the observation point. Use this information to calculate the retarded potential for the dynamic dipole as a function of r (where $r \gg a$).

40. Demonstrate that the energy density associated with the uniform electrostatic field E_0 within a parallel plate capacitor is $\varepsilon_0 E_0^2/2$ by considering the work performed by a battery that charges the capacitor at constant voltage.

41. At a particular region in space, the electric field is found to be $(E_0/a^2)(x^2\hat{x} + y^2\hat{y} + z^2\hat{z})$. Determine the local charge density.

42. How would Maxwell's equations have to be modified if magnetic monopoles existed?

43. Assume that solutions to the electromagnetic wave equations have the form of Eq. (1.52). Present an argument similar to that which leads to Eq. (1.53) wherein E and B are shown to be perpendicular, starting from Maxwell equation IV.

44. Derive the wave equation for B starting from Maxwell equation IV.

45. What would Eq. (1.48) look like if $\nabla \cdot \mathbf{E} \neq 0$?

46. Show that the plane wave in a vector field $\mathbf{F}$ is transverse if $\nabla \cdot \mathbf{F} = 0$ and is longitudinal if $\nabla \times \mathbf{F} = 0$.

47. An electromagnetic wave carries momentum in addition to energy. Starting from the characteristics of the Poynting vector, derive an expression for the momentum per unit area per unit time that is delivered to a surface at normal incidence provided the surface reflects 50% of the incident optical flux density. What is the magnitude of the momentum density delivery rate in a 1 W/cm² beam?

48. Prove that the Poynting vector is equal to the optical flux density. (Start by forming the divergence of $\mathbf{S}$.)

49. A 100-W light bulb is in the geometrical center of a cubical room with sides of 3 m. Find the time-averaged total energy within the room from the light bulb, assuming that no light is absorbed or reflected by the walls. (The walls are perfectly transparent.)

50. Determine the magnitude of the magnetic induction in an electromagnetic wave that carries a power density of 100 mW/cm². How much current would be required in a long straight wire so that a magnetic induction of this same magnitude would be generated at a distance of 1 mm from the wire?

51. The flux density 1 m from a point source is 15 mW/cm². Determine the total flux emitted by the source. Estimate the flux density at a distance of 2 m from the source.

52. Derive an expression for the flux density in terms of the Poynting vector $\mathbf{S}(r)$ as a function of the radial distance r from a point source whose strength is A (where A has the dimensions of electric field times length). Repeat the calculation for an infinitely long line source where r is the radial distance far from the source and A has the dimensions of electric field times (length)$^{1/2}$.

53. A light beam of circular cross section with a diameter of 5 mm carries 200 mW of power. The beam intersects a wall such that the angle between the wall and the incoming beam is 20°. Quantitatively determine the size and shape of the illuminated spot and calculate the power density on the wall assuming that the beam is uniform.

54. In Foucault's rotating mirror experiment, what would the distance between the rotating mirror and the retroreflector mirror need to be if the mirror had an angular velocity of 100 rad/s and a deflection of 1° was desired for the return reflection?

55. How is it possible for a wire grating, such as was used by Hertz, to polarize an electromagnetic wave with a wavelength of 5.4 m?

2 Optics of Planar Interfaces

We have seen that the fundamental laws of geometrical optics, including the two that involve interaction with matter (the laws of reflection and refraction), can be derived from either Fermat's principle or from Huygens' construction. Neither of these approaches, however, is capable of providing information about the fraction of the radiant energy that ends up in the reflected or transmitted beams. We need a more complete theory to describe these phenomena as well as the interaction of light with absorbing, rather than transparent, materials.

In principle, we may start from the quantum theory of matter, but it is more practical to parameterize the behavior of the medium in the presence of electric and magnetic fields. Once the parameters are identified theoretically or experimentally, the electromagnetic theory of light is sufficient to explain the reflection and refraction characteristics of an interface. Ultimately, the quantum theory is required to justify the details of these parameters.

With plane interfaces the geometry of the modification of ray directions and wavefronts is particularly simple. The description of the changes in amplitude and phase of the optical fields is the subject matter of the optics of planar interfaces.

2.1 Light Waves in Matter

The starting point for the treatment of light in matter from the electromagnetic perspective is contained within the four equations of Maxwell, which we recall here for convenience:

$$\text{I} \quad \nabla \cdot \bar{\mathbf{E}} = \frac{\rho}{\varepsilon_0} \tag{1.39}$$

$$\text{II} \quad \nabla \cdot \bar{\mathbf{B}} = 0 \tag{1.41}$$

$$\text{III} \quad \nabla \times \bar{\mathbf{E}} = -\frac{\partial \bar{\mathbf{B}}}{\partial t} \tag{1.43}$$

$$\text{IV} \quad \nabla \times \bar{\mathbf{B}} = \mu_0 \left(\bar{\mathbf{J}} + \varepsilon_0 \frac{\partial \bar{\mathbf{E}}}{\partial t} \right) \tag{1.45}$$

The microscopic form for these equations has been averaged over a volume element $\Delta \mathscr{V}$ so as to eliminate the dramatic variations due to the atomic structure of matter. Thus these equations already are in terms of the macroscopic fields $\bar{\mathbf{E}}$ and $\bar{\mathbf{B}}$ and the macroscopic charge density $\bar{\rho}$ and current density $\bar{\mathbf{J}}$.

We need to separate out the contributions to the source terms $\bar{\rho}$ and $\bar{\mathbf{J}}$ that come from the static and moving charges associated with the electrons and ions of the matter. By doing this, we can lump the influence of the matter into easy-to-handle parameters and, at the same time, produce a form of Maxwell equations from which a wave equation can be derived. This will tell us how the electromagnetic wave propagates within matter and how the characteristics of the propagation depend on the parameters in Maxwell's equations.

We must remember that the fields we are talking about are high-frequency oscillating optical fields and that the response of the electrons and ions must be considered at these high frequencies. This is easy for many students to forget, because much introductory electromagnetic theory is concerned with static electric and magnetic fields. The material parameters are functions of the optical frequency. In quantum mechanical terms, this means that the elementary excitations of the matter depend on the energies of the individual photons. Unless otherwise specified, we will assume that the light under consideration is monochromatic with frequency $v = \omega/2\pi$, thus representing photons of energy hv.

A. Bound Charges and Currents

1. Polarization. Macroscopic regions of matter can contain a net charge or current density. For instance, a pith ball can be charged electrostatically, or a metallic conductor can be carrying a macroscopic current. These are called *free charges* and *free current*, respectively. They should be distinguished from *bound charge* and *bound current*. The argument for bound charge is summarized as follows.

A given molecule may have a net charge; if so, it contributes to the free charge density ρ_f. Whether or not the molecule has a net charge, it may be polarized; that is, there may be a relative separation of positive charges (atomic nuclei) and negative charges (electrons) within it. Let $\mathbf{r}_{+j}$ be the position vector of the jth positive charge ($+q_j$) with respect to, say, the center of mass of the molecule, and let $\mathbf{r}_{-i}$ be the position vector of the ith negative charge ($-q_i$). Then the electric dipole moment of the molecule is defined to be

$$\mathbf{p} = \sum_j q_j \mathbf{r}_{+j} - \sum_i q_i \mathbf{r}_{-i} \tag{2.1}$$

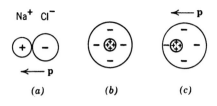

Fig. 2.1 (a) Permanent dipole.
(b) Unpolarized atom (zero dipole).
(c) Polarized atom (induced dipole).

In the quantum mechanical picture of a molecule, the position vectors $\mathbf{r}_{+j}$ and $\mathbf{r}_{-i}$ must represent all regions of space at which there is some probability of finding an electron or a nucleus. We cannot uniquely identify the exact position of each point change. This problem can be solved by interpreting the electron contribution as an integral:

$$-\sum q_i \mathbf{r}_{-i} \rightarrow \frac{-|e| \iint\!\int P(\mathbf{r})\mathbf{r}\, d\mathscr{V}}{\iint\!\int P(\mathbf{r})\, d\mathscr{V}}$$

where $-|e|$ is the charge on an electron and $P(\mathbf{r})$ is the quantum mechanical probability of finding an electron at position $\mathbf{r}$. A similar interpretation can be made for $\sum q_j \mathbf{r}_{+j}$, if necessary.

Generally, we distinguish between permanent electric dipoles, as is the case with an NaCl molecule, which is composed of two oppositely charged ions, and induced dipoles resulting from charge displaced in an external field. These two cases are illustrated in Fig. 2.1.

The polarization vector $\mathbf{P}$ is defined to be the averaged dipole moment per unit volume:

$$\mathbf{P} = \frac{1}{\Delta\mathscr{V}}\sum \mathbf{p} \tag{2.2}$$

where the sum is over all dipoles in the macroscopically small but microscopically large volume $\Delta\mathscr{V}$.

Bound charge is defined to be the macroscopic charge in matter resulting from the presence of macroscopic polarization $\mathbf{P}$. When $\mathbf{P}$ is constant, each macroscopic volume element contains the same number of positive and negative bound charges, and the bound charge density ρ_b is zero. When $\mathbf{P}(\mathbf{r})$ is spatially varying, a nonzero bound-charge density can appear. This is shown schematically in Fig. 2.2a. The mean dipole moment $\langle \mathbf{p} \rangle$ increases gradually from zero in plane A to a net value in the $+x$ direction in plane B. We see that the B plane tends to cut through some dipoles, whereas the A plane does not. There is thus a net negative (bound) charge in the volume between planes A and B.

The general formula for the bound-charge density is

$$\rho_b = -\boldsymbol{\nabla}\cdot\mathbf{P} \equiv -\frac{\partial P_x}{\partial x} - \frac{\partial P_y}{\partial y} - \frac{\partial P_z}{\partial z} \tag{2.3}$$

We will not derive Eq. (2.3) here, but instead we will make it plausible by reference to Fig. 2.2. Part (b) of the figure shows the individual molecular dipoles lined up,

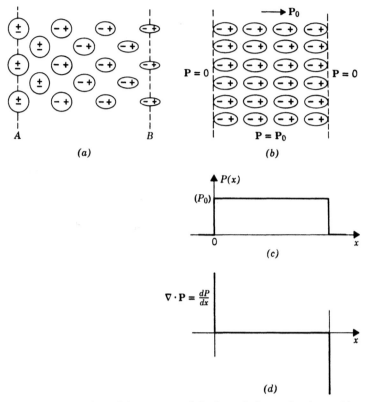

Fig. 2.2 Illustration of the concept of the bound-charge density and its relationship to the dipole moment per unit volume spatial dependence of the dipole moment per unit volume.

producing a uniform polarization $\mathbf{P}_0$ inside a dielectric slab. The function $P(x)$ gives the magnitude of $\mathbf{P}$ and is sketched in part (c). Then $\nabla \cdot \mathbf{P} = dP/dx$ is sketched in part (d), which consists of two localized "spikes." From part (b) we see that there is a net positive surface charge density on the right surface and a negative surface charge on the left. In this example, the spikes represent the bound surface charge density.

2. Bound Current. Bound current results from motion of the charges bound in matter. The resulting current density is

$$\mathbf{J}_b \equiv \frac{1}{\Delta \mathscr{V}} \sum \left(q_{+i} \frac{dr_{+i}}{dt} - q_{-i} \frac{d\mathbf{r}_{-i}}{dt} \right) \tag{2.4}$$

$$= \frac{d\mathbf{P}}{dt}$$

In magnetic material, account must also be taken of the magnetization $\mathbf{M}$, which is defined to be the average magnetic dipole moment per unit volume:

$$\mathbf{M} = \frac{1}{\Delta \mathscr{V}} \sum \mathbf{m} \tag{2.5}$$

The magnetization of material comes from atomic currents resulting from either the spinning of electrons or the electron motion within the atoms. The general relationship is

$$\mathbf{J}_m = \nabla \times \mathbf{M} \tag{2.6}$$

As with Eq. (2.3), we will provide a plausibility argument to justify this expression. Consider a square cross section of a material that possesses a net magnetic moment along the z-axis as in Fig. 2.3a. The atomic currents that give rise to the net moment cancel everywhere except on the surface. If an adjacent region of the material has a different magnetization, then the circulating current around that region will differ in magnitude from the first region (Fig. 2.3b).

In this example $\nabla \times \mathbf{M} = (dM_z/dy)\hat{\mathbf{x}}$, which is related to $\mathbf{J}_m = J_m \hat{\mathbf{x}}$ by Eq. (2.6). To see this more clearly, consider the approximation for the magnetic dipole moment of one of the small regions, $m_z = I \Delta x \Delta y$. Within a region the magnetization is

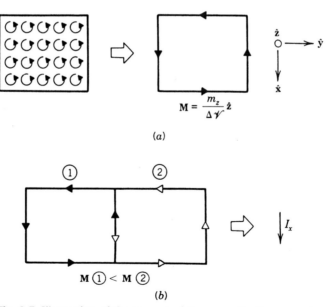

Fig. 2.3 Illustration of the concept of the magnetization current density and its relationship to the spatial dependence of the magnetic moment per unit volume.

$$M_z = \frac{m_z}{\Delta \mathcal{V}} \cong \frac{I}{\Delta z} \tag{2.7}$$

and the derivative with respect to y is approximately

$$\frac{dM_z}{dy} \cong \frac{\Delta I}{\Delta y \Delta z} \tag{2.8}$$

But the change in the surface current is, from Fig. 2.3b,

$$\Delta I = I_x$$

thus,

$$\frac{dM_z}{dy} \cong \frac{I_x}{\Delta y \Delta z} \cong J_x \tag{2.9}$$

When the orientation and spatial variation of $\mathbf{M}$ is more complicated, then the other components of $\nabla \times \mathbf{M}$ are needed to specify the complete magnetization current density.

This modification to Eq. (2.4) leads to the most general form for the bound-current density,

$$\mathbf{J}_b = \frac{d\mathbf{P}}{dt} + \nabla \times \mathbf{M} \tag{2.10}$$

Free charges and currents are then defined by the differences

$$\rho_f = \bar{\rho} - \rho_b \tag{2.11}$$

$$\mathbf{J}_f = \bar{\mathbf{J}} - \mathbf{J}_b \tag{2.12}$$

3. Maxwell's Equations in Matter. In Eqs. (2.11) and (2.12), we have separated the total density $\bar{\rho}$ and current density $\bar{\mathbf{J}}$, which appear in Maxwell equations I and IV respectively (Eqs. 1.30 and 1.45), into parts arising from the structure of matter and parts resulting from free charges and currents. This makes it possible to clarify the role of matter in defining the electromagnetic fields.

a. Field Equations. Using Eqs. (2.11) and (2.3), Maxwell's equation I (Gauss' law for $\mathbf{E}$) becomes

$$\nabla \cdot \bar{\mathbf{E}} = \frac{1}{\varepsilon_0} (\rho_f + \rho_b) = \frac{\rho_f}{\varepsilon_0} - \frac{\nabla \cdot \mathbf{P}}{\varepsilon_0} \tag{2.13}$$

Through Eqs. (2.12) and (2.10), Maxwell equation IV (displacement current version of Ampere's law) is changed also:

$$\nabla \times \bar{\mathbf{B}} = \mu_0 \left(\bar{\mathbf{J}}_f + \bar{\mathbf{J}}_b + \varepsilon_0 \frac{\partial \bar{\mathbf{E}}}{\partial t} \right)$$

$$= \mu_0 \left(\bar{\mathbf{J}}_f + \frac{d\mathbf{P}}{dt} + \nabla \times \mathbf{M} + \varepsilon_0 \frac{\partial \bar{\mathbf{E}}}{\partial t} \right) \tag{2.14}$$

Our goal is to transform these into forms that look like the original Maxwell equations I and IV. This we can do by introducing the *electric displacement* **D**:

$$\mathbf{D} \equiv \varepsilon_0 \mathbf{E} + \mathbf{P} \tag{2.15}$$

We have dropped the bar over **E** in Eq. (2.15). *From now on it is understood that* **E** *and* **B** *refer to macroscopic fields*. We also introduce the *magnetic field* **H**:

$$\mathbf{H} \equiv \frac{\mathbf{B}}{\mu_0} - \mathbf{M} \tag{2.16}$$

Equation (2.13) then may be written

$$\nabla \cdot \mathbf{D} = \rho_f \tag{2.17}$$

and Eq. (2.14) may be written

$$\nabla \times \mathbf{H} = \mathbf{J}_f + \frac{\partial \mathbf{D}}{\partial t} \tag{2.18}$$

Along with Maxwell equations II and III (Gauss' law for **E** and Faraday's law), which remain unchanged, Eqs. (2.17) and (2.18) give the four Maxwell equations in the presence of matter:

$$\text{I} \quad \nabla \cdot \mathbf{D} = \rho_f \tag{2.17}$$

$$\text{II} \quad \nabla \cdot \mathbf{B} = 0 \tag{1.41}$$

$$\text{III} \quad \nabla \times \mathbf{E} = \frac{-\partial \mathbf{B}}{\partial t} \tag{1.43}$$

$$\text{IV} \quad \nabla \times \mathbf{H} = \mathbf{J}_f + \frac{\partial \mathbf{D}}{\partial t} \tag{2.18}$$

The source equations now contain only free charges and currents.

b. Energy Equations. In the presence of matter it becomes necessary to redefine the energy density as

$$U = \tfrac{1}{2}[\mathbf{D} \cdot \mathbf{E} + \mathbf{B} \cdot \mathbf{H}] \tag{2.19}$$

The Poynting vector is now

$$\mathbf{S} = \mathbf{E} \times \mathbf{H} \tag{2.20}$$

B. Response Functions

1. Parameterization of Matter. To proceed with our development of a wave equation for light in matter, we must express **D** and **H** as functions of **E** and **B**. This is particularly simple if the relationships are *local*, *isotropic*, and *linear*.

The response of the medium to an applied field is local if the response can be determined by examining the applied field at the same point. In the case of the electric field, the response may be evaluated by the polarization. As shown in Fig.

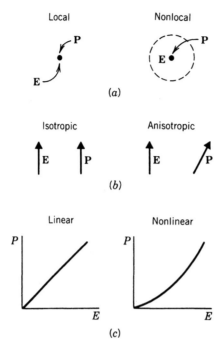

Fig. 2.4 Schematic representation of the parameterization characteristics of matter that are of importance in identifying material parameters. Here they are expressed as relationships between **E** and **P**.

2.4*a*, the polarization depends on the nature of the electric field in a region surrounding the point of interest in nonlocal media. This is a complication that is more important in nonhomogeneous systems.

In an isotropic medium the direction of the applied field is preserved in the response. As shown in Fig. 2.4*b*, the polarization direction may be different from the electric field in anisotropic material. This has important consequences when considering the polarization characteristics of the light wave as it travels through the material. We will discuss this in detail in Chapter 9. Here we will only be concerned with isotropic systems in which the direction of the applied field is not important.

If, in addition, the medium is linear, then the magnitude of the response is directly proportional to the magnitude of the stimulus. As shown in Fig. 2.4*c*, a nonlinear medium may demonstrate gain. This important characteristic of nonlinear optics is the basis of laser operation. Frequently, nonlinearity is coupled with anisotropy.

For the local, isotropic, linear case, the polarization is related to the total electric field by

$$\mathbf{P} = \varepsilon_0 \chi \mathbf{E} \tag{2.21}$$

where χ is the electric susceptibility. This means that

$$\mathbf{D} = \varepsilon_0 \mathbf{E} + \mathbf{P} = \varepsilon_0 (1 + \chi)\mathbf{E}$$

To simplify the notation, two alternative functions are frequently introduced: The permittivity of the medium is

$$\varepsilon = \varepsilon_0(1 + \chi) \tag{2.22}$$

which in the SI system has the units of ε_0, $C^2 N^{-1} m^{-2}$. Alternatively, the relative permittivity or dielectric function may be used

$$\kappa = \frac{\varepsilon}{\varepsilon_0} = (1 + \chi) \tag{2.23}$$

This is a dimensionless number. Any of these material parameters are functions of the optical frequency. The relation between the displacement and the total electric field becomes

$$\mathbf{D} = \varepsilon_0 \kappa \mathbf{E} = \varepsilon \mathbf{E} \tag{2.24}$$

In magnetic material that conforms to our simplifications, the relationship between the magnetization and the magnetic field is

$$\mathbf{M} = \chi_m \mathbf{H} \tag{2.25}$$

so that

$$\mathbf{B} = \mu_0(\mathbf{H} + \mathbf{M}) = \mu_0(1 + \chi_m)\mathbf{H}.$$

If we define the permeability as

$$\mu = \mu_0(1 + \chi_m) \tag{2.26}$$

which has units of μ_0, N sec^2 C^{-2}, or the relative permeability

$$\kappa_m = \mu/\mu_0 = 1 + \chi_m \tag{2.27}$$

the magnetic induction $\mathbf{B}$ will have a simple connection to the magnetic field intensity $\mathbf{H}$.

$$\mathbf{B} = \mu_0 \kappa_m \mathbf{H} = \mu \mathbf{H} \tag{2.28}$$

The material-dependent parameters χ and χ_m are, in general, complex numbers. This means that $\mathbf{D}$ may be out of phase with $\mathbf{E}$, and $\mathbf{B}$ may be out of phase with $\mathbf{H}$.

In the approach chosen here, all the properties of the medium are wrapped up in the complex parameters χ and χ_m. An alternative procedure would be to consider the electric susceptibility to be a real number. The out-of-phase, or imaginary, part of the response would have to be associated with another material-dependent parameter. This is usually accomplished by introducing Ohm's law and treating the motion of free electrons in metallic media as leading to a contribution to the free-current density.

2. Reformulated Field Equations. The characteristics of the medium are now contained in the parameters χ (or ε) and χ_m (or μ), which are functions of the optical frequency. They include the information necessary to describe dispersion, absorption, and high-frequency magnetic response. We will consider problems in which no free charge or free current is found. Only bound charge and current, which are associated with the electrons and ions of the atoms in the optical medium, appear in the source-dependent equations. However, Eqs. (2.17) and (2.18) are simplified under these conditions. Because the bound sources have been reexpressed in terms of associated fields, the reformulated Maxwell equations look like Eq. (1.46):

$$\text{I} \quad \varepsilon \mathbf{V} \cdot \mathbf{E} = 0$$

$$\text{II} \quad \mathbf{V} \cdot \mathbf{B} = 0$$

$$\text{III} \quad \mathbf{V} \times \mathbf{E} = -\frac{\partial \mathbf{B}}{\partial t}$$

$$\text{IV} \quad \mathbf{V} \times \mathbf{B} = \mu \varepsilon \frac{\partial \mathbf{E}}{\partial t}$$

(2.29)

C. Plane Waves in Matter

1. The Wave Equation. Through the similarity between the sets of Eqs. (2.29) and (1.46) we can immediately write down the wave equations for $\mathbf{E}$ and $\mathbf{B}$ in matter.

$$\nabla^2 \mathbf{E} - \mu \varepsilon \frac{\partial^2 \mathbf{E}}{\partial t^2} = 0; \quad \nabla^2 \mathbf{B} - \mu \varepsilon \frac{\partial^2 \mathbf{B}}{\partial t^2} = 0$$

(2.30)

These same equations hold for $\mathbf{D}$ and $\mathbf{H}$. They differ from the free-space equations by the substitutions

$$\varepsilon_0 \to \varepsilon$$

and

$$\mu_0 \to \mu$$

where the new values are complex quantities.

Plane wave solutions to Eqs. (2.30) exist in the form of Eqs. (1.52).

$$\mathbf{E} = \mathbf{E}_0 e^{i(\omega t - \mathbf{k} \cdot \mathbf{r} + \varphi_E)}$$

(2.31)

$$\mathbf{B} = \mathbf{B}_0 e^{i(\omega t - \mathbf{k} \cdot \mathbf{r} + \varphi_B)}$$

We can still use the operator shortcuts of Eqs. (1.29) and (1.30). When applied to Eq. (2.30), we find

$$k^2 = \mu \varepsilon \omega^2$$

(2.32)

This shows that k is, in general, complex.

We define a complex index of refraction

$$\tilde{n}^2 \equiv (n - i\mathscr{k})^2 \equiv \mu \varepsilon c^2 = \frac{\mu}{\mu_0} \frac{\varepsilon}{\varepsilon_0}$$

(2.33)

such that

$$k = \tilde{n} \frac{\omega}{c}$$

(2.34)

Some authors write $\tilde{n} = n(1 - i\kappa)$, where κ is called the attenuation index. Others use a time dependence $e^{-i\omega t}$, which changes the sign of i in all our equations. For example, the index of refraction would be written $n + i\ell$. However, n and ℓ would be the same functions as defined here.

Applying the operator shortcuts to Eq. (2.29) III, we arrive at a relationship between **E** and **B**, similar to Eq. (1.53).

$$\mathbf{B} = \frac{\mathbf{k} \times \mathbf{E}}{\omega} = \frac{\tilde{n}}{c} \hat{\mathbf{s}} \times \mathbf{E} \tag{2.35}$$

Thus **E** and **B** are still perpendicular to each other and mutually perpendicular to the direction of propagation $\hat{\mathbf{s}}$. But they are no longer in phase with each other for the most general case that still meets our restrictions of locality, isotropy, and linearity.

2. Attenuation. In the cases where **k** is a complex quantity, as it is for metals and other absorbing media, the optical fields become attenuated with distance. Substituting Eqs. (2.34) into the first of Eqs. (2.31) (with $\varphi_E = 0$) shows that

$$\mathbf{E} = \mathbf{E}_0 \exp\left[-\ell \frac{\omega}{c} \hat{\mathbf{s}} \cdot \mathbf{r}\right] \exp\left[i\left(\omega t - n\frac{\omega}{c} \hat{\mathbf{s}} \cdot \mathbf{r}\right)\right] \tag{2.36}$$

Assuming that the direction of propagation is such that

$$\hat{\mathbf{s}} \cdot \mathbf{r} = r$$

this leads to the simplification

$$\mathbf{E} = \mathbf{E}_0 e^{-r/\delta} e^{i\omega(t - nr/c)} \tag{2.37}$$

where

$$\delta = \frac{c}{\ell \omega} = \frac{\lambda_0}{2\pi\ell} \tag{2.38}$$

is the classical *skin depth* and is a measure of the length that light will penetrate through the attenuating medium. (λ_0 is the wavelength of the light as measured in a vacuum.)

3. Dispersion. If **k** is purely real, as it is for optically transparent media, in conjunction with real values for ε and μ, then Eq. (2.37) becomes

$$\mathbf{E} = \mathbf{E}_0 e^{i\omega(t - nr/c)} \tag{2.39}$$

The electromagnetic wave described by this formula propagates with a phase velocity of

$$v = \frac{c}{n} = \frac{1}{\sqrt{\mu\varepsilon}} \tag{2.40}$$

This confirms our definition of the refractive index and shows the consistency of the electromagnetic theory with the early ideas surrounding the speed of light in matter.

Since the optical properties of a transparent medium are functions of the frequency, the angle of deviation described by Snell's law of refraction must also change with frequency. This characteristic of matter is called *dispersion*.

4. Electromagnetic Energy in a Medium. We know that, in general, **E** and **B** will not be in phase with each other. To see this explicitly form

$$\mathbf{B} = \frac{\tilde{n}}{c}\,\hat{\mathbf{s}} \times \mathbf{E}_0 e^{-r/\delta} e^{i\phi}, \quad \phi = \omega(t - nr/c) \tag{2.41}$$

which can be written as

$$\mathbf{B} = \frac{|\tilde{n}|}{c}\,\hat{\mathbf{s}} \times \mathbf{E}_0 e^{-r/\delta} e^{i(\phi - \varphi_{EB})} \tag{2.42}$$

where

$$\tan \varphi_{EB} = \frac{\ell}{n}$$

Equation (2.20) is an expression for the rate of energy propagation per unit area due to the optical plane wave in the medium. We must exercise care in calculating the vector product. To do so, let us express **E** and **H** in trigonometric form. In what follows we will assume that μ *is real*, an excellent assumption for most optical materials. Then,

$$\mathbf{H} = \frac{\mathbf{B}}{\mu} = \frac{|\tilde{n}|}{\mu c}\,\hat{\mathbf{s}} \times \mathbf{E}_0 e^{-r/\delta} \cos(\phi - \varphi_{EB}) \quad \text{and} \quad \mathbf{E} = \mathbf{E}_0 e^{-r/\delta} \cos \phi$$

The Poynting vector is

$$\mathbf{S} = \frac{|\tilde{n}|}{c\mu}\,|E_0|^2 e^{-2r/\delta} \cos \phi \cos(\phi - \varphi_{EB})\hat{\mathbf{s}}$$

The time average involves

$$\langle \cos \phi \cos(\phi - \varphi_{EB}) \rangle = \langle \cos^2 \phi \cos \varphi_{EB} + \cos \phi \sin \phi \sin \varphi_{EB} \rangle$$
$$= \tfrac{1}{2} \cos \varphi_{EB}$$

But since $\tan \varphi_{EB} = \ell/n$, then

$$\cos \varphi_{EB} = \frac{n}{|\tilde{n}|}$$

Therefore, the time-averaged energy-flux density in matter is

$$\langle \mathbf{S} \rangle = \frac{n}{c\mu} \frac{|E_0|^2}{2} e^{-Kr} \hat{\mathbf{s}} \tag{2.43}$$

where the *absorption coefficient* is

$$K \equiv \frac{2}{\delta} = \frac{4\pi\ell}{\lambda_0} \tag{2.44}$$

There is exponential decay of the optical signal with r. The optical power is reduced to e^{-1} of its starting value in a distance equal to K^{-1}. Note that when $\ell = 1$ then $K^{-1} = \lambda_0/4\pi$, so that the penetration is roughly one order of magnitude less than λ_0.

In nonmagnetic media (often the case) $\mu = \mu_0$ and $(c\mu_0)^{-1} = c\varepsilon_0$. Thus

$$\langle \mathbf{S} \rangle = n\left[\frac{\varepsilon_0 c}{2}|\mathbf{E}_0|^2 \hat{\mathbf{s}}\right]e^{-Kr} \tag{2.45}$$

where the terms in brackets would be the time-averaged energy-flux density in a vacuum.

5. Optical Properties. The functions n and ℓ versus frequency (or wavelength) represent the characteristics of a local, isotropic, linear medium in the electromagnetic theory of light. These may be measured through careful observation of the reflection and/or transmission of light through a sample of the material of interest. They may also be calculated from microscopic theories of the optical interactions with the electrons and ions of the material. We will cover this in more detail in section 2.4.

Once determined, the optical properties may be tabulated for future use in solving problems in optical design. These data are also useful as experimental verifications of the microscopic theory. For both reasons, the optical properties of matter are important.

Table 2.1 gives the complex indices of refraction for two characteristic materials. The optical properties of a typical metal are shown in Fig. 2.15a; there we find a large value of ℓ.

2.2 Reflection and Transmission of Light at an Interface

In this section, we derive and discuss the coefficients that describe the changes in field amplitude and irradiance that light is subject to on reflection and transmission at an interface between two media having different optical properties. In addition, we will prove the laws of reflection and refraction (Snell's law) once again.

We assume that there is an abrupt interface between two otherwise isotropic and homogeneous local and linear media. In general, the indices of refraction may be complex.

A. Boundary Conditions

The basis for the electromagnetic approach to this problem is the establishment of boundary conditions that impose restrictions on the relationships between fields on either side of the interface.

Table 2.1. Characteristic Optical Properties

Wavelength (micrometers)	Fused Silica[a]	Silicon[b]	
	n^c	n	k
0.28	1.49416	3.244	5.230
0.30	1.48779	4.894	3.938
0.32	1.48273	4.983	3.272
0.34	1.47865	5.234	3.018
0.36	1.47528	6.147	2.976
0.38	1.47248	6.510	0.881
0.40	1.47011	5.619	0.341
0.42	1.46809	5.130	0.191
0.44	1.46634	4.824	0.130
0.46	1.46483	4.608	0.093
0.48	1.46350	4.454	0.069
0.50	1.46232	4.281	0.055
0.52	1.46128	4.207	0.042
0.54	1.46034	4.112	0.035
0.56	1.45949	4.049	0.029
0.58	1.45873	3.987	0.025
0.60	1.45803	3.933	0.022
0.62	1.45739	3.892	0.019
0.64	1.45681	3.854	0.017
0.66	1.45626	3.816	0.015
0.68	1.45576	3.782	0.014
0.70	1.45529	3.755	0.012

[a] I. H. Malitson. *J. Opt. Soc. Am. 55*, 1205, 1965.

[b] G. E. Jellison and F. A. Modine. *Oak Ridge National Laboratory Report* TM-8002, 1982.

[c] The imaginary part of the index of refraction of fused silica is so small in this wavelength range that it is commonly neglected.

Maxwell's four equations are presented in integral form in Eqs. (1.38), (1. (1.42), and (1.44). After our introduction of the response of matter and parameterization for the local isotropic linear approximation, these equati become

$$\text{I} \quad \oiint \varepsilon \mathbf{E} \cdot d\mathbf{A} = 0$$

$$\text{II} \quad \oiint \mathbf{B} \cdot d\mathbf{A} = 0 \qquad\qquad (2$$

$$\text{III} \quad \oint \mathbf{E} \cdot d\mathbf{l} = \frac{-\partial}{\partial t} \iint \mathbf{B} \cdot d\mathbf{A}$$

$$\text{IV} \quad \oint \frac{\mathbf{B}}{\mu} \cdot d\mathbf{l} = \frac{\partial}{\partial t} \int\!\!\int \varepsilon \mathbf{E} \cdot d\mathbf{A}$$

where, in general, ε is complex. We will assume from now on that μ is real.

1. Continuity of Normal Components of $\varepsilon\mathbf{E}$ and $\mathbf{B}$. Consider the "pillbox" shaped geometrical region pictured in Fig. 2.5a. The height h is assumed infinitesimally small. The area $\mathscr{A}$ is finite but arbitrary. The volume $\mathscr{A}h = \mathscr{V}$ is infinitesimally small. From Maxwell equation I applied to this region we have

$$\int\!\!\int \varepsilon E_n \, d\mathscr{A} = 0$$

Because h is very small, there is no finite contribution to the integral from the sides. What remains is

$$\int\!\!\int_{\mathscr{A}_1} \varepsilon E_n \, d\mathscr{A} + \int\!\!\int_{\mathscr{A}_2} \varepsilon E_n \, d\mathscr{A} = \int\!\!\int_{\mathscr{A}} (\varepsilon_1 E_{1n} - \varepsilon_2 E_{2n}) \, d\mathscr{A} = 0$$

where $\varepsilon_1 E_{1n}$ and $\varepsilon_2 E_{2n}$ represent the outward normal components of the displacement fields just to the left of and to the right of the interface. Because the surface $\mathscr{A}$ is arbitrary, we are at liberty to choose a small enough area to ensure that E_n remains constant.

We may conclude that

$$\varepsilon_1 E_{1n} = \varepsilon_2 E_{2n} \qquad (2.47)$$

The normal component of $\varepsilon\mathbf{E}$ is continuous across the interface.

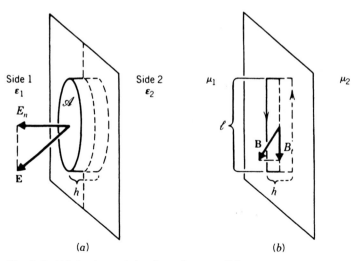

Side 1 Side 2
ε_1 ε_2
E_n
$\mathbf{E}$
μ_1 μ_2
ℓ
$\mathbf{B}$ B_t
h h
$\mathscr{A}$

(a) (b)

Fig. 2.5 Aids in determining boundary conditions.

The same derivation applies to B. Hence

$$B_{1n} = B_{2n} \qquad (2.48)$$

The normal component of **B** *is also continuous.*

2. Continuity of Tangential Components of B/μ and E. Consider now the "ribbon-shaped" surface shown in Fig. 2.5b. The width h is infinitesimally small and the length ℓ is finite but arbitrary. The area of the ribbon is negligible because h is very small. Thus, when applying Maxwell equation IV to this surface, the right side of the equation will be zero. The line integral is to be taken around the periphery of the ribbon, but the ends contribute nothing.

Thus we conclude that

$$\int_{\ell_1} \frac{B_t}{\mu} \, d\ell + \int_{\ell_2} \frac{B_t}{\mu} \, d\ell = \int_{\ell} \left(\frac{B_{1t}}{\mu_1} - \frac{B_{2t}}{\mu_2} \right) d\ell = 0$$

where B_{1t}/μ_1 and B_{2t}/μ_2 are the tangential components of the magnetic field on either side of the interface. The ribbon is arbitrary, so we may choose a small enough length to avoid variations in the fields on a given side of the interface. Thus,

$$\frac{B_{1t}}{\mu_1} = \frac{B_{2t}}{\mu_2} \qquad (2.49)$$

or, *the tangential component of* **B**/μ *is continuous across the interface.*

An analogous derivation yields

$$E_{1t} = E_{2t} \qquad (2.50)$$

So E_t, *the tangential component of* **E** *must also be continuous across the interface.*

B. The Laws of Geometrical Optics at Interfaces

Let the interface lie in the xy plane, and let the plane of incidence be the xz plane as shown in Fig. 2.6. The fields **E** and **B** may be resolved into their components in the plane of incidence and their y-components normal to the plane of incidence. Two independent cases then arise: the case where **E** is perpendicular to the plane of incidence (σ-case) with **B** in the plane of incidence, and the case where **E** is in the plane of incidence (π-case) with **B** perpendicular to it.

The spatial dependence of plane waves that represent the incident, the reflected, and the transmitted beams are $e^{-i\mathbf{k} \cdot \mathbf{r}}$, $e^{-i\mathbf{k}'' \cdot \mathbf{r}}$, and $e^{-i\mathbf{k}' \cdot \mathbf{r}}$, respectively. The wave vectors will have components

$$\mathbf{k} = (k_x, 0, k_z) = \tilde{n}\left(\frac{\omega}{c}\right)(\sin\theta, 0, \cos\theta) \qquad (2.51a)$$

$$\mathbf{k}'' = (k_x'', k_y'', k_z'') \quad \text{with} \quad k_x''^2 + k_y''^2 + k_z''^2 = \tilde{n}^2\left(\frac{\omega}{c}\right)^2 \qquad (2.51b)$$

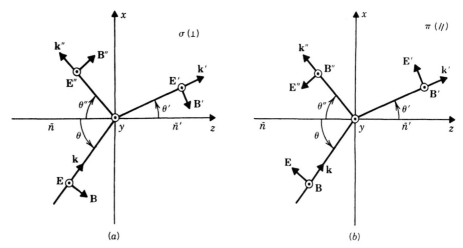

Fig. 2.6 Geometry that establishes the conventions for the optics at an interface: (a) $\mathbf{E}$ perpendicular to the plane of incidence; (b) $\mathbf{E}$ parallel to the plane of incidence.

$$\mathbf{k}' = (k_x', k_y', k_z') \quad \text{with} \quad k_x'^2 + k_y'^2 + k_z'^2 = \tilde{n}'^2 \left(\frac{\omega}{c}\right)^2 \qquad (2.51c)$$

where $\tilde{n}$ and $\tilde{n}'$ are the complex refractive indices in the two media.

The continuity requirements on the tangential parts of $\mathbf{E}$ and $\mathbf{B}/\mu$ across the interface at $z = 0$ then take the form

$$E_t e^{-ixk_x} + E_t'' e^{-i(xk_x'' + yk_y'')} = E_t' e^{-i(xk_x' + yk_y')}$$

$$\left(\frac{B_t}{\mu}\right) e^{-ixk_x} + \left(\frac{B_t''}{\mu}\right) e^{-i(xk_x'' + yk_y'')} = \left(\frac{B_t'}{\mu'}\right) e^{-(xk_x' + yk_y')}$$

The fields E_t, E_t'', E_t', B_t, B_t'', and B_t' in the previous expressions are the *amplitudes* of the respective field components. The time dependence $e^{i\omega t}$ is the same for all fields, so it cancels out. (We have eliminated the subscript "0" to simplify the notation. Later, when we are dealing exclusively with the electric field, we will use A, A'', and A' for the *magnitude* of the field amplitudes $\mathbf{E}_0$, $\mathbf{E}_0''$, and $\mathbf{E}_0'$ or for the strength of the electric field in a spherical wave.) These equations must hold for all x and y and will be valid if and only if the phases are everywhere the same.

One consequence is the condition

$$k_y'' = k_y' = 0 \qquad (2.52a)$$

This says that the reflected and transmitted wave vectors must lie in the plane of incidence (xz plane). Other requirements are

$$k_x'' = k_x \qquad (2.52b)$$

and

$$k_x' = k_x \qquad (2.52c)$$

Combining Eqs. (2.51) with Eqs. (2.52b and c) leads to $k''_x = \tilde{n}(\omega/c) \sin \theta$ and $k'_x = \tilde{n}(\omega/c) \sin \theta$. From Fig. 2.6 these can also be written as

$$\tilde{n}\left(\frac{\omega}{c}\right) \sin \theta'' = \tilde{n}\left(\frac{\omega}{c}\right) \sin \theta \tag{2.53}$$

or

$$\sin \theta'' = \sin \theta$$

and

$$\tilde{n}'\left(\frac{\omega}{c}\right) \sin \theta' = \tilde{n}\left(\frac{\omega}{c}\right) \sin \theta \tag{2.54}$$

or

$$\tilde{n}' \sin \theta' = \tilde{n} \sin \theta$$

Equation (2.53) is equivalent to $\theta'' = \theta$, the law of reflection, which holds for any value of $\tilde{n}$ and $\tilde{n}'$. Equation (2.54) is equivalent *in form* to Snell's law, which is used to find the direction of the transmitted beam in dielectrics. Although Eq. (2.54) remains valid when the indices are complex, the direction of the transmitted beam is not equal to θ' unless both $\tilde{n}$ and $\tilde{n}'$ are real (as they are for optically transparent media).

C. Amplitude Relations

With the phase-matching conditions thus properly accounted for, the boundary conditions become conditions on the amplitudes of the fields

$$E_y + E''_y = E'_y \tag{2.55a}$$

$$B_x + B''_x = \frac{\mu}{\mu'} B'_x \tag{2.55b}$$

$$B_y + B''_y = \frac{\mu}{\mu'} B'_y \tag{2.56a}$$

$$E_x + E''_x = E'_x \tag{2.56b}$$

From here on we will assume that the media are nonmagnetic (an excellent approximation in most cases), so that $\mu = \mu_0$ in all the equations.

1. σ Case: E Perpendicular to the Plane of Incidence. The boundary conditions are expressed in Eq. (2.55). For the σ case, E has only a y component. Then B must lie in the xz plane. Equation (2.35) gives B in terms of E

$$\mathbf{B} = \frac{\mathbf{k} \times \mathbf{E}}{\omega}$$

Equation (2.55b), which now says that B_x is continuous, then becomes

$$(\mathbf{k} \times \mathbf{E})_x + (\mathbf{k}'' \times \mathbf{E}'')_x = (\mathbf{k}' \times \mathbf{E}')_x \tag{2.57}$$

Because the E's have only y components, we may write

$$E_y \equiv E, \quad E_y'' \equiv E'', \quad E_y' \equiv E'$$

The k's have only x and z components; thus, $(\mathbf{k} \times \mathbf{E})_x = -k_z E_y$. Using these specific relationships, we obtain from Eq. (2.55a) (which now says that E_y is continuous)

$$E + E'' = E' \tag{2.58a}$$

and from Eq. (2.57)

$$k_z E + k_z'' E'' = k_z' E'$$

But $k_z'' = -k_z$. Thus

$$E - E'' = \frac{k_z'}{k_z} E' \tag{2.58b}$$

Equations (2.58a and b) are readily solved

$$E' = \frac{2}{1 + \left(\dfrac{k_z'}{k_z}\right)} E \equiv \tau_\sigma E \tag{2.59a}$$

$$E'' = \frac{1 - \left(\dfrac{k_z'}{k_z}\right)}{1 + \left(\dfrac{k_z'}{k_z}\right)} E \equiv \rho_\sigma E \tag{2.59b}$$

From Eq. (2.54) we have

$$a \equiv \frac{k_z'}{k_z} = \frac{\tilde{n}' \cos \theta'}{\tilde{n} \cos \theta} = \frac{\tan \theta}{\tan \theta'} \tag{2.60}$$

the last equality following from Snell's law. Using Eq. (2.60) and the trigonometric identity

$$\cos \theta \sin \theta' + \sin \theta \cos \theta' = \sin(\theta + \theta')$$

we find for the *transmission coefficient*

$$\tau_\sigma = \frac{2}{1 + a} = \frac{2\tilde{n} \cos \theta}{\tilde{n} \cos \theta + \tilde{n}' \cos \theta'} \tag{2.61a}$$

$$= \frac{2 \sin \theta' \cos \theta}{\sin(\theta + \theta')} \tag{2.61b}$$

and similarly for the *reflection coefficient*

$$\rho_\sigma = \frac{1 - a}{1 + a} = \frac{\tilde{n} \cos \theta - \tilde{n}' \cos \theta'}{\tilde{n} \cos \theta + \tilde{n}' \cos \theta'} \tag{2.62a}$$

$$= -\frac{\sin(\theta - \theta')}{\sin(\theta + \theta')} \tag{2.62b}$$

The last equalities for τ_σ and ρ_σ in Eqs. (2.61) and (2.62) are known as the Fresnel relations for the σ case (after Augustin Fresnel who was first to derive their form from the mechanical-elastic theory of light). They describe the magnitudes of the reflected and transmitted electric fields in terms of the incident field magnitude. In addition, since τ_σ and ρ_σ are, in general, complex numbers, these coefficients express how the phase of the electric field changes on encountering an interface. Careful reference should be made to the conventions established in Fig. 2.6 for the directions of the fields that have been defined to be positive and in phase with each other. For instance, in Eq. (2.62b) if the ratio of the sine functions is positive the reflection coefficient is negative. This means that there is a phase shift of 180° for the optical electric field on reflection in this case.

At this point it is useful to compare the preceding results with those that hold when the roles of incident and transmitted waves are interchanged. Let us therefore consider the situation when the light travels backward along $-\mathbf{k}'$. The incident angle is now θ'. The reflected light appears at $\theta'' = \theta'$, and the transmitted light is found at θ in what is now the transmitting medium. The ratio a becomes

$$a' = \frac{\tilde{n} \cos \theta}{\tilde{n}' \cos \theta'} = \frac{1}{a}$$

and the reflection coefficient obeys

$$\rho_\sigma' = \frac{1 - a'}{1 + a'} = \frac{1 - a^{-1}}{1 + a^{-1}} = \frac{-(1 - a)}{1 + a} = -\rho_\sigma$$

The minus sign in the relation

$$\rho_\sigma' = -\rho_\sigma \tag{2.63}$$

corresponds to a 180° phase difference when the reflection coefficients from opposite sides of an interface are compared.

If we make a similar comparison for the transmitted fields, the transmission coefficient from the opposite side is

$$\tau_\sigma' = \frac{2}{1 + a'} = \frac{2}{1 + a^{-1}} = \frac{2a}{1 + a} = a\tau_\sigma$$

or

$$\tau_\sigma \tau_\sigma' = a\tau_\sigma^2 = \frac{4a}{(1 + a)^2}$$

This should be compared with

$$1 - \rho_\sigma^2 = \frac{(1 + a)^2 - (1 - a)^2}{(1 + a)^2} = \frac{4a}{(1 + a)^2}$$

Thus

$$\tau_\sigma \tau_\sigma' = 1 - \rho_\sigma^2 \tag{2.64}$$

Equations (2.63) and (2.64) also hold for the π case. They are important relationships that will be used later in the book.

2. π Case: E in the Plane of Incidence. In this case **B** has only a y-component. The resulting directions of **E**, **E''**, and **E'** are as shown in Fig. 2.6b. This figure suggests that we adopt for the scalar quantities

$$B_y \equiv B, \ B_y'' \equiv B''$$

and

$$B_y' \equiv B'$$

Because from Eq. (2.35) $B = \tilde{n}E/c$, the boundary condition for B in the π case [Eq. (2.56a)] is

$$\tilde{n}E + \tilde{n}E'' = \tilde{n}'E' \tag{2.65}$$

To generate a modification of Eq. (2.56b) that is similar to Eq. (2.57), as the modified form of Eq. (2.55a), we need to apply the plane-wave operator shortcuts to the Maxwell equation (2.29) IV. This leads to

$$-i\mathbf{k} \times \mathbf{B} = \mu_0 \varepsilon i \omega \mathbf{E}$$

The tangential or x-component of E, which must be continuous across the interface, is then

$$E_x = \frac{-c^2}{\omega} \frac{(\mathbf{k} \times \mathbf{B})_x}{\tilde{n}^2} = \frac{c^2}{\omega} \frac{k_z B}{\tilde{n}^2} \tag{2.66}$$

where the last equality follows from the fact that $k_y = 0$. Using Eq. (2.66) in Eq. (2.56b) yields

$$\frac{k_z \tilde{n} E}{\tilde{n}^2} + \frac{k_z'' \tilde{n} E''}{\tilde{n}^2} = \frac{k_z' \tilde{n}' E'}{\tilde{n}'^2}$$

or, since $k_z'' = -k_z$

$$\tilde{n}(E - E'') = \left(\frac{\tilde{n}}{\tilde{n}'}\right)^2 \left(\frac{k_z'}{k_z}\right) \tilde{n}' E' \tag{2.67}$$

The solutions to Eqs. (2.65) and (2.67) are

$$E' = \tau_\pi E \qquad E'' = \rho_\pi E$$

with

$$\tau_\pi = \frac{2(\tilde{n}/\tilde{n}')}{1 + b}, \quad \rho_\pi = \frac{1 - b}{1 + b} \tag{2.68}$$

and

$$b = \left(\frac{\tilde{n}}{\tilde{n}'}\right)^2 \left(\frac{k_z'}{k_z}\right) = \left(\frac{\tilde{n}}{\tilde{n}'}\right)^2 a \tag{2.69}$$

Alternative versions may be derived, namely,

$$\tau_\pi = \frac{2\tilde{n}\cos\theta}{\tilde{n}'\cos\theta + \tilde{n}\cos\theta'} = \frac{2\sin\theta'\cos\theta}{\sin(\theta + \theta')\cos(\theta - \theta')} \tag{2.70a}$$

$$\rho_\pi = \frac{\tilde{n}'\cos\theta - \tilde{n}\cos\theta'}{\tilde{n}'\cos\theta + \tilde{n}\cos\theta'} = \frac{\tan(\theta - \theta')}{\tan(\theta + \theta')} \tag{2.70b}$$

The last equalities in Eqs. (2.70a and b) express the Fresnel relations for the π-case.

Through a development similar to that surrounding Eqs. (2.63) and (2.64) we may readily demonstrate that

$$\rho'_\pi = -\rho_\pi$$

and

$$\tau_\pi \tau'_\pi = 1 - \rho_\pi^2$$

Thus the general expressions are most appropriately written as

$$\rho' = -\rho \tag{2.71}$$

and

$$\tau\tau' = 1 - \rho^2 \tag{2.72}$$

D. Energy Reflection and Transmission Coefficients

The average Poynting vector $\langle S \rangle$ gives the time-averaged power per unit area; for a plane wave this is a maximum on a surface normal to the direction of propagation. From Eq. (2.45) we have at $r = 0$

$$\langle \mathbf{S} \rangle = n\frac{\varepsilon_0 c}{2}|\mathbf{E}_0|^2\hat{\mathbf{s}}$$

where n is the real part of $\tilde{n}$.

The average power crossing a unit area of a plane surface parallel to the interface, the irradiance, is given by the magnitude of the z-component of $\langle S \rangle$, which is $\langle S \rangle_z = \langle S \rangle \cos\theta$. Thus we have for the three waves

$$\text{Incident}\quad \langle S \rangle_z = \frac{\varepsilon_0 c}{2} n|E|^2 \cos\theta$$

$$\text{Reflected}\quad \langle S \rangle_z'' = \frac{\varepsilon_0 c}{2} n|E''|^2 \cos\theta$$

$$\text{Transmitted}\quad \langle S \rangle_z' = \frac{\varepsilon_0 c}{2} n'|E'|^2 \cos\theta'$$

The reflectance R is defined by the ratio

$$R \equiv \frac{\langle S \rangle_z''}{\langle S \rangle_z} = \left|\frac{E''}{E}\right|^2 = |\rho|^2 \tag{2.73}$$

and the transmittance T is defined by the ratio

$$T \equiv \frac{\langle S \rangle'_z}{\langle S \rangle_z} = \frac{|E'|^2 n' \cos \theta'}{|E|^2 n \cos \theta} = |\tau|^2 \frac{n' \cos \theta'}{n \cos \theta} \tag{2.74}$$

We can readily show that for both σ and π cases we have

$$T + R = 1 \tag{2.75}$$

which must follow from conservation of energy, because the power per unit area at the interface in the incident beam equals the power in the reflected beam plus the power in the transmitted beam.

E. Dielectric Media

When the materials on either side of the interface are dielectrics, the indices $\tilde{n}$ and $\tilde{n}'$ are real. At normal incidence the distinction between the σ and π cases is lost. We have a and $b^{-1} = n'/n$ and hence

$$\rho_\sigma = \frac{n - n'}{n + n'} = -\rho_\pi \tag{2.76a}$$

$$\tau_\sigma = \frac{2n}{n + n'} = \tau_\pi \tag{2.76b}$$

The sign difference in Eq. (2.76a) is necessary because our convention in Fig. 2.6b shows the incident and reflected optical electric field in opposite directions at normal incidence.

For a simple numerical example, take $n'/n = 1.5$. Then

$$\rho_\sigma = -\rho_\pi = -0.2$$

and the reflectance is

$$R = R_\sigma = R_\pi = \rho_\sigma^2 = 0.04$$

The transmission coefficient is

$$\tau_\sigma = \tau_\pi = 0.8$$

and the transmittance is

$$T = T_\sigma = \tau^2 n'/n = 0.96 = 1 - R$$

The next cases to be discussed involve a nonzero angle of incidence.

1. Oblique Incidence Where $n < n'$ (External Reflection). This would be the case for light incident from air onto a glass surface, for instance. In the σ case the parameter a is given by

$$a = \frac{n' \cos \theta'}{n \cos \theta} = \frac{n'\sqrt{1 - \sin^2 \theta'}}{n \cos \theta} = \frac{\sqrt{n'^2 - n^2 \sin^2 \theta}}{n \cos \theta}$$

This is real and positive for all values of θ. Furthermore, from Snell's law—$\sin \theta' = (n/n') \sin \theta$—we find that for all values of θ between zero and $90°$ the inequality $\theta' < \theta$ is obeyed. Thus,

$$\rho_\sigma = \frac{-\sin(\theta - \theta')}{\sin(\theta + \theta')}$$

is always negative, indicating that in the σ case the electric field undergoes a $180°$ phase shift on external reflection.

At grazing incidence, when $\theta \rightarrow 90°$ we have $a \rightarrow + \infty$ and hence $\rho_\sigma \rightarrow -1$. The overall behavior of ρ_σ and $R_\sigma = \rho_\sigma^2$ as a function of θ is shown in Fig. 2.7 for the case $n'/n = 1.5$.

For the π case, we must consider the parameter $b = (n/n')^2 a$ and a reflection coefficient given by

$$\rho_\pi = \frac{1 - b}{1 + b} = \frac{\tan(\theta - \theta')}{\tan(\theta + \theta')}$$

The limiting value at grazing incidence is $\rho_\pi(90°) = -1$. At normal incidence ρ_π is positive

$$\rho_\pi(0°) = \frac{n' - n}{n' + n} [= 0.2 \text{ for our numerical example}]$$

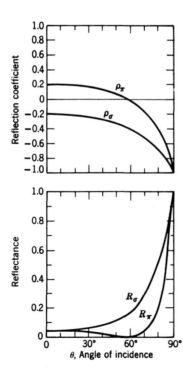

Fig. 2.7 Reflection coefficients and reflectances for a dielectric interface with $n'/n = 1.5$.

[Note that the convention in Fig. 2.6 requires that a positive ρ_π as associated with a 180° phase shift at normal incidence.] As θ ranges from 0° to 90°, ρ_π decreases monotonically from 0.2 to -1. It equals zero when

$$\theta + \theta' = 90° \tag{2.77}$$

because then $\tan(\theta + \theta')$ becomes infinite.

When θ satisfies Eq. (2.77), then the angle of incidence is at *Brewster's angle* θ_B. If Snell's law is used, Eq. (2.77) gives (at $\theta = \theta_B$),

$$\tan \theta_B = \frac{n'}{n} \tag{2.78}$$

For our numerical example of $n'/n = 1.5$, $\theta_B = 56.3°$. The corresponding angle of refraction is 33.7° under the Brewster condition. Because ρ_π is zero at Brewster's angle, this effect can be used as an efficient way of reducing the light to one linearly polarized component. The σ component has a reflection coefficient

$$\rho_\sigma(\theta_B) = -\sin(\theta_B - \theta_B'') = \sin 22.6° = -0.384$$

and the reflectance $R(\theta_B)$ is $0.147 \approx 15\%$.

When the material has a complex index of refraction, the reflectance versus incident angle is similar to the dielectric medium example except that there is, in general, no angle at which Brewster's relation is satisfied. Instead, a pseudo-Brewster angle is demonstrated at which the π-polarized component goes through a *minimum* (Fig. 2.15b).

2. Oblique Incidence Where $n > n'$ (Internal Reflection). An example of this situation is the light incident on the surface of a pool of water from beneath the water. Snell's law shows that $\theta' > \theta$ until $\theta = 90°$. This defines the critical angle of incidence θ_c:

$$\sin \theta_c = \frac{n'}{n} \tag{2.79}$$

For $n/n' = 1.5$, $\theta_c = 41.8$.

At angles of incidence greater than the critical angle, we have $\sin \theta' > 1$; this means that the angle θ' becomes imaginary. In this subsection we will assume that the angle of incidence is less than the critical angle. Then θ' is real, as is

$$a = \frac{n' \cos \theta'}{n \cos \theta} = \frac{\sqrt{n'^2 - n^2 \sin^2 \theta}}{n \cos \theta}$$

At normal incidence we now have a positive value for

$$\rho_\sigma = -\rho_\pi(0°) = \frac{n - n'}{n + n'} \left[= +0.2 \text{ for } \frac{n}{n'} = 1.5 \right]$$

At the critical angle we have $a = b = 0$, so that

$$\rho_\sigma(\theta_c) = \rho_\pi(\theta_c) = 1$$

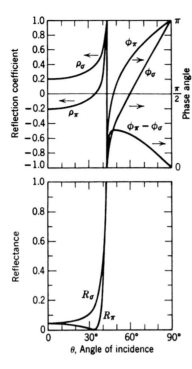

Fig. 2.8 Reflection coefficients, phase shifts, and reflectances for a dielectric interface with $n/n' = 1.5$.

The coefficient ρ_π now increases monotonically from -0.2 to $+1$ as θ increases from $0°$ to θ_c. Now ρ_π passes through zero at the new Brewster's angle given by

$$\tan \theta_B = \frac{n'}{n}$$

For $n/n' = 1.5$, $\theta_B = 33.7°$. The resulting behavior of ρ_σ, ρ_π, R_σ, and R_π is shown in Fig. 2.8.

3. Total Internal Reflection. In the case where $n > n'$ and the angle of incidence is greater than the critical angle defined by Eq. (2.79), the parameter a becomes purely imaginary. We therefore introduce a new real parameter γ so that

$$a = \frac{k'_z}{k_z} = -i\gamma \equiv -i\frac{\sqrt{n^2 \sin^2 \theta - n'^2}}{n \cos \theta} \tag{2.80}$$

The choice of the minus sign in Eq. (2.80) is dictated by the need to have decreasing exponential dependence in the z-direction for the transmitted wave. We will see in a few lines that this is necessary because the transmitted wave does not propagate. The complex notation for the transmitted field is then given by

$$E'(x, z, t) = E' e^{i(\omega t - k' \cdot r)} = E' e^{i(\omega t - x k_x)} e^{-\gamma k_z z} \tag{2.81}$$

with $k_x = n\omega (\sin \theta)/c$, $k_z = n\omega (\cos \theta)/c = 2\pi n (\cos \theta)/\lambda_0$ and with λ_0 the vacuum wavelength. The power flux density in the second medium is proportional to

$$|E'|^2 e^{-2\gamma k_z z} = |E'|^2 e^{-2z/\delta}$$

with a skin depth given by

$$\delta = \frac{1}{k_z \gamma} = \frac{\lambda_0}{2\pi\sqrt{n^2 \sin^2 \theta - n'^2}} \tag{2.82}$$

A wave such as (2.81) that has a real exponential decay in one spatial direction is called an inhomogeneous plane wave or an *evanescent wave*.

The reflection coefficients become

$$\rho_\sigma = \frac{1 + i\gamma}{1 - i\gamma}; \quad \rho_\pi = \frac{1 + i\left(\dfrac{n}{n'}\right)^2 \gamma}{1 - i\left(\dfrac{n}{n'}\right)^2 \gamma}$$

The reflectances are unity:

$$R_\sigma = |\rho_\sigma|^2 = \left|\frac{1 + i\gamma}{1 - i\gamma}\right|^2 = \frac{1 + \gamma^2}{1 + \gamma^2} = 1; \quad R_\pi = |\rho_\pi|^2 = 1$$

The light is totally reflected as long as the incident angle is greater than the critical angle. This is called *total internal reflection*.

Since $R_\sigma = R_\pi = 1$, the transmittances T_σ and T_π are zero. There is no flow of energy in the second medium in a direction normal to the interface. This can be directly demonstrated by a calculation of $\langle S \rangle_{tz}$. However, there is a nonzero component of $\langle S \rangle$ in the x-direction. This means that the evanescent wave moves along the surface at the intersection of the plane of incidence.

The coefficients ρ_σ and ρ_π are complex and impose a phase change on the electric field that is a fractional part of 180°. This remains true even though the amplitude of the reflected field remains the same at all angles greater than the critical angle. To better discuss this, we introduce the angles ϕ_σ and ϕ_π as defined by

$$\rho_\sigma = e^{-i\phi_\sigma} \quad \text{and} \quad \rho_\pi = e^{-i\phi_\pi} \tag{2.83}$$

This requires

$$\tan \frac{\phi_\sigma}{2} = \gamma \quad \text{and} \quad \tan \frac{\phi_\pi}{2} = \frac{n^2}{n'^2} \gamma$$

The phase angles ϕ_σ and ϕ_π are plotted in the right part of the top of Fig. 2.8.

If the incident wave is linearly polarized such that the orientation of the optical field has both σ and π components, then there will be a relative phase difference between the reflected σ and π components. This results in elliptical polarization (see section 9.2).

4. Frustrated Total Internal Reflection. The exponential decay of $E \sim e^{-z/\delta}$ of the fields in the second medium can be detected if a second interface is placed at a

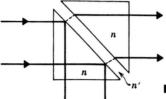

Fig. 2.9 Geometry in which frustrated internal reflection can be demonstrated.

certain depth d in that medium. Often the third medium has the same refractive index as the first, as in the example in Fig. 2.9. If d is somewhat larger than the skin depth δ, we can neglect multiple reflection effects—the first multiple-reflected beam will travel a distance $3d$ and hence will be attenuated by the factor $e^{-3d/\delta}$, which we assume is much less than $e^{-d/\delta}$.

Just inside the second interface, the amplitude of the transmitted field for, say, the π component will be given by

$$E'_\pi = E_\pi \tau_\pi \tau'_\pi e^{(-d/\delta)}$$

where E_π is the field amplitude just outside the first interface. There is a similar expression for the σ component. Our previous results for the τ's hold with the parameter a replaced by $i\gamma$. In particular, we have

$$\tau_\pi \tau'_\pi = 1 - \rho_\pi^2 = 1 - e^{2i\phi_\pi}$$

The coefficient for energy transmission across the gap is then given by

$$T_\pi = 2e^{-2d/\delta}(1 - \cos 2\phi_\pi) \quad (d > \delta) \tag{2.84}$$

and, because there are no dissipative, loss-producing mechanisms, by the principle of energy conservation, the beam reflected by the $n - n' - n$ "sandwich" would be described by a reflectance

$$R_\pi = 1 - T_\pi$$

When d is not greater than a few times the skin depth, these results must be modified to take multiple reflections into account or, equivalently, to take the proper boundary conditions into account.

The phenomenon of frustrated total internal reflection is exactly analogous to the quantum-mechanical phenomenon of penetrating or tunneling of a plane wave through a one-dimensional rectangular barrier.

2.3 Applications in Planar Surface Optics

Through our understanding of the electromagnetic theory of light we are able to solve many problems in planar surface optics. We present here a small collection of examples.

A. Dielectrics

When the absorption coefficient of the material is zero, $K = 4\pi k/\lambda_0 = 0$, then the indices of refraction are real. Snell's law can be used to define the direction of the transmitted ray uniquely. This is also the case for which total internal reflection is well-defined.

1. Prism Refraction. The angle of refraction will be a function of the wavelength of light through the indices of the media comprising an interface. This phenomenon, called *dispersion*, is such that for many optical materials (such as glass) the refractive index is a decreasing function of λ (as illustrated in Table 2.1). This may be a problem in many optical design situations because it leads to *chromatic aberration*. The effect may also be used to advantage to produce monochromatic light, provided all but the desired radiation can be physically blocked after it is dispersed. A convenient shape for dispersion analysis and exploitation is the prism (see Fig. 2.10).

The prism is the heart of the *prism monochromator*. Although such instruments have been largely replaced by *grating monochromators* (in which an interference grating effects dispersion), the analysis of dispersion in a prism still provides an instructive example for the application of Snell's law.

Following the convention established in Fig. 2.10, we see that the angle of deviation at P is $\theta - \theta'$. At R the deviation is $\theta_F - \theta_{IN}$. The total deviation is therefore

$$\theta_D = \theta + \theta_F - \theta' - \theta_{IN}$$

By extrapolating the normals to where they cross at Q we define a small triangle PQR. An exterior angle, at Q, is also identified, which is equal to the apex angle α of the prism. Because the sum of the other two interior angles must equal the exterior angle, we have

$$\alpha = \theta' + \theta_{IN} \tag{2.85}$$

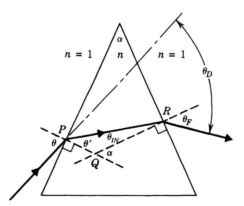

Fig. 2.10 Refraction through a prism.

The total deviation becomes

$$\theta_D = \theta + \theta_F - \alpha \tag{2.86}$$

The final angle θ_F is related to θ and α through Snell's law by

$$\sin \theta_F = n \sin \theta_{IN} \tag{2.87}$$

$$\sin \theta = n \sin \theta' \tag{2.88}$$

and Eq. (2.85).

The angular dispersion is

$$\frac{d\theta_D}{d\lambda} = \frac{d\theta_D}{dn} \frac{dn}{d\lambda} \tag{2.89}$$

If θ is fixed, then

$$\frac{d\theta_D}{dn} = \frac{d\theta_F}{dn} \tag{2.90}$$

Differentiating Eq. (2.87) leads to

$$\frac{d\theta_F}{dn} \cos \theta_F = n \cos \theta_{IN} \frac{d\theta_{IN}}{dn} + \sin \theta_{IN} \tag{2.91}$$

From Eq. (2.85)

$$\frac{d\theta_{IN}}{dn} = \frac{-d\theta'}{dn}$$

and from Eq. (2.88)

$$0 = n \cos \theta' \frac{d\theta'}{dn} + \sin \theta'$$

Thus

$$\frac{d\theta_{IN}}{dn} = \frac{1}{n} \tan \theta' \tag{2.92}$$

Combining Eqs. (2.91) and (2.92) leads to

$$\frac{d\theta_F}{dn} = \frac{\cos \theta_{IN} \tan \theta' + \sin \theta_{IN}}{\cos \theta_F} \tag{2.93}$$

But $\cos \theta_{IN} = \cos(\alpha - \theta') = \sin \alpha \sin \theta' + \cos \alpha \cos \theta'$, and $\sin \theta_{IN} = \sin(\alpha - \theta') = \sin \alpha \cos \theta' - \cos \alpha \sin \theta'$. If we substitute these into Eq. (2.93), we get

$$\frac{d\theta_F}{dn} = \frac{\sin \alpha}{\cos \theta_F \cos \theta'} \tag{2.94}$$

The dispersive characteristics of a prism can be used to measure the index of refraction of the prism with high precision. Assume that the incident light is

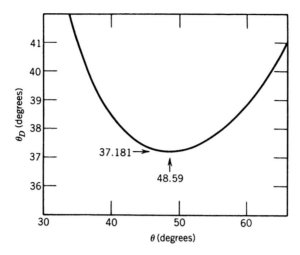

Fig. 2.11 Total deviation through a prism, whose apex angle is $\alpha = 60°$, which is constructed of a transparent material with index $n = 1.5$, as a function of the incident angle.

monochromatic. As θ is varied, we will find that θ_D changes, passing through a minimum at a well-defined angle (Fig. 2.11). Let us examine the conditions for minimum deviation.

For an extremum in θ_D, we must have [from Eq. (2.86)]

$$\frac{d\theta_D}{d\theta} = 1 + \frac{d\theta_F}{d\theta} = 0$$

or

$$\frac{d\theta_F}{d\theta} = -1 \tag{2.95}$$

The differential forms of Eqs. (2.87) and (2.88) are

$$\cos\theta_F \, d\theta_F = n\cos\theta_{IN} \, d\theta_{IN}$$

and

$$\cos\theta \, d\theta = n\cos\theta' \, d\theta'$$

also from Eq. (2.85) we get

$$d\theta' = -d\theta_{IN}$$

Using the last three equations in Eq. (2.95) produces

$$\frac{d\theta_F}{d\theta} = \frac{\left(\dfrac{-n\cos\theta}{\cos\theta_F}\right)}{\left(\dfrac{n\cos\theta'}{\cos\theta}\right)} = -1$$

This requires that

$$\frac{\cos\theta_{IN}}{\cos\theta_F} = \frac{\cos\theta'}{\cos\theta}$$

This will be true if

$$\theta = \theta_F$$

and

$$\theta_{IN} = \theta'$$

Using this information in Eqs. (2.85) and (2.86) establishes the following conditions at minimum deviation

$$\theta' = \theta_{IN} = \frac{\alpha}{2} \tag{2.96}$$

and

$$\theta = \theta_F = \frac{\alpha + \theta_D}{2} \tag{2.97}$$

These expressions can be reintroduced into Snell's law for either interface to yield

$$\sin\left(\frac{\alpha + \theta_D}{2}\right) = n \sin\left(\frac{\alpha}{2}\right)$$

or

$$n = \frac{\sin\left(\dfrac{\alpha + \theta_D}{2}\right)}{\sin\left(\dfrac{\alpha}{2}\right)} \tag{2.98}$$

The incident wavelength can be changed. At each wavelength the deviation where $\theta_F = \theta_{IN}$ (or minimum deviation) can be measured. Then from Eq. (2.98) the index of refraction versus λ can be identified.

2. Dielectric Waveguides. One of the most prominent applications of optical technology is in the field of fiber optic communications. Thin ($\approx 50 \ \mu m$) cylindrical glass or plastic fibers can be used to carry signals in place of metal wires. The big advantage is in the available bandwidth when the carrier is a light wave. This makes it possible for one fiber to carry many more different independent signals than can a wire. There are also advantages in weight and independence from limited resources. In the laboratory, short-length optical fibers are also very useful; for example, you can isolate a sensitive photoelectric device from electronic noise by shielding the device and routing the light signal to it via a fiber.

A proper treatment of electromagnetic wave propagation in an optical fiber requires a solution of the wave equation subject to the boundary conditions imposed by the fiber. We will not cover this topic. Our approach is to consider the geometrical optics of a ray that remains in a plane containing the optical axis. This will enable us to discuss some of the fundamental ideas associated with optical fibers without the complications of the complete wave theory.

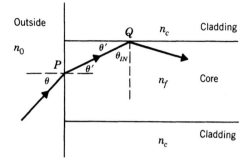

Fig. 2.12 Relevant angles in the ray treatment in the meridional plane of an optical fiber.

The principle behind the ray theory of an optical fiber is total internal reflection. The light is multiply-reflected and therefore is guided down the fiber.

a. *Acceptance Angle.* Our model fiber is shown in cross section in Fig. 2.12. The refractive indices n_c and n_f are functions of wavelength. The behavior that we now discuss holds for a limited range of wavelengths for a fiber system composed of given materials. We must have total internal reflection at point Q. By symmetry, all subsequent reflections will then also conform to this condition. This means

$$\theta_{IN} \geqslant \theta_c = \sin^{-1}\left(\frac{n_c}{n_f}\right) \tag{2.99}$$

The internal angle θ_{IN} is related to the transmitted angle θ' at P by $\theta' = 90° - \theta_{IN}$. This in turn is connected to the exterior angle θ by Snell's law.

$$n_o \sin \theta = n_f \sin \theta'$$

Substituting this information into Eq. (2.99) leads to

$$\theta_{IN} = 90° - \sin^{-1}\left(\frac{n_o \sin \theta}{n_f}\right) \geqslant \sin^{-1}(n_c/n_f)$$

or

$$90° - \sin^{-1}\left(\frac{n_c}{n_f}\right) \geqslant \sin^{-1}\left(\frac{n_o \sin \theta}{n_f}\right)$$

Take the sine of both sides of the inequality to get

$$\sin\left[90° - \sin^{-1}\left(\frac{n_c}{n_f}\right)\right] \geqslant \frac{n_o \sin \theta}{n_f}$$

The left side is

$$\sin\left[90° - \sin^{-1}\left(\frac{n_c}{n_f}\right)\right] = \cos\left[\sin^{-1}\left(\frac{n_c}{n_f}\right)\right]$$

$$= \frac{\sqrt{n_f^2 - n_c^2}}{n_f}$$

Therefore,

$$n_o \sin \theta \leqslant \sqrt{n_f^2 - n_c^2} \qquad (2.100)$$

This shows that there is a maximum angle of incidence outside of which entering rays will not be totally reflected within the fiber. We call the quantity on the left side of Eq. (2.100) the *numerical aperture* of the fiber. It is a measure of the acceptance cone for the fiber in the outside medium of index n_o.

For the largest acceptance cone, we would like to arrange for the index of refraction of the cladding to be as small as possible. This is achieved if there is no cladding at all. However, this leads to other problems associated with the loss of intensity.

b. *Attenuation.* In communications applications optical fibers must be many kilometers long. Over this distance even a very small attenuation per unit length can be costly in terms of signal strength.

To evaluate this parameter, consider Fig. 2.13, which shows a small section of a fiber that is carrying a ray inclined at θ' with respect to the fiber axis. The length of one trip between reflections is ℓ_1, corresponding to a length down the fiber of L_1. These are related by

$$\ell_1 = \frac{L_1}{\cos \theta'}$$

or

$$\ell_1 = \frac{L_1}{\left[1 - \dfrac{n_0^2}{n_f^2}\sin^2\theta\right]^{1/2}}$$

Over the total length L of the fiber the actual path of the ray traverses a length

$$\ell = \frac{L}{\cos \theta'} \qquad (2.101)$$

The transmission efficiency of the fiber will be related to the absorption coefficient of the fiber core material through Eq. (2.45), which is equivalent to

$$T = \exp\left(-\frac{KL}{\cos \theta'}\right) \qquad (2.102)$$

In addition to the absorption loss, there will be reflection losses if the surface of the core is not clean. This can be prevented by introducing cladding to protect the

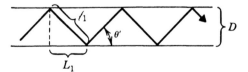

Fig. 2.13 Analysis of the propagation path through an optical fiber in the meridional plane.

sensitive optical surface from contamination. In the absence of cladding, the reflectance for one reflection may be 99.9% rather than 100%.

The number of reflections will be equal to the ratio L/L_1, where from Fig. 2.13 $L_1 = D/\tan \theta'$. Thus,

$$N = L\frac{\tan \theta'}{D} \tag{2.103}$$

and the reflection loss will lead to an efficiency

$$T = R_1^N \tag{2.104}$$

As an example of these concepts, consider an optical fiber in air whose index of refraction is 1.5. Assume that its diameter is 50 μm and it is 50 cm long. The core material is such that the absorption coefficient is 0.01 cm^{-1}. If the incident angle is 10°, then by Snell's law the value of θ' is 6.65°. For a clean fiber there is no reflection loss. The transmission loss under these conditions leads to an efficiency by Eq. (2.102) of 60%. If the fiber is dirty such that $R_1 = 99.9\%$, the loss increases dramatically. Under our conditions using Eq. (2.103) there will be 1165 reflections. Combining the reflection efficiency of Eq. (2.104) with the transmission efficiency gives a total efficiency for the dirty fiber of only 18%. This is why cladding is used, even if the resulting configuration has a smaller acceptance angle.

3. Attenuated Total Reflection (ATR). An interesting modification of the phenomenon of total internal reflection occurs when the second medium is somewhat "lossy," that is, when its index $\tilde{n}'$ is not entirely a real quantity. The exponential penetration of light into the second medium is now accompanied by an irreversible loss of energy. The resulting reflectance will be less than unity, and we say that the otherwise totally internally reflected beam has been attenuated.

Consider the second medium to have a complex refractive index

$$\tilde{n}' = n' - i k'$$

so that

$$\tilde{n}'^2 = n'^2 - k'^2 - i2n'k'$$

The parameter γ is now complex and is given by

$$\tilde{\gamma} = \frac{(n^2 \sin^2 \theta - n'^2 + k'^2 + i2n'k')^{1/2}}{n \cos \theta} \tag{2.105}$$

The discussion is simplified by assuming that the absorption properties of the second medium are weak, then $n'^2 \gg k'^2$ and $n^2 \sin^2 \theta - n'^2 \gg 2n'k'$. We can now write $\tilde{\gamma} = \gamma_r + i\gamma_i$ where

$$\gamma_r \approx \frac{\sqrt{n^2 \sin^2 \theta - n'^2}}{n \cos \theta} \tag{2.106a}$$

and

$$\gamma_i \approx \frac{\gamma_r n'}{n^2 \sin^2 \theta - n'^2} k' \tag{2.106b}$$

such that

$$\gamma_i \ll \gamma_r$$

This gives us for the reflection coefficient in the σ case

$$\rho_\sigma = \frac{1 + i\tilde{\gamma}}{1 - i\tilde{\gamma}} = \frac{(1 - \gamma_i) + i\gamma_r}{(1 + \gamma_i) - i\gamma_r}$$

and for the reflectance

$$R_\sigma = |\rho_\sigma|^2 = \frac{(1 - \gamma_i)^2 + \gamma_r^2}{(1 + \gamma_i)^2 + \gamma_r^2} \approx \frac{1 + \gamma_r^2 - 2\gamma_i}{1 + \gamma_r^2 + 2\gamma_i} \approx 1 - \frac{4\gamma_i}{1 + \gamma_r^2}$$

The reduction in reflectance from the case of total internal reflection is

$$1 - R_\sigma = \frac{4\gamma_i}{1 + \gamma_r^2} \tag{2.107a}$$

The π case can be treated by replacing $\tilde{\gamma}$ by $(n/n')^2\tilde{\gamma}$ with the result

$$1 - R_\pi = \frac{4\gamma_i\left(\dfrac{n}{n'}\right)^2}{1 + \gamma_r^2\left(\dfrac{n}{n'}\right)^4} \tag{2.107b}$$

Often the imaginary part k' owes its origin to the presence of a relatively dilute concentrate of absorbing atoms or molecules in the second medium. For $k' \ll 1$, the effect of these atoms or molecules on $\tilde{n}'$ would be very slight, so that for a fixed angle of incidence θ and a limited range of wavelength, we could treat γ_r and the coefficient of k' in Eq. (2.106b) as constants. Hence, Eqs. (2.106) and (2.107) may be used to give k' from measurements of $(1 - R)$.

Experiments using the ATR technique are often performed as indicated in Fig. 2.14. The medium with index n is used as a tool to study the medium with index $\tilde{n}'$. The former medium must be highly transparent in the desired wavelength range and of large refractive index. The interface between the two media should be of good optical quality, and the two surfaces should be in good optical contact if one wants the expressions derived here to apply quantitatively; otherwise, each interface is actually a complicated $\tilde{n}'$-air-n sandwich. The multiple reflection

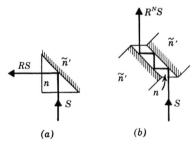

(a)   (b)

Fig. 2.14 Attenuated internal reflection: (a) single reflection; (b) multiple reflection.

technique of Fig. 2.14*b* is used to amplify the effects of a small reduction in reflectance. If $(1 - R) \ll 1$, then after N reflections, we have

$$R^N = [1 + (R - 1)]^N \approx 1 + N(R - 1)$$

and the reduction from the case of total internal reflection is

$$1 - R^N \cong N(R - 1)$$

giving an N-fold amplification.

As an example, consider the case where $n = 2.2$ and $n' = 1.5$. Then $\theta_c = 43.1°$. Let the angle of incidence be $\theta = 50°$. We then find the following:

$$\gamma_r = 0.54, \qquad \gamma_i = 1.38k'$$

$$1 - R_\sigma = 4.3k'$$

$$1 - R_\pi = 5.1k'$$

if k' is 10^{-2}, then

$$1 - R_\sigma = 0.043$$

$$1 - R_\pi = 0.052$$

In a transmission experiment the measured absorption coefficient would be (for $\lambda_0 = 1 \ \mu m$)

$$K = \frac{4\pi k'}{\lambda_0} = 1.25 \times 10^3 \ \text{cm}^{-1}$$

To measure this in transmission, we need a thin sample. Suppose that we can measure an attenuation by a factor of 100. Then we set $e^{-K\Delta z} = 10^{-2}$ and find that the resulting value of Δz would need to be $3.7 \cdot 10^{-3} \ \text{cm} = 37 \ \mu m$. It is often impractical to prepare samples this thin. The ATR technique has the advantage of providing the same information with thick samples provided that surface preparation is of sufficiently high quality. Even if it were not, the ATR technique could be used to locate conveniently the positions of maxima in k' as a function of wavelength.

The ATR technique has been used in practice mainly in the infrared spectral region with relatively strongly absorbing samples. In the longer wavelength infrared regions, the requirements on surface quality are not as stringent as they are in the visible regions.

B. Nontransparent Media

This section discusses some of the optical properties of nontransparent media. Such materials have a large value of k, the imaginary part of the complex index of refraction. Light incident at all angles is then strongly attenuated. It is useful to discuss two limiting cases that have sizable values for k, although it is true that most real systems represent an intermediate case.

In one limit the *dielectric function* is real, but *negative*. The phenomenon is very similar to total internal reflection. The index of refraction $\tilde{n}'$ is purely imaginary, as are the parameters a and b. The reflectance R_σ or R_π is unity. The transmission coefficients τ_σ and τ_π have magnitudes *of order* unity. Thus, the fields E and H just inside the surface are relatively large. They do not propagate as a wave, however. The components of E and H that contribute to the normal or z component of the Poynting vector $\mathbf{S}$ are $180°$ out of phase, so that the time average of S_z then behaves like $\langle \cos \omega t \sin \omega t \rangle = 0$. This phenomenon has its counterpart in other regions of the electromagnetic spectrum, for instance, the reflection of radio waves in the ionosphere.

In the other limiting case, the dielectric constant is complex with a positive real part, but with a sizable imaginary part. The refractive index $\tilde{n}'$ is large in absolute value compared with n; $|a|$ is large compared with unity, and $|b|$ is small. Then R_σ and R_π are close to unity. The transmission coefficients τ_σ and τ_π are both small in absolute value because of the large mismatch $|\tilde{n}'| \gg n$, and the fields E, H just inside the interfaces are quite small. The power lost to dissipative absorption processes is small, not because the processes are weak, but because the fields are weak.

We can calculate the reflectance of an absorbing material by Eq. (2.73) using Eq. (2.62b) for the σ case and Eq. (2.70b) for the π case. The results are shown in Fig. 2.15 for an air-metal interface. Note the similarity with Fig. 2.7. In this case, there is a "pseudo-Brewster angle" where the π-polarized reflectance goes through a minimum.

The form of Eq. (2.54) remains valid as an expression of the conservation of the x component of the complex wavevector. However, it cannot be used to define the direction of a ray because the meaning of a plane wave must be reinterpreted inside an absorbing material.

To explore these concepts in mathematical form, consider the incident medium to be nonabsorbing, so that the components of the transmitted complex wavevector in the (x, y, z) directions are

$$\mathbf{k}' = \frac{\omega}{c} (n \sin \theta, 0, \tilde{n}' \cos \theta') \tag{2.108}$$

See Fig. 2.16.

The z component can be reexpressed in terms of the incident angle

$$\tilde{n}' \cos \theta' = \sqrt{\tilde{n}'^2 - n^2 \sin^2 \theta} \equiv \eta e^{-i\beta} \tag{2.109}$$

where η and β are determined by solving the pair of equations

$$\eta^2 \cos 2\beta = n'^2 - \kappa'^2 - n^2 \sin^2 \theta \tag{2.110a}$$

and

$$\eta^2 \sin 2\beta = 2n'\kappa' \tag{2.110b}$$

Thus the transmitted wave vector is

$$\mathbf{k}' = \frac{\omega}{c} [n \sin \theta, 0, (\eta \cos \beta - i\eta \sin \beta)] \tag{2.111}$$

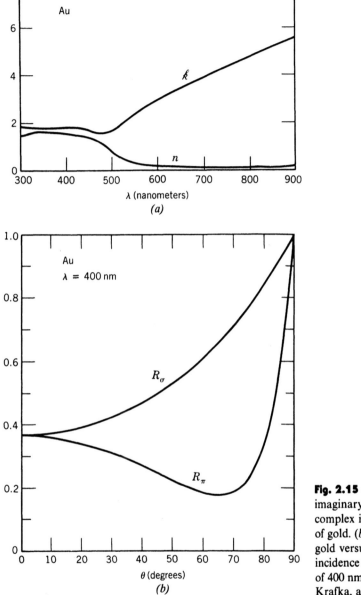

Fig. 2.15 (*a*) Real and imaginary parts of the complex index of refraction of gold. (*b*) Reflectivity of gold versus angle of incidence at a wavelength of 400 nm. (After Weaver, Krafka, and Lynch.)

If we substitute this into the form for the spatial dependence of the plane wave, we find

$$e^{-i\mathbf{k}' \cdot \mathbf{r}} = e^{-i(\omega/c)(n \sin \theta \, x + \eta \cos \beta \, z)} \, e^{-(\omega/c)\eta \sin \beta \, z} \tag{2.112}$$

This shows that the surfaces of constant amplitude are given by $z = $ constant.

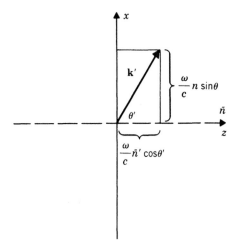

Fig. 2.16 Representation of the complex wavevector within a material with nonnegligible $\mathscr{k}'$.

These are planes parallel to the interface. The surfaces of constant real phase are defined by

$$n \sin \theta \, x + \eta \cos \beta \, z = \text{const} \tag{2.113}$$

These two surfaces are obviously not coincident in the case where $\mathscr{k}' \neq 0$.

We can identify the normal to the surfaces of constant real phase (as we might in the case where we want to identify a "ray") from Eq. (2.113). This will be along

$$n \sin \theta \, \hat{\mathbf{x}} + \eta \cos \beta \, \hat{\mathbf{z}} = A\hat{\mathbf{x}} + B\hat{\mathbf{z}} \tag{2.114}$$

If we define θ_T by

$$\sin \theta_T \equiv A(A^2 + B^2)^{-1/2} \tag{2.115}$$

and then let

$$n_T \equiv \sqrt{n^2 \sin^2 \theta + \eta^2 \cos^2 \beta} \tag{2.116}$$

we recover the functional form of Snell's law from Eq. (2.115)

$$\sin \theta_T = \frac{n \sin \theta}{n_T}$$

However, in this case the angle θ_T depends on the angle of incidence in a complicated way.

2.4 Introduction to the Optical Properties of Matter

Phenomenologically, the presence of matter results in a dielectric function, $\kappa \neq 1$ (and sometimes in a nonzero relative permeability, $\kappa_m \neq 1$). How then is κ determined from the microscopic properties of matter? More basically, can we see directly how the presence of atoms and molecules modifies the propagation law for light waves? These are some of the questions to be discussed in this section. To simplify, the discussion will be limited to nonmagnetic media where $\kappa_m = 1$ and $\mathbf{B} = \mu_0 \mathbf{H}$. Magnetic media may be described by theories somewhat analogous to the ones to be developed in this section.

A key role is played by the polarization $\mathbf{P}$ because it yields the bound charge $\rho_b = -\nabla \cdot \mathbf{P}$ and the bound current density $\mathbf{J} = d\mathbf{P}/dt$. If we assume that the "free" currents and charges $\mathbf{J}_f$ and ρ_f are known, then once $\mathbf{P}$ is known, all the sources of $\mathbf{E}$ and $\mathbf{B}$ are known, and $\mathbf{E}$ and $\mathbf{B}$ may be determined (in principle at least) by integration of Maxwell's equations or of equations derived from them.

Following Eqs. (2.21) and (2.22) for linear, local, isotropic media, we express the polarization in terms of the field at the same point and the susceptibility or, alternatively, the dielectric function.

$$\mathbf{P}(\mathbf{r}) = \varepsilon_0 \chi \mathbf{E}(\mathbf{r}) \tag{2.117}$$

$$\kappa = 1 + \chi \tag{2.118}$$

A. Model for a Dilute Nonpolar Gas

Consider a medium composed of molecules. The polarization $\mathbf{P}$ is the average dipole moment per unit volume. We assume that in a field-free environment the molecules have no net dipole moment; any dipoles must therefore be induced as in Fig. 2.1. Such a medium is called *nonpolar*.

The induced dipole moment on the ith molecule will be given in a linear approximation by

$$\mathbf{p}_i = \alpha \mathbf{E}_m(\mathbf{r}_i) \tag{2.119}$$

Here $\mathbf{E}_m(\mathbf{r}_i)$ is the local microscopic electric field at the center of mass $\mathbf{r}_i$ of the molecule. We hereby neglect variations in the external field over the spatial extent of the molecule; this is a good assumption for visible light and for molecules of ordinary size. For polymers or crystals, for which, technically speaking, the molecules have macroscopic dimensions, the argument must be applied to a small subunit, or cell, that is repeated through the structure. The function α in Eq. (2.119) is called the *polarizability*.

The microscopic local field $\mathbf{E}_m(\mathbf{r}_i)$ depends not only on $\mathbf{r}_i$ but also on the positions $\mathbf{r}_j$ of all the other molecules in the system and on the orientation and magnitude of the dipoles induced on these molecules. If we average the microscopic field $\mathbf{E}_m$ over all the molecules in a macroscopically small but microscopically large volume element at $\mathbf{r}$, we obtain the average local field $\mathbf{E}_l(\mathbf{r})$ acting on a molecule at $\mathbf{r}$. This local field should be contrasted with the ordinary macroscopic field $\mathbf{E}(\mathbf{r})$, which is the average of the microscopic field over *all* positions in a small volume element at $\mathbf{r}$, not merely over positions occupied by molecules.

Two tasks must now be performed if the theory is to be carried further: (1) a model for a molecule must be constructed and α calculated, and (2) the local field $\mathbf{E}_l$ must be related to the macroscopic field $\mathbf{E}$.

1. Simple Molecular Model. A simple model of a nonpolar molecule is provided by a harmonic oscillator (Fig. 2.17). Let a charge q of mass m be attached to another charge $-q$ of essentially infinite mass (for simplicity) by a harmonic "spring" with spring constant C. The equilibrium separation $\mathbf{r}$ of the two charges is zero if no external forces are applied.

For frequencies corresponding to the visible region of the spectrum, the major contribution to α and hence ultimately to κ comes from oscillating electrons. Then q would be minus the electronic charge $(-|e|)$ and m would be the mass of an electron. In the infrared region of the spectrum the oscillating charges can be ions or parts of molecules having much larger values of m. Fortunately, the present derivation can apply to a variety of oscillating systems.

The equation of motion for the position vector of the movable charge will be written

$$m\frac{d^2\mathbf{r}}{dt^2} = -C\mathbf{r} - \frac{m}{\tau}\frac{d\mathbf{r}}{dt} + \mathbf{F} \tag{2.120}$$

Here the external force $\mathbf{F}$ is given by

$$\mathbf{F} = q\mathbf{E}_m \tag{2.121}$$

where $\mathbf{E}_m$ is the microscopic electric field at the molecule. We have included a damping force $-(m/\tau)\, d\mathbf{r}/dt$ in Eq. (2.120), which is proportional to the velocity, so that we can describe irreversible losses caused by collisions with other molecules or to radiation damping. This term may not always yield the correct quantitative behavior resulting from collisions, but it does give the main qualitative effects. The constant τ can be interpreted as a damping or collision time. It is a measure of the mean time that an oscillation will decay in the system.

The natural resonance frequency ω_0 of the undamped oscillator is given by

$$\omega_0 = \sqrt{\frac{C}{m}} \tag{2.122}$$

We assume a single angular frequency in the driving field and write

$$\mathbf{E}_m = \mathbf{E}_{m,0}\, e^{i\omega t} \quad \text{and} \quad \mathbf{r} = \mathbf{r}_0\, e^{i\omega t}$$

where $\mathbf{E}_{m,0}$ and $\mathbf{r}_0$ are complex, time-dependent amplitudes. Equations (2.120),

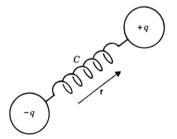

Fig. 2.17 Simple model of a diatomic molecule.

(2.121), and (2.122) are combined and use is made of the operator shortcut [Eq. (1.28] to obtain

$$\frac{d\mathbf{r}}{dt} = i\omega\mathbf{r}, \quad \frac{d^2\mathbf{r}}{dt^2} = -\omega^2\mathbf{r}$$

and

$$\mathbf{r} = \frac{\dfrac{q}{m}}{\omega_0^2 - \omega^2 + \dfrac{i\omega}{\tau}} \mathbf{E}_m \tag{2.123}$$

The induced dipole moment is then $\mathbf{p} = q\mathbf{r}$.

Equation (2.119) then becomes simply $\mathbf{p} = \alpha\mathbf{E}_m$ with

$$\alpha = \frac{\dfrac{q^2}{m}}{\omega_0^2 - \omega^2 + \dfrac{i\omega}{\tau}} \tag{2.124}$$

The polarization is then the average dipole moment per unit volume or

$$\mathbf{P}(\mathbf{r}) = Nq\langle\mathbf{r}\rangle = \frac{N\dfrac{q^2}{m}}{\omega_0^2 - \omega^2 + \dfrac{i\omega}{\tau}} \langle\mathbf{E}_m\rangle = \frac{N\dfrac{q^2}{m}}{\omega_0^2 - \omega^2 + \dfrac{i\omega}{\tau}} \mathbf{E}_l(\mathbf{r}) \tag{2.125}$$

Here N is the mean number of molecules per unit volume, and $\mathbf{E}_l(\mathbf{r})$ is, as explained earlier, the average field $\langle\mathbf{E}_m\rangle$ acting on a molecule at $\mathbf{r}$. Thus,

$$\mathbf{P}(\mathbf{r}) = N\alpha\mathbf{E}_l(\mathbf{r}) \tag{2.126}$$

Since $i = \sqrt{-1}$ appears in the denominator of Eqs. (2.124) and (2.125), α will be a complex function of ω. Write it in the form

$$\alpha = \alpha_r - i\alpha_i = \frac{\left(\dfrac{q^2}{m}\right)(\omega_0^2 - \omega^2)}{(\omega_0^2 - \omega^2)^2 + \dfrac{\omega^2}{\tau^2}} - i\frac{\left(\dfrac{q^2}{m}\right)\left(\dfrac{\omega}{\tau}\right)}{(\omega_0^2 - \omega^2)^2 + \dfrac{\omega^2}{\tau^2}}$$

The real, physical electric field is given by

$$\mathbf{E}_l(\text{real}) = \text{Re}\{\mathbf{E}_{l,0}\, e^{i\omega t}\} = \mathbf{E}_{l,0} \cos \omega t$$

Similarly, the real, physical polarization will be

$$\mathbf{P}(\text{real}) = \text{Re}\{N\alpha\mathbf{E}_{l,0}\, e^{i\omega t}\} = N\mathbf{E}_{l,0}\alpha_r \cos \omega t + N\mathbf{E}_{l,0}\alpha_i \sin \omega t$$

The first term is oscillating in phase with the local field, the second term is 90° out of phase.

2. Very Dilute Gas. Having determined the polarizability, we must now decide what to use for the local field E_l. For a very dilute system, the molecules have such a small effect on the fields that $E_l(r)$ can be approximated well enough by the macroscopic field $E(r)$. We will later make a better approximation for E_l and will therefore be in a position to assess the applicability of our present assumption that $E_l = E$.

With $E_l = E$, the susceptibility becomes

$$\chi = \frac{N\alpha}{\varepsilon_0}$$

and the dielectric function is given by

$$\kappa = 1 + \chi = 1 + \frac{\dfrac{Nq^2}{m\varepsilon_0}}{\omega_0^2 - \omega^2 + \dfrac{i\omega}{\tau}} \tag{2.127}$$

If we separate real and imaginary parts via $\kappa = \kappa_r - i\kappa_i$, we find

$$\kappa_r - 1 = \frac{\left(\dfrac{Nq^2}{m\varepsilon_0}\right)(\omega_0^2 - \omega^2)}{(\omega_0^2 - \omega^2)^2 + \dfrac{\omega^2}{\tau^2}} \tag{2.128}$$

$$\kappa_i = \frac{\left(\dfrac{Nq^2}{m\varepsilon_0}\right)\left(\dfrac{\omega}{\tau}\right)}{(\omega_0^2 - \omega^2)^2 + \dfrac{\omega^2}{\tau^2}} \tag{2.129}$$

The low-density assumption allowing us to set $E_l = E$ will be shown later to hold provided that

$$\tfrac{1}{3}|\chi| \ll 1$$

This inequality will be used to simplify the calculation of the complex index of refraction $\tilde{n} = \kappa^{1/2}$. Use of the approximation

$$(1 + \delta)^{1/2} \approx 1 + \tfrac{1}{2}\delta \quad \text{for small } \delta \text{ leads to}$$

$$\tilde{n} = (1 + \chi)^{1/2} \simeq 1 + \frac{1}{2}\chi = 1 + \tfrac{1}{2}(\kappa - 1)$$

$$= 1 + \frac{\left(\dfrac{Nq^2}{2m\varepsilon_0}\right)\left(\omega_0^2 - \omega^2 - \dfrac{i\omega}{\tau}\right)}{(\omega_0^2 - \omega^2)^2 + \dfrac{\omega^2}{\tau^2}} \tag{2.130}$$

If we split $\tilde{n}$ into real and imaginary parts by writing $\tilde{n} = n - i\ell$, we obtain the expressions

$$n - 1 = \tfrac{1}{2}(\kappa_r - 1) = \frac{\left(\dfrac{Nq^2}{2m\varepsilon_0}\right)(\omega_0^2 - \omega^2)}{(\omega_0^2 - \omega^2)^2 + \dfrac{\omega^2}{\tau^2}} \tag{2.131}$$

$$\mathscr{k} = \tfrac{1}{2}\kappa_i = \frac{\left(\dfrac{Nq^2}{2m\varepsilon_0}\right)\left(\dfrac{\omega}{\tau}\right)}{(\omega_0^2 - \omega^2)^2 + \dfrac{\omega^2}{\tau^2}} \tag{2.132}$$

The constant

$$\omega_p \equiv \left(\frac{Nq^2}{m\varepsilon_0}\right)^{1/2}$$

will appear often in the equations that follow. Here ω_p has the dimensions of angular frequency and is called the angular *plasma frequency*. It also has a physical significance of its own. The plasma frequency is the oscillation frequency that results if one part of a gas of free particles of charge q, mass m, and number density N is displaced with respect to the rest of the gas and with respect to a fixed oppositely charged background charge distribution. In terms of ω_p, Eqs. (2.131) and (2.132) become

$$n - 1 = \tfrac{1}{2}(\kappa_r - 1) = \frac{\omega_p^2 \dfrac{(\omega_0^2 - \omega^2)}{2}}{(\omega_0^2 - \omega^2)^2 + \dfrac{\omega^2}{\tau^2}} \tag{2.133}$$

$$\mathscr{k} = \tfrac{1}{2}\kappa_i = \frac{\omega_p^2 \dfrac{\omega}{2\tau}}{(\omega_0^2 - \omega^2)^2 + \dfrac{\omega^2}{\tau^2}} \tag{2.134}$$

According to Eq. (2.45), the flux density or other energy-related quantity in the beam decays exponentially. For example,

$$S = S_0 \exp\left(-2\omega\mathscr{k}\,\frac{r}{c}\right)$$

The energy lost from the light beam goes into the irreversible processes responsible for the collision or damping time τ. For instance, if collisions between molecules provide the main mechanism that limits τ, then these collisions provide a means of converting electromagnetic energy into energy associated with the translational motion of the molecules, that is, into heat.

3. Characteristics of the Complex Index of Refraction. Plots of n and $\mathscr{k}$ versus ω are shown in Fig. 2.18 for the case $\omega_0\tau \gg 1$, which is the usual one for not too dense

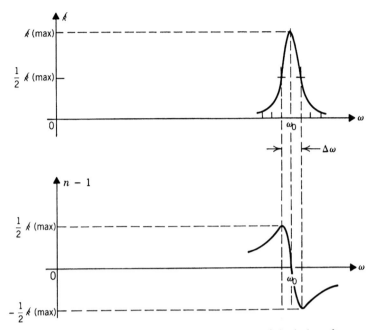

Fig. 2.18 Behavior of the imaginary and real parts of the index of refraction for the simple molecule in Fig. 2.17.

a gas. In this limit, for ω close to the resonance frequency ω_0, we may replace ω by ω_0 everywhere in Eqs. (2.133) and (2.134) except in the expression $(\omega^2 - \omega_0^2)$, which we can approximate as follows:

$$\omega^2 - \omega_0^2 = (\omega - \omega_0)(\omega + \omega_0) \approx 2\omega_0(\omega - \omega_0) \tag{2.135}$$

Thus, *near resonance* the real and imaginary parts of the index of refraction are

$$n = 1 + \frac{\omega_p^2 \dfrac{(\omega_0 - \omega)}{4\omega_0}}{(\omega_0 - \omega)^2 + \dfrac{1}{(2\tau)^2}} \tag{2.136}$$

$$= 1 + \tfrac{1}{2}(\kappa_r - 1)$$

$$\kappa = \frac{\dfrac{\omega_p^2}{8\tau\omega_0}}{(\omega_0 - \omega)^2 + \dfrac{1}{(2\tau)^2}} \tag{2.137}$$

$$= \tfrac{1}{2}\kappa_i$$

At resonance $(\omega = \omega_0)$, κ takes on its maximum value

$$\kappa_{\max} = \tfrac{1}{2}\kappa_i(\max) = \frac{\omega_p^2\tau}{2\omega_0} = \frac{Nq^2\tau}{2m\omega_0\varepsilon_0}$$

and both n and κ_r equal 1. Above resonance, n is less than 1, below resonance it is greater than 1. In Eq. (2.137), ℓ is expressed as a *Lorentzian* function of ω. When $\omega = \omega_0 \pm 1/2\tau$, ℓ is reduced to half its maximum value. The full width at half maximum is thus

$$\Delta\omega = 2\frac{1}{2\tau} = \frac{1}{\tau} \tag{2.138}$$

Because we have assumed $\omega_0\tau \gg 1$, the width is very much less than the resonance frequency.

At the half-maximum points of ℓ the real part of the index of refraction n takes on its maximum and minimum values of

$$n(\text{max}) = 1 + \frac{\omega_p^2\tau}{4\omega_0} = 1 + \frac{\ell(\text{max})}{2} \quad \text{when } \omega = \omega_0 - \frac{\Delta\omega}{2} \tag{2.139}$$

and

$$n(\text{min}) = 1 - \frac{\omega_p^2\tau}{4\omega_0} = 1 - \frac{\ell(\text{max})}{2} \quad \text{when } \omega = \omega_0 + \frac{\Delta\omega}{2} \tag{2.140}$$

4. Numerical Examples of Optical Properties. As an illustration, take $\omega_0 = 3 \times 10^{15}$ rad/sec (associated with a wavelength of light at resonance of 628 nm), $N = 10^{17}/\text{cm}^3$ (corresponding to a gas at room temperature at 2.5 torr pressure), with $q = -1.602 \times 10^{-19}$ C and $m = 9.109 \times 10^{-31}$ kg (for the electron). Then

$$\omega_p^2 = 3.18 \times 10^{26} \text{ (rad/sec)}^2$$

and

$$\ell(\text{max}) = 0.53$$

This is too large a value to be consistent with the assumption $|\chi| \ll 3$. If we want to make this assumption for the preceding example, we must stay away from the peak ℓ, that is, we must be certain that ω is far enough from ω_0 so that $\ell \ll 1$ and $|n - 1| \ll 1$.

Equations (2.136) and (2.137) represent a slightly simplified version of Eqs. (2.133) and (2.134) near resonance. Away from resonance the second term, ω^2/τ^2, in the denominators of the latter equations is small and can be neglected, in comparison with $(\omega_0^2 - \omega^2)$, which will be comparable with ω_0^2. Thus, away from resonance we can write

$$n - 1 = \frac{\dfrac{\omega_p^2}{2}}{\omega_0^2 - \omega^2} \tag{2.141a}$$

$$\ell = \frac{\left(\dfrac{\omega_p^2}{2}\right)\left(\dfrac{\omega}{\tau}\right)}{(\omega_0^2 - \omega^2)^2} = (n - 1)\left(\frac{\omega^2}{\omega_0^2 - \omega^2}\right)\frac{1}{\omega\tau} \tag{2.141b}$$

Because $\omega_0 \tau \gg 1$ and $(\omega_0^2 - \omega^2)$ is not too close to zero, the factor $|\omega^2/(\omega_0^2 - \omega^2)|(1/\omega\tau)$ is $\ll 1$. This means that $\ell \ll |n - 1|$, so that away from resonance n can be considered to be purely real. For the preceding numerical example the off-resonance value of n will be given by

$$n = 1 + 1.77 \times 10^{-5}\left(1 - \frac{\omega^2}{\omega_0^2}\right)^{-1}.$$

Note that above resonance n is less than unity, and below resonance it is greater than unity. As the frequency tends to zero the index tends to the static limit

$$n(0) = 1 + \frac{\omega_p^2}{2\omega_0^2} \qquad [= 1 + 1.77 \times 10^{-5} \text{ for our example}]$$

Note that when damping or collisions are neglected so that ℓ is allowed to go to infinity, the complex quantities α, χ, κ, and $\tilde{n}$ all go to infinity at resonance. A finite value of τ keeps these quantities finite at $\omega = \omega_0$, but they still have a large absolute value there. Specifically, the real parts of α and χ are zero at resonance, and the real parts of κ and $\tilde{n}$ are unity there, whereas the imaginary parts of all four functions become very large. We say that these functions have a "near singularity" at $\omega^2 = \omega_0^2$.

In the case of higher number densities the functions κ, χ, and $\tilde{n}$ still have near singularities, but not at ω_0^2, where the polarizability still has its near singularity. Then κ and χ keep the same functional form they had at low densities. In particular, the equations for κ_r and κ_i look exactly like the ones written previously above, with ω_0^2 replaced by a new frequency parameter $\bar{\omega}^2$. The real and imaginary parts of the index of refraction become more complicated at higher densities—so complicated that we often choose to give our results in terms of the dielectric function. Of course, when κ_i is very small, so is ℓ, and we may write simply

$$n = \sqrt{\kappa_r}$$

The discussion of high-density dielectrics will be postponed until we cover the following discussion of wave propagation in conducting media.

B. Conducting Media

In a conducting medium such as a metal or an ionized gas there is a cloud of (nearly) free electrons moving through a background of relatively stationary positive charge. The molecular model of the preceding section may be modified to describe this new case by setting the force constant C and, hence, the resonance frequency ω_0 equal to zero. We can argue (with some success, it turns out) that for a free electron gas the local field acting on a given electron is more likely to be simply the macroscopic field, even at high densities, than for a gas or liquid composed of molecules. Molecules have a well-defined volume that they occupy to the exclusion of other molecules. This exclusion affects the distribution of other molecules about a given molecule—it is not the same as the distribution of molecules about an arbitrary point in space. Thus the average electric field acting on a given molecule is likely to be different from the average field at an arbitrary point in space, especially at high molecular densities. Electrons, on the other hand, occupy no such well-defined volume, and the average field seen by one of them is much more likely to be well approximated by the average field at an arbitrary point in space.

The classical calculation of the dielectric function and the index of refraction of the electron gas proceeds by setting $\omega_0 = 0$ and $\mathbf{E}_l = \mathbf{E}$. Then,

$$\kappa = 1 + \frac{\dfrac{Nq^2}{m\varepsilon_0}}{(i\omega/\tau) - \omega^2} = 1 - \frac{\omega_p^2}{\omega^2 - (i\omega/\tau)} \tag{2.142}$$

where again the plasma frequency

$$\omega_p = \left(\frac{Nq^2}{m\varepsilon_0}\right)^{1/2}$$

has been introduced.

For the numerical example discussed previously N was $10^{17}/cm^3$ and $\omega_p \approx 1.8 \times 10^{13}$ rad/sec. For the ionized gas layers in the ionosphere, N is about $10^5/cm^3$, and hence $\omega_p \approx 2 \times 10^7$ rad/sec, which corresponds to a frequency of 3 MHz. For a metal, N is about $10^{22}/cm^3$, and hence $\omega_p \approx 6 \times 10^{15}$ rad/sec.

As a first approximation at high frequencies where we can often go to the limit $\tau \rightarrow \infty$, the assumption is then equivalent to assuming that the electron gas is *collisionless*. Then

$$\kappa = 1 - \frac{\omega_p^2}{\omega^2} \tag{2.143}$$

A major consequence of this result is that the index of refraction, $n = \sqrt{\kappa}$, is purely imaginary for $\omega < \omega_p$ and purely real for $\omega > \omega_p$. In the former case, we have

$$\tilde{n} = -i\ell, \quad \ell = \frac{\sqrt{\omega_p^2 - \omega^2}}{\omega} \tag{2.144}$$

A medium with a purely imaginary index of refraction cannot propagate electromagnetic waves. We have already seen that under these conditions the waves are completely reflected at the boundary of the medium.

Despite the fact that there is no propagation of waves, there is penetration of fields into the conducting medium. Equation (2.37) is applicable to this case and gives for the physical electric field

$$\mathbf{E} = \mathbf{E}_0 \, e^{-r/\delta} \cos(\omega t)$$

Here the skin depth is

$$\delta = \frac{c}{\ell\omega} = \frac{c}{\sqrt{\omega_p^2 - \omega^2}} \tag{2.145}$$

When $\omega > \omega_p$ (and $\tau \to \infty$), the dielectric function is positive and $\tilde{n}$ is purely real, $\tilde{n} = n$,

$$n = \frac{\sqrt{\omega^2 - \omega_p^2}}{\omega} \tag{2.146}$$

Electromagnetic waves then propagate with no losses with phase velocity

$$v = \frac{\omega}{k} = \frac{c}{n}$$

For a metal, ω_p corresponds to frequencies in the near ultraviolet region of the optical spectrum. We thus expect visible and infrared light to be strongly reflected, because for such light $\omega < \omega_p$, but ultraviolet light for which $\omega > \omega_p$ should be transmitted. This is indeed the case; for instance, sodium metal is relatively transparent for wavelengths below 210 nm. For the ionosphere, where the plasma frequency of about 3 MHz is in the radio frequency region, we expect higher frequencies to be transmitted and lower frequencies to be reflected. This is, roughly, what actually happens, but the real situation is much more complicated, in part because of the presence of the earth's magnetic field.

In the more general case of finite damping, the dielectric function κ from Eq. (2.142) when separated into real and imaginary parts gives

$$\kappa_r = 1 - \frac{\omega_p^2}{\omega^2 + \tau^{-2}} = \frac{\omega^2 + \tau^{-2} - \omega_p^2}{\omega^2 + \tau^{-2}}$$

$$\kappa_i = \left[\frac{\omega_p^2}{\omega^2 + \tau^{-2}} \right] \frac{1}{\omega\tau} \tag{2.147}$$

Because

$$\kappa = \kappa_r - i\kappa_i = \tilde{n}^2 = (n - ik)^2 = n^2 - k^2 - i2nk$$

we find

$$n^2 - k^2 = \kappa_r$$

$$2nk = \kappa_i \tag{2.148}$$

The real and imaginary parts of the dielectric function and of the index of refraction are plotted in Fig. 2.19 for $\omega_p = 6 \times 10^{15}$ rad/sec and $1/\tau = 3 \times 10^{13}$ rad/sec, typical for a simple metal such as sodium. The critical frequency ω_c at the onset of transparency is now defined to be the frequency at which $\kappa_r = 0$. Thus,

$$\omega_c^2 = \omega_p^2 - \left(\frac{1}{\tau} \right)^2 \tag{2.149}$$

The model functions of Fig. 2.19 are qualitatively similar to the real quantities for "free-electron-like" metals such as gold. Compare Fig. 2.19b with the data in Fig. 2.15a.

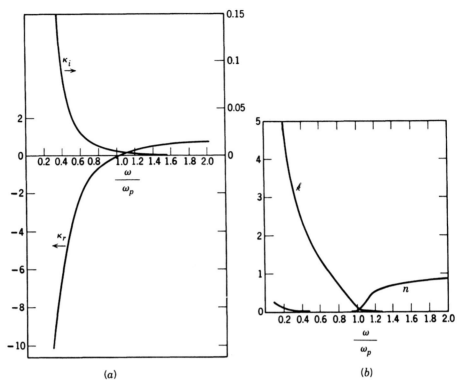

Fig. 2.19 (a) Real and imaginary parts of the dielectric function for a metal. (b) Real and imaginary parts of the index of refraction for a metal. Here $\omega_p = 6 \times 10^{15}$ rad/sec and $1/\tau = 3 \times 10^{13}$ rad/sec.

C. Denser Dielectrics

Return now to the model of a dielectric medium as a collection of molecules, and consider higher densities so that the local field cannot be expected to be the same as the macroscopic field. In many cases it is a good approximation to write

$$E_l = E + \frac{1}{3\varepsilon_0} P \qquad (2.150)$$

The term

$$\frac{P}{3\varepsilon_0}$$

is called the *Lorentz local field correction*.

1. Lorentz Local Field. Some of the arguments that lead to the Lorentz local field will now be given with the help of Fig. 2.20. The molecules are represented by circles, their centers by dots, and random positions in the medium by X's. Each

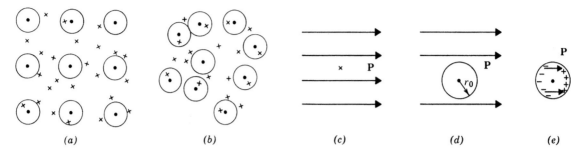

Fig. 2.20 Aids in the determination of the Lorentz local field.

molecule has an induced dipole moment, which is not shown. Two extremes of molecular arrangements are depicted—a completely ordered crystalline array in (*a*) and a disordered fluid or amorphous solid in (*b*).

The microscopic field $E_m(r)$ is the external field from sources outside the medium plus the field from the dipoles. The macroscopic field E is the average of E_m as r ranges over random positions in a limited volume, that is, over the X's. This will be the same as the field inside a dielectric continuum having macroscopic polarization P, as indicated schematically in Fig. 2.20*c*. The actual value of E will depend on the value of the external field, on the position (and time) dependence of P, and on the shape of the medium; fortunately it is not necessary to produce an explicit formula for E.

The local field E_l is the average of $E_m(r$, as r ranges over all the black dots in Fig. 2.20. In this case, the "self-field" from the molecule at the dot in question must be excluded. Because each molecule occupies a well defined volume, the spacing between dots is at least twice the molecular radius r_m; the nearest molecules contributing to the field must be at least this distance $2r_m$ away. Furthermore, the distribution of molecules about a given molecule will have inversion symmetry and be isotropic for a fluid or amorphous solid composed of spherical molecules. For certain simple crystal structures, such as cubic structures (but not for all crystal structures), the symmetry is high enough to be also considered isotropic in this context. These two features of the molecular distribution about a given molecule—a minimum separation $2r_m$ and an isotropic distribution—will be preserved during the averaging process that defines E_l. The result is that the dipoles become essentially smeared into a dielectric continuum with polarization P. The minimum separation is preserved by having a cavity in this continuum of molecular dimensions centered about r, and the isotropic distribution can be assured by making the cavity spherical. Then E_l is the field at the center of this cavity, shown in Fig. 2.20*d*.

Two valid approximations greatly simplify the calculation of E_l. The cavity has molecular dimensions and hence is much smaller than the wavelength of visible light. This means that P can be treated as a constant in the region of the cavity. This also means that the time for light to propagate across the region of the cavity is very much less than the period of oscillation of the polarization field, so that retardation effects resulting from the finite velocity of electromagnetic waves may also be neglected.

If the cavity were to be filled with a uniformly polarized sphere having polarization **P**, the field at its center would then be the macroscopic field **E**. Let $\mathbf{E}_s$ be the field at the center of an isolated uniformly polarized sphere. Then we must have

$$\mathbf{E}_l + \mathbf{E}_s = \mathbf{E}$$

or

$$\mathbf{E}_l = \mathbf{E} - \mathbf{E}_s$$

The field $\mathbf{E}_s$ (Fig. 2.20e) is readily shown to be

$$\mathbf{E}_s = -\frac{\mathbf{P}}{3\varepsilon_0}$$

This gives Eq. (2.150),

$$\mathbf{E}_l = \mathbf{E} + \frac{\mathbf{P}}{3\varepsilon_0}$$

2. Calculation of χ and κ. Equation (2.150) leads to

$$\mathbf{P} = N\alpha\mathbf{E}_l = N\alpha\mathbf{E} + \frac{N\alpha}{3\varepsilon_0}\mathbf{P}$$

The electric susceptibility is then

$$\chi = \frac{P}{\varepsilon_0 E} = \frac{N\alpha/\varepsilon_0}{1 - N\alpha/(3\varepsilon_0)} \qquad (2.151)$$

and the dielectric constant is given by

$$\kappa = 1 + \chi = 1 + \frac{N\alpha/\varepsilon_0}{1 - N\alpha/(3\varepsilon_0)} \qquad (2.152)$$

The earlier low-density results were

$$\chi = \frac{N\alpha}{\varepsilon_0}$$

$$\kappa = 1 + \frac{N\alpha}{\varepsilon_0}$$

They hold when the second term in the denominator in Eqs. (2.151) and (2.152) can be neglected, or when

$$\left|\frac{N\alpha}{3\varepsilon_0}\right| \ll 1$$

Equation (2.152) may be rearranged to give

$$\frac{\kappa - 1}{\kappa + 2} = \frac{N\alpha}{3\varepsilon_0} \tag{2.153}$$

This is known as the Clausius-Mosotti-Lorentz-Lorenz equation, which is named after the four people that derived various versions of it. There are other ways of writing it. For instance, we can put

$$N = N_A n_m$$

where N_A is Avogadro's number and n_m is the number of moles per unit volume. Then

$$\frac{\kappa - 1}{\kappa + 2} = \frac{1}{3\varepsilon_0} \alpha_m n_m$$

where

$$\alpha_m = N_A \cdot \alpha$$

is the polarizability per mole (molar polarizability).

For simplicity we will consistently use Eq. (2.153), and for brevity we call it the Clausius–Mosotti equation. It says that the quantity

$$\frac{1}{N} \frac{(\kappa - 1)}{(\kappa + 2)}$$

should be a constant independent of the number density N, the constant being

$$\frac{\alpha}{3\varepsilon_0}$$

If the Clausius–Mossotti formula is taken seriously, we should be able to calculate α from the refractive index in the gas phase and then predict κ in the liquid phase at much higher density. This often works very well (within a few percent) for nonpolar molecules in the transparent regions of the optical spectrum. In the absorptive regions the molecular polarizability can change at high densities because of changes in the damping constant τ resulting from short-range interactions between molecules—essentially τ becomes much shorter.

3. Interpretation of the Dense Dielectric Result. Consider for simplicity the case of a single oscillator so that

$$\alpha(\omega) = \frac{\dfrac{q^2}{m}}{\omega_0^2 - \omega^2 + \dfrac{i\omega}{\tau}}$$

We say that α has a "near singularity" at $\omega^2 = \omega_0^2$ because its absolute value is maximum there. Specifically, at resonance we have

$$\alpha(\omega_0) = \frac{q^2\tau}{im\omega_0}$$

and in absolute value this is $\omega_0\tau$ times larger than the zero-frequency value of

$$\alpha(0) = \frac{q^2}{m\omega_0^2}$$

In the low-density limit, when it is a good approximation to write

$$\kappa = 1 + \frac{N\alpha}{3\varepsilon_0}$$

this near singularity of α at $\omega^2 = \omega_0^2$ is shared by the complex dielectric function and the complex index of refraction. The near singularity appears as a sharply peaked function $k(\omega) = \frac{1}{2}\kappa_i(\omega)$ having a maximum at resonance and a characteristic dispersive wiggle for $n(\omega) - 1 = \frac{1}{2}[\kappa_r(\omega) - 1]$, that sends it through zero at resonance. Optical measurements on such a dilute dielectric medium would yield a value for ω_0.

In dense media where the Lorentz local field correction applies, this is no longer the case. Equations (2.152) and (2.124) give

$$\kappa = 1 + \frac{\omega_p^2}{\left(\omega_0^2 - \frac{\omega_p^2}{3}\right) - \omega^2 + \frac{i\omega}{\tau}} \tag{2.154}$$

where

$$\omega_p = \left(\frac{Nq^2}{m\varepsilon_0}\right)^{1/2}$$

represents the plasma frequency that the oscillating charges would have if they were released from their restoring "springs."

Now κ has a near singularity at

$$\bar{\omega}^2 \equiv \omega_0^2 - \frac{\omega_p^2}{3}$$

not at

$$\omega^2 = \omega_0^2$$

The resulting dispersive behavior in

$$\kappa_r - 1 = \frac{\omega_p^2(\bar{\omega}^2 - \omega^2)}{(\bar{\omega} - \omega^2)^2 + \frac{\omega^2}{\tau^2}} \tag{2.155}$$

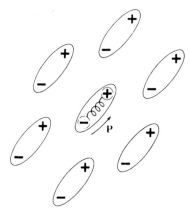

Fig. 2.21 Physical picture of the action of the local field.

and peak of

$$\kappa_i = \frac{\dfrac{\omega_p^2 \omega}{\tau}}{(\bar{\omega}^2 - \omega^2)^2 + \dfrac{\omega^2}{\tau^2}} \tag{2.156}$$

now occur at $\omega^2 = \bar{\omega}^2$. Any optical measurements would suggest that $\bar{\omega}$, not ω_0, is the resonance frequency of the system; indeed this is precisely the case—the resonance *has* shifted to ω.

The resonance frequency of isolated molecules thus shifts downward when the molecules are assembled into dense matter. The shift results from the smoothed-out effect of long-range dipole–dipole interactions as introduced via the Lorentz local field term $\mathbf{P}/3\varepsilon_0$. That the shift should be downward can be understood from Fig. 2.21. If the molecules are vibrating together, a given molecule sees an electric field, from its neighbors that tends to increase its dipole moment, that is, stretch its "spring." This weakens the restoring effect of the "spring," lowers the effective spring constant, and gives a lower resonance frequency for the coupled system.

In quantum mechanics a shift in the frequency of an oscillator corresponds to a shift in the energy of the transition that corresponds to it. Such energy shifts are common in dense, many-particle systems and go by the name of "renormalization."

4. Characteristics of the High-Density Result. The high-density expressions Eqs. (2.154), (2.155), and (2.156) are formally the same as the low-density expressions Eqs. (2.127), (2.128), and (2.129) for κ_r and κ_i, with ω_0 replaced by $\bar{\omega}$, but it is no longer true that

$$(n - 1) = \tfrac{1}{2}(\kappa_r - 1) \quad \text{and} \quad \ell = \tfrac{1}{2}\kappa_i$$

Instead, we must use Eq. (2.148):

$$n^2 - \ell^2 = \kappa_r \quad \text{and} \quad 2n\ell = \kappa_i$$

The resulting expressions for n and ℓ are complicated and will not be given here in analytical form. Typical behavior near resonance is shown in Fig. 2.22.

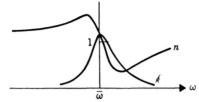

Fig. 2.22 Real and imaginary parts of the index of refraction near a resonance for high-density media.

All earlier low-density results for $\kappa_r - 1$ and κ_i may now be written with $\bar{\omega}$ in place of ω_0. For instance, near resonance

$$\kappa_r - 1 = \frac{\dfrac{\omega_p^2(\bar{\omega} - \omega)}{2\bar{\omega}}}{(\bar{\omega} - \omega)^2 + \dfrac{1}{(2\tau)^2}} \qquad (2.157)$$

$$\kappa_i = \frac{\dfrac{\omega_p^2}{4\tau\bar{\omega}}}{(\bar{\omega} - \omega)^2 + \dfrac{1}{(2\tau)^2}} \qquad (2.158)$$

Here κ_i is described by a Lorentzian function having a maximum

$$\kappa_i(\text{max}) = \frac{\omega_p^2\tau}{\bar{\omega}}$$

and a width

$$\Delta\omega = \frac{1}{\tau}$$

The imaginary part κ_i, as approximated by Eq. (2.158), has integrated area

$$A = \int_0^\infty \kappa_i(\omega)\, d\omega = \frac{\pi\omega_p^2}{\bar{\omega}} = \frac{\pi N q^2}{m\bar{\omega}\varepsilon_0} \qquad (2.159)$$

independent of τ. This expression is accurate when $\Delta\omega \ll \bar{\omega}$, for then all the contribution to κ_i comes when $|\omega - \bar{\omega}| \ll \bar{\omega}$, that is, when Eq. (2.157) is a good approximation. As τ increases, the peak for κ_i rises and narrows, but it keeps constant area. This is seen by comparing Figs. 2.23a and 2.23b. The maximum and minimum of κ_r still occur at the half-maximum points of κ_i and may be readily shown to be

$$\kappa_r(\text{max}) = 1 + \frac{\omega_p^2\tau}{2\bar{\omega}} = 1 + \tfrac{1}{2}\kappa_i(\text{max}) \qquad \text{when } \omega = \bar{\omega} - \frac{\Delta\omega}{2}$$

$$\kappa_r(\text{min}) = 1 - \frac{\omega_p^2\tau}{2\bar{\omega}} = 1 - \tfrac{1}{2}\kappa_i(\text{max}) \qquad \text{when } \omega = \bar{\omega} + \frac{\Delta\omega}{2} \qquad (2.160)$$

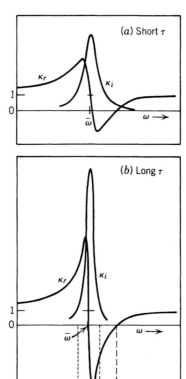

Fig. 2.23 Real and imaginary parts of the dielectric function for two values of the lifetime τ. In part (b), T denotes the transparent regions, A the region of absorption, whose boundaries are ill-defined, and M the region of "metallic" reflection.

Except for the region $\omega - (\Delta\omega/2) \leq \omega \leq \omega + (\Delta\omega/2)$, κ_r increases with increasing frequency. The real part of the index of refraction, n, has similar behavior, although the frequency range is not quite the same (see Fig. 2.22). This region where n is not increasing with ω, called the region of *anomalous dispersion*, coincides with the region of greatest absorption where k is large. For the usual or "normal" dispersion, n increases with increasing frequency (decreasing wavelength), so that for visible light, n is greater in the blue region of the spectrum than in the red region.

Note from Fig. 2.23b that when κ_i has a strong, narrow resonance peak, κ_r necessarily is negative over a frequency range just above resonance. The significance of this behavior is most apparent when τ is very long. Then $\Delta\omega$ is very small, and except for ω very close to $\bar{\omega}$ (within a few multiples of $\Delta\omega$), we may neglect the imaginary part κ_i of κ and write

$$\kappa \cong \kappa_r = 1 + \frac{\omega_p^2}{\bar{\omega}^2 - \omega^2} = \frac{\bar{\omega}^2 + \omega_p^2 - \omega^2}{\bar{\omega}^2 - \omega^2} \tag{2.161}$$

Unlike Eq. (2.157), this expression is valid for all values of ω not too close to $\bar{\omega}$.

According to Eq. (2.161), κ is negative for

$$\bar{\omega}^2 < \omega^2 < \bar{\omega}^2 + \omega_p^2$$

To be consistent with our neglect of κ_i, in the left-hand inequality we should keep ω^2 greater than $(\bar{\omega} + M\Delta\omega)^2 \cong \bar{\omega}^2 + 2M\bar{\omega}\Delta\omega$, where M is of the order of 5 or 10. Thus we require

$$\bar{\omega}^2 + 2M\bar{\omega}\Delta\omega < \omega^2 < \bar{\omega}^2 + \omega_p^2 \qquad (2.162)$$

as the condition for both a negative value of κ_r and a very small value of κ_i. If Eq. (2.162) is to hold for a nonzero range of ω, it is necessary that

$$\omega_p^2 > 2M\bar{\omega}\Delta\omega = \frac{2M\bar{\omega}}{\tau}$$

In this case we find

$$\kappa_i(\text{max}) = \frac{\omega_p^2 \tau}{\bar{\omega}} > 2M \qquad (2.163)$$

The greater the maximum value of κ_i exceeds this inequality, the greater the range of ω over which Eq. (2.162) is satisfied.

Thus, under the appropriate conditions just stated, we have a dielectric function given by

$$\kappa = -\left(\frac{\bar{\omega}^2 + \omega_p^2 - \omega^2}{\omega^2 - \bar{\omega}^2}\right) < 0 \qquad (2.164)$$

The index of refraction $\tilde{n} = \sqrt{\kappa}$ then is purely imaginary:

$$\tilde{n} = -i\left(\frac{\bar{\omega}^2 + \omega_p^2 - \omega^2}{\omega^2 - \bar{\omega}^2}\right)^{1/2} \qquad (2.165)$$

As discussed in connection with Eq. (2.144), a medium with a purely imaginary refractive index will totally reflect light incident onto it from outside. The exponential penetration of the fields into the medium is accompanied by no energy loss.

5. Multiple Oscillators. Consider a medium that is a homogeneous mixture of more than one type of molecule, and let the jth type have number density N_j and molecular polarizability α_j. The polarization is then written

$$\mathbf{P} = \sum_j N_j \alpha_j \mathbf{E}_l$$

The use of the Lorentz local field with this formula leads to a modified Clausius--Mossotti equation that says that the quotient $(\kappa - 1)/(\kappa + 2)$ is additive:

$$\frac{\kappa - 1}{\kappa + 2} = \frac{1}{3\varepsilon_0} \sum_j N_j \alpha_j \qquad (2.166)$$

This formula holds very well for mixtures of nonpolar gases at *not too high densities.*

Still another change may be made. Suppose that a given molecule contains several types of oscillators. Each such oscillator contributes a term to the molecular polarizability

$$\alpha_j = \frac{f_j \dfrac{q^2}{m_j}}{\omega_j^2 - \omega^2 + \dfrac{i\omega}{\tau_j}}$$

The total polarizability is

$$\alpha = \sum_j \alpha_j \tag{2.167}$$

The constant f_j is called the *oscillator strength* of the jth oscillator. It is introduced in an ad hoc fashion in the classical theory of dispersion. A quantum-mechanical calculation of the polarizability leads to exactly the same result with explicit expressions for the f_j in terms of "matrix elements" for intramolecular dipole transitions. Each transition is identified with a classical oscillator, and the transition energy is given by Planck's constant h times the oscillator circular frequency $v_j = (\omega_j/2\pi)$.

If Eq. (2.167) is used with the Lorentz local field, we obtain

$$\frac{\kappa - 1}{\kappa + 2} = \frac{1}{3\varepsilon_0} N \sum_j \alpha_j \tag{2.168}$$

which is formally the same as Eq. (2.166) for a mixture of different molecules if Nf_j is replaced by N_j.

Suppose that the molecules when isolated have several widely spaced resonance frequencies $\omega_1^2 \ll \omega_2^2 \ll \omega_3^2$, and so on. Then the contribution of the jth oscillator to the molecular polarizability

$$\alpha_j(\omega) = \frac{f_j \dfrac{q^2}{m_j}}{\omega_j^2 - \omega^2 + \dfrac{i\omega}{\tau_j}}$$

is nearly constant near the resonance frequency ω_l of the lth oscillator. For oscillators j at lower frequencies than $\omega_l (\omega_j^2 \ll \omega_l^2)$ we have for ω near ω_l

$$\alpha_j(\omega) \approx \frac{-f_j q^2}{m_j \omega^2}$$

and for oscillators at higher frequencies $(\omega_j^2 \gg \omega_l^2)$

$$\alpha_j \approx \frac{f_j q^2}{m_j \omega_j^2}$$

The contribution of one of the oscillators, say the lth, to κ may be described as follows using Eq. (2.168). Define κ_{0l} by

$$\frac{\kappa_{0l} - 1}{\kappa_{0l} + 2} = \frac{N}{3\varepsilon_0} \sum_{j \neq l} \alpha_j \tag{2.169}$$

Here κ_{0l} is the dielectric function in the absence of the lth oscillator. The dielectric function in the presence of the lth oscillator is given by

$$\frac{\kappa - 1}{\kappa + 2} = \frac{N}{3\varepsilon_0} \sum_{j \neq l} \alpha_j + \frac{N}{3\varepsilon_0} \alpha_l = \frac{\kappa_{0l} - 1}{\kappa_{0l} + 2} + \frac{N}{3\varepsilon_0} \alpha_l \tag{2.170}$$

with

$$\alpha_l = \frac{f_l \dfrac{q^2}{m_l}}{\omega_l^2 - \omega^2 + \dfrac{i\omega}{\tau_l}}$$

Define

$$\Delta\kappa = \kappa - \kappa_{0l}$$

Then it is easy to show that

$$\Delta\kappa = \frac{f_l \omega_{pl}^2 \left(\dfrac{\kappa_{0l} + 2}{3}\right)^2}{\omega_l^2 - \left(\dfrac{\omega_{pl}^2}{3}\right)\left(\dfrac{\kappa_{0l} + 2}{3}\right) f_l - \omega^2 + \dfrac{i\omega}{\tau_l}} \tag{2.171}$$

with

$$\omega_{pl}^2 = \frac{Nq^2}{m_l \varepsilon_0}$$

Here $\Delta\kappa$ shows a resonance at

$$\bar{\omega}_l^2 = \omega_l^2 - \frac{\omega_{pl}^2}{3}\left(\frac{\kappa_{0l} + 2}{3}\right) f_l \tag{2.172}$$

This is shifted down from ω_l^2 because of interactions among dipoles. Roughly speaking, the dipoles resonating at $\bar{\omega}_l$ interact within a medium having a dielectric function, hence the presence of κ_{0l} in Eq. (2.172).

Far below resonance, $\Delta\kappa$ tends to a constant

$$\Delta\kappa_l \equiv \frac{f_l \omega_{pl}^2}{\bar{\omega}_l^2}\left(\frac{\kappa_{0l} + 2}{3}\right)^2 \tag{2.173}$$

and far above it $\Delta\kappa$ tends to zero. In both cases, these limits are essentially attained before the next isolated resonance frequency ω_{l-1} or ω_{l+1} is reached, because of the assumed inequalities $\omega_{l-1}^2 \ll \omega_l^2 \ll \omega_{l+1}^2$. As the resonance is transversed from

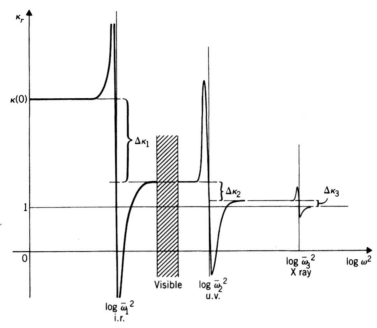

Fig. 2.24 Representation of the dielectric function (real part only) for a system whose molecules display three resonances in different regions of the spectrum. This behavior is typical of nonmetals.

above, the dielectric function increases by $\Delta\kappa_l$. To the extent that the m_i's are approximately the same, so are the plasma frequencies ω_{pl}^2. The dielectric function shift $\Delta\kappa_l$ is then approximately proportional to $(\bar{\omega}_l)^{-2}$, according to Eq. (2.173). This will be quite small for large resonance frequencies.

The resulting behavior of κ is shown semiquantitatively in Fig. 2.24. A logarithmic scale is used as an abscissa so that the condition $\bar{\omega}_1^2 \ll \bar{\omega}_2^2 \ll \bar{\omega}_3^2$ can be satisfied and all interesting behavior shown on one plot. Only the real part κ_r is plotted; the imaginary part κ_i will be nonzero essentially only in the regions of rapid change in κ_r—the regions of anomalous dispersion.

The type of behavior shown in Fig. 2.24 is exhibited by many real nonconducting systems. The lowest resonance frequency $\bar{\omega}_1$ might typically be about $2 \cdot 10^{14}$ rad/sec ($\lambda = 10\ \mu m$) and would correspond to the infrared-active internal vibrations of one part of the molecule with respect to another. It is often called the fundamental infrared absorption frequency. Several such vibrations often occur, but we show only one. For diatomic molecules, such as O_2 and N_2, and for monatomic molecules, such as He, there is no infrared-active mode; thus the ω_1 resonance is missing.

In this section we have painted a rough picture of the dispersion properties of most transparent materials. The frequency dependence of κ_r or n is intimately connected with the existence of regions of large absorption. These regions are associated with various characteristic excitation processes. Measurement of the optical properties of a medium provide one important method of studying these processes.

REFERENCES

Abeles, F., ed. *Optical Properties of Solids*. North Holland, Amsterdam, 1972.

Balkanski, Minko, ed. *Optical Properties of Solids*. North Holland, Amsterdam, 1980.

Bloembergen, N. *Nonlinear Optics*. Benjamin, New York, 1965.

Born, Max, and Emil Wolf. *Principles of Optics*. Pergamon Press, Oxford, 1980.

Cook, David M. *The Theory of the Electromagnetic Field*. Prentice-Hall, Englewood Cliffs, N.J., 1975.

Driscoll, Walter G., and William Vaughan. *Handbook of Optics*. McGraw-Hill, New York, 1978.

Fowles, Grant R. *Introduction to Modern Optics*. Holt, Rinehart and Winston, New York, 1968.

Garbuny, Max. *Optical Physics*. Academic Press, New York, 1965.

Harrick, N. J. *Internal Reflection Spectroscopy*. Interscience, New York, 1967.

Hodgson, John Noel. *Optical Absorption and Dispersion in Solids*. Chapman and Hall, London, 1970.

Jackson, John D. *Classical Electrodynamics*. Wiley, New York, 1975.

Kline, Morris, and Irwin W. Kay. *Electromagnetic Theory and Geometrical Optics*. Interscience, New York, 1965.

Lavin, E. P. *Specular Reflection*. Elsevier, New York, 1971.

Levenson, Marc D. *Introduction of Nonlinear Spectroscopy*. Academic Press, New York, 1982.

Marcuse, D. *Theory of Dielectric Optical Waveguides*. Academic Press, New York, 1974.

O'Neill, E. L. *Introduction to Statistical Optics*. Addison-Wesley, Reading, Mass, 1963.

Okoshi, Takanori. *Optical Fibers*. Academic Press, London, 1982.

Seraphin, B. O., ed. *Optical Properties of Solids: New Developments*. North Holland, Amsterdam, 1976.

Shen, Y. R., *The Principles of Nonlinear Optics*. Wiley, New York, 1984.

Weaver, J. H., C. Krafka, and D. W. Lynch. *Optical Properties of Metals*. Fach-Informations-Zentrum, Karlsruhe, West Germany, 1981.

Wooten, Fredrick. *Optical Properties of Solids*. Academic Press, New York, 1972.

Yariv, A. *Quantum Electronics*. Wiley, New York, 1975.

Zernike, F., and J. E. Midwinter. *Applied Nonlinear Optics*. Wiley, New York, 1973.

PROBLEMS

Section 2.1   Light Waves in Matter

1. Identify examples of free charge, free current, bound charge, and bound current.

2. Draw qualitative diagrams for the electric field **E**, the displacement **D**, and the polarization as functions of

position perpendicular to the plates of a parallel plate capacitor that is filled with a dielectric material whose permitivity is ε. Show the results for positions both inside and outside the plates of the capacitor.

3. A local, isotropic, linear, uniform dielectric cylinder is found with its polarization vector parallel to the axis of the cylinder. If the ends of the cylinder have radii a and the cylinder is b long, find the electrostatic potential as a function of position along the axis of the cylinder a distance r from the center of the cylinder (where $r \gg b, a$). Repeat the calculation, except this time assume that the dielectric is in the shape of a uniform sphere of radius a.

4. Derive the dispersion equation for light in matter, Eq. (2.32).

5. Imagine a plane wave propagating in a linear, isotropic, local, and homogeneous medium that is conducting ($\mathscr{k} \neq 0$). At a given instant in time, how much does the amplitude of E change from one maximum to the next as r changes?

6. An indoor swimming pool is illuminated with red lights during a special effects presentation at a water show. The overhead lights are provided with filters that select a narrow band of wavelengths centered on 600 nm. If the index of refraction of the water is 1.33, what is the wavelength of the light under water? For swimmers under the water, what wavelength would they observe?

7. Within a homogeneous, linear, isotropic medium a 1-cm^2 light beam travels 1 cm. If the absorption coefficient is $3 \times 10^{-4} \text{ cm}^{-1}$ and the initial value of the irradiance of the light beam is 0.5 W/cm², find the rate at which energy is deposited within the 1 cm³ volume.

8. What is the magnitude of the absorption coefficient (in cm^{-1}) that would cause the power in a light beam to be attenuated to $1/e$ of its starting value after traveling 20 nm in the medium? If the vacuum wavelength of the light is 514.5 nm, find the imaginary part of the index of refraction in this example.

9. The complex index of refraction of a material is $4 - i2$. Find the complex dielectric function and the classical skin depth (the latter expressed as a fraction of the vacuum wavelength). What is the phase relationship between the electric and the magnetic fields? Assume that the medium is nonmagnetic.

10. Derive expressions for the real and the imaginary parts of the index of refraction in terms of the real and imaginary parts of the dielectric function, assuming the medium is nonmagnetic.

Section 2.2   Reflection and Transmission of Light at an Interface

11. Show that the equations used to determine the reflection and transmission coefficients satisfy the requirements of continuity of the normal components of **B** and **D**.

12. Derive the Fresnel relations Eqs. (2.61b), (2.62b), and the second forms of Eqs. (2.70a) and (2.70b).

13. Prove energy conservation, Eq. (2.75), for σ and π polarization.

14. Derive Eqs. (2.71) and (2.72), which relate the amplitude reflection and transmission coefficients to the situation where the roles of incident and transmitted waves are interchanged for the explicit case that the radiation is π polarized.

15. Create a quantitatively accurate plot of the amplitude reflection coefficient as a function of the angle of incidence for both σ and π polarization in the case of internal reflection from a medium with an index of refraction of 1.7 with the external medium index of refraction 1.3.

16. Present a step-by-step derivation of the Brewster's angle equation

$$\tan \theta_B = \frac{n'}{n}$$

17. In the reflection of a laser beam from flint glass ($n = 1.7$) in the external configuration find the value of the σ polarized reflectance at an angle of incidence equal to the Brewster angle.

18. Under what conditions at normal incidence would the reflectance and the transmittance be equal to each other? What is the value of R and T under these conditions?

19. Find the reflectance and the transmittance for external reflection at the interface between air and water ($n = 1.33$) when the angle of incidence is 50°. Perform your calculation for both states of linear polarization in the incident beam.

20. Use a Taylor series expansion to approximate the reflectance of an interface between two transparent dielectrics that are very similar such that $n'/n = 1 + \eta \approx 1$. Perform the calculation at normal incidence.

21. Consider light at normal incidence onto a metal surface, the incident electric field being E. What is the electric field just inside the surface for silicon at 700 nm? Repeat the calculation for 500 nm. Use the optical properties in Table 2.1.

22. You are given a pure sample with a good polished surface of an unknown opaque material with a high reflectance at a certain wavelength and at normal incidence. How could you tell whether the large value of R was due to a large value of k or to a large value of n?

23. If there is an equal mix of σ and π light in the incident beam, what will the reflectance be in terms of the reflectance of the pure components?

24. Unpolarized light reflects from the surface of water ($n = 1.33$) at an angle of $60°$ with respect to the surface normal. Draw a quantitatively accurate polar plot of the reflectance as a function of the angle, measured with respect to the plane of incidence of polarizing filter that is used to analyze the characteristics of the reflected light.

25. Consider the case of transmission through a parallel plate dielectric with multiple reflections taken into account. Assume that the thickness is large enough that interference phenomena are not important. Show that the transmittance of the plate is given by

$$\frac{1 - R}{1 + R}$$

where R is the reflectance of a single interface between air and the dielectric medium.

26. Show by direct calculation of the Poynting vector that for total internal reflection the normal component has zero time average in the transmitting medium.

27. Consider total internal reflection at a dielectric interface, concentrating on the inhomogeneous plane wave in the transmitting medium. Find the surfaces of constant amplitude, the surfaces of constant phase, and the phase velocity.

28. Find the phase difference between the reflected wave and the incident wave for the case of internal reflection at an incident angle of $70°$ from a dielectric medium with an index of refraction of 1.6. Perform the calculation for both σ and π polarizations.

29. In the configuration demonstrating frustrated total

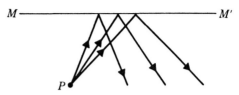

Fig. 2.25

internal reflection, as in Fig. 2.9, find the size of the air gap that produces a transmission coefficient of 50% if the prisms are constructed of glass with an index of refraction of 1.6 and the incident vacuum wavelength is 632.8 nm with π polarization.

Section 2.3 Applications in Planar Surface Optics

30. Show that the rays from P in Fig. 2.25 that are reflected by the plane mirror MM' appear to be coming from the image point P'. Locate P'.

31. (a) Using Fig. 2.26, prove that a ray reflected from both faces of a corner mirror is sent back parallel to its original direction.

(b) Locate the image point P' from which the doubly reflected rays appear to be coming if P is a point source.

32. Two plane mirrors are positioned so as to contact each other on a line, thus making a wedge whose apex angle is $40°$. A person is located between the mirrors on an imaginary plane that contains the line of contact of the mirrors and that is an angular distance of $10°$ from one of the mirrors. Find the number of self images that the person will see and identify their locations.

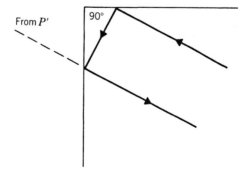

Fig. 2.26

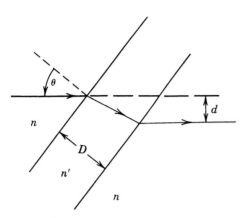

Fig. 2.27

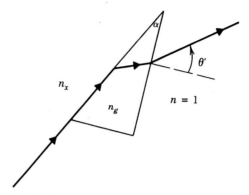

Fig. 2.28

33. How accurately must the apex and the angle of minimum deviation be known if the index of refraction is to be determined to four significant digits?

34. A prism with an apex angle of 40° is constructed from an unknown transparent material. The angle of incidence at minimum deviation is measured to be 37.32°. What is the index of refraction of the prism material?

35. Show that for a thin prism (small apex angle) the angle of minimum deviation θ_D is related to the prism index of refraction n and the prism angle α by

$$\theta_D \cong (n-1)\alpha$$

36. Calculate the deviation d produced by the plane parallel slab in Fig. 2.27 as a function of n, n', D, and θ.

37. A ray with glancing incidence onto an Abbe prism from a medium of refractive index n_x will emerge at angle θ' (see Fig. 2.28).

(a) If n_g is the refractive index of the glass and α the prism angle, show that

$$n_x = \sin \alpha (n_g^2 - \sin^2 \theta')^{1/2} + \cos \alpha \sin \theta'$$

(This effect is used to measure indices of refraction.)

(b) Find n_x for $n_g = 1.600$ and $\alpha = 45°$ if $\theta' = +15°$ and if $\theta' = -15°$.

38. Light from a distant star will be bent by refraction in the atmosphere so as to distort the apparent angle of observation (measured with respect to the zenith directly overhead). Consider that the atmosphere is a uniform layer of constant thickness with an index of refraction of 1.000292. Derive an expression from which the true angle can be calculated.

(a) Assume that the earth and the atmosphere are both flat.

(b) Use a more realistic model that allows for the curvature of the earth.

39. You are able to determine distance by the angle that your eyes measure between two rays coming from the same object. Using this concept, find the apparent height of an object above the surface of the water if you are below the water looking up at the object. Let the actual height above the water be 1 m and the index of refraction of the water be 1.33.

40. A source P embedded in a medium with an index of refraction n emits a narrow bundle of rays with an infinitesimal angular spread $\Delta\theta$ about a center ray

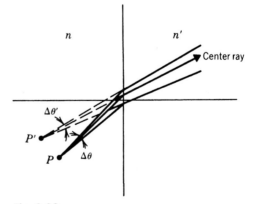

Fig. 2.29

having an angle of incidence onto a plane interface with a medium whose index of refraction is n' (Fig. 2.29). Find the position of the image point P' from which the rays appear to be diverging and calculate the angular spread $\Delta\theta'$.

(a) Consider only rays in the plane of the illustration.

(b) Consider only a fan of rays about the center ray in a plane perpendicular to the plane of the illustration.

41. A small particle is embedded at the center of a glass cube of index 3/2. An observer is located on the extension of the cube diagonal. Sketch a quantitatively accurate diagram that illustrates the view of the cube from the observer's perspective, including all images of the particle.

42. A fiber-optic element is under design. If the end will be in air and the cladding material is to have an index of 1.3, then what will the index of the fiber core have to be so that all light, regardless of the incident angle, is captured by the fiber?

43. An optical fiber has a core diameter of 15 μm. If the indices of refraction of the core and the cladding are 1.7 and 1.5, respectively, find the minimum radius of a bend in the fiber that will not destroy the internal reflection condition. Perform the analysis with meridional rays only (rays that stay in the plane of the bend).

44. A circular raft with a radius of 5 ft floats on water ($n = 1.33$). Assuming an overcast but bright sky, find the shape and volume of water beneath the raft that will not be illuminated. Neglect scattering and reflection of light within the water.

45. In an ATR experiment from an element with index of refraction 1.71 the π-polarized reflectance is found to be 0.924 when the internal angle of incidence is 60°. If the external medium has an index of refraction 1.33, find the absorption coefficient of that medium.

46. A plane light wave with a wavelength of 500 nm is incident on an absorbing medium at an angle of 45°. If the complex index of refraction is $3 - i0.2$, determine the orientation and speed of the surfaces of constant real phase just inside the surface. Repeat the calculation for an angle of incidence of 60°. How do these results compare with the situation where the index of refraction is 3, purely real?

Section 2.4 Introduction to the Optical Properties of Matter

47. Calculate the induced *electrostatic* field at the center of a spherical cavity within a local, isotropic, linear, and uniformly polarized dielectric medium that is otherwise infinite in extent. Express your answer in terms of the uniform polarization **P**. The total local field in this case would be the applied field **E** plus the induced field resulting from the polarization of the medium.

48. In a material within which the molecular dipoles are free to rotate, the energy U_d of a single dipole is related to the local electric field

$$U_d = -\mathbf{p} \cdot \mathbf{E}$$

The net dipole moment per unit volume can be calculated as a statistical average with the weighting factor given by the Boltzmann relation

$$\exp\!\left(\frac{-U_d}{k_B T}\right)$$

By symmetry, in a material containing N dipoles per unit volume the net polarization must be along the direction of the local electric field (where θ is the angle between a given dipole and the local field). The statistically averaged result for **p** is

$$\mathbf{p} = \frac{\displaystyle\int_0^\pi \exp\!\left(\frac{-U_d}{k_B T}\right) p \cos\theta \; d\Omega}{\displaystyle\int_0^\pi \exp\!\left(\frac{-U_d}{k_B T}\right) d\Omega}$$

Carry out the integration to determine the result for the net polarization of this material. ($d\Omega$ = differential solid angle = $2\pi \sin\theta \, d\theta$.)

49. A normalized Lorentzian function may be written

$$L(x; \gamma) = \frac{\dfrac{\gamma}{\pi}}{x^2 + \gamma^2}$$

Its maximum value of $1/(\pi\gamma)$ occurs at $x = 0$. The half-width at half maximum is γ. The area under the peak is unity:

$$\int_{-\infty}^{\infty} L(x; \gamma) \, dx = 1$$

(a) Prove the preceding statements.

(b) Consider a dilute system of N oscillators per unit volume embedded in a dielectric medium. The oscillators have charge q, mass m, natural frequency ω_0, oscillator strength f and damping time τ, where $\omega_0\tau \gg 1$. The medium itself has no absorption near ω_0, so that its index of refraction n may be treated as a real constant number for ω near ω_0. Show that for ω near ω_0, the absorption coefficient is well approximated by

$$K(\omega) = \frac{\pi Nq^2}{2mc\varepsilon_0}\frac{(n^2+2)^2}{9n}fL\left[(\omega-\omega_0);\frac{1}{2\tau}\right]$$

(c) Change variables to $1/\lambda = \omega/2\pi c$ (wavenumbers). Show that

$$\int K\left(\frac{1}{\lambda}\right)d\left(\frac{1}{\lambda}\right) = f\frac{(n^2+2)^2}{9n}\frac{\pi}{\lambda_p^2}$$

where $\lambda_p = 2\pi c/\omega_p$ represents the vacuum wavelength of light at the plasma frequency

$$\omega_p = \sqrt{\frac{Nq^2}{m\varepsilon_0}}$$

The integral is to be taken over the wavenumber range about the resonance peak where K is nonzero.

50. Consider an oscillator O with strength f, mass m, isolated resonance frequency ω_0, a long damping time τ, and concentration N imbedded in a medium with index of refraction $n_\infty(\omega)$; that is, the dielectric function would be $[n_\infty(\omega)]^2$ if the oscillators in question were not present. Assume that n_∞ may be treated as a constant for all frequencies from the vicinity of ω_0 to zero. This means that the oscillators that contribute to n_∞ have their resonances at frequencies much higher than ω_0.

(a) Write the analogs of Eqs. (2.157) and (2.158) for κ_r and κ_i and Eq. (2.159)

$$\int_0^{\bullet(above\ \omega_0)}\kappa_i(\omega)\,d\omega$$

(b) Let ω_t be the frequency of the oscillator O in the medium at which κ_i attains its maximum value (resonance frequency in the medium). Assume N and τ are large enough so that κ_r passes through zero near resonance. Let ω_l be the frequency at which κ_r becomes positive again. Assume that $\omega_l - \omega_t \gg 1/\tau$. Show that

$$\omega_l^2 - \omega_t^2 = f\frac{Nq^2}{m\varepsilon_0}\frac{(n_\infty^2+2)^2}{9n_\infty^2}$$

and that

$$\frac{\omega_l^2}{\omega_t^2} = \frac{\kappa(0)}{n_\infty^2}$$

This last equation is known as the Lyddane-Sachs-Teller relation.

51. There is a class of electronic impurity centers in crystals called "color centers." The absorption coefficient produced by one such center, the "F center" in a sodium chloride crystal, is shown in Fig. 2.30. Treat this center as a harmonic oscillator with the mass of an electron imbedded in the crystal (of refractive index $n = 1.55$). Assume the experimental lineshape to be Lorentzian, and from it estimate the lifetime of the oscillator and the oscillator strength. Calculate a value for the maximum change in the real part of the index of refraction caused by the presence of the F centers in this sample.

52. For the model of a metal assumed for Fig. 2.19, the plasma frequency $\omega_p = 6 \times 10^{15}$ rad/sec corresponds to a wavelength $\lambda_p = 314$ nm. Calculate numerical values for the penetration depth $1/K$ for light at normal incidence at these wavelengths: 349 nm, 314 nm, and 286 nm.

53. Consider a substance in the "metallic" reflection region where κ_r is negative and κ_i quite small.

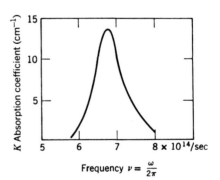

Fig. 2.30

(a) Show that to first order in κ_i, the reflectance R at normal incidence obeys

$$1 - R = \frac{2\kappa_i/\ell}{1 + \ell^2}$$

(b) Calculate $(1 - R)$ for a metal with $\omega_p\tau = 100$ when $\omega/\omega_p = 0.8$ and 0.95.

54. A useful approximate equation to describe the index of refraction of transparent media in the visible is Cauchy's equation

$$n = A + \frac{B}{\lambda^2}$$

Show that this approximation may be derived from equations such as Eqs. (2.155) and (2.156) under the assumption that the frequency of the light is much less than the frequencies ω_j of all the important oscillators.

55. When light propagates through a transparent medium that has a resonant absorption at a frequency less than that of the light, the dielectric function is less than unity. The phase velocity $v = c/n$ will be greater than the velocity of light in a vacuum. This is not incon-sistent with the principle of special relativity because, in order to carry information, there must be imposed on an ideal purely monochromatic sinusoidal light wave some form of modulation or change in amplitude. As a result the light signal must be considered to be a superposition of several monochromatic components, each traveling with the velocity v. However, the modulation must travel at the "group velocity," which is defined to be

$$\frac{1}{v_g} = \frac{dk}{d\omega}$$

Show that

$$\frac{1}{v_g} = \frac{1}{v} + \frac{\omega}{c}\frac{dn}{d\omega}$$

Demonstrate that the group velocity is less than the velocity of light if the index of refraction is of the following form

$$n^2 = 1 + \frac{B}{\omega_0^2 - \omega^2}$$

in the region $\omega > \omega_0$.

3 Geometrical Optics

Geometrical optics is that part of optics involving image formation and related phenomena that can be discussed within the framework of the three "laws" of *reflection*, *refraction*, and *rectilinear propagation*. We will be primarily concerned with the reflection from a spherical mirror (which would usually be covered with a metallic coating) and refraction at a spherical interface between two optically transparent materials ($\tilde{n}$ for both media purely real). We will also discuss the combination of optical elements into optical instruments and describe the details of several examples.

3.1 Ray Tracing

The light rays associated with a plane electromagnetic wave are lines parallel to the propagation wavevector $\mathbf{k}$. These are also perpendicular to the surfaces of constant phase. For the more general wavefront, we can still identify a surface on which the phase is constant. For a wave emanating from a point source for instance, these would be concentric spheres. In this example the rays would be radial lines directed outward from the point source. In the general case for local isotropic, linear media, the rays will be everywhere perpendicular to the constant phase surfaces and parallel to the wavevector.

Most optical systems consist of a series of refracting or reflecting surfaces to provide the required deviation of the light rays together with appropriate stops and apertures to limit the angular and spatial extent of the rays. The optical designer must often trace the path of rays through the system. This is done by repeated applications of the laws of geometrical optics.

This is simple in principle, but in practice it can become a formidable task if high accuracy is required. In this section we indicate the type of calculation necessary to perform accurate ray tracing through a series of coaxial spherical refracting surfaces. In subsequent work (sections 3.2–3.5) we will need only approximate versions of these formulas, which will be derived as we go along.

A. Refraction and Reflection

Let us first present the general expressions for the directions of the refracted and reflected rays in vector form. We already know that the wavevectors for the incident, refracted, and reflected waves will have the forms

$$\mathbf{k} = n\frac{\omega}{c}\hat{\mathbf{s}} \tag{3.1a}$$

$$\mathbf{k}' = n'\frac{\omega}{c}\hat{\mathbf{s}}' \tag{3.1b}$$

and

$$\mathbf{k}'' = n\frac{\omega}{c}\hat{\mathbf{s}}'' \tag{3.1c}$$

The $\hat{\mathbf{s}}$, $\hat{\mathbf{s}}'$, and $\hat{\mathbf{s}}''$ are unit vectors, and the refractive indices n and n' are both real (see Fig. 3.1). These must all lie in a plane of incidence. This plane contains the surface normal $\hat{\mathbf{\eta}}$, which here is defined as a unit vector outward into the transmitting medium. (Do not confuse n and n' with $\hat{\mathbf{\eta}}$.) The plane of incidence also contains $\hat{\mathbf{t}}$, (see Fig. 3.2a), the unit vector tangent to the interface and perpendicular to $\hat{\mathbf{\eta}}$. Angles θ, θ' and θ'' are given by

$$\hat{\mathbf{\eta}} \cdot \hat{\mathbf{s}} = \cos\theta \tag{3.2a}$$

$$\hat{\mathbf{\eta}} \cdot \hat{\mathbf{s}}' = \cos\theta' \tag{3.2b}$$

$$\hat{\mathbf{\eta}} \cdot \hat{\mathbf{s}}'' = \cos\theta'' \tag{3.2c}$$

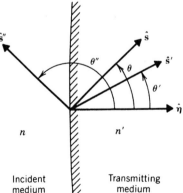

Fig. 3.1 Ray direction convention at an interface. Note that all angles are measured in the same sense from the surface normal.

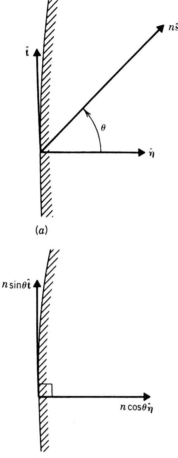

Fig. 3.2 Decomposition of a ray at the interface into normal and tangential components.

Note that the reflected angle as used here is defined as the supplement of the angle usually used in describing reflection. Our definition here is more consistent with a generalized formalism wherein all the angles are measured with respect to the surface normal in the same sense of rotation.

The phase-matching conditions of Chapter 2 lead to relationships among the k's in Eq. (3.1). These involve the ray directions and the appropriate refractive index. Because ω/c is the same for all three Eqs. labeled (3.1), we must then concentrate on relationships among the different $n\hat{s}$. With this in mind, we express $n\hat{s}$ in terms of $\hat{\eta}$ and $\hat{t}$ (see Fig. 3.2b):

$$n\hat{s} = n \cos \theta\, \hat{\eta} + n \sin \theta\, \hat{t} \tag{3.3}$$

1. Refraction. Starting from the form of Eq. (3.3) in the transmitting medium and applying Snell's law to the second term on the right yields

$$n'\hat{s}' = n' \cos \theta'\, \hat{\eta} + n \sin \theta\, \hat{t}$$

We would like this to be written in terms of $n\hat{s}$ in the incident medium.

$$n\hat{s} = n \cos \theta \, \hat{\eta} + n \sin \theta \, \hat{t}$$

Combining the previous two equations leads to the desired result.

$$n'\hat{s}' = n\hat{s} + (n' \cos \theta' - n \cos \theta)\hat{\eta} \tag{3.4}$$

We can determine $\cos \theta'$ in terms of $\cos \theta$ by Snell's law.

$$n' \cos \theta' = \sqrt{n'^2 - n^2 \sin^2 \theta} = \sqrt{n'^2 - n^2 + n^2 \cos^2 \theta} \tag{3.5}$$

Thus if we are initially given only the vectors $\hat{s}$ and $\hat{\eta}$ and the indices n and n', we first determine $\cos \theta$ by Eq. (3.2a) and then $\cos \theta'$ by Eq. (3.5). This allows us to calculate $\hat{s}'$ with Eq. (3.4).

2. Reflection. Using Eq. (3.3) in the incident medium for the incident and reflected rays leads to

$$\hat{s} = \cos \theta \, \hat{\eta} + \sin \theta \, \hat{t} \tag{3.6a}$$

$$\hat{s}'' = \cos \theta'' \, \hat{\eta} + \sin \theta'' \, \hat{t} \tag{3.6b}$$

But the law of reflection *in our ray-tracing convention* becomes

$$\theta'' = 180° - \theta \tag{3.7}$$

which means that

$$\cos \theta'' = - \cos \theta$$

and Eqs. (3.6a and b) can be combined to produce

$$\hat{s}'' = \hat{s} - 2 \cos \theta \, \hat{\eta} \tag{3.8}$$

B. Image Formation

1. Image Criteria. Equations (3.4) and (3.8) form the basis of a vector approach to geometrical optics. They describe the changes in direction of a light ray when it encounters an interface.

In many cases our goal will be to exploit the conditions that lead to image formation. An object that acts as the source for an optical system may be considered to be a collection of point sources, each having its own location and brightness. To understand the production of an image of an extended object, we must first study how a point source is imaged.

Figure 3.3 illustrates the image formation process for a source at point P. All the rays contributing to the image at P' (of which rays A and D are examples) must arrive at P' in phase with one another. (If they do not, then random interference would lead to a very small total amplitude for the optical field.) One of the ways to accomplish this is to provide a mechanism whereby the optical path length is increased over the ordinary path length by an amount that is more for rays B and C

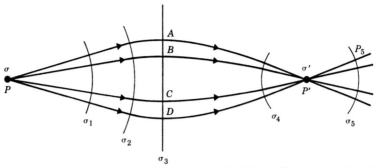

Fig. 3.3 Ray paths and wavefronts associated with a point source at P and a real point image at P'.

than for rays A and D. This could be done by requiring all the rays to pass through a glass plate whose thickness varied, being thinner at the edges than at the center. This, of course, is just a common lens. We can understand its operation in terms of Fermat's principle, which shows that all the rays must travel equivalent optical path lengths if they are to be simultaneously allowed.

An image can also be formed by a combination of several reflections and refractions and/or by the continuous bending of light rays in a medium where the index of refraction is a smoothly varying function of position. We say that P' is a *real image* of P if all the rays in a bundle leaving P and subtending a finite (rather than infinitesimal) solid angle come together at P' (Fig. 3.3). If the medium near P is uniform and isotropic, the wave surface σ_1 near P will be spherical. Similarly, the wave surface σ_5 near P' will be spherical, if the medium there is uniform and isotropic. (The intermediate wave surfaces σ_2 and σ_3 may be distorted, however.) At P and P' themselves, the wave surfaces σ and σ' degenerate to a point.

The rays from P may not actually cross at P'; they may diverge as though they met there. This forms what is known as a *virtual image* at P' (Fig. 3.4). The region of divergence must occur in uniform media so that the wave surface σ_5 is spherical. Otherwise there is no way to be sure that the rays near σ_5 appear to be coming from P'.

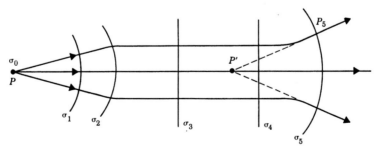

Fig. 3.4 Virtual image at P'. The rays *appear* to be diverging from P'.

The optical path from P to σ_5 is the same for all rays. The apparent (dashed) path from P' to σ_5 is the same for all rays, and so is the apparent optical path from P to P'. This must evidently be defined by the relation

$$\text{OPL}\,(PP') \equiv \text{OPL}\,(PP_5) - \text{OPL}\,(P'P_5)$$

$$= \text{OPL}\,(PP_5) - n'\overline{P'P_5} \tag{3.9}$$

The minus sign occurs because the rays at P_5 point away from P'. This equation also holds for the real image illustrated in Fig. 3.3. In both cases, n' is the index of refraction near σ_5.

2. Cartesian Surfaces. If we require that the task of focusing be performed through a single encounter with a reflecting or refracting surface in otherwise homogeneous media, then it is a matter of simple geometry to determine the contour which that surface must have. The condition can be established for monochromatic incident light from one object point at a fixed distance from the surface. Such a surface is called a *Cartesian surface*.

The Cartesian surfaces for reflection are conic sections. We illustrate them in Fig. 3.5 but leave to the reader the proofs that they are indeed what we claim.

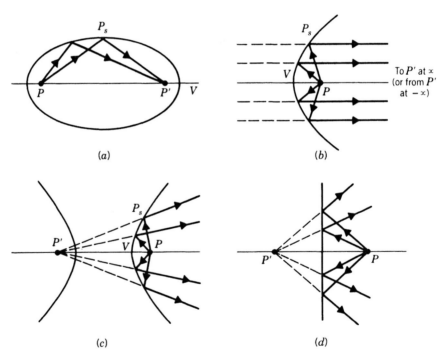

(a) (b)

(c) (d)

Fig. 3.5 Cartesian surfaces for reflection: (*a*) Ellipsoid; P and P' at foci. (*b*) Paraboloid; P at focus. (*c*) Hyperboloid; P and P' at foci. (*d*) Plane.

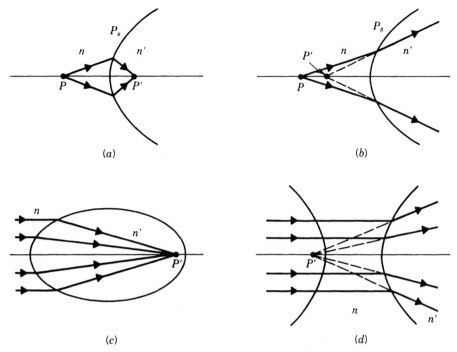

Fig. 3.6 Cartesian surfaces for refraction: (*a*) Real image. (*b*) Virtual image. (*c*) Real image, object at infinity. (Ellipsoid eccentricity $e = n/n' < 1$.) (*d*) Virtual image, object at infinity. (Hyperboloid.)

The Cartesian surfaces for refraction are generally not simple curves. In Fig. 3.6*a* the case of a real image is shown. In this case $n\overline{PP_s} + n'\overline{P_sP'} = $ constant, and the curve is a Cartesian ovoid of revolution. Figure 3.6*b* illustrates virtual image formation for which the equation, based on Eq. (3.9)

$$\text{OPL } (PP') = n\overline{PP_s} - n'\overline{P_sP'} \tag{3.10}$$

holds. The refracting surface can sometimes be spherical in this case.

If the object or image is at infinity, then the Cartesian surfaces may be conic sections, as indicated in Figs. 3.6*c* and 3.6*d*.

C. Refraction and Reflection at Spherical Surfaces

Good Cartesian surfaces are difficult to produce. The overwhelming majority of focusing operations in optical systems are performed by surfaces that are sections of spheres. The resulting images are, by necessity, only approximate.

1. Refraction at Spherical Surfaces. Let us apply the vector form of the law of refraction, Eq. (3.4), to the case where the interface is described by a section of a sphere. The medium on the incident side of the interface is characterized by index of

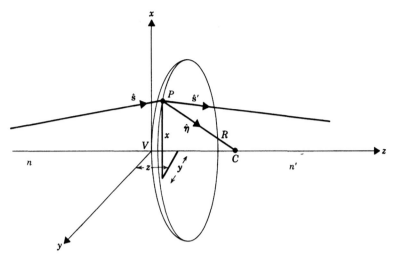

Fig. 3.7 Geometry for refraction at a spherical interface.

refraction n and on the transmitting side of the interface by n'. The optical materials are otherwise local, isotropic, linear, and nonabsorbing. The situation is depicted in Fig. 3.7.

The vertex is at point V where the surface intersects the z axis at the origin. Other important points are identified by vectors leading from the origin. These are specified in terms of their components along x, y, and z. The spherical refracting surface has radius of curvature R and center of curvature at point C. A vector identifying C has components

$$\mathbf{C}{:}(0, 0, R) = (0, 0, \pm |R|)$$

The incident ray meets the surface at the point P, identified by vector

$$\mathbf{P}{:}\left[x, \, y, \left(R \mp \sqrt{R^2 - x^2 - y^2}\right)\right]$$

Use the upper (lower) sign when the z component of $\vec{c}$ is positive (negative).

The normal to the surface at P is a unit vector pointing from P toward C. It is given by

$$\hat{\boldsymbol{\eta}} = \frac{\mathbf{C} - \mathbf{P}}{\pm |\mathbf{C} - \mathbf{P}|} = \frac{\mathbf{C} - \mathbf{P}}{R}$$

with components

$$\hat{\boldsymbol{\eta}}{:}\left[\frac{-x}{R}, \, \frac{-y}{R}, \, \left(\frac{\sqrt{R^2 - x^2 - y^2}}{|R|}\right)\right] \tag{3.11}$$

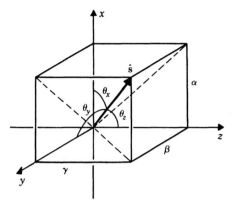

Fig. 3.8 Standard geometry showing a ray direction unit vector with its direction cosines.

We now introduce the components (called *direction cosines*) of the unit vectors $\hat{s}$ and $\hat{s}'$ (see Fig. 3.8):

$\hat{s}$ has components (α, β, γ) with length $\alpha^2 + \beta^2 + \gamma^2 = 1$ (3.12a)

$\hat{s}'$ has components $(\alpha', \beta', \gamma')$ with length $\alpha'^2 + \beta'^2 + \gamma'^2 = 1$ (3.12b)

We can write the x and y components of Eq. (3.4) by using Eqs. (3.11) and (3.12)

$$n'\alpha' = n\alpha - \left[\frac{(n'\cos\theta' - n\cos\theta)}{R}\right]x \tag{3.13a}$$

$$n'\beta' = n\beta - \left[\frac{(n'\cos\theta' - n\cos\theta)}{R}\right]y \tag{3.13b}$$

$$n'\gamma' = n\gamma + \left[\frac{(n'\cos\theta' - n\cos\theta)}{|R|}\right]\sqrt{R^2 - x^2 - y^2} \tag{3.13c}$$

[Note that γ and γ' can always be determined from the normalization condition, Eq. (3.12)]

Since $\cos\theta = (\hat{s} \cdot \hat{n})$, we may use Eqs. (3.11) and (3.12) to determine $\cos\theta$ in terms of the direction cosines α, β of $\hat{s}$ and the transverse coordinates x, y of P:

$$\cos\theta = \left\{\left[1 - \left(\frac{x}{R}\right)^2 - \left(\frac{y}{R}\right)^2\right](1 - \alpha^2 - \beta^2)\right\}^{1/2} - \left(\frac{x\alpha}{R}\right) - \left(\frac{y\beta}{R}\right) \tag{3.14a}$$

We then determine $\cos\theta'$ by Eq. (3.5)

$$n'\cos\theta' = \sqrt{n'^2 - n^2 + n^2\cos^2\theta} \tag{3.14b}$$

and α' and β' by use of Eqs. (3.13a, b). We also have the obvious relation

$$x' = x \tag{3.13d}$$

$$y' = y \tag{3.13e}$$

between the transverse coordinates of the ray just after refraction and its coordinates just before refraction.

Regarding sign conventions, we note that the implicit assumption has been made that the refracting surface is convex toward the incident side, that is, that V is on the incident side of C. Equations (3.5), (3.13), and (3.14) hold also for a concave surface, that is, one with C on the incident side of V, provided that R is defined to be negative in that case and provided $\cos \theta$ and $\cos \theta'$ are defined to remain positive. These conventions require some rewriting of intermediate steps of the derivation; the results remain valid.

Equations (3.13) can be considered as the equations of a transformation, denoted symbolically by $\mathscr{R}$, that the direction cosines α, β and coordinates x, y of a ray undergo during a single refraction.

2. Reflection at Spherical Surfaces. Now consider reflection off a vertical spherical surface with vertex V and center of curvature C, both on the z-axis, as shown in Fig. 3.9. The radius of curvature, R, will be considered positive if V is on the incident side of C (as it was for refraction). The incident ray, $\hat{s}$, meets the surface at the point P specified by vector

$$\mathbf{P}{:}(x, y, z) = \left[x, y, \left(R \mp \sqrt{R^2 - x^2 - y^2} \right) \right]$$

at which point the surface normal $\hat{\eta}$ has components

$$\hat{\eta}{:}\left[-\frac{x}{R}, -\frac{y}{R}, \left(\frac{\sqrt{R^2 - x^2 - y^2}}{|R|} \right) \right]$$

as before.

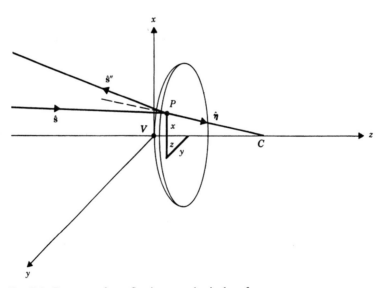

Fig. 3.9 Geometry for reflection at spherical surface.

The reflected ray direction, $\hat{s}''$ is given by Eq. (3.8). We arrive at the reflection transformation at a spherical surface:

$$\mathcal{R}'' \begin{cases} \alpha'' = \alpha + \left[\dfrac{2 \cos \theta}{R} \right] x & \text{(3.15a)} \\[2mm] \beta'' = \beta + \left[\dfrac{2 \cos \theta}{R} \right] y & \text{(3.15b)} \\[2mm] \gamma'' = \gamma - \left[\dfrac{2 \cos \theta}{|R|} \right] \sqrt{R^2 - x^2 - y^2} & \text{(3.15c)} \\[2mm] x'' = x & \text{(3.15d)} \\[2mm] y'' = y & \text{(3.15e)} \end{cases}$$

We also have the auxiliary relation to determine $\cos \theta$ from Eq. (3.14a).

3. Translation between Spherical Surfaces. Before and after refraction or reflection at an interface, the rays will travel undeviated in straight lines until they meet other surfaces. It will be convenient to describe this linear propagation as a mathematical transformation in a way that is compatible with our prescription for refraction and reflection.

As shown in Fig. 3.10, we consider two spherical refracting surfaces whose

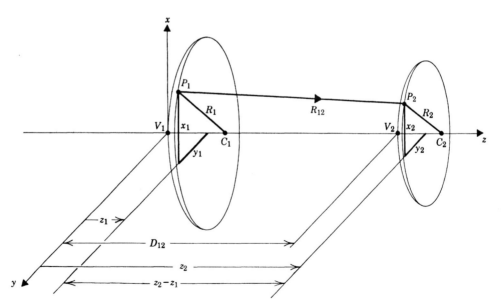

Fig. 3.10 Translation between two spherical interfaces.

vertices are on the z-axis located at the origin and at D_{12}, respectively. Let the intersection of a ray with the first surface be at P_1 specified by vector

$$\mathbf{P}_1 : \left[x_1, y_1, \left(R_1 \mp \sqrt{R_1^2 - x_1^2 - y_1^2} \right) \right] \tag{3.16a}$$

The direction of the ray on the transmitting side of P_1 is

$$\hat{\mathbf{s}}_1' : (\alpha_1', \beta_1', \gamma_1') \tag{3.16b}$$

The corresponding quantities for the second surface are

$$\mathbf{P}_2 : \left[x_2, y_2, \left(D_{12} + R_2 \mp \sqrt{R_2^2 - x_2^2 - y_2^2} \right) \right] \tag{3.17a}$$

and

$$\hat{\mathbf{s}}_2 : (\alpha_2, \beta_2, \gamma_2) \tag{3.17b}$$

The primes in Eqs. (3.16a and b) are an indication that the intersection and ray direction are evaluated after the optical interaction with the interface. The coordinates in Eqs. (3.17a and b) are without primes because these are evaluated before the encounter with the second interface.

At P_2 the ray direction is the same as after refraction at P_1, thus

$$\alpha_2 = \alpha_1' \tag{3.18a}$$

and

$$\beta_2 = \beta_1' \tag{3.18b}$$

The distance between P_1 and P_2 is given by R_{12}, where

$$\begin{aligned}
R_{12}^2 = {} & (x_2 - x_1')^2 + (y_2 - y_1')^2 \\
& + \left[D_{12} + R_2 - R_1 \mp \sqrt{R_2^2 - x_2^2 - y_2^2} \right. \\
& \left. \pm \sqrt{R_1^2 - x_1'^2 - y_1'^2} \right]^2
\end{aligned} \tag{3.19}$$

In our convention, the sign of R_{12} must be the same as the sign of $(z_2 - z_1)$.

The translation transformation can be expressed in vector form as

$$\mathbf{P}_2 = \mathbf{P}_1 + R_{12}\hat{\mathbf{s}}_1' \tag{3.20}$$

or in component form as

$$x_2 = x_1' + R_{12}\alpha_1' \tag{3.18c}$$

and

$$y_2 = y_1' + R_{12}\beta_1' \tag{3.18d}$$

Equations (3.18) together with the distance Eq. (3.19) form the translation transformation from spherical surface 1 to spherical surface 2. This we may symbolize as $\mathscr{T}$.

4. Multiple-Interface Systems. The preceding mathematical formalism may be applied repeatedly as the ray makes its way through the optical system. Application of the procedure to a selection of incident rays (for instance, those emanating from an object at different directions) will provide exact information about the locations of the resulting rays. This type of calculation is most conveniently performed by interactive computer algorithms that allow adjustment of the parameters describing the optical surfaces. Most modern lens design is performed in this way.

Table 3.1 summarizes the transformations for refraction, reflection, and translation that apply to spherical interfaces.

3.2 Paraxial Optics

Although the ray-tracing equations are exact, they provide little physical insight into the image-formation process. To reach a simplified analytical result, some approximations are necessary. The first-order case is useful as a starting point for more exact calculations, but it is often good enough to be used for many applications as it stands. This is valid when an optical axis can be defined for the system under study and when all light rays and all surface normals to refracting or reflection surfaces make small angles with the axes. Such rays are called *paraxial rays*. It was essentially the paraxial approximation that Kepler used when he first formulated the theory of the telescope and the magnifier.

A. Refraction

1. Approximations. The first-order theory that approximates the exact ray-tracing formulas of section 3.1 follows from the condition that the direction cosines α and β remain small and γ is close to unity. This is equivalent to the requirement that θ_z in Fig. 3.8 remain small. Similarly, all transverse dimensions x and y are treated as small compared with the radii of curvature R. Within these limitations we may replace the exact coordinates of the intercept of the ray with the optical surface at P by $(x, y, 0)$. The surface normal at that point becomes $(-x/R, -y/R, 1)$. The angles θ and θ' are also considered to be small, so that Snell's law becomes

$$n\theta = n'\theta' \tag{3.21}$$

and

$$\cos \theta = 1, \quad \cos \theta' = 1 \tag{3.22}$$

The equations of the refraction transformation (3.13) then become the following linear equations:

Table 3.1. Ray-Tracing Transformations: Spherical Interfaces

Refraction $\mathscr{R}$

$$x' = x$$

$$y' = y$$

$$n'\alpha' = n\alpha - \left[\frac{(n' \cos \theta' - n \cos \theta)}{R}\right]x$$

$$n'\beta' = n\beta - \left[\frac{(n' \cos \theta' - n \cos \theta)}{R}\right]y$$

$$n'\gamma' = n\gamma + \left[\frac{(n' \cos \theta' - n \cos \theta)}{R}\right]\sqrt{R^2 - x^2 - y^2}$$

Reflection $\mathscr{R}''$

$$x'' = x$$

$$y'' = y$$

$$\alpha'' = \alpha + \left[\frac{2 \cos \theta}{R}\right]x$$

$$\beta'' = \beta + \left[\frac{2 \cos \theta}{R}\right]y$$

$$\gamma'' = \gamma - \left[\frac{2 \cos \theta}{R}\right]\sqrt{R^2 - x^2 - y^2}$$

where $\quad \cos \theta = \left\{\left[1 - \left(\frac{x}{R}\right)^2 - \left(\frac{y}{R}\right)^2\right](1 - \alpha^2 - \beta^2)\right\}^{1/2} - \left(\frac{x\alpha}{R}\right) - \left(\frac{y\beta}{R}\right)$

and $\quad n' \cos \theta' = \sqrt{n'^2 - n^2 + n^2 \cos^2 \theta}$

Translation $\mathscr{T}_{12}$

$$x_2 = x'_1 + R_{12}\alpha'_1$$

$$y_2 = y'_1 + R_{12}\beta'_1$$

$$\alpha_2 = \alpha'_1$$

$$\beta_2 = \beta'_1$$

$$\gamma_2 = \gamma'_1$$

where $\quad R_{12}^2 = (x_2 - x'_1)^2 + (y_2 - y'_1)^2 + (z_2 - z'_1)^2$

with $\quad (z_2 - z'_1) = (D_{12} + R_2 - R_1 - \sqrt{R_2^2 - x_2^2 - y_2^2} + \sqrt{R_1^2 - x'^2_1 - y'^2_1})$

sign of $\quad R_{12} = \text{sign of } (z_2 - z'_1)$

$$\mathscr{R} \begin{cases} n'\alpha' = n\alpha + \dfrac{n - n'}{R} x \quad * \\[2ex] n'\beta' = n\beta + \dfrac{n - n'}{R} y \\[2ex] x' = x \qquad\qquad * \\[2ex] y' = y \end{cases} \tag{3.23}$$

Note that these equations remain valid in the limit $R \to \infty$.

The translation equations also simplify under these approximations. Equation (3.19) gives simply $R_{12} = D_{12}$. The translation transformation of Eq. (3.18) becomes

$$\mathscr{T}_{12} \begin{cases} \alpha_2 = \alpha'_1 \qquad\qquad * \\[2ex] \beta_2 = \beta'_1 \\[2ex] x_2 = x'_1 + D_{12}\alpha'_1 \quad * \\[2ex] y_2 = y'_1 + D_{12}\beta'_1 \end{cases} \tag{3.24}$$

Note that in Eqs. (3.23) and (3.24) x and α at a given point depend only on x and α at other points, not at all on y and β. In other words, the pairs of variables (x, α) and (y, β) are decoupled from one another and may be treated independently. Still another way of saying this is that the projections of the rays on the xz- or yz-planes behave independently. (This is not true in the exact theory.) Because of this independence, it is no longer necessary to perform calculations of both projections simultaneously. We choose here to concentrate on the projection in the xz-plane. The projections behave as though the rays were actually lying in the xz-plane, that is, as though y and β were zero. Such rays that lie in a single plane containing the z-axis are called *meridional rays*.

When we make the assumption of meridional rays lying in the xz-plane, we are led to the two-dimensional situation shown in Figs. 3.11 and 3.12 for refraction and translation, respectively.

The direction cosine α is in general defined to be the cosine of the angle the ray makes with the x-axis. If the rays lie in the xz-plane, this also becomes the sine of the angle the ray makes with the z-axis. In the paraxial ray approximation, the sine of this angle equals the angle itself as shown in Figs 3.11 and 3.12. That is, $\alpha = \cos \theta_x \cong \sin \theta_z \cong \theta_z$. The relevant parts of the transformations in Eqs. (3.23) and (3.24) are marked with asterisks ($*$).

Successive application of these approximations allow us to express the transformation from a general point P_0 on the incident side of a refracting surface to a final point P_2 in the transmitting medium. Figure 3.13a shows the actual situation and Fig. 3.13b the linearized approximation of the same situation. We need to translate from P_0 to P_1 where the ray is transformed by refraction to

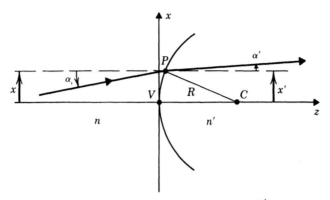

Fig. 3.11 Refraction in the paraxial limit for meridional rays. The separation between the x-axis and point P is negligible.

assume a new angle. At the distance D_{12} the translation transformation will tell us at what height the final ray is found. We have,

$$\mathcal{T}_{01} \begin{cases} x_1 = x_0 + D_{01}\alpha_0 \\ \alpha_1 = \alpha_0 \end{cases} \text{(New height)}$$

$$\mathcal{R}_1 \begin{cases} x_1' = x_1 \\ \alpha_1' = \dfrac{n}{n'}\alpha_1 + \left[\dfrac{(n - n')}{Rn'}\right]x_1 \end{cases} \text{(New angle)}$$

$$\mathcal{T}_{12} \begin{cases} x_2 = x_1' + D_{12}\alpha_1' \\ \alpha_2 = \alpha_1' \end{cases} \text{(New height)}$$

or, putting these together in a single transformation,

$$x_2 = \left[\frac{(n - n')D_{12}}{n'R_1} + 1\right]x_0 + \left[D_{01} + \frac{nD_{12}}{n'} + \frac{(n - n')D_{01}D_{12}}{n'R_1}\right]\alpha_0 \quad (3.25a)$$

$$\alpha_2 = \left[\frac{n - n'}{n'R_1}\right]x_0 + \left[\frac{n}{n'} + \frac{(n - n')D_{01}}{n'R_1}\right]\alpha_0 \quad (3.25b)$$

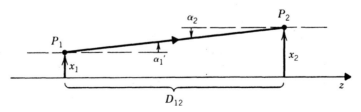

Fig. 3.12 Translation in the paraxial limit.

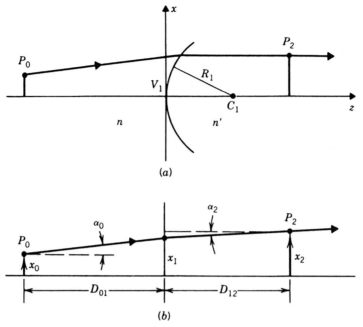

Fig. 3.13 Refraction plus translation near a spherical interface: (a) physical situation; (b) paraxial approximation.

Note the linear dependence of x_2 and α_2 on x_0 and α_0. This follows from the linear dependence of each individual transformation and will be exploited when we introduce the matrix formalism a bit later.

2. Image Formation. (a) *Location of Image.* If P_2 is to be an image of P_0, then any ray leaving P_0, regardless of the value of α_0 (as long as it is small), should arrive at P_2. This means that x_2 must be independent of α_0; that is, the coefficient of α_0 in Eq. (3.25a) must vanish if P_2 is to be an image of P_0. The imaging condition thus becomes

$$D_{01} + \frac{nD_{12}}{n'} + \frac{(n - n')D_{01}D_{12}}{n'R_1} = 0$$

This can be rearranged to yield the well-known equation

$$\frac{n'}{D_{12}} + \frac{n}{D_{01}} = \frac{n' - n}{R_1} \tag{3.26}$$

The *sign convention* we have been using is as follows.

1. Light rays travel from left to right (in the $+z$ direction).
2. R_1 is positive if V_1 is to the left of C_1, that is, if the refracting surface is convex toward the left; otherwise, R_1 is negative.

3. D_{01} is positive if P_0 is to the left of V_1; otherwise, D_{01} is negative.

4. D_{12} is positive if P_2 is to the right of V_1; otherwise, D_{12} is negative.

5. The angles α_0, α_1, and so on are positive if the ray direction is obtained by rotating the $+z$ axis counterclockwise through an acute angle; otherwise, the angles are negative.

6. The distances x_0, x_1, and so on are positive if up, negative if down.

It is useful to note that when $D_{01} = \infty$, the image will be located at the point

$$D_{12} \equiv f_1' = \frac{n'R_1}{n' - n} \tag{3.27a}$$

and if $D_{12} = +\infty$, the object must have been located at the point

$$D_{01} \equiv f_1 = \frac{nR_1}{n' - n} \tag{3.27b}$$

These distances are called the *image* and *object focal lengths*, respectively. Using them, we can write Eq. (3.26) in the form

$$\frac{n'}{D_{12}} + \frac{n}{D_{01}} = \frac{n' - n}{R_1} = \frac{n'}{f_1'} = \frac{n}{f_1} \tag{3.28}$$

(b) *Lateral Magnification.* Now that we have made x_2 in Eq. (3.25a) independent of α_0, we are left with an equation of the form $x_2 = m_x x_0$, where the lateral magnification m_x is given by

$$m_x = \frac{n - n'}{R_1} \cdot \frac{D_{12}}{n'} + 1 = \frac{D_{12}}{n'}\left(\frac{n - n'}{R_1} + \frac{n'}{D_{12}}\right)$$

This is the usual magnification that gives image size in terms of object size. We use Eq. (3.26) to express D_{12} in terms of D_{01} and obtain

$$m_x = \frac{x_2}{x_0} = \frac{D_{12}}{n'}\left(\frac{n - n'}{R_1} + \frac{n' - n}{R_1} - \frac{n}{D_{01}}\right) = -\frac{n}{n'}\frac{D_{12}}{D_{01}} \tag{3.29}$$

Hence, if D_{01} and D_{12} are both positive, m_x must be negative; this means that with x_0 positive, x_2 will be negative, or P_2 will be below the z axis if P_0 is above it. This is shown in Fig. 3.14.

(c) *Ray-Angle Magnification.* Consider two rays leaving P_0 with angular separation $\Delta\alpha_0$ and arriving at P_2 with angular separation $\Delta\alpha_2$ as shown in Fig. 3.15a. One of the rays could equivalently cross the z-axis, and the other could be the axis itself, as shown in Fig. 3.15b. The ray angle magnification m_α is given by

$$m_\alpha = \frac{\Delta\alpha_2}{\Delta\alpha_0} = -\frac{D_{01}}{D_{12}} \tag{3.30}$$

as can be seen from the illustrations. This result can also be obtained from Eq. (3.25b) by taking differences for fixed x_0. The ray-angle magnification is a useful concept when the angular spread of rays forming the image must be considered.

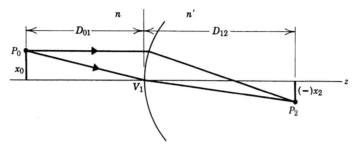

Fig. 3.14 Image inversion when both object and image distances are positive.

From Eqs. (3.29) and (3.30), we obtain a useful relation called the *Lagrange* or *Smith-Helmholtz equation*:

$$\frac{x_2}{x_0} \cdot \frac{\Delta\alpha_2}{\Delta\alpha_0} = \frac{n}{n'} \quad \text{or} \quad n'x_2\Delta\alpha_2 = nx_0\Delta\alpha_0 \quad \text{or} \quad n'x_2\alpha_2 = nx_0\alpha_0 \qquad (3.31)$$

Another version of this is

$$m_x m_\alpha = \frac{n}{n'} \qquad (3.31')$$

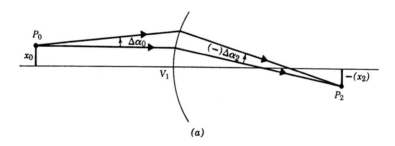

(a)

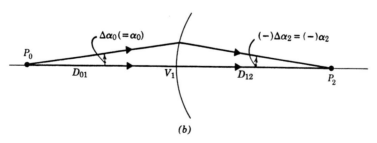

(b)

Fig. 3.15 Ray angle magnification: *(a)* general ray angular separation; *(b)* angular separation from the optical axis.

which says that large linear magnification can be obtained only at the expense of a large demagnification of the ray angles, that is, only by having much smaller angular divergence at the image than at the object.

When the image formation condition (3.26) is satisfied, we may use Eqs. (3.27), (3.29), and (3.30) in Eqs. (3.25a and b) to obtain for the ray-tracing transformation from object point P_0 to paraxial image point P_2 the following simple result:

$$x_2 = m_x x_0 \tag{3.32a}$$

$$\alpha_2 = -\frac{1}{f'_1} x_0 + m_\alpha \alpha_0. \tag{3.32b}$$

(d) *Virtual Object and Image.* The preceding formulas (3.25–3.32) hold when D_{01} and/or D_{12} is negative. A negative value for D_{12} means that the image is to the left of V_1; the light rays in the second medium are diverging as though they were coming from an image P_2 to the left of the vertex V_1, as shown in Fig. 3.16a. This is called the virtual image.

If D_{01} is negative, the rays appear to be converging at P_0 to the right of the vertex. This is called a *virtual object* and is shown in Fig. 3.16b. An auxiliary lens (shown dashed) is necessary to prepare the converging rays for such an object.

B. Reflections

1. Approximations. We now make the same paraxial ray assumptions that we used in the previous section and apply them to the reflection equations of section 3.1C2. The result is that Eqs. (3.15) become

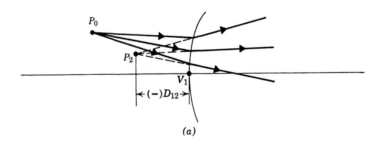

(a)

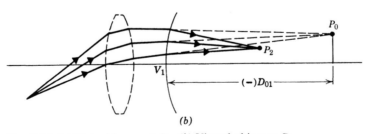

(b)

Fig. 3.16 (a) Virtual image at P_2. (b) Virtual object at P_0.

$$\mathscr{R}_1'' \begin{cases} \alpha_1'' = \alpha_1 + \dfrac{2}{R_1}x_1 & \text{(3.33a)} \\[3mm] \beta_1'' = \beta_1 + \dfrac{2}{R_1}y_1 & \text{(3.33b)} \\[3mm] \gamma_1'' = \gamma_1 - \dfrac{2\sqrt{R_1^2 - x_1^2 - y_1^2}}{R_1} \approx -1 & \text{(3.33c)} \\[3mm] x_1'' = x_1 & \text{(3.33d)} \\[2mm] y_1'' = y_1 & \text{(3.33e)} \end{cases}$$

The translation equations remain the same as in the paraxial refraction case. The pairs of variables (x, α) and (y, β) are uncoupled from one another and, as before, we choose to work with the pair (x, α). This is equivalent to working with meridional rays lying in the xz plane. The translation transformation $\mathscr{T}_{01}$ and $\mathscr{T}_{12}$ and the reflection transformation $\mathscr{R}_1''$ that describe the situation shown in Fig. 3.17 are given by

$$\mathscr{T}_{01} \begin{cases} x_1 = x_0 + D_{01}\alpha_0 \\ \alpha_1 = \alpha_0 \end{cases}$$

$$\mathscr{R}_1'' \begin{cases} x_1'' = x_1 \\[2mm] \alpha_1'' = \dfrac{2x_1}{R_1} + \alpha_1 \\[2mm] \gamma_1'' = -1 \end{cases}$$

$$\mathscr{T}_{12} \begin{cases} x_2 = x_1'' + D_{12}\alpha_1'' \\ \alpha_2 = \alpha_1'' \end{cases}$$

Note that, in Fig. 3.17, the radius of curvature R is negative, as is the reflected angle α_1''.

The transformations just presented can be combined to give the overall transformation from the pair variables (x_0, α_0) to the pair (x_2, α_2):

$$x_2 = \left(1 + \frac{2D_{12}}{R_1}\right)x_0 + \left(D_{01} + D_{12} + \frac{2D_{01}D_{12}}{R_1}\right)\alpha_0 \qquad \text{(3.34a)}$$

$$\alpha_2 = \frac{2}{R_1}x_0 + \left(\frac{2D_{01}}{R_1} + 1\right)\alpha_0 \qquad \text{(3.34b)}$$

2. Image Formation.   (a) *Location of Image.* We continue to use the line of reasoning employed in the refraction case. The condition for image formation is that x_2 be independent of α_0 or

$$\frac{1}{D_{12}} + \frac{1}{D_{01}} = -\frac{2}{R_1} \qquad \text{(3.35)}$$

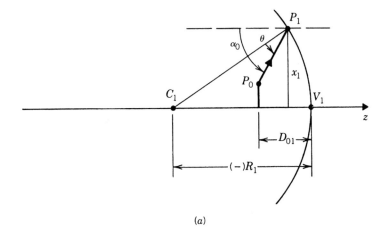

(a)

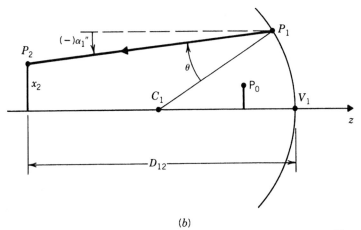

(b)

Fig. 3.17 Paraxial reflection of meridional rays: (a) incident ray; (b) reflected ray.

The focal length f is defined by

$$f = -\frac{R_1}{2} \tag{3.36}$$

If $D_{12} = \pm \infty$, then $D_{01} = f$, and vice versa.

The *sign convention* for *reflection* is as follows.

1. Positive ray direction is from left to right before reflection and from right to left after reflection.
2. R_1 is positive if V_1 is to the left of C_1.
3. D_{01} is positive if P_0 is to the left of V_1.
4. D_{12} is positive if P_2 is to the left of V_1.
5. The angles α_0, α_1 are positive if the ray direction is obtained by rotating the $+z$ axis counter clockwise through an acute angle.

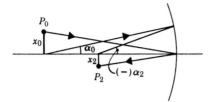

Fig. 3.18 Lateral magnification and ray-angle magnification in reflection.

6. The angles α_1'', α_2 are positive if the ray direction is obtained by rotating the $-z$ axis clockwise through an acute angle.

(The sign conventions in (5) and (6) are designed to preserve the interpretation of α_0, α_1, α_1'', α_2, as x components of unit vectors along the rays.)

(b) *Lateral Magnification.* Once the imaging condition (3.35) is satisfied, we may compute the lateral magnification by

$$m_x = \frac{x_2}{x_0} = -\frac{D_{12}}{D_{01}} \tag{3.37}$$

Again, for both a real image and a real object (D_{12} and D_{01} both positive), the image and object cannot both be erect or both be inverted (Fig. 3.18).

(c) *Ray-Angle Magnification.* We may derive an expression for the ray-angle magnification from the second of Eqs. (3.34) if we put $x_0 = 0$ (Fig. 3.18).

$$m_\alpha = \frac{\alpha_2}{\alpha_0} = -\frac{D_{01}}{D_{12}} \tag{3.38}$$

The Lagrange or Smith–Helmholtz equations then becomes

$$\frac{x_2 \alpha_2}{x_0 \alpha_0} = 1$$

(d) *Virtual Object and Image.* Our sign convention is still such that the object is real when D_{01} is positive, and the image is real when D_{12} is positive. The object is virtual when D_{01} is negative, and the image is virtual when D_{12} is negative.

3.3 Matrix Methods

The ray-tracing equations of paraxial optics are linear in the variables x and α. The individual refraction and translation equations are linear, and the resulting overall transformation is also linear. The equations of a linear transformation are very conveniently written using matrices, and the results of several consecutive linear transformations can be compactly written in terms of the product of the individual matrices. The matrix method makes it possible to discuss the paraxial optical theory of simple and compound lenses and of lens systems with a minimum of mathematical manipulation. It is also an expedient method for demonstrating some important theorems about paraxial optical systems.

A. Transformation Matrices

1. Refraction. At a spherical surface (identified as surface 1) between media n (on the left) and n' (on the right) with radius of curvature R_1, Eqs. (3.23) provide the paraxial refraction transformation

$$n'\alpha'_1 = n\alpha_1 + \left(\frac{n - n'}{R_1}\right)x_1 = n\alpha_1 - \mathscr{P}_1 x_1 \tag{3.39a}$$

$$x'_1 = x_1 \tag{3.39b}$$

where,

$$\mathscr{P}_1 \equiv \left(\frac{n' - n}{R_1}\right) \tag{3.40}$$

is called the *power* of the interface.

Define ray-column matrices

$$\mathbf{r}_1 \equiv \begin{pmatrix} n\alpha_1 \\ x_1 \end{pmatrix} \tag{3.41a}$$

and

$$\mathbf{r}'_1 \equiv \begin{pmatrix} n'\alpha'_1 \\ x'_1 \end{pmatrix} \tag{3.41b}$$

associated with the left and right sides of the interface, respectively. Also define a refraction matrix

$$\mathbf{R}_1 \equiv \begin{pmatrix} 1 & -\mathscr{P}_1 \\ 0 & 1 \end{pmatrix} \tag{3.42}$$

The refraction transformation can then be written in matrix form as

$$\mathbf{r}'_1 = \mathbf{R}_1 \mathbf{r}_1$$

2. Translation. We are concerned with the translation operation that transforms a ray from the right side of one optical element through a homogeneous space characterized by index of refraction n, to the left side of a second optical element. From Eqs. (3.24) we have

$$\alpha_2 = \alpha'_1$$

$$x_2 = x'_1 + D_{12}\alpha'_1$$

Which can be written in the equivalent matrix form as

$$\begin{pmatrix} n\alpha_2 \\ x_2 \end{pmatrix} = \begin{pmatrix} 1 & 0 \\ \dfrac{D_{12}}{n} & 1 \end{pmatrix} \begin{pmatrix} n\alpha'_1 \\ x'_1 \end{pmatrix}$$

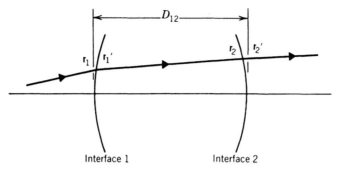

Fig. 3.19 Ray matrices.

thus defining the translation matrix

$$\mathbf{T}_{12} = \begin{pmatrix} 1 & 0 \\ \dfrac{D_{12}}{n} & 1 \end{pmatrix} \tag{3.43}$$

3. Combined Operations. These transformations may be combined to give the overall transformation through several refracting and transmitting elements. For instance, in Fig. 3.19 the ray $\mathbf{r}_1$ is changed into $\mathbf{r}_2'$ by successive steps:

$$\mathbf{r}_1 \rightarrow \mathbf{r}_1' \rightarrow \mathbf{r}_2 \rightarrow \mathbf{r}_2'$$

Using the matrix formalism, this can be written as

$$\mathbf{r}_2' = \mathbf{R}_2 \mathbf{T}_{12} \mathbf{R}_1 \mathbf{r}_1 \equiv \mathbf{M} \mathbf{r}_1$$

where

$$\mathbf{M} \equiv \mathbf{R}_2 \mathbf{T}_{12} \mathbf{R}_1 \tag{3.44}$$

is the *system matrix* that transforms the ray from the left side of interface 1 to the right side of interface 2.

In the general case of an arbitrary number of optical elements, within the paraxial approximation, the transformation would look like a generalization of Eq. (3.44), namely

$$\mathbf{r}' = \mathbf{M} \mathbf{r} \tag{3.45}$$

Here $\mathbf{M}$ is the matrix product of all the $\mathbf{R}$ matrices for refracting elements and the $\mathbf{T}$ matrices for translation between elements written in reverse order. That is, because the light travels from left to right through the system, the sequence of matrices must be arranged from right to left. This is necessary because of the character of the matrix multiplication rules.

Note that the determinant of the refraction, translation, or system matrix will always be unity:

$$\det \mathbf{R} = 1 \tag{3.46a}$$

$$\det \mathbf{T} = 1 \tag{3.46b}$$

$$\det \mathbf{M} = 1 \tag{3.46c}$$

4. Conjugate Planes. The system matrices that we have been discussing were derived with the idea that the rays in question need to be identified at the refracting interfaces within the system. The formalism is more general than that. We can employ the translation matrix to move our reference planes away from the optical interfaces. This is particularly important when considering *conjugate planes*. These are *images of each other*. In Fig. 3.20, P and P' are *conjugate points*. The geometry and refracting characteristics of the overall optical setup are contained within the transformation matrix. The details of this matrix will be revealed later. For now, let us just identify the optical influence of the shaded region as contained in matrix $\tilde{\mathbf{M}}$. We use here the tilde notation over the matrix to indicate that the matrix connects conjugate planes.

The general transformation takes on the form

$$\begin{pmatrix} n'\alpha' \\ x' \end{pmatrix} = \begin{pmatrix} \tilde{M}_{11} & \tilde{M}_{12} \\ \tilde{M}_{21} & \tilde{M}_{22} \end{pmatrix} \begin{pmatrix} n\alpha \\ x \end{pmatrix}$$

from which we can identify the role of each of the terms of the system matrix.

$$n'\alpha' = \tilde{M}_{11} n\alpha + \tilde{M}_{12} x$$

or

$$\alpha' = \left[\tilde{M}_{11}\left(\frac{n}{n'}\right) \right]\alpha + \left[\frac{\tilde{M}_{12}}{n'} \right]x \tag{3.47}$$

and

$$x' = [\tilde{M}_{21} n]\alpha + [\tilde{M}_{22}]x \tag{3.48}$$

If P and P' are conjugate points, then all the rays that start at P must end up at

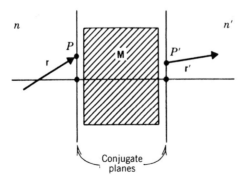

Fig. 3.20 Conjugate planes connected by system matrix **M**.

P'. This means that, in Eq. (3.48), x' must be independent of the angle of the original ray, α. This condition is met if

$$\tilde{M}_{21} = 0 \tag{3.49}$$

Then $x' = \tilde{M}_{22}x$, and we can define the *lateral magnification*

$$m_x \equiv \frac{x'}{x} = \tilde{M}_{22} \tag{3.50}$$

as the linear scale factor between points on the conjugate planes.

Let us also define a ray-angular magnification as

$$m_\alpha \equiv \frac{\Delta\alpha'}{\Delta\alpha} \tag{3.51}$$

Then from Eq. (3.47) we can see that

$$m_\alpha = \tilde{M}_{11}\left(\frac{n}{n'}\right) \tag{3.52}$$

The overall transformation matrix between conjugate planes now becomes

$$\tilde{\mathbf{M}} = \begin{pmatrix} m_\alpha \dfrac{n'}{n} & \tilde{M}_{12} \\ 0 & m_x \end{pmatrix} \tag{3.53}$$

This must still satisfy Eq. (3.46c), thus

$$m_x m_\alpha \frac{n'}{n} = 1 \tag{3.54'}$$

When written in terms of the components of the rays this becomes

$$nx\,\Delta\alpha = n'x'\,\Delta\alpha' \tag{3.54}$$

These equations should be recognized as the Lagrange equations that were first presented in Eq. (3.31) for a single refracting surface.

B. Single Lens

1. General Formulation. A lens consists of two refracting surfaces in series. Fig. 3.19 is a prototypical example. In the class of lenses we will consider, the interfaces are spherical sections. However, aspheric lenses can be built for specialized purposes (such as focusing a monochromatic laser beam). In the paraxial limit, we can use the matrix formalism to describe the behavior of a general ray that passes through both surfaces of the lens. Fig. 3.21 shows the annotated model system. The intercepts of the front and rear surfaces of the lens with the optical axis are at the vertices V and V', respectively. The indices of refraction need not be the same on either side of the lens, and the thickness D_ℓ in the general case, need not be small.

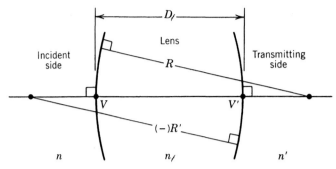

Fig. 3.21 Single lens of refractive index n_ℓ sandwiched between media having indices n and n'.

From Eq. (3.40) we identify the powers of the left and right surfaces as

$$\mathscr{P} = \frac{n_\ell - n}{R} \tag{3.55a}$$

$$\mathscr{P}' = \frac{n' - n_\ell}{R'} \tag{3.55b}$$

where the appropriate sign conventions must be followed. (See section 3.2A2a for a review) A common case is the double convex lens in air, for which we have

$$\mathscr{P} = \frac{n_\ell - 1}{|R|}$$

$$\mathscr{P}' = \frac{n_\ell - 1}{|R'|}$$

The system matrix follows from the form of Eq. (3.44), where, from Eq. (3.42) and (3.43), the component matrices are

$$\mathbf{R} = \begin{pmatrix} 1 & -\mathscr{P} \\ 0 & 1 \end{pmatrix}$$

$$\mathbf{R}' = \begin{pmatrix} 1 & -\mathscr{P}' \\ 0 & 1 \end{pmatrix}$$

$$\mathbf{T} = \begin{pmatrix} 1 & 0 \\ \dfrac{D_\ell}{n_\ell} & 1 \end{pmatrix}$$

When we multiply this out we get

$$\mathbf{M} = \begin{pmatrix} 1 - \left(\dfrac{\mathscr{P}' D_\ell}{n_\ell}\right) & -\mathscr{P} - \mathscr{P}' + \left(\dfrac{\mathscr{P}\mathscr{P}' D_\ell}{n_\ell}\right) \\ \dfrac{D_\ell}{n_\ell} & 1 - \left(\dfrac{\mathscr{P} D_\ell}{n_\ell}\right) \end{pmatrix} \tag{3.56}$$

2. The Thin Lens. A lens is classified as thin if the thickness is negligible with respect to the object and the image distances and the focal length. In our development this means that D_ℓ is allowed to become zero. In this approximation the system matrix of Eq. (3.56) becomes

$$\mathbf{M}_{thin} = \begin{pmatrix} 1 & -\mathcal{P}_{thin} \\ 0 & 1 \end{pmatrix} \tag{3.57}$$

where $\mathcal{P}_{thin} = \mathcal{P} + \mathcal{P}'$ or

$$\mathcal{P}_{thin} = \frac{n_\ell - n}{R} + \frac{n' - n_\ell}{R'} \tag{3.58}$$

Note that $\mathbf{M}_{thin}$ has the same form as the refraction matrix for a single spherical interface, Eq. (3.42).

For a thin lens in air, Eq. (3.58) simplifies to

$$\mathcal{P}_{thin} = (n_\ell - 1)\left(\frac{1}{R} - \frac{1}{R'}\right) \tag{3.59}$$

This is commonly referred to as the *lens maker's equation.*

C. Principal Planes

1. General Transformation. In the thin lens all the refraction is considered to occur at a single plane located at the center of the lens. The rays travel undeviated up to, and away from, this plane. The system matrix takes on a simple form, as presented in Eq. (3.57). In general, an optical system will have a more complicated system matrix if, as for the thick lens in Eq. (3.56), the reference planes are associated with the location of the vertices (the physical intersection of the front and back refracting surfaces with the optical axis).

Consider the optical system pictured in Fig. 3.22. The details of the shaded region are not specified. However, this may be a thick lens or a compound lens. A compound lens is made up of several thick or thin lenses. The refracting surfaces are not necessarily close together. Thick lenses are necessary to obtain short focal lengths. Compound lenses are necessary if the aberrations, or deviations from

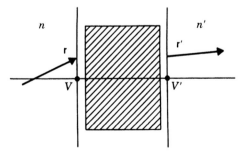

Fig. 3.22 Optical system with matrix $\mathbf{M}_{VV'}$ referred to the vertices.

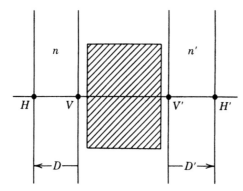

Fig. 3.23 Convention for locating the principal planes with positive distances D and D'.

paraxial behavior, of single thin lenses are to be corrected. The initial and final optical elements intersect the optical axis at the vertices V and V', respectively. The incident refractive index is n, and the refractive index of the final medium is n'. The system matrix that transforms rays from a plane at V to the plane at V' has the general form

$$\mathbf{M}_{VV'} = \begin{pmatrix} M_{11} & M_{12} \\ M_{21} & M_{22} \end{pmatrix} \tag{3.60}$$

subject to the requirement that

$$\det \mathbf{M} = 1$$

We now ask if it is possible to find new reference planes instead of those at V and V' for which the converted system matrix will take on the form of that for a thin lens. These will turn out to be the *principal planes* that intersect the optical axis at H and H' in Fig. 3.23. They are identified by translations through distances D and D', respectively as shown.

We need $\tilde{\mathbf{M}}_{HH'}$. We have $\mathbf{M}_{VV'}$. The conversion from the second form to the first is accomplished by

$$\tilde{\mathbf{M}}_{HH'} = \mathbf{T}'\mathbf{M}_{VV'}\mathbf{T}$$

The translations $\mathbf{T}$ and $\mathbf{T}'$ are

$$\mathbf{T} = \begin{pmatrix} 1 & 0 \\ \dfrac{D}{n} & 1 \end{pmatrix}$$

$$\mathbf{T}' = \begin{pmatrix} 1 & 0 \\ \dfrac{D'}{n'} & 1 \end{pmatrix}$$

The new system matrix becomes

$$\tilde{\mathbf{M}}_{HH'} = \begin{pmatrix} M_{11} + \dfrac{M_{12}D}{n} & M_{12} \\ \dfrac{M_{11}D'}{n'} + \dfrac{M_{12}DD'}{nn'} + M_{21} + \dfrac{M_{22}D}{n} & M_{22} + \dfrac{M_{12}D'}{n'} \end{pmatrix} \quad (3.61)$$

Now it is also possible independently to establish what the system matrix must look like when referred to the principal planes. If the matrix is to have the same form as that associated with the thin lens, then it must resemble Eq. (3.57). For this to be true we must have

$$\tilde{\mathbf{M}}_{HH'} = \begin{pmatrix} 1 & M_{12} \\ 0 & 1 \end{pmatrix} \quad (3.62)$$

The (2, 1) term in Eq. (3.62) is zero. This is the same as the conjugate plane requirement of Eq. (3.49). Therefore, the principal planes at H and H' are images of each other. Since the (2, 2) element is unity, then by Eq. (3.50) we see that we have lateral magnification of unity between the principal planes. Figure 3.24 illustrates the concept. Rays 1, 2, and 3 undergo apparent refractions only at the principal planes. In between the principal planes the rays are mathematically translated without changing the distance from the optical axis. These planes are mathematical entities only. The actual refractions take place, of course, at the physical surfaces within the optical system. The locations of the principal planes could be within or outside of the image forming system. The order of H and H' could be interchanged. The point to be made here is that these are mathematical planes, and the rays *behave* as though they were deviated as shown in Fig. 3.24.

The forms of Eqs. (3.61) and (3.62) must be equivalent. Thus from the (1, 1) element and the (2, 2) element we find

$$M_{11} + M_{12}\frac{D}{n} = 1$$

and

$$M_{22} + M_{12}\frac{D'}{n'} = 1$$

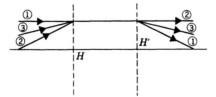

Fig. 3.24 Behavior of rays intersecting a pair of points on principal planes.

which can be rewritten as

$$D = \left(\frac{n}{M_{12}}\right)(1 - M_{11}) \tag{3.63}$$

$$D' = \left(\frac{n'}{M_{12}}\right)(1 - M_{22}) \tag{3.64}$$

where M_{11}, M_{12}, and M_{22} are the appropriate terms in the matrix $\mathbf{M}_{VV'}$ which is assembled from the physical arrangement of optical components in the form of Eq. (3.60).

Equations (3.63) and (3.64) tell us how to locate the principal planes given the system matrix referred to the vertices V and V'.

$$D > 0 \text{ if } H \text{ is on the left side of } V$$

$$D' > 0 \text{ if } H' \text{ is on the right side of } V'$$

These are conventions established by the geometry of Fig. 3.23.

Equations (3.63) and (3.64) are only meaningful if the condition

$$M_{12} \neq 0 \tag{3.65}$$

is satisfied. This then becomes the requirement that our general paraxial system be image-forming like a "thin lens" and that a finite location for H and H' can be found.

It is convenient to identify the power of the general system as

$$-\mathscr{P}_{\text{syst}} = M_{12} \tag{3.66}$$

Then the system matrix takes on the simplified form:

$$\tilde{\mathbf{M}}_{HH'} = \begin{pmatrix} 1 & -\mathscr{P}_{\text{syst}} \\ 0 & 1 \end{pmatrix} \tag{3.67}$$

Also from the comparison of Eqs. (3.61) and (3.62) in the (2, 1) element, we require that

$$\frac{M_{11}D'}{n'} + \frac{M_{12}DD'}{nn'} + M_{21} + \frac{M_{22}D}{n} = 0$$

which, together with Eq. (3.66) means that the system power is

$$\mathscr{P}_{\text{syst}} = -M_{12} = \frac{nM_{11}}{D} + \frac{n'M_{22}}{D'} + \frac{nn'M_{21}}{DD'} \tag{3.68}$$

This must be nonzero for an image-forming optical system.

2. Application to Thick Lenses. As an example of the use of these techniques we return to the single lens of section B. We wish to find the lens power and the locations of the principal planes with respect to the vertices V and V'. From this information we will be able to compute the image characteristics for a given object with relatively little effort, because when referred to the principal planes, a complicated optical system acts like a thin lens.

The system matrix for the thick lens is the same as that found in Eq. (3.56). Thus,

$$M_{11} = 1 - \frac{\mathscr{P}'D_\ell}{n_\ell}$$

$$M_{12} = \left(\frac{\mathscr{P}\mathscr{P}'D_\ell}{n_\ell}\right) - \mathscr{P} - \mathscr{P}'$$

$$M_{21} = \frac{D_\ell}{n_\ell}$$

$$M_{22} = 1 - \left(\frac{\mathscr{P}D_\ell}{n_\ell}\right)$$

From Eq. (3.68) (where the system in this case is just the single thick lens) we can evaluate the power.

$$\mathscr{P}_{\text{syst}} = \mathscr{P}_\ell = \mathscr{P} + \mathscr{P}' - \left(\frac{\mathscr{P}\mathscr{P}'D_\ell}{n_\ell}\right) \tag{3.69}$$

From Eqs. (3.63) and (3.64) the locations of the principal planes can be identified.

$$D = \frac{n\mathscr{P}'D_\ell}{-\mathscr{P}_\ell n_\ell} \tag{3.70}$$

$$D' = \frac{n'\mathscr{P}D_\ell}{-\mathscr{P}_\ell n_\ell} \tag{3.71}$$

The paraxial optical characteristics of the lens, referred to the principal planes, are now contained within the previous three equations and

$$\tilde{\mathbf{M}}_{HH'} = \begin{pmatrix} 1 & -\mathscr{P}_\ell \\ 0 & 1 \end{pmatrix} \tag{3.72}$$

The separation of the principal planes is given by

$$t = D_\ell + D + D' = D_\ell\left(1 - \frac{\mathscr{P}n'}{\mathscr{P}_\ell n_\ell} - \frac{\mathscr{P}'n}{\mathscr{P}_\ell n_\ell}\right) \tag{3.73a}$$

In the common case where the lens is in air this becomes

$$t = D_\ell\left(1 - \frac{(\mathscr{P} + \mathscr{P}')}{n_\ell\mathscr{P}_\ell}\right) \tag{3.73b}$$

There are times when we wish to treat even a thin lens as having a nonzero thickness to locate its principal planes. When D_ℓ is small compared with R and R', but not entirely negligible, we have

$$\mathscr{P}_\ell = \mathscr{P}_{\text{thin}} = \mathscr{P} + \mathscr{P}' \tag{3.74a}$$

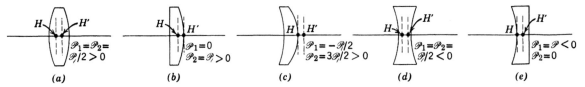

Fig. 3.25 Examples of principal plane locations for different lens shapes. Here $n = 1.5$ and $t = D/3 =$ principal plane separation.

and

$$t = D_\ell \left(1 - \frac{1}{n_\ell} \right) \tag{3.74b}$$

$$D = \left(\frac{-\mathscr{P}'}{\mathscr{P} + \mathscr{P}'} \right) \frac{D_\ell}{n_\ell} \tag{3.74c}$$

$$D' = \left(\frac{-\mathscr{P}}{\mathscr{P} + \mathscr{P}'} \right) \frac{D_\ell}{n_\ell} \tag{3.74d}$$

When $n_\ell = 1.5$ (a typical value), we have $t = D/3$. Some examples are shown in Fig. 3.25.

3. Combination of Two Image-Forming Systems. A system of two simple lenses separated by distance d is a prototype for several important optical instruments. In the general case, each "lens" might be a complicated lens system. We model the optical characteristics of each component system by means of its power and the location of its principal planes. (If the systems are simple thin lenses, then both principal planes for a given lens are located at the center of that lens.) We will develop the more general case.

The situation is defined in Fig. 3.26. The power of the first system is $\mathscr{P}_1$; that of

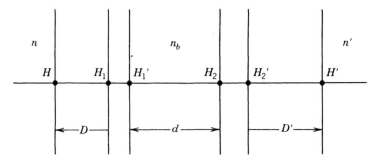

Fig. 3.26 Conventions for two image-forming systems. The overall principal planes are at H and H'.

the second, $\mathscr{P}_2$. The transformation matrix between the principal planes of the first system is

$$\tilde{\mathbf{M}}_{H_1 H_{1'}} = \begin{pmatrix} 1 & -\mathscr{P}_1 \\ 0 & 1 \end{pmatrix} \tag{3.75}$$

Between the principal planes of the second system the transformation matrix is

$$\tilde{\mathbf{M}}_{H_2 H_{2'}} = \begin{pmatrix} 1 & -\mathscr{P}_2 \\ 0 & 1 \end{pmatrix} \tag{3.76}$$

The overall transformation matrix between the first principal plane of the first system and the second principal plane of the second system is identical in form to **M** in Eq. (3.56) and is given

$$
\begin{aligned}
\mathbf{M}_{H_1 H_2'} &= \tilde{\mathbf{M}}_{H_2 H_2'} \begin{pmatrix} 1 & 0 \\ \dfrac{d}{n_b} & 1 \end{pmatrix} \tilde{\mathbf{M}}_{H_1 H_1'} \\[2mm]
&= \begin{pmatrix} 1 - \left(\dfrac{\mathscr{P}_2 d}{n_b} \right) & -\mathscr{P}_1 - \mathscr{P}_2 + \left(\dfrac{\mathscr{P}_1 \mathscr{P}_2 d}{n_b} \right) \\[3mm] \dfrac{d}{n_b} & 1 - \left(\dfrac{\mathscr{P}_1 d}{n_b} \right) \end{pmatrix}
\end{aligned}
\tag{3.77}
$$

This matrix now plays the role of the general system matrix

$$\mathbf{M} = \begin{pmatrix} M_{11} & M_{12} \\ M_{21} & M_{22} \end{pmatrix} \tag{3.78}$$

for which we already know how to determine the overall system power $\mathscr{P}_{\text{syst}}$ and the locations of the principal planes H and H' of the overall system.

From Eq. (3.66) the power is

$$\mathscr{P}_{\text{syst}} = \mathscr{P}_1 + \mathscr{P}_2 - \left(\frac{\mathscr{P}_1 \mathscr{P}_2 d}{n_b} \right) \tag{3.79}$$

From Eq. (3.70)

$$D = \frac{n \mathscr{P}_2 d}{-\mathscr{P}_{\text{syst}} n_b} \tag{3.80}$$

and from Eq. (3.71)

$$D' = \frac{n' \mathscr{P}_1 d}{-\mathscr{P}_{\text{syst}} n_b} \tag{3.81}$$

In Eq. (3.80), D is the distance of the first principal plane of the overall system to the left of the first principal plane of the first component, and in Eq. (3.81) D' is the distance of the second principal plane of the overall system to the right of the second principal plane of the second component. The component principal planes are defined in terms of their distance from the vertices of the component optical parts. For the case of two thin lenses separated by distance d, D and D' become the distance to the left of the first lens and to the right of the second lens, respectively. As a unit, the combination will have a system matrix of the form described by Eq. (3.67), with $\mathscr{P}_{\text{syst}}$ given by Eq. (3.79).

Thus we see that in this case also, the reduced optical equivalent in the paraxial approximation behaves like a thin lens when referred to the overall system principal planes.

3.4 Image Formation

A. General Concepts of Image Formation

From the information in the previous two sections (B and C) a configuration of optical elements can be associated with a single overall system matrix. This matrix will bear the form of

$$\tilde{\mathbf{M}}_{HH'} = 1\begin{pmatrix} 1 & -\mathscr{P} \\ 0 & 1 \end{pmatrix}$$

and is referred to the principal planes of the system. In constructing this matrix, the necessary relationships that serve to locate the principal planes with respect to the physical surfaces in the system are derived along the way.

We now wish to explore the relationship between a general object and its image. The situation is illustrated in Fig. 3.27. For each point on the object plane there is a conjugate point on the image plane. Fig. 3.27 is a more detailed expansion of Fig. 3.20, where the characteristics of conjugate planes were first described. Here

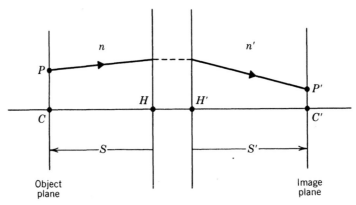

Fig. 3.27 General object–image relationship.

we can include the object distance S and the image distance S'. These are measured from the principal planes of the overall optical system, as shown in Fig. 3.27.

We use the translation matrices to create a matrix that transforms points on the plane at C into points on the plane at C'.

$$\tilde{\mathbf{M}}_{CC'} = \begin{pmatrix} 1 & 0 \\ S'/n' & 1 \end{pmatrix} \underbrace{\begin{pmatrix} 1 & -\mathscr{P} \\ 0 & 1 \end{pmatrix}}_{\tilde{\mathbf{M}}_{HH'}} \begin{pmatrix} 1 & 0 \\ S/n & 1 \end{pmatrix}$$

When this is multiplied out we find that

$$\tilde{\mathbf{M}}_{CC'} = \begin{pmatrix} 1 - \dfrac{\mathscr{P}S}{n} & -\mathscr{P} \\ \dfrac{S'}{n'} + \dfrac{S}{n} - \dfrac{\mathscr{P}SS'}{nn'} & 1 - \dfrac{\mathscr{P}S'}{n'} \end{pmatrix} \tag{3.82}$$

Because this is a transformation between conjugate planes, Eq. (3.82) must have the form of Eq. (3.53). Equating these matrices element by element yields, from the (2, 1) elements

$$\frac{S'}{n'} + \frac{S}{n} - \frac{\mathscr{P}SS'}{nn'} = 0 \tag{3.83}$$

from the (1, 1) elements

$$1 - \frac{\mathscr{P}S}{n} = m_\alpha \frac{n'}{n} \tag{3.84}$$

and from the (2, 2) elements

$$1 - \frac{\mathscr{P}S'}{n'} = m_x \tag{3.85}$$

The first of these relations can be rewritten in the form

$$\frac{n}{S} + \frac{n'}{S'} = \mathscr{P} \tag{3.86}$$

This is the familiar functional form of the "thin-lens equation." We see now that, provided S and S' are measured from the principal planes, the imaging behavior of a complicated optical system is the same, in the paraxial limit, as that of a thin lens.

If the object distance is taken to infinity, so that the incident rays are all parallel to the optical axis, as in Fig. 3.28a, then the image distance identifies the *image focal length, f', and the image focal point, F'*.

$$S' \underset{\text{Lim } S \to \infty}{=} \frac{n'}{\mathscr{P}} \equiv f' \tag{3.87}$$

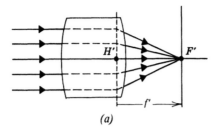

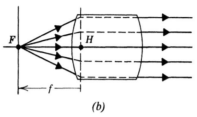

Fig. 3.28 (a) Image focal point. (b) Object focal point.

When the image is at infinity, as in Fig. 3.28b, the object distance identifies the *object focal length, f*, and the *object focal point, F*.

$$S \bigg|_{\text{Lim} S' \to \infty} = \frac{n}{\mathscr{P}} \equiv f \tag{3.88}$$

Note that these are related through the system power such that

$$\frac{n}{f} = \frac{n'}{f'} \tag{3.89}$$

When the index of refraction is the same on both sides of the optical system, the object and image focal lengths are equal, and Eq. (3.86) becomes

$$\frac{1}{S} + \frac{1}{S'} = \frac{1}{f} \tag{3.90}$$

This is the elementary form of the thin-lens equation.

The magnifications that are associated with the image plane referred to the object plane can be extracted from Eqs. (3.84) and (3.85) by using the result of Eq. (3.86). For the angular magnification we have

$$m_\alpha = -\frac{S}{S'} \tag{3.91}$$

and for the lateral magnification

$$m_x = -\frac{n}{n'}\frac{S'}{S} \tag{3.92}$$

Another important performance parameter is the longitudinal magnification, m_z. This is defined as

$$m_z \equiv \frac{dS'}{dS} \tag{3.93}$$

Differentiating Eq. (3.83) produces

$$\frac{1}{n'}\frac{dS'}{dS} + \frac{1}{n} - \frac{\mathscr{P}}{nn'}\left[S\frac{dS'}{dS} + S'\right] = 0$$

which can be rearranged to

$$\frac{dS'}{dS}\left[1 - \frac{\mathscr{P}S}{n}\right] = \frac{n'}{n}\left[\frac{\mathscr{P}S'}{n'} - 1\right]$$

Using Eqs. (3.84) and (3.85), this can be written in terms of m_x and m_α.

$$\frac{dS'}{dS} = - \frac{\dfrac{n'}{n}m_x}{m_\alpha \dfrac{n'}{n}}$$

Finally, employing the Lagrange relation, Eq. (3.54'), this becomes

$$m_z = \frac{dS'}{dS} = -\frac{n'}{n}m_x^2 \tag{3.94}$$

This tells us how the image will be shifted along the optical axis in response to small changes in the object distance.

B. Graphical Construction of Image Formation

We can adopt a graphical technique for identifying the location of the image provided we know the locations of the principal planes and the focal points. The object focal point is F and the image focal point is F'. The situation is shown in Fig. 3.29, where P' is the image of P. In general, the indices of refraction on either side of the system are not equal.

Rays 1 and 2 are constructed using the concept described in Fig. 3.28. Rays that are parallel to the optical axis before (or after) the optical interaction pass through (or appear to pass through) the image (or object) focal point. In Fig. 3.29 ray 1 is parallel to the optical axis prior to its encounter with the lens. After passing through the lens, it is deviated so as to pass through the image focal point, F'. Because ray 2 passes through the object focal point, it must become parallel to the optical axis after the refraction. These two rays cross at only two points, P and P'. Thus, given the location of P, the conjugate point P' can be found by construction.

To add another ray that is easy to construct so as to overdetermine the

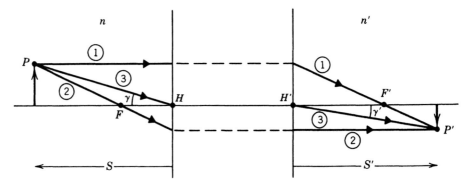

Fig. 3.29 Graphical construction for locating the image.

location of the image, we can draw a ray from P to the optical axis at H. This ray makes the angle γ with the optical axis at H (where it has zero height). The transformation matrix from H to H' has the form of Eq. (3.67). If we compare that matrix with the general form of Eq. (3.53) for conjugate points, we find that the $(1, 1)$ terms in Eq. (3.53) must be unity. This means that the angular magnification from H to H' is given by

$$m_\alpha = \frac{n}{n'} \tag{3.95}$$

At $x = 0$, this relationship is equivalent to

$$\frac{\gamma'}{\gamma} = \frac{n}{n'}$$

in Fig. 3.29. Thus we can construct ray 3 on the transmitting side with the information

$$\gamma' = \gamma\left(\frac{n}{n'}\right) \tag{3.96}$$

The three rays 1, 2, and 3 all cross at P and at P'.

If the optical system is a single thin lens, then H and H' become the same point in Fig. 3.29, which is located at the center of the thin lens. If the indices of refraction are the same on both sides of the optical system, then the angles γ and γ' are equal. Fig. 3.30 illustrates the case where both of the previous simplifications are in effect.

When the lens is concave, as in Fig. 3.31a, the focal length, by Eq. (3.87) or (3.88), is negative. This places the image focal point to the left of the lens and the object focal point to the right of the lens. The result is that, *for a real object*, there will be no position where the rays cross. The image is virtual. As viewed from the right side of the lens, the image would appear to be located on the object side as shown and would be reduced in size. Rays 1, 2, and 3 have been drawn following the same procedure as for Fig. 3.29. The apparent extensions of ray 1 back into the

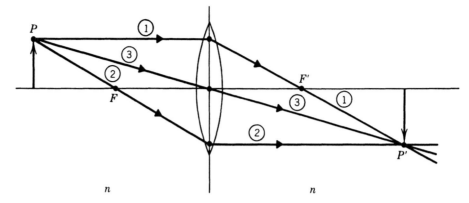

Fig. 3.30 Graphical construction for a thin lens with the refractive index on the incident side equal to that on the transmitting side.

region to the left of the lens and ray 2 forward to the object focal point are required to determine from where the rays are diverging.

For a virtual object location (object on the right side of the lens) the situation would be similar to Fig. 3.31b. Here other optical elements (not shown) would be required to create the convergence of rays that, if the lens were not present, would produce an image at P. The effect of the positive focal length lens is to cause a more abrupt convergence. The real image is formed at P'. In this case, apparent extensions of the incident rays toward the virtual object are required to identify the intercepts and angles for rays 1 and 2 properly.

An additional useful expression can be derived from a consideration of Fig. 3.32, which is a simplified form of Fig. 3.29. Here we have defined the object location by the use of the distance X, the algebraic difference between the conventional object distance and the focal length.

$$X = S - f \tag{3.97}$$

Likewise X' is the redefined image distance.

$$X' = S' - f' \tag{3.98}$$

These change sign as the object moves toward the optical system past F, or as the image moves toward the system past F'.

Triangles I and II are similar; thus

$$\frac{-x'}{x} = \frac{f}{X}$$

Also, triangles I′ and II′ are similar, leading to

$$\frac{x}{-x'} = \frac{f'}{X'}$$

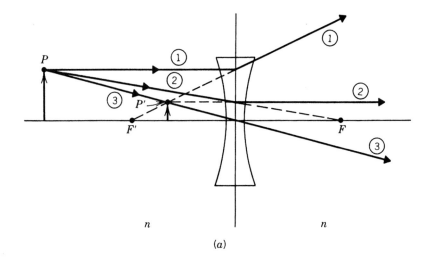

(a)

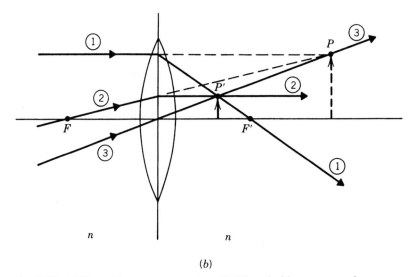

(b)

Fig. 3.31 (a) Virtual image construction. (b) Virtual object construction.

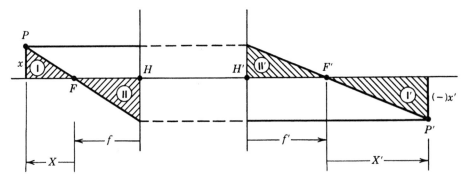

Fig. 3.32 Object and image geometry leading to the Newtonian form of the thin-lens equation.

Multiplication of the previous two equations yields the *Newtonian* form of the object–image relationship.

$$XX' = ff' \qquad (3.99)$$

This is equivalent to the image relationship in Eq. (3.86).

A summary of the object–image distance relationship is presented in Fig. 3.33, which is a plot of Eq. (3.99). For a converging lens, where both f and f' are positive, the ordinate shows the normalized value of X' versus the normalized value of X. This shows how the image distance increases as the object moves in toward the lens, approaching F (upper branch of the hyperbola in Fig. 3.33). Whenever the object is on the right side of F, the image is found on the left side of F', as shown by the lower branch of the hyperbola in Fig. 3.33.

For diverging lenses, with negative values of f and f', F is on the right side of the system and F' is on the left side. Any real object will be on the positive X (thus the negative X/f) side of the abscissa. Access to the values of X/f near the origin will be restricted for real objects because F is on the "wrong" side of the lens. This shows that the image will be on the positive X' side of F' and quite close to F'. Because F' is also on the "wrong" side of the lens, this means that the image will be virtual.

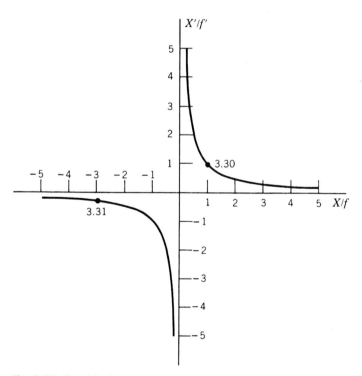

Fig. 3.33 Graphical representation of Eq. (3.99). Points representing the situations in Figs. 3.30 and 3.31 are identified.

Figs. 3.30 and 3.31 are represented on the diagram in Fig. 3.33 by the identified points. The virtual object with a converging lens and the virtual image with the diverging lens are associated with the same point as shown.

3.5 Examples of Paraxial Optics

A. Image-Forming Systems

1. The Human Eye. The eye as an optical instrument is pictured in Fig. 3.34. Here only the principal planes are identified with respect to the front surface of the cornea. In actuality, the aqueous humor, the crystalline lens, and the vitreous humor all must be taken into account when the optical properties of the eye are considered. Because the refractive indices are different on the incident and transmitting sides of the focusing elements, the focal lengths f and f' are not equal.

These data represent the relaxed eye. The image is formed on the retina that is at F'. The relaxed eye is focused on an infinitely distant object. The thickness and curvature of the crystalline lens can change to allow for objects that are nearer than infinity. For younger persons, the eye can "accommodate" for objects quite close to the eye. This ability decreases with age. A convenient design parameter is 250 mm for the near point for an object at which comfortable viewing can still take place.

Many optical instruments are designed to be used by direct viewing. This means that the final image must appear on the retina under comfortable circumstances. The eye should therefore be relaxed; otherwise, the accommodation required will cause strain during continual viewing.

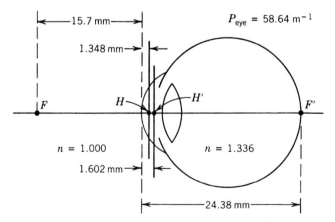

Fig. 3.34 Optical equivalent of the human eye. An image must be formed on the retina.

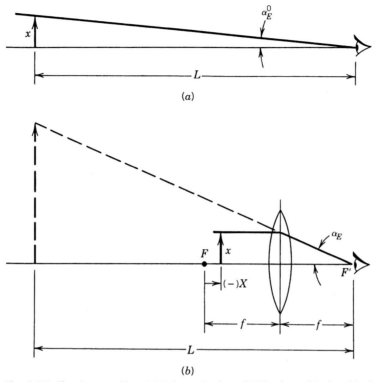

Fig. 3.35 Simple magnifier. (*a*) Direct viewing. (*b*) Viewing with the aid of a simple lens.

2. Magnifier. The simple magnifier (Fig. 3.35) is used to create a virtual erect, enlarged image that is located at the near point of the eye. The rays coming from the magnifier appear to originate on this virtual object. The conventional definition of the magnifying power of a simple instrument like this, where the rays are nearly parallel as they enter the eye, is

$$M \equiv \frac{\alpha_E}{\alpha_E^0} \tag{3.100}$$

where α_E and α_E^0 are the angles that a ray from the edge of the object makes with the optical axis at the eye with and without the magnifier, respectively. For this comparison to be valid the object, in the case without the magnifier, and the virtual image, in the case with the magnifier, must be at the same distance L from the eye, as shown in Fig. 3.35.

If the object has a lateral size of x, then

$$\alpha_E^0 = \frac{x}{|L|} \tag{3.101}$$

The eye of the observer will be located at about the image focal point F'. This means that the image distance (as measured from F') is $|L|$, and the object distance (as measured from F) should be

$$X = -\frac{f^2}{|L|} \tag{3.102}$$

The minus sign shows that the object must be closer to the lens than F, as shown in Fig. 3.35b.

A ray that meets the optical axis at F' would have to have been parallel to the optical axis before the lens. From the geometry of Fig. 3.35b, we find

$$\alpha_E = \frac{x}{f} \tag{3.103}$$

Thus the magnifying power is (from Eqs. 3.100, 3.101, and 3.103)

$$M = \frac{|L|}{f} \tag{3.104}$$

or

$$M = \frac{250 \text{ mm}}{f}$$

if the comparison is made at the near point.

With this definition we can rewrite the expression for the required location of the object such that the image will be located at the near point. From Eq. (3.102)

$$X = -\frac{f}{M} \tag{3.105}$$

To increase the magnifying power over that achieved with a thin lens, we could decrease the focal length as suggested by Eq. (3.104). This implies that the radii of curvature of the lens surfaces must be small so that the power is large. This cannot be done without making the lens thick. Fig. 3.36 shows the simple magnifier geometry with a single thick lens. By comparing this diagram with that of Fig. 3.35b we can readily see that Eqs. (3.102) through (3.104) will remain valid, provided the distances are measured with respect to the principal planes.

To obtain large power in a single lens, its surfaces must have small radii of curvature, and the ray "bending" is through such a large angle (α_E in Fig 3.35b) that the small-angle constraints of the paraxial theory are not satisfied. When this is the case, departures from our straightforward image-formation theory are found, and a good quality image is not produced.

3. Doublet. An efficient way of achieving short focal lengths and thus large magnification with a better quality image is through the use of two lenses that form a doublet.

   a. General Considerations. A doublet consists of two lenses with principal

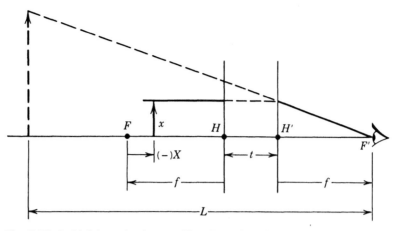

Fig. 3.36 A thick-lens simple magnifier. Only the principal planes of the lens are shown.

plane separation d, as in Fig. 3.37. Here we want to identify the separation of focal points l such that

$$d = f'_1 + f_2 + l \qquad (3.106)$$

In this definition, l has a sign that is positive provided F'_1 is to the left of F_2. The power of the doublet can be written [from Eq. (3.79)]

$$\mathscr{P} = -\mathscr{P}_1\mathscr{P}_2 l \qquad (3.107)$$

We set $n_b = n' = 1$ in Fig. 3.37, but anticipate use of a doublet as an immersion microscope where $n \neq 1$. From Eq. (3.87), the image focal length of the doublet will be

$$f' = -\frac{f'_1 f_2}{l} \qquad (3.108)$$

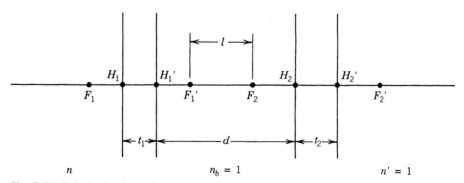

Fig. 3.37 Principal points of the doublet.

and the object focal length will follow from Eq. (3.89)

$$f = nf' \qquad (3.109)$$

The principal planes of the doublet, thus defined, will be found at a distance to the left of H_1 given by

$$D = \frac{f_1' dn}{l} \qquad (3.110a)$$

and at a distance to the right of H_2' given by

$$D' = \frac{f_2 d}{l} \qquad (3.110b)$$

This means that the principal planes are separated by a distance

$$t = D + D' + d + t_1 + t_2 = (l + f_1' n + f_2)\left(1 + \frac{f_1' n}{l} + \frac{f_2}{l}\right) + t_1 + t_2 \quad (3.111)$$

b. Microscope. In the microscope (Fig. 3.38) the focal lengths of the first element, the *objective*, and that of the second element, the *eyepiece*, are small compared with the distance that separates the focal points l. Often l is a standard 160 mm. If we consider the doublet as a single unit with focal length given by Eq. (3.108), then the focal length f' is small. Substituting f' from Eq. (3.108) into Eq. (3.104) for the magnifying power, we find

$$M = \left(\frac{L}{f_2}\right)\left(-\frac{l}{f_1'}\right) \qquad (3.112)$$

which can be rewritten as

$$M = M_E \, m_{x0} \qquad (3.113)$$

where, as before, M_E is the magnifying power of the eyepiece and m_{x0} is a good approximation of the transverse magnification of the objective alone.

To see this more clearly, consider Fig. 3.38. From Eq. (3.105) we would like to arrange for the intermediate image to be located at $X_2 = -f_2/M_E$. Here it will be the object for the eyepiece. Then the eyepiece will create a final virtual image at $X_2' = -|L|$. This will provide a valid comparison with the magnifying power of the simple magnifier. Because the focal points of the two lenses are separated by l, the intermediate image will be located at a distance to the right of F_1' given by

$$X_1' = l - X_2 = l\left(\frac{1 - X_2}{l}\right) \qquad (3.114)$$

Using this in Eq. (3.99), we can find the object position

$$X_1 = \frac{f_1 f_1'}{X_1'} = \frac{n(f_1')^2}{l(1 - X_2/l)} \qquad (3.115)$$

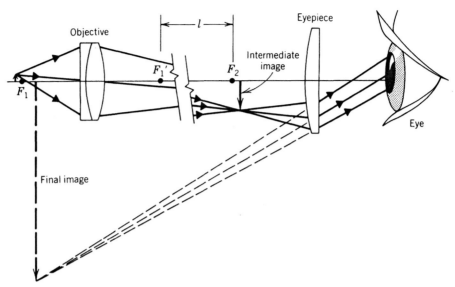

Fig. 3.38 Compound microscope.

The lateral magnification of the objective is given by Eq. (3.92):

$$m_{x0} = -\frac{nS'_1}{S_1} = -n\frac{X'_1 + f'_1}{X_1 + nf'_1}$$

Using Eqs. (3.114) and (3.115), this becomes

$$m_{x0} = -\frac{l}{f'_1}(1 - (X_2/l)) \tag{3.116}$$

If $X/l \ll 1$, as in most cases, the terms involving this ratio can be neglected in Eq. (3.116). Then the lateral magnification of the objective is very nearly

$$m_{x0} \simeq -\frac{l}{f'_1} \tag{3.117}$$

as stated earlier.

In a typical example, $f_1 = f'_1 = f_2 = 16$ mm, $n = 1$, and $M = 250/16 = 15.625$. In the microscope configuration the new magnifying power is $M = (15.625)(-160/16) = -156.25$, considerably larger than that of the single lens alone.

4. Telephoto Lens. The purpose of a camera lens is to produce a real, inverted image on the focal plane where the film is located. Adjustment of the position of the lens or the power of a compound lens is required to accommodate a variety of object distances. A simple lens used in such a capacity is shown in Fig. 3.39a. The relationship between the object distance and the image distance is contained in Eq.

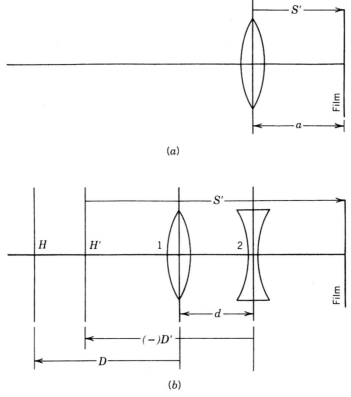

Fig. 3.59 (a) Camera with a simple lens. (b) Telephoto lens. The effect is to increase the image distance without extending the physical location of the lenses an unreasonable amount away from the film.

(3.38). If the object distance is very large, say $S = L$, then

$$S' = a \approx f \tag{3.118}$$

where a is the distance from the lens to the film.

The image magnification, from Eq. (3.92), will be

$$m_x = -\frac{S'}{L} = -\frac{a}{L} \simeq -\frac{f}{L} \tag{3.119}$$

If a larger image is desired without changing the distance from the camera to the object (here approximately equal to L), then a different lens that has a larger focal length must be used. However, from Eq. (3.118) we can see that this necessitates that the lens be located a greater distance from the film. This may be impractical.

The same optical effect can be achieved with a compound lens. This is the motivation for the telephoto lens. In the compound lens the image distance is measured from H'. If we can arrange to move H' a considerable distance away from the film by suitable choices of the optical elements, then the magnification will increase.

Most telephoto lenses are multielement devices that are optimized for small aberrations. We can study the concept by an assembly consisting of two lenses (Fig. 3.39b). We require that D' be considerably negative. This means that, by Eq. (3.81), the focal length of the first lens should be positive. We also require that H be to the left of H', thus

$$D > D' - d$$

and the focal length of the second lens should be negative, as shown in Fig. 3.39b.

Now the image distance is $S' = |D'| + a$. This means that the magnification is now

$$m_x \simeq - \frac{|D'| + a}{L} \tag{3.120}$$

which is greater in absolute magnitude than m_x with the simple lens. To gain practicality with this design, we need $|D'| > d$. Then the physical location of the components will be closer to the camera body than a simple single lens would be to obtain the same magnification. Typical values for a are in the order of 50 mm while $|D'|$ can be several hundreds of millimeters.

B. Telescopic Systems

1. General Considerations. In section 3.3A4 we explored the general conditions obeyed by an optical matrix that connects conjugate planes. We found that rays that start at a point on one plane must end up at a certain point on the other plane independent of their original angle. Later, in section 3.3C1 we showed that M_{12} in the system matrix had to be nonzero for image formation to take within the "thin lens" formalism.

Here we examine the situation where the angles of the final rays are directly proportional to the angles of the corresponding incident rays, independent of the ray height. This means the M_{12} element must be zero. The ray angles obey

$$n'\alpha' = M_{11}n\alpha$$

An optical system of this type has zero power.

Now, in general, it is true that the (1, 2) matrix element is invariant under translations; that is, the (1, 2) element of the matrix

$$\begin{pmatrix} 1 & 0 \\ \dfrac{Z'}{n'} & 1 \end{pmatrix} \begin{pmatrix} M_{11} & M_{12} \\ M_{21} & M_{22} \end{pmatrix} \begin{pmatrix} 1 & 0 \\ \dfrac{Z}{n} & 1 \end{pmatrix}$$

is M_{12}. Thus, if M_{12} is zero, it will remain zero unless refracting surfaces are added to the system.

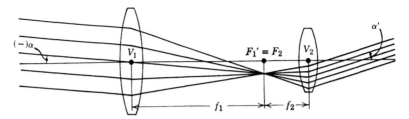

Fig. 3.40 Astronomical telescope.

2. Astronomical Telescope. One of the simplest examples of a telescopic system is the astronomical telescope (Fig. 3.40). This consists of two positive lenses, the objective lens with a focal length f_1 and the eye lens, or eyepiece, with a focal length f_2 separated by a distance $d = f_1 + f_2$ in air. Then the total power of the combination is given by Eq. (3.79) as

$$\mathscr{P} = \frac{1}{f_1} + \frac{1}{f_2} - \frac{d}{f_1 f_2} = 0$$

A parallel incident beam emerges from the telescope as another parallel beam making a different angle. The transformation matrix from the first lens to the second lens is

$$\mathbf{M}_{V_1 V_2} = \begin{pmatrix} \dfrac{-f_1}{f_2} & 0 \\ f_1 + f_2 & \dfrac{-f_2}{f_1} \end{pmatrix} \tag{3.121}$$

The transformation equations become

$$\alpha' = -\frac{f_1}{f_2} \alpha \tag{3.122a}$$

$$x' = -\frac{f_2}{f_1} x + (f_1 + f_2)\alpha \tag{3.122b}$$

The angular magnification will be

$$m_\alpha = -\frac{f_1}{f_2} \tag{3.123}$$

This is the factor by which the telescope magnifies the angular separation of two distant objects. Note also that these concepts remain valid when f_2 is negative. In this case we have a *Galilean telescope* (Fig. 3.41). It gives a positive angular magnification $f_1/(-f_2)$.

3. General Telescopic System. The general telescopic system is quite similar to the two-lens system we have just discussed. The general form for the transformation

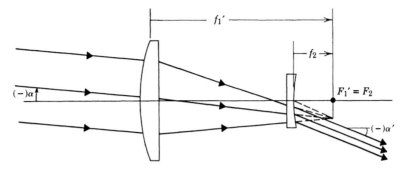

Fig. 3.41 Galilean telescope (fashioned after the first telescope).

matrix will then be

$$\mathbf{M} = \begin{pmatrix} M_{11} & 0 \\ M_{21} & M_{22} \end{pmatrix}$$

This must obey Eq. (3.46c)

$$\det M = 1 = M_{11}M_{22}$$

If we continue to define the angular magnification by

$$m_\alpha = M_{11}\left(\frac{n}{n'}\right) \tag{3.124}$$

we then obtain

$$\mathbf{M} = \begin{pmatrix} m_\alpha(n'/n) & 0 \\ M_{21} & \dfrac{1}{m_\alpha(n'/n)} \end{pmatrix} \tag{3.125}$$

Even though the telescope is most commonly employed to form a virtual image (which is then further processed into a real image within the eye or a camera), we can still find conjugate planes. The locations of these planes will not be given by the simple thin-lens equations, but we can examine the situation by the following development.

An object–image relation is obtained between planes that intersect the optical axis at C and C' (Fig. 3.42) when

$$\mathbf{M}_{CC'} = \begin{pmatrix} 1 & 0 \\ Z'/n' & 1 \end{pmatrix}\begin{pmatrix} m_\alpha(n'/n) & 0 \\ M_{21} & \dfrac{1}{m_\alpha(n'/n)} \end{pmatrix}\begin{pmatrix} 1 & 0 \\ Z/n & 1 \end{pmatrix}$$

$$= \begin{pmatrix} m_\alpha(n'/n) & 0 \\ \dfrac{Z'm_\alpha}{n} + \dfrac{Z}{n'm_\alpha} + M_{21} & \dfrac{1}{m_\alpha(n'/n)} \end{pmatrix} \tag{3.126}$$

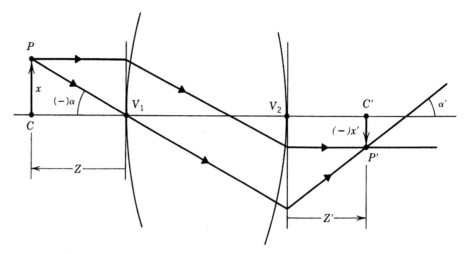

Fig. 3.42 General telescope system; image formation.

We require that P and P' be conjugate points; thus Eq. (3.126) must have the form of Eq. (3.53).

$$\tilde{\mathbf{M}} = \begin{pmatrix} m_\alpha(n'/n) & \tilde{M}_{12} \\ 0 & m_x \end{pmatrix} \qquad (3.53)$$

This shows that the lateral magnification must be

$$m_x = \left(\frac{m_\alpha n'}{n}\right)^{-1}$$

as required by the Lagrange equation. It also proves that an object–image relationship exists when

$$\frac{Z' m_\alpha}{n} + \frac{Z}{n' m_\alpha} + M_{21} = 0$$

or

$$Z' m_\alpha + Z m_x = -M_{21} n \qquad (3.127)$$

The distances Z and Z' are measured from the telescope vertices.

If the telescope is the simple astronomical system *in air*, then the telescope is described by the matrix of Eq. (3.121).

$$M_{21} = f_1 + f_2$$

$$m_\alpha = -\frac{f_1}{f_2}$$

$$m_x = -\frac{f_2}{f_1}$$

The object–image relationship becomes

$$Z' = -\left(\frac{f_2}{f_1}\right)^2 Z + f_2\left(1 + \frac{f_2}{f_1}\right) \tag{3.128}$$

REFERENCES

Arfken, G. *Mathematical Methods for Physicists.* Academic Press, New York, 1970.

Born, Max, and Emil Wolf. *Principles of Optics.* Pergamon Press, Oxford, 1980.

Brouwer, W. *Matrix Methods in Optical Instrument Design.* Benjamin, New York, 1964.

Conrady, A. E. *Applied Optics and Optical Design.* Dover, New York, 1957.

Cosslett, V. E. *Modern Microscopy.* Cornell University Press, Ithaca, N.Y., 1966.

Cox, Arthur. *Photographic Optics.* Focal Press, London, 1966.

Driscoll, Walter G., and William Vaughan. *Handbook of Optics.* McGraw-Hill, New York, 1978.

Graham, C. H., ed. *Vision and Visual Perception.* Wiley, New York, 1965.

Herzberger, Max. *Modern Geometrical Optics.* Interscience Publishers, New York, 1958.

Hopkins, George W. "Basic Algorithms for Optical Engineering," in Robert R. Shannon and James C. Wyant, eds. *Applied Optics and Optical Engineering.* Academic Press, New York, 1983, vol. 9, p. 1.

Kingslake, Rudolf. *Lens Design Fundamentals.* Academic Press, New York, 1978.

Kuiper, G. P. and B. M. Middlehurst, ed. *Telescopes.* University of Chicago Press, Chicago, 1960.

Levi, Leo. *Applied Optics.* Wiley, New York, 1968.

Martin, L. C. *The Theory of the Microscope.* Elsevier, New York, 1965.

Smith, Warren J. *Modern Optical Engineering.* McGraw-Hill, New York, 1966.

Verdeyen, Joseph T. *Laser Electronics.* Prentice-Hall. Englewood Cliffs, N.J., 1981.

Welford, W. T. *Geometrical Optics.* North Holland, Amsterdam, 1962.

PROBLEMS

Section 3.1   Ray Tracing

1. Consider an interface between two different transparent media that we will call the xy plane. The indices of refraction are 1 and 1.5 on the negative and the positive z sides of the interface, respectively. A ray is found passing through the point $(x, y, z) = (-10 \text{ cm}, 0, -10 \text{ cm})$ traveling in the direction $\hat{\mathbf{s}} = 0.34\hat{\mathbf{x}} + 0.94\hat{\mathbf{z}}$. Find the z coordinate at the point where the ray intercepts the yz plane.

2. A refracting surface is defined by the normal $\hat{\mathbf{\eta}} = -1/\sqrt{3}(\hat{\mathbf{x}} + \hat{\mathbf{y}} + \hat{\mathbf{z}})$. If the incident ray is along $\hat{\mathbf{s}} = -\hat{\mathbf{x}}$ and the indices of refraction on the incident and the transmitting sides are 1.0 and 1.33, respectively, find the directions of the reflected and the transmitted rays.

3. A corner cube mirror is formed by surfaces whose normals are $\hat{\mathbf{\eta}}_1 = -\mathbf{x}$, $\hat{\mathbf{\eta}}_2 = -\hat{\mathbf{y}}$, $\hat{\mathbf{\eta}}_3 = -\hat{\mathbf{z}}$ (Fig. 3.43). Prove that the reflected ray will have the opposite direction as the incident ray ($\hat{\mathbf{s}}'' = -\hat{\mathbf{s}}$) independent of the direction of $\hat{\mathbf{s}}$ by applying the vector form of the law of reflection.

4. An ellipsoidal reflector exists for the positive z part of the equation for its surface

$$\frac{x^2}{a^2} + \frac{y^2}{a^2} + \frac{z^2}{b^2} = 1$$

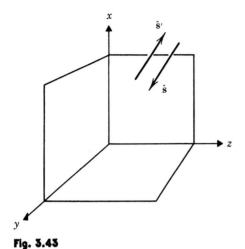

Fig. 3.43

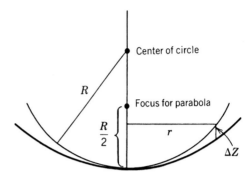

Fig. 3.44

where $a = 10$ cm and $b = 20$ cm. An incident ray parallel to the positive z-axis passes through the point $x = 2$ cm, $y = 2$ cm, $z = -2$ cm. Use the vector form of the law of reflection to find the direction of the reflected ray.

5. Consider the problem of exact ray tracing through a surface defined by $f = 0 = z + a(x^2 + y^2)$, where $a = 1$ cm^{-1}. Let the incident ray start at $(0, 0, -10$ cm$)$ inclined $20°$ with respect to the z-axis and in the xz plane. The incident medium is air and the transmitting medium has an index of refraction of 1.5. Find the intersection of the transmitted ray with the xy plane.

6. Derive an expression for the deviation between a spherical surface and a paraboloid of revolution that

contacts the sphere at one point and whose radius of curvature is less than the sphere radius. Express your answer as a distance, measured parallel to the symmetry axis of the paraboloid at a particular distance perpendicular to the symmetry axis (Fig. 3.44).

7. A 6 in. mirror with a radius of curvature of 96 in. is a reasonable approximation to a paraboloid with a focal length of 48 in. If the incident light has a wavelength of 550 nm, find the error between the spherical mirror and the paraboloid at the edge of the mirror in terms of the fraction of the wavelength of the incident light. This is a measure of the error involved in focusing a distant star (whose light meets a spherical telescope mirror as a plane wave).

8. Prove that the conic surfaces of revolution in Fig. 3.5 are Cartesian surfaces for reflection.

9. Prove that the spherical surface shown in Fig. 3.45 is Cartesian for refraction with a virtual image formed at

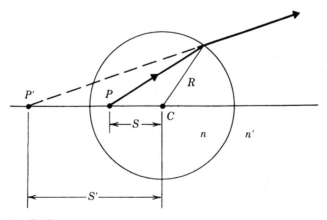

Fig. 3.45

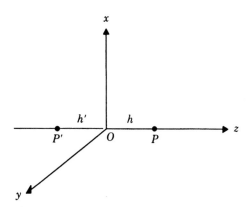

Fig. 3.46

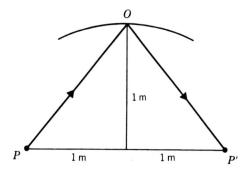

Fig. 3.48

P' provided that S and S' satisfy $n^2S = n'^2S' = nn'R$. (Use Eq. 3.9 and establish equality of the appropriate optical paths.)

10. (a) What is the nature of the Cartesian reflecting surface to be inserted at O so that a virtual image of P is formed at P' (Fig. 3.46)?

  (b) Which of the following surfaces will give approximate virtual imaging of P at P' in Fig. 3.46?

  (i) $z = a(x^2 + y^2)$
  (ii) $z = ax^2 + by^2, a \neq b$
  (iii) $z = ax^3 + by^3$
  (iv) $z = a(x^3 + y^3) + cy^4$
  (v) $z = a(x^4 + y^4)$

Explain.

11. Show that the ellipsoidal surface of revolution in Fig. 3.47 will give perfect image formation with refraction for the parallel beam at the focus F' of the ellipse provided that the eccentricity $e = c/a$ satisfies $e = n/n'$.

12. Consider Fig. 3.48 in which an approximate image of P is desired at P' after reflection from a toroidal mirror at O. What should be the values of the two principal radii of curvature? (One principal curve lies in the plane of the figure, the other lies in a perpendicular plane.)

13. Design an off-axis paraboloid reflector that will take a collimated beam with a diameter of 0.5 cm and focus it to a spot 10 cm from the point where the center of the beam meets the reflector (Fig. 3.49).

14. A plane wave is incident from air onto a spherical glass bead of radius R that has an index of refraction of n. Derive an expression for the wave surface after refraction. Express your answer in terms of the radial distance from the optical axis that is defined as the

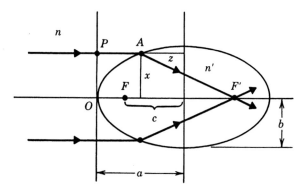

Fig. 3.47

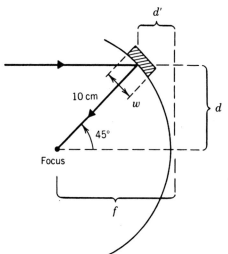

Fig. 3.49

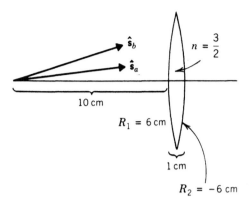

Fig. 3.50

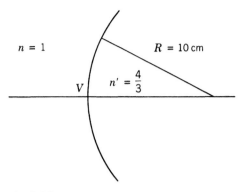

Fig. 3.51

original propagation direction of the plane wave through the point of first contact of a wavefront with the bead.

15. Repeat the derivation of Eqs. (3.13a) and (3.13b) for the specific case that the refracting surface is concave.

16. Use the ray-tracing formulas given in Table 3.1 to follow a ray that starts on the optical axis in front of the lens shown in Fig. 3.50 and that is inclined with respect to the z-axis by **(a)** $5°$ **(b)** $10°$. Follow the ray until it recrosses the z-axis.

17. Use the exact ray-tracing formulas to determine the apparent size of the bore in a glass capillary tube that is 10 mm in diameter with a 2-mm bore if the index of refraction of the glass is 1.562.

Section 3.2 Paraxial Optics

18. Derive the next higher-order correction to Eqs. (3.23) by expanding the exact ray-tracing formulas in terms of x/R and y/R.

19. Derive the lateral magnification relation Eq. (3.29) by means of the ray $P_0 V_1$ as in Fig. 3.14.

20. Derive the ray-angle magnification relation, Eq. (3.30), from Eq. (3.25b).

21. (a) Find the left and right focal lengths of the system in Fig. 3.51.

   (b) Find the image position and size for a 1-cm-high erect object located: **(i)** 50 cm to the left of V; **(ii)** 30 cm to the left of V; **(iii)** 20 cm to the left of V; **(iv)** 20 cm to the right of V (virtual object). (If the image is

located at $\pm \infty$, give, instead of image size, a value for the angle the parallel exit rays make with the z-axis.)

22. Locate the images of P_1, P_2, P_3 in Fig. 3.52. Find the lateral magnification $\overline{P_2' P_1'}/\overline{P_2 P_1}$ and the longitudinal magnification $\overline{P_3' P_1'}/\overline{P_3 P_1}$. Assume $\overline{P_2 P_1} \ll 10$ cm, $\overline{P_3 P_1} \ll 10$ cm.

23. Find the location and size of the final images of the objects at O_1 and O_2 as shown in Fig. 3.53.

24. (a) Calculate where a parallel beam of light incident from the left in Fig. 3.54 would be focused.

   (b) Turn the "lens" around and repeat the calculation.

25. Rework Problem 17 using the paraxial approximations.

26. In terms of the actual depth, what is the apparent depth of a swimming pool as viewed from directly above? (The index of refraction of water is 1.33.) Repeat the calculation at a viewing angle of $45°$.

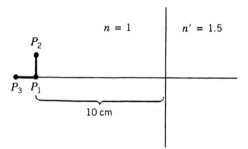

Fig. 3.52

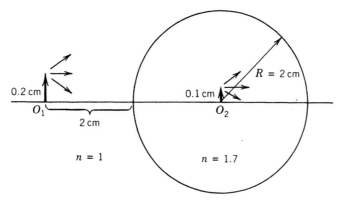

Fig. 3.53

27. An imperfection that is 1.5 cm below the surface of a glass block appears in its proper orientation but magnified by a factor of 1.2. If the index of refraction is 1.65, determine the curvature of the surface of the block.

28. A 3-cm-long goldfish is at the center of a spherical fishbowl filled with water ($n = 1.33$). Find the location and size of the image of the fish if the fishbowl is 12 in. in diameter.

If the fish is oriented so as to be pointing directly at the observer and the center of the fish is at the center of the fishbowl, how long does the fish appear?

29. Locate the position and size of the image of a 1-cm-high object formed by a spherical mirror under the following conditions: **(i)** $R = -20$ cm, $D_{01} = 30$ cm; **(ii)** $R = -20$ cm, $D_{01} = 15$ cm; **(iii)** $R = -20$ cm, $D_{01} = 8$ cm; **(iv)** $R = -20$ cm, $D_{01} = -15$ cm; **(v)** $R = +20$ cm, $D_{01} = -8$ cm. Illustrate each case with a rough sketch.

30. An incident ray passes a point 20 cm to the left of a reflecting spherical surface (of radius -30 cm) at a height of -2 cm below the z-axis (which passes

through the center of curvature) and with an inclination angle with the z-axis of 0.05 rad. Find the height and inclination angle of the reflected ray 10 cm to the left of the surface.

31. Find the location and size of the self-image that the driver of a car sees by reflection from the front windshield if the driver's head is 20 in. from the glass and the glass has a radius of curvature of 12 ft.

32. Describe the characteristics of the image of the sun that reflects off a spherical drop of honey 3 mm in diameter.

33. Plot the relationship between D_{12}/R_1 and D_{01}/R_1 for the range of D_{01}/R_1 from -3 to $+3$ in reflection from a spherical surface. Identify the region associated with real and virtual objects and images. Can this graph be used for both convex and concave mirrors?

34. Design a spherical reflector that is to be used as a collimator for a projection system where the lamp-to-mirror distance is to be 2 in. Will this be an efficient way of producing "parallel" light?

35. A small sphere is placed along the axis of a concave spherical mirror. Find the location(s) along the axis where the image of the sphere will appear spherical without distortion in the direction of the optical axis.

36. Light from a point source on the axis of a spherical concave mirror a distance 10 in. in front of the mirror illuminates a spot 0.5 in. in diameter on the surface of the mirror. The rest of the mirror is blocked by a mask. The reflected light converges to a focus with a cone of 70 mrad. Identify the characteristics of the mirror.

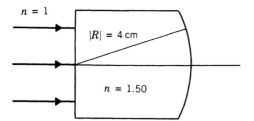

Fig. 3.54

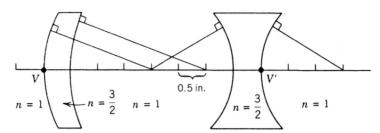

Fig. 3.55

Section 3.3 Matrix Methods

37. Prove the associative law of matrix multiplication:

$$(AB)C = A(BC)$$

38. Find 2×2 matrices **A** and **B** for which $AB \neq BA$.

39. Prove for two 2×2 matrices **A** and **B** that $\det(AB) = \det(A)\det(B)$.

40. Compose the system matrix for the optical configuration shown in Fig. 3.55.

41. Find the system matrix appropriate for transforming rays from the front face of the block to the back face of the block in Fig. 3.56.

42. The system matrix that relates an object that is 6 ft. tall to an image is given by

$$\begin{pmatrix} -20 & -20\,\text{m}^{-1} \\ 0 & -0.05 \end{pmatrix}$$

The object is in air, the image is in water ($n = 1.33$). What size is the image? A cone of light 2 mrad at the object will be converted to a cone that converges at the image with what angle?

43. Glass with an index of refraction of 1.58 is to be used to construct a thin lens that will be planar on the back side. For a power of $2\,\text{m}^{-1}$, what will the front surface curvature need to be? How will the power change if the lens is made symmetrical with the same curvature (but negative in sign) on the back as on the front?

44. Verify the locations of the principal planes in Fig. 3.25 for a lens index of refraction of 1.5.

45. Repeat the ray-tracing solution to Problem 3.16 only this time use the matrix method in the paraxial approximation.

46. Design a matrix formalism that would be useful for the paraxial treatment of multiple *reflections* from spherical surfaces.

47. Find the power and the location of the principal planes of the "lenses" discussed in Figs. 3.53 and 3.54.

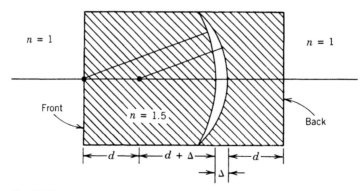

Fig. 3.56

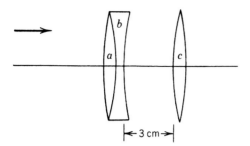

Fig. 3.57

48. Given the system matrix

$$\begin{pmatrix} 0.5 & -0.125 \text{ cm}^{-1} \\ 3 \text{ cm} & 1.25 \end{pmatrix}$$

which characterizes a multicomponent lens. Find the principal planes and the system power. Draw a sketch of the location of the principal planes with respect to the front and rear surfaces of the system. Assume that these surfaces are 9 cm apart.

49. Consider the combination of lenses shown in Fig. 3.57. $n_a = 1.5$, $n_b = 1.3$, $n_c = 1.6$. Find the system matrix for the combination. $|R| = 15$ cm for all surfaces. $D = 0.5$ cm for all lens thicknesses. From the system matrix find the location of the system principal planes with respect to the front and rear refracting surfaces and identify the system power.

50. Use the matrix technique to find the power and the locations of the principal planes for a combination of two thin lenses each with the same power $\mathscr{P}$ greater than zero separated by a distance d:
   (a) Where $d = 1/\mathscr{P}$.
   (b) Where $d = (3/4)(1/\mathscr{P})$.

Section 3.4 Image Formation

51. Locate the image graphically for a positive focal length thin lens when
   (a) $0 < S < f$.
   (b) $-f < S < 0$ (virtual object).
   (c) $S < -f$.

52. Use the graphical technique to find the location and size of the image produced by the negative focal length lens, $f = -2$ cm, when an object 1.5 cm tall is in front of the lens a distance of 3 cm.

53. For a negative lens we have $f < 0$.
   (a) Find the focal length of such a lens assumed thin when $R_1 = -40$ cm, $R_2 = 50$ cm, and $n = 1.75$.
   (b) Locate graphically and analytically the position and size of an image when $S = 2f$, $S = f$, $S = f/2$, $S = -f/2$.

54. Two thin lenses, one with focal length $+f$ the other with focal length $-f$, are mounted a distance f apart. Find the principal planes and focal planes of the combination. Now repeat the calculation, but the focal lengths are $+f$ and $-f/6$ and the lens separation is $(2/3)f$. For both cases, analytically determine the location of the image of an object a distance $+f$ in front of the first lens surface. Express your answer in terms of the distance from the last lens surface. Find the magnification in each case.

55. Find the object focal point for a combination of two thin lenses whose individual object focal points coincide.

56. A thin positive lens of focal length f is placed between a point source and a screen. Let the distance between the source and the screen be fixed at L. Derive an expression for all the possible locations of the lens (measured from the source) that will lead to a real image on the screen.

57. A lens of focal length 100 cm is near a 2-cm-tall object that is located 30 cm to the left of the object focal point. Using exclusively the Newtonian form of the thin lens equations, find the location and size of the image. What is the longitudinal magnification in this example. If the lens is constructed of glass whose index of refraction is 1.5 and water is the transmitting medium ($n = 1.33$), how do these results change?

58. Prove that the lateral magnification for two lenses in series is given by

$$m_x = \frac{f_1 S_2'}{d(S_1 - f_1) - S_1 f_1}$$

where f_1 is the focal length of the first lens, d is the separation between lenses, S_1 is the object distance from the first lens, and S_2' is the image distance from the second lens.

59. Two lenses having focal lengths $f_1 = +9$ cm and $f_2 = -18$ cm are placed 3 cm apart. If an object 2.5 cm high is located 20 cm in front of the first lens, calculate

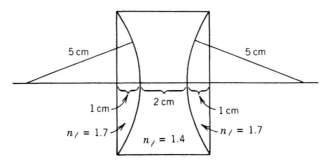

Fig. 3.58

(a) The position of the final image.

(b) The size of the final image.

(c) Check your solution graphically.

60. A parallel beam of light enters a clear plastic bead 2.5 cm in diameter and having an index of refraction of 1.440. At what point beyond the bead are these rays brought to a focus?

61. A thick lens with radii $R_1 = -4.5$ cm and $R_2 = -3.6$ cm has a thickness of 3.0 cm and an index of 1.56.

(a) Find the power of the lens.

(b) Find the distances from the vertices to the principal planes.

(c) Find the distances from the vertices to the focal points.

(d) If an object is placed 24 cm in front of the first vertex of the lens, find the location of the image with respect to the second vertex.

(e) Graph your solution.

62. Use the matrix method to find the locations of the principal planes and the focal planes for the optical system in Fig. 3.58. Express your measurements with respect to the vertices. With respect to the second vertex, locate the image of an object that is 20 cm in front of the first vertex.

63. The thick lens shown in Fig. 3.59 is exposed to parallel light.

(a) Treat the lens as thin with the location of the thin lens at the center of the actual lens and in this way calculate the focal point for the rays.

(b) Repeat this exercise treating the lens as thick yet still in the paraxial limit.

(c) Use the exact ray-tracing formalism to find the optical axis intercept of rays that are 0.25 in. and 0.5 in. from the optical axis on the incident side.

Section 3.5   Examples of Paraxial Optics

64. In a nearsighted eye the image focal point for an infinitely distant object is in front of the retina. Assume that the eye has the characteristics of Fig. 3.34 except that F' is 22 mm from the front of the cornea rather than 24.38 mm as required. Find the power of a corrective lens that could be placed 14 mm in front of the cornea that will place F' on the retina. Repeat your calculation for a contact lens.

65. For a simple magnifier that has a magnifying power of $M = 20$ when the image is at the near point when used as in Fig. 3.35, find the focal length of the lens and the distance at which the object should be placed as measured from the lens.

66. Design a microscope objective lens for a standard microscope that will yield a magnification of 200 when used in combination with an eye lens with a focal length of 20 mm.

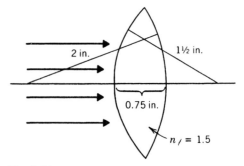

Fig. 3.59

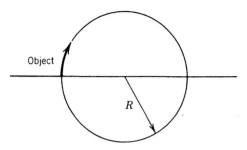

Fig. 3.60

67. Consider a small glass sphere with index n, radius R. Find a value of n for which a virtual image is formed at infinity. What then is the resulting microscope magnification? See Fig. 3.60.

68. Show that the combination of two lenses having equal and opposite powers a finite, positive distance d apart has a net positive power $\mathscr{P}$, and find $\mathscr{P}$ as a function of d.

69. A microscope is assembled with elements having focal planes as shown in Fig. 3.61.

(**a**) Find the focal length and the location of the principal planes of the combination. This microscope is used to project a real image a distance 500 mm to the right of F'_2.

(**b**) How far to the left of F_1 must the object be placed?

(**c**) What is the overall lateral magnification?

(**d**) Find the answers to (**b**) and (**c**) when the image is virtual and 250 mm to the left of F'_2.

70. A telescope consists of two positive lenses, $f_1 = 30$ cm for the objective and $f_2 = 2$ cm for the eye lens. If the object is 250 m away from the objective, how far apart should the lenses be placed if the image is to be located at the near point of the standard eye? Find the location of the intermediate image.

71. An astronomical telescope (Fig. 3.40) is made of two thin positive lenses such that $m_\alpha = -10$ and $f_1 + f_2 = 1$ m. It is used with a lens separation $d = 1$ m. An observer looks through the telescope at the eye lens.

(**a**) What is the closest distance an object may be to the objective and still be conveniently seen? (Image farther from the observer than 250 mm).

(**b**) Find the answer to (**a**) when a Galilean telescope is used with $m_\alpha = +10$, $f_1 + f_2 = 0.8$ m, $d = 0.8$ m.

72. The telescope in problem 71(**a**) is used as a doublet to observe an object 2 m to the left of the objective. The lens separation d is adjusted to give a virtual image located 0.3 m to the left of the eye lens. Find d and the overall magnification.

73. Consider a telephoto lens consisting of a Galilean telescope of angular magnification m_α together with a lens of focal length f. Show that the focal length of the combination is $m_\alpha f$ and locate the principal planes (see Fig. 3.41).

74. A wide-angle camera lens has a smaller effective focal length than a normal lens while being placed in the same position with respect to the film (rear element to film separation constant in the comparison). Following arguments similar to those that surround Eq. (3.120) for the telephoto lens, derive an expression for the lateral magnification of a wide-angle lens.

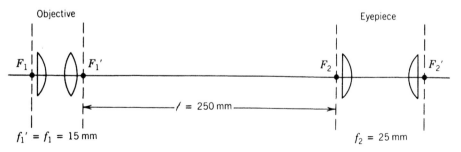

Fig. 3.61

4 Practical Geometrical Optics

4.1 Stops and Apertures

It is not enough to be able to predict the image position and size for an optical system. Other important properties of an image include its brightness and the size of the field of view. These topics require a study of the limitations of the spatial and angular extent of light beams by stops and apertures. To make such a study, we will have to introduce some new definitions and concepts. Even if the optical system under consideration is designed to accept light rays that are not paraxial, it is usually sufficient to treat stops and apertures using the methods of paraxial optics.

A. Aperture Stop and Pupils

Consider an object located on the axis of an optical system. Imagine the object to be emitting a cone of light of varying apex angle. For a small angle, all the light rays will pass through the system, but as the angle is increased, some of the rays will be stopped by the rim of a lens or a mechanical aperture somewhere in the system. The *aperture stop* (AS) is defined to be that stop or lens rim which physically limits the solid angle of rays passing through the system from an on-axis object point.

1. Single Thin Lens. *a. Aperture Stop on Incident Side.* A simple example of an aperture stop is shown in Fig. 4.1a. Here the hole in the screen limits the solid angle of rays from the object at P_0 that can pass through the system. It is the aperture stop.

The rays are cut off at A and B. The images of A and B are A' and B'. To an

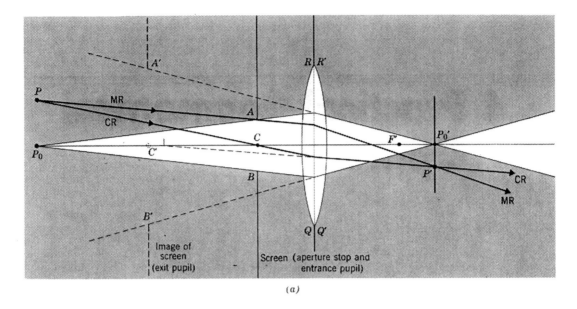

(a)

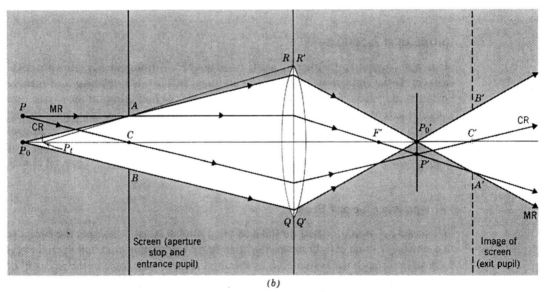

(b)

Fig. 4.1 Ray diagrams for the case of the aperture stop on the incident side of the lens: (a) virtual exit pupil; (b) real exit pupil.

observer looking back through the lens from a position near P_0', it will appear as if A' and B' are cutting off the rays.

If we move the screen to the left of F, we have the situation shown in Fig. 4.1b. The screen still is the aperture stop, but the images A', B' of A and B are now to the right of the image P_0'. To an observer who moves sufficiently far to the right, it still appears as if the rays are being cut off by A' and B'.

The image of the aperture stop that is formed by the light after it goes through the lens is called the *exit pupil*. In Fig. 4.1*a* the exit pupil is virtual, in Fig. 4.1*b* it is real. To an observer examining the light after its encounter with the lens, the exit pupil is the opening that defines the solid angle for the rays that *converge* to the on-axis *image* point.

By analogy, we can define an *entrance pupil*. This is the opening that an observer would identify as the limitation on the solid angle for rays *diverging* from an on-axis *object point*. In Fig. 4.1 the aperture stop acts as the entrance pupil. The aperture stop is physically located between the object and the lens. In this position the aperture stop limits the solid angle of the rays coming from the object.

The term "pupil" has to do with the design of an optical system intended for use with visual observation. It is most desirable for the exit pupil to coincide approximately with the pupil of the observer's eye. This could happen in Fig. 4.1*b* but not in Fig. 4.1*a*.

A ray from an off-axis object P through the center of C of the aperture stop is called a *chief ray* (CR). Its initial portion (extended if necessary) passes through the center of the entrance pupil. The conjugate ray $P'C'$ will (appear to) pass through the center C' of the exit pupil.

A ray such as PA from an object point through the edge of the aperture stop is called a *marginal ray* (MR). The conjugate ray $P'A'$ appears to pass through the edge of the exit pupil. The marginal rays drawn from an on-axis object identify the maximum solid angle that can be passed by the system.

If the object plane in Fig. 4.1*b* were somewhat closer to the screen, the rays from P_0 would cease to be limited by AB but instead would be limited by the lens mount QR. Hence the latter would become the aperture stop. The transition point occurs for P_0 at P_t, where P_t, A, and R are colinear. For P_0 closer to the lens than this, its rim is the aperture stop.

b. Aperture Stop on Transmitting Side. Two simple cases where the aperture stop encounters the light after the lens are shown in Figs. 4.2*a* and *b*. It is then its own exit pupil. The primes indicate that the light at the aperture stop has already gone through the lens.

The entrance pupil is the real image of the aperture stop in Fig. 4.2*a* and the virtual image of the aperture stop in Fig. 4.2*b*. In either case, this is the opening that an observer would identify as the limitation for rays diverging from the object point P_0.

2. Multilens System. In a multilens system one can always identify an aperture stop using the same definitions as were presented in section 4.1A1. This is the physical opening that limits the amount of light that can pass through the system. The entrance pupil will be the opening or the appropriate image of an opening toward which the marginal rays from an on-axis object appear to diverge. The exit pupil will be the opening or image of an opening from which these same marginal rays appear to converge toward the image. In Fig. 4.3 the aperture stop is $\overline{AB}$, while the entrance pupil is AB and the exit pupil is $A'B'$.

A systematic method of finding the entrance pupil is to image all stops and lens

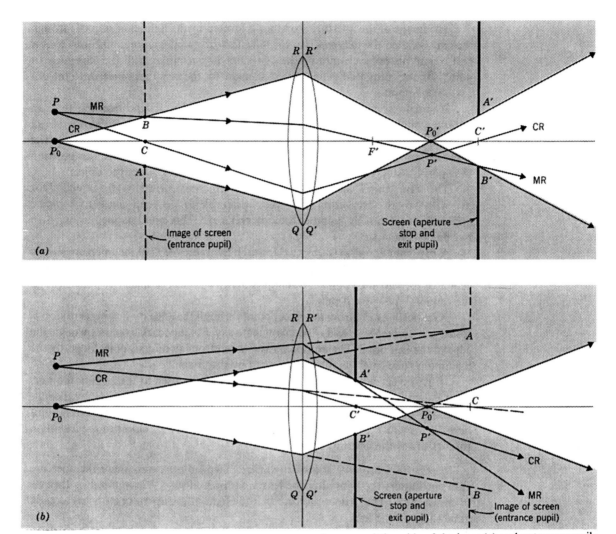

Fig. 4.2 Ray diagrams for the case of the aperture stop on the transmitting side of the lens: (*a*) real entrance pupil; (*b*) virtual entrance pupil.

rims to the left (in the direction opposite to that which is followed by the rays from the object) through all intervening refracting elements of the system. Find the solid angle subtended by each image at P_0. The one with the smallest magnitude is the entrance pupil, and the physical object corresponding to it is the aperture stop. Alternatively, the exit pupil may be found by imaging all stops and lens rims to the right (in the same direction as that followed by the light from the object) through all intervening refracting elements of the system. Determine the solid angle subtended by each image at P_0'. The one with the smallest solid angle is the exit pupil, and the corresponding real physical object is the aperture stop. These two procedures must, of course, identify the same object as the aperture stop.

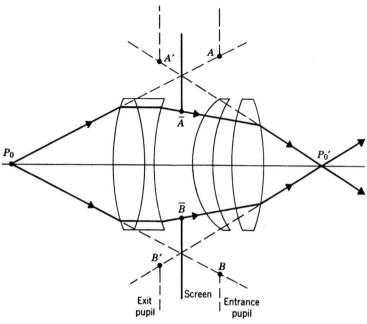

Fig. 4.3 Marginal rays in a multilens system.

3. Matrix Method of Locating the Aperture Stop. *a. General Discussion.* The matrix method may be useful in locating the aperture stop, especially when a one-dimensional use of angles rather than solid angles is sufficient. Let the object be at the plane that cuts the axis at A on Fig. 4.4. Then a ray from A that just misses the rim of the aperture stop will be the ray that just passes through the system. Let α_0 be the angle it makes with the axis. If $\mathbf{M}_{AB}$ represents the transformation matrix from the plane at A to the plane of the aperture stop, then in general we have

$$\begin{pmatrix} n'\alpha' \\ x' \end{pmatrix} = \mathbf{M}_{AB} \begin{pmatrix} n\alpha \\ x \end{pmatrix}$$

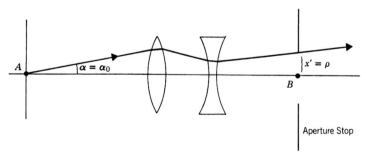

Fig. 4.4 Marginal ray illustrating the relationship between the angle at the object and the size of the aperture stop.

or, if the lens system is in air,

$$\alpha' = M_{11}\alpha + M_{12}x \qquad (4.1a)$$

$$x' = M_{21}\alpha + M_{22}x \qquad (4.1b)$$

Now for the ray in question we have

$$x = 0, \quad \alpha = \alpha_0, \quad \text{and } x' = \rho$$

where ρ is the radius of the aperture stop. Equation (4.1b) then gives

$$\alpha_0 = \frac{\rho}{M_{21}} \qquad (4.2)$$

Thus the aperture stop has the *smallest* ratio of radius ρ to M_{21} for all the stops or lens rims in the system.

b. Example. As a simple example, we consider the case illustrated in Fig. 4.5. The issue to be decided is whether L_1 or the image of L_2 subtends the smallest angle at A. Now L_1 subtends the half-angle

$$\alpha_1 = 0.1 \text{ rad}$$

The matrix method requires the calculation of the (2, 1) element of the matrix

$$\mathbf{M}_{AL_2} = \begin{pmatrix} 1 & 0 \\ D_{12} & 1 \end{pmatrix}\begin{pmatrix} 1 & -1/f_1 \\ 0 & 1 \end{pmatrix}\begin{pmatrix} 1 & 0 \\ D_{01} & 1 \end{pmatrix} = \begin{pmatrix} -1 & -0.2 \\ 4 & -0.2 \end{pmatrix}$$

This is $M_{21} = 4$ cm. Hence by Eq. (4.2) we have

$$\alpha_2 = 0.25 \text{ rad}$$

For the system of Fig. 4.5, the rim of L_1 is the aperture stop and the entrance pupil.

The direct method would be to find the image of L_2 as formed by L_1. It is located a distance $S' = 30$ cm to the left of L_1 and is $|S'/S| \times 1$ cm = 5 cm high. This also shows the angle subtended at A to be 0.25 rad.

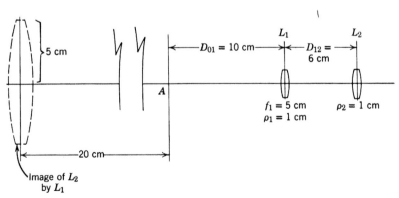

Fig. 4.5

B. Field Stop and Windows

1. Definitions. It should be clear that the aperture stop determines the illumination of an on-axis point image, because the greater the solid angle of the transmitted light cone, the greater the transmitted light flux. The flux density in the off-axis parts of an image depends in part on the size and location of another stop, the field stop.

Both chief rays and marginal rays will leave an off-axis object, but not all such rays may pass through to the image. The *field stop* is that stop or lens rim that limits the solid angle formed by chief rays. The image of the field stop formed by optical elements on the incident side of the field stop is called the *entrance window* (En.W.). The image of the field stop formed by all optical elements after the field stop is the *exit window* (Ex.W.). If no additional refracting elements exist between the field stop and the object or image, then the field stop itself acts as the entrance or exit window, respectively.

If the entrance window is near the object plane, it tends to limit the *field of view*, that is, the lateral extent of the object that will be imaged by the system. To an observer examining the image, the exit window appears to limit the area of the image, just as a window limits the area that an observer can see when he or she looks outdoors. The exit pupil, on the other hand, tends to limit the solid angle of rays that converge to each point of the image.

We have used the term "tends to" in the preceding paragraph, because in many optical systems the field of view is not sharply limited by the field stop; instead, there is a gradual loss of light as we move farther off axis. In addition to this effect, and related to it, is another effect: A cone of light rays from an off-axis source to the entrance pupil will not necessarily be transmitted entirely. It can be *partially* cut off by the field stop or by other stops or lens rims in the system. This is called *vignetting*.

2. Example. To illustrate these ideas, we consider an optical system consisting of two thin lenses of the same focal length f, where

$$\overline{P_0 L_1} = 2f, \quad \overline{L_1 L_2} = 4f, \quad \overline{L_2 P_0'} = 2f$$

There is an aperture stop right behind L_1 (Fig. 4.6). An on-axis object P_0 is imaged first at P_0' and then at P_0''. The aperture stop and entrance pupil are at L_1. From the lens equation applied to the imaging of the aperture stop by L_2,

$$\frac{1}{S} + \frac{1}{S'} = \frac{1}{f}$$

with $S = 4f$, we find

$$\frac{S'}{S} = \frac{4}{3}\frac{f}{4f} = \frac{1}{3}$$

Thus the exit pupil will be one-third as large as the aperture stop.

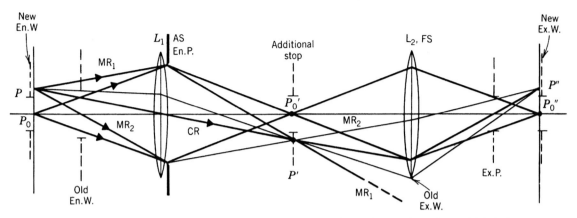

Fig. 4.6 Vignetting in a symmetrical two-lens system. Ray MR_1 misses the opening of L_2, which is the field stop. This causes a loss of image brightness for off-axis parts of the image.

Lens L_2 is the field stop and also the exit window in this example. Because of the symmetric placement of the lenses, the entrance window is one-third the size of L_2 and located a distance $(4/3)f$ to the left of L_1.

As P_0 moves off axis, rays such as MR_1 are lost by *vignetting*. If a uniformly illuminated screen were placed in the object plane at P, its image would be nonuniformly illuminated and would be roughly half as bright at P'' as at P_0''.

One can put an additional stop at the intermediate image point P_0' as shown by the dashed lines. If it subtends a smaller angle, as seen from the center of the aperture stop L_1, than does L_2, it will become the field stop. The new entrance window will now be in the object plane, and (in this example) it will be the same size as the field stop (both are shown dashed in Fig. 4.6). As P_0 moves off axis toward P, the chief rays will now be cut off first by this new field stop, but some of the marginal rays will still miss L_2, and there will still be loss of brightness in the image plane. The *field of view* in the object plane is now sharply limited by the new entrance window, but we still have vignetting.

3. Field Lens. The vignetting in the example of section 3.1B2 can be completely eliminated by putting a *field lens* L_3 at P_0' (Fig. 4.7). Because of its location, it will not affect the imaging of light from the on-axis point P. It will deviate the whole cone of light originating at an off-axis point P. This deviation will be most effective if L_3 focuses L_1 onto L_2, for a marginal ray leaving the bottom of L_1 will be deflected to the top of L_2, whereas without the field lens L_3, this marginal ray would completely miss L_2. If L_3 is stronger than necessary to focus L_1 onto L_2, then a marginal ray leaving the bottom of L_1 will be deviated so much that it misses the top of L_2.

In this optimum case, an entire cone of rays from an off-axis point in the object plane to the full entrance pupil is either completely successful in passing through the rest of the system or completely unsuccessful, depending on whether or not P is

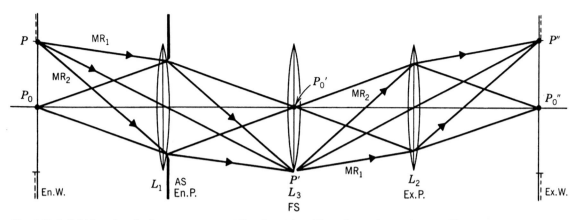

Fig. 4.7 A field lens bends the rays to correct for vignetting. There is no change in magnification resulting from the introduction of the field lens.

within the entrance window. The field stop is at L_3, the entrance window is in the object plane, the field of view is sharply limited by the entrance window, and the brightness of the image matches, in a first approximation, that of the object. This will be discussed in detail in a later section.

The insertion of a field lens at key places in an optical system can make a marked difference in the ability of the system to transmit a useful amount of light. Of course, this can be done in a system made up of well-separated individual components, but not for a compound lens.

The concept of the field stop is often not particularly useful in discussing a compound lens, such as that in Fig. 4.3. For such a lens, the vignetting of a cone of rays from an off-axis point can be described by vignetting diagrams, as shown in Fig. 4.8. We first consider each ray-limiting aperture or lens rim in turn. Using those refracting elements that are placed to the left of each, we form an image of each ray-limiting object. Usually there are only three important limiting objects, the first lens rim (FLR), the aperture stop (AS), and the last lens rim (LLR). The FLR is already the left-most element, so no image of it need be formed. The image of the AS is the entrance pupil (En.P). Finally, the image of the last lens rim (ILLR) involves refraction by all lenses in the system except the last lens.

When the object is on-axis, the limiting cone of light is identified by rays that head toward the entrance pupil. For off-axis object points, the rays must be subject to additional limitations imposed by the FLR and the LLR. To determine this, draw rays headed from the object point toward the FLR and the ILLR. If either of these define a cone of light smaller than that defined by the En.P., then vignetting occurs. The top marginal rays are prone to limitation by the FLR, and the bottom marginal rays are prone to limitation by the LLR. The cone of light that passes through the system in Fig. 4.8 is shown bright in the vertical sections of the left half of each part of the figure.

To illustrate the limitations imposed on rays out of the plane of the drawing,

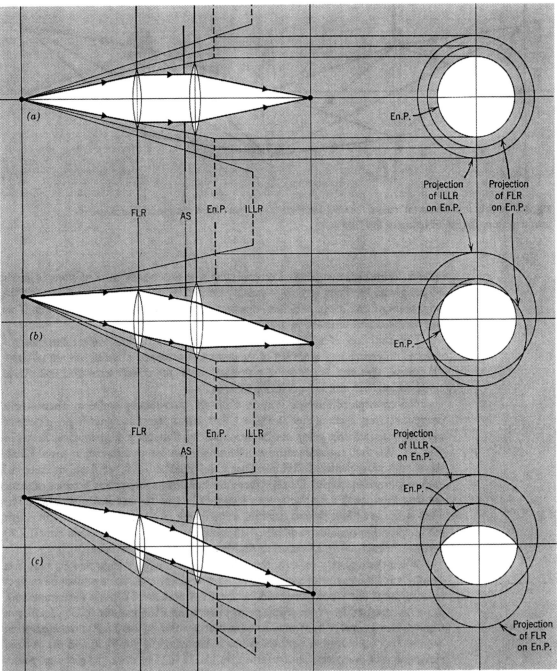

Fig. 4.8 Vignetting diagrams. Ray limiting apertures are imaged to the left. From an object point, cones of potential rays are drawn to the images of the apertures. The intersection of these identify the useful cone for a particular object point. (*a*) On-axis. (*b*) Off-axis by a small amount. (*c*) Off-axis by a larger amount.

we can project the FLR and the ILLR from the object point onto the plane of the entrance pupil (En.P.). If the angles are not too large, these projections can be approximated by circles. (They are really ellipses.) We then find the area common to all three projections, shown bright on the right half of each part of Fig. 4.8. The light passing through the system consists of all rays from the object point that are headed for this common (bright) area in the plane of the entrance pupil.

The effect of "stopping down" the lens can be seen by decreasing the size of the circle that represents the entrance pupil. To obtain a uniform decrease in illumination (and, ordinarily, to minimize aberrations), it is desirable to have the center of the entrance pupil correspond to the center of the bright area. This is not the case in Fig. 4.8, because the aperture stop is not centered between the two lenses, which in this example have equal power.

4.2   Radiometry and Photometry

A real optical system must not only produce an image of the desired size at the right place, but the image must be bright enough to be useful. Image brightness will depend on object brightness and on the size and location of stops within the optical system.

In optics there are two types of units for energy-related quantities, physical and psychophysical. In our notation, the subscript e (for energy) will denote physical units, and v (for visual) psychophysical units.

A. Physical or Radiometric Nomenclature

1. Definitions. We will discuss the physical system first, using the mks system or a hybrid system in which area is measured in square centimeters.

Radiant energy, Q_e in joules, refers to the total amount of energy emitted, transferred, or collected in a radiation process. The *radiant energy density,* U_e in joules/m^3, is the radiant energy contained in a unit volume of space. These two are related by

$$U_e = dQ_e/d\mathcal{V} \tag{4.3}$$

where $d\mathcal{V}$ is an arbitrary differential volume element. We have already seen in Chapter 1 that the energy density due to an electromagnetic field as given by Eq. (1.54) is

$$U_e = \varepsilon_0 |\mathbf{E}|^2 \tag{4.4a}$$

or in the presence of matter

$$U_e = \varepsilon |\mathbf{E}|^2 \tag{4.4b}$$

We are usually interested in time-averaged quantities because our detectors cannot follow the rapid variation of the electromagnetic field, thus we need

$$\langle U_e \rangle = \frac{\varepsilon}{2}|\mathbf{E}|^2 \tag{4.4c}$$

In the rest of this section we will simplify the notation by eliminating the time-average brackets. However, it should be understood that all quantities are averaged over a number of oscillations of the fields. Radiant power or *radiant flux* Φ_e in watts is the time rate of change, or rate of transfer, of radiant energy. This can be expressed as

$$\Phi_e = dQ_e/dt \tag{4.5}$$

If we identify a closed surface that defines a volume as shown in Fig. 4.9, then the total radiant flux out through the closed surface is

$$\Phi_e = \oiint \mathbf{S} \cdot d\mathbf{A} \tag{4.6}$$

where $d\mathbf{A}$ is the differential surface area element vector taken positive outward; $\mathbf{S}$ is the radiant energy flux density as defined in Eq. (2.20)

$$\mathbf{S} = \mathbf{E} \times \mathbf{H}$$

After the time average in nonabsorbing, local, linear, isotropic, nonmagnetic media this becomes [Eq. (2.45)]

$$\mathbf{S} = \frac{nc\varepsilon_0}{2}|E_0|^2 \hat{\mathbf{s}} \tag{4.7}$$

In the general case, the net $\mathbf{S}$ from all contributions must be determined at each spot on the surface so that the integral in Eq. (4.6) can be completed. [In most cases this will turn out to be a vector sum of the time-averaged $\mathbf{S}$'s from each source. However, if the fields from the different sources are coherent (same frequency and bearing constant phase relationships among them), then the *fields* must be vectorially combined before $\mathbf{S}$ can be determined (more about this in Chapter 5).]

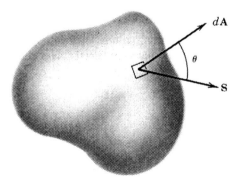

Fig. 4.9 Closed surface used for calculating the total radiant flux.

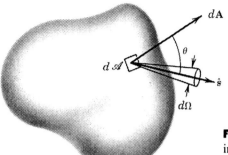

Fig. 4.10 Geometry for the radiant flux integral in the case where the light possesses a range of directions.

The combination of flux density from different incoherent sources can be expressed as

$$\Phi_e = \oiint [\mathbf{S}_1 + \mathbf{S}_2 + \mathbf{S}_3 + \dots] \cdot d\mathbf{A}$$

or as

$$\Phi_e = \oiint [S_1 \cos \theta_1 + S_2 \cos \theta_2 + \dots] \, d\mathscr{A} \qquad (4.8)$$

When a continuous range of directions for **S** is present, this is more conveniently written as

$$\Phi_e = \oiint \left[\int \frac{dS}{d\Omega} \cos \theta \, d\Omega \right] d\mathscr{A} \qquad (4.9)$$

This is illustrated in Fig. 4.10. The factor within the bracket is a solid angle integral over all directions. The increment $d\Omega$ is defined in Fig. 4.11 as

$$d\Omega = \frac{d\mathscr{A}'}{R^2} \qquad (4.10)$$

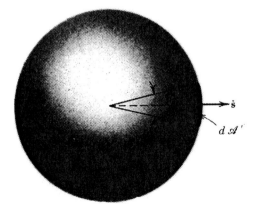

Fig. 4.11 Defining relationships for the solid-angle differential element.

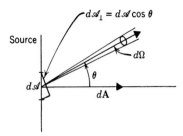

Fig. 4.12 Radiance from a source of area $d\mathscr{A}$ in the direction θ with respect to the surface normal of the source and for the solid angle $d\Omega$.

This is described by a cone, centered about a direction specified by unit vector $\hat{s}$, which intercepts a portion of a sphere. Solid angles are measured in *steradians* (sr). For a complete sphere, $\Omega = 4\pi$ sr.

The differential angular dependence of the radiant flux density $dS/d\Omega$ is given the name *radiance* and is signified by L_e. Thus,

$$L_e = \frac{dS}{d\Omega} = \frac{d^2\Phi_e}{\cos\theta\, d\Omega\, d\mathscr{A}} \tag{4.11}$$

The radiance is given in W m^{-2} sr^{-1}.

We are often concerned with the characteristics of sources or images in an optical system. This is best described by the radiance, because the properties of the source or image usually depend on angle ($\hat{s}$) and position. Figure 4.12 illustrates how to specify the radiance of a source. The radiant flux emitted by $d\mathscr{A}$ of the source into an element of solid angle $d\Omega$ about the direction $\hat{s}$ is written

$$d^2\Phi_e = L_e\, d\Omega\, d\mathscr{A}_\perp = L_e\, d\Omega\, d\mathscr{A}\cos\theta = L_e\, d\Omega(\hat{s}\cdot dA) \tag{4.12}$$

In principle, the radiance L_e may be measured by means of the apparatus illustrated in Fig. 4.13. The small opening or aperture has area $d\mathscr{A}_0$. Project this area onto the source using straight lines from a point on the detector. The projected area at the source is then $d\mathscr{A}_p = d\mathscr{A}_0(l+d)^2/l^2$. The solid angle subtended by the detector as seen from the source is $d\Omega = d\mathscr{A}_{det}(l+d)^2$. Thus, $d\mathscr{A}_p\, d\Omega = d\mathscr{A}_0\, d\mathscr{A}_{det}/l^2$. Hence, if the detector is calibrated to read in watts, we take its output $d^2\Phi_e$ and obtain the radiance from the equation

$$L_e = \frac{d^2\Phi_e}{d\mathscr{A}_p\, d\Omega} = \frac{d^2\Phi_e l^2}{d\mathscr{A}_0\, d\mathscr{A}_{det}}$$

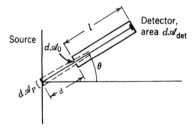

Fig. 4.13 Device with which the radiance from a source could be measured.

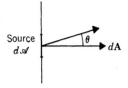

Fig. 4.14 The radiant exitance is the power per unit area leaving $d\mathcal{A}$ and going to the right. It includes all propagation directions θ from 0 to 90°.

One of the most important physical quantities in radiometry is the flux of radiant energy flowing across unit area of a real or imaginary surface. The accepted convention makes a distinction between source surfaces and other surfaces. The *radiant exitance M_e* (in W/m²) is the total emitted flux density (Fig. 4.14). Similarly the *irradiance E_e* (in W/m²) is the incident radiant flux density on a real or imaginary surface. Thus we have for the flux to or from area $d\mathcal{A}$

$$d\Phi_e = M_e \, d\mathcal{A} \qquad \text{(out from } d\mathcal{A}\text{)}$$
$$d\Phi_e = E_e \, d\mathcal{A} \qquad \text{(onto } d\mathcal{A}\text{)} \qquad (4.13)$$

These quantities are related to the radiance through an angular integration. For instance, the radiant exitance of a source may be obtained by integration of Eq. (4.12) for fixed $d\mathcal{A}$ (at a given spot on the source).

$$d\Phi_e = \int_{(1/2)} d^2\Phi_e = d\mathcal{A} \int_{(1/2)} L_e \cos\theta \, d\Omega$$

Hence

$$M_e = \int_{(1/2)} L_e \cos\theta \, d\Omega \qquad (4.14)$$

Here

$$\int_{(1/2)} (\ldots) \, d\Omega$$

refers to an angular integration over one-half of the full solid angle, that is, over all outward directions from the surface.

Another important quantity is the *radiant intensity I_e* (in W/s). This represents the flux per unit solid angle radiated by an entire source in a given direction (represented by ŝ). This is shown in Fig. 4.15. The intensity is of interest if the source

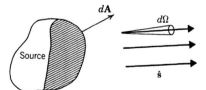

Fig. 4.15 The radiant intensity $I(\hat{s})$ represents the total power per unit solid angle radiating in direction ŝ from a finite source.

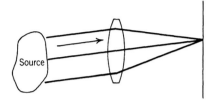

Fig. 4.16 All rays in the same direction $\hat{s}$ are focused at the same point by the lens.

is far away and if one has a small detector so that all rays from any part of the source to any part of the detector are essentially parallel. It is also of interest if the light is studied in the focal plane of a lens, for then all rays from the source with a given direction $\hat{s}$ are focused at the same point (Fig. 4.16).

If we consider a cone of rays making a solid angle $d\Omega$ about $\hat{s}$ (Fig. 4.15), then the flux is given by

$$d\Phi_e = I_e \, d\Omega \tag{4.15}$$

The intensity I_e will, in general, depend on the direction of $\hat{s}$. The intensity I_e may be obtained by integration of Eq. (4.12) over area, for fixed $\hat{s}$ and $d\Omega$:

$$d\Phi_e = \int d^2\Phi_e = d\Omega \iint L_e \cos\theta \, d\mathscr{A} = \hat{s} \, d\Omega \cdot \iint L_e \, d\mathbf{A}$$

Hence

$$I_e = \iint L_e \cos\theta \, d\mathscr{A} = \hat{s} \cdot \iint L_e \, d\mathbf{A} \tag{4.16}$$

Here the integral extends over that part of the surface that may radiate in direction $\hat{s}$, for example, over the shaded area in Fig. 4.15.

2. Special Cases. *a. Isotropic Source.* For a uniform spherical source of radius R_0 emitting total flux $\Phi_{e,\text{tot}}$, we obtain, by integrating over the entire surface,

$$\Phi_{e,\text{tot}} = \iint M_e \, d\mathscr{A} = 4\pi R_0^2 M_e \tag{4.17a}$$

Then a small detector of area $d\mathscr{A}_{\text{det}}$ placed as shown in Fig. 4.17 will receive flux

$$d\Phi_e = \Phi_{e,\text{tot}} \frac{d\mathscr{A}_{\text{det}}}{4\pi R^2} = \frac{\Phi_{e,\text{tot}}}{4\pi} \frac{d\mathscr{A}_{\text{det}}}{R^2} = \frac{\Phi_{e,\text{tot}}}{4\pi} \, d\Omega \tag{4.17b}$$

By comparison with Eq. (4.15) we obtain

$$I_e = \frac{\Phi_{e,\text{tot}}}{4\pi} \tag{4.18}$$

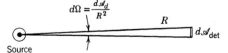

Fig. 4.17

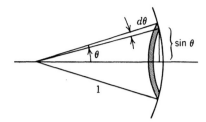

Fig. 4.18 Annular solid angle differential element.

for an isotropic source. For such a source, assumed spherical, we also obtain, using (4.17a),

$$I_e = R_0^2 M_e \qquad (4.19)$$

b. Lambert Source. A Lambert source is one that has the same directional properties as a small hole in a cavity. Within a cavity, the light is completely randomized. It propagates everywhere in all directions with equal radiance. It is also uniform across the area of the hole. Thus L_e is not a function of ŝ or of position on the source. For a Lambert source, the apparatus in Fig. 4.13 would produce a detector output independent of position and angle θ, as long as the field of view of the detector is limited by the opening of area $d\mathscr{A}_0$ and not by the edge of the source.

For a Lambert source we may take L_e outside of the integral in Eq. (4.14). It is then useful to redefine the solid angle element $d\Omega$ such that $\cos \theta$ will be a constant on the element. The annulus shown in Fig. 4.18 satisfies this requirement. Then $d\Omega = 2\pi \sin \theta \, d\theta$. For a *Lambert source* we then have

$$M_{e,L} = L_e \int_0^{\pi/2} \cos \theta (2\pi \sin \theta) \, d\theta = \pi L_e \int_0^{\pi/2} d(\sin^2 \theta) = \pi L_e \qquad (4.20)$$

3. Spectral Quantities. The preceding definitions refer to total radiation at all wavelengths. A spectral version of each may be defined. For example, a *spectral radiant flux* $\Phi_e(\lambda) \, d\lambda$ represents the flux in a wavelength range between λ and $\lambda + d\lambda$. Usually wavelengths are specified in terms of nanometers. Thus $\Phi_e(\lambda)$ would be in watts/nanometer.

We obtain by integration

$$\Phi_e = \int_0^\infty \Phi_e(\lambda) \, d\lambda \qquad (4.21)$$

Other spectral quantities are $Q_e(\lambda)$, $U_e(\lambda)$, $M_e(\lambda)$, $E_e(\lambda)$, $I_e(\lambda)$, and $L_e(\lambda)$.

It is also of interest in cases involving the interaction of radiation with matter to be able to specify the photon flux in an optical beam. The spectral quantities identify a unique wavelength that, through the quantum theory, is associated with photons of a given energy. The energy per photon is hc/λ (where $h = 4.135 \times 10^{-15}$ eV sec is Planck's constant). Thus all the spectral quantities involving energy can be converted to quantities involving photons by dividing the radiant expressions by hc/λ. Thus the photon flux (photons per second) in a range of wavelengths from λ to $\lambda + d\lambda$ is

$$N(\lambda)\, d\lambda = \frac{\lambda \Phi_e(\lambda)}{hc}\, d\lambda \qquad (4.22)$$

4. Numerical Examples of Radiometric Quantities. One of the most intense incoherent sources is a super-high-pressure mercury lamp. This has a radiance in the visible region of the spectrum of about 250 W/(cm² sr). Continuous gas lasers have a relatively low total power output, but it is highly directional. A typical example is furnished by a 4-mW helium–neon laser emitting light at 632.8 nm. Its output can be focused into a spot $r = 0.1$ mm with the beam making a cone of half-angle $\theta = 2$ mrad. The area of the spot is about 3.1×10^{-4} cm². The solid angle that the cone makes is $d\Omega \cong (\theta R)^2 \pi / R^2 \cong \theta^2 \pi \simeq 1.3 \times 10^{-5}$ sr. The product $d\mathscr{A}\, d\Omega$ is 4×10^{-9} cm² sr. The radiance L_e is the power divided by $d\mathscr{A}\, d\Omega$ or about 10^6 W/(cm² sr). This is 400 times brighter than the mercury lamp.

B. Psychophysical or Photometric Nomenclature

We now turn to photometric quantities recorded by a human observer using visible light. In the science of photometry, comparisons between different sources are made by an observer. The comparison is ultimately related to one or more standard sources. A system of photometric units has been developed that parallels exactly the physical units discussed earlier. This is shown in Table 4.1. The actual equivalence between physical and psychophysical units depends on several variables: the conditions of observation, the age and experience of the observer, where the light is falling on the retina of the observer's eye, and the wavelengths present in the light.

1. Definitions. The unit of *luminous flux* Φ_v is the lumen, and that of *luminous intensity* I_v is the candle or lumen/steradian. One candle (1 cd) corresponds to a flux of one lumen (1 lm) through a solid angle of one steradian (1 sr). Hence an isotropic source with an intensity of 1 cd will radiate a total luminous flux of 4π lm.

Luminous exitance and *illuminance* flux density (in lm/m² or lux) are quantities representing luminous flux density. Lux are also called meter-candles. Other units are the foot-candle (lm/ft²) and the phot (lm/cm²):

$$1\text{-ft-cd} = 10.764 \text{ m-cd (lux)} = 1.0764 \times 10^{-3} \text{ phot}$$

$$1 \text{ phot} = 10^4 \text{ lux}$$

Like its physical counterpart, the radiance L_e, the *luminance* L_v [in lm/(m² sr)] is flux per unit solid angle per unit projected area of source. Since 1 lm/sr = 1 cd, the mks units of L_v are also cd/m². 1 cd/m² is sometimes called a nit. Other units are the *stilb* or lm/(cm² sr) or cd/cm² $= 10^4$ cd/m²; the *lambert* $= (1/\pi)$ cd/cm² $= (10^4/\pi)$ cd/m²; the *millilambert* $= (1/1000\ \pi)$ cd/cm² $= (10/\pi)$ cd/m²; and the *foot-lambert* $= 0.0003426$ cd/cm² $= 3.426$ cd/m².

Table 4.1. Correspondence of Radiometric
and Photometric Nomenclature

Radiometry	Physical Symbol	mks Units
Radiant energy	Q_e	joule
Radiant density	U_e	J/m^3
Radiant flux	Φ_e	Watt
Radiant exitance	M_e	W/m^2
Irradiance	E_e	W/m^2
Radiant intensity	I_e	W/sr
Radiance	L_e	$W/sr\text{-}m^2$

Photometry	Psychophysical Symbol	mks Units
Luminous energy	Q_v	talbot
Luminous density	U_v	$talbot/m^3$
Luminous flux	Φ_v	lumen
Luminous exitance	M_v	lm/m^2
Illuminance	E_v	lm/m^2 (lux)
Luminous intensity	I_v	lm/sr (candle)
Luminance	L_v	$lm/sr\text{-}m^2$

The *primary standard* of the photometric system of units is the luminous exitance of the surface of a black body radiator at the freezing temperature of platinum (2043.5°K), which is *defined* to be $60 \text{ cd/cm}^2 = 60 \text{ stilb} = 60 \text{ lm/(cm}^2 \text{ sr)}$.

2. Conversion Between Luminous and Radiant Energy Units. The most important variable in the conversion is the wavelength present in the light. Sources of equal spectral radiance will appear to an observer to have different psychophysical spectral luminance as the wavelength varies. Results of experiments with many observers have been combined to yield the *standard luminosity curve* shown in Fig. 4.19. This curve is a function of λ, which we will denote y_λ; it has a peak value of unity at $\lambda = 555$ nm. This means that light of wavelength λ gives a factor of y_λ as much psychological sensation as light of wavelength 555 nm. The conversion at $\lambda = 555$ nm is standardized to be

$$K_m = 680 \text{ lm/W} \tag{4.23}$$

This means that 1 W of flux at 555 nm gives the same physical sensation as 680 lm. For other wavelengths the conversion factor is

$$K = K_m y_\lambda = 680 y_\lambda \text{ lm/W} \tag{4.24}$$

that is, for $\lambda \neq 555$ nm, 1 W produces less than 680 lm.

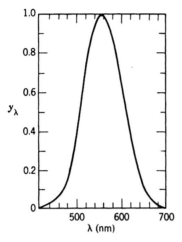

Fig. 4.19 Standard luminosity curve.

We may use Eq. (4.24) to convert any spectral physical quantity in Table 4.1 to the corresponding psychophysical quantity. For instance, if we have a spectral flux $\Phi_e(\lambda)$, then the luminous flux is given by

$$\Phi_v = 680 \int_0^\infty y_\lambda \Phi_e(\lambda)\, d\lambda \ [\text{lm}] \tag{4.25}$$

The average conversion factor $\bar{K}$ is defined by

$$\Phi_v = \bar{K}\Phi_e \tag{4.26}$$

where

$$\Phi_e = \int \Phi_e(\lambda)\, d\lambda$$

$\bar{K}$ is given by

$$\bar{K} = \frac{680 \int_0^\infty y_\lambda \Phi_e(\lambda)\, d\lambda}{\int_0^\infty \Phi_e(\lambda)\, d\lambda} \tag{4.27}$$

If the intensity from the same source is desired, the conversion would be

$$I_v = 680 \int_0^\infty y_\lambda I_e(\lambda)\, d\lambda \ [\text{cd}] = \bar{K} I_e \tag{4.28}$$

where

$$I_e = \int_0^\infty I_e(\lambda)\, d\lambda$$

and

$$\bar{K} = \frac{680 \int_0^\infty y_\lambda I_e(\lambda)\, d\lambda}{\int_0^\infty I_e(\lambda)\, d\lambda} \left[\frac{\text{cd}}{\text{W/sr}}\right] \tag{4.29}$$

Equations (4.27) and (4.29) are equivalent, because for a given source and given geometry, $I_e(\lambda)$ will be proportional to $\Phi_e(\lambda)$.

The luminous incidence from the sun at noon at the equator is about 10^5 lux, which is 500,000 times greater than that from the full moon under similar circumstances.

The luminous flux density from a "60-W" light is about 50 lux = 50 lm/m^2 at a distance of 1 m. The total flux would then be 50 lm/m^2 × 4π m^2 = 600 lm or about 1 W visible radiant flux.

A fully adapted human eye can see about 10^{-9} lux. If we assume that the pupil of the eye is 2 mm in radius, then the area of the eye receiving this light is $\pi \cdot 2^2 \cdot 10^{-6}$ m^2 $\sim 13 \times 10^{-6}$ m^2, which means that about 10^{-14} lm can be detected by a human eye. This amounts to about 1.5×10^{-17} W of visible power or about 44 phot/sec.

A photomultiplier tube has a sensitivity as limited by "dark current" at room temperature of about 5×10^{-10} lm. This can be reduced to as low as 10^{-15} lm by cooling the tube.

A reasonably fast photographic film (ASA 400) will develop a very noticeable gray area if exposed to 10^{-2} lux for 1 s. If the resolution amounted to 30 lines/mm, then we could distinguish the darkening in an area of $1[(1/30) \cdot 10^{-3}]^2 \approx 10^{-9}$ m^2 produced in 1 sec by $10^{-2} \cdot 10^{-9} = 10^{-11}$ lm of light.

C. Examples of Radiometry

We now drop the subscripts e and v so that discussion may be applied equally well to physical or psychophysical quantities. We use radiant terms in what follows, but the relationships are valid for photometric terms and luminous terms.

1. Simple Sources. In most cases, calculations involve complicated angular and surface integrals. Sometimes the geometry is particularly simple. Further simplication comes from the assumption that the sources behave like Lambert sources. We wish to calculate the irradiance E at a point P a distance R from the center of the source.

a. Disk. The geometry for a disk is shown in Fig. 4.20. Consider a ring-shaped element of the source of area $d\mathscr{A}_s$ that is equal to

$$d\mathscr{A}_s = 2\pi r\, dr = \pi d(r^2) = \pi d(\rho^2)$$

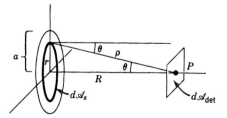

Fig. 4.20 Geometry associated with a disk-shaped source.

where $\rho^2 = R^2 + r^2$ is taken to be the dependent variable in the integration that will cover the entire disk. The flux reaching area $d\mathscr{A}_{det}$ on the detector at P that comes from this ring will be [Eq. (4.12)]

$$d^2\Phi = Ld\Omega \, d\mathscr{A}_s \cos\theta$$

Here $d\Omega$ is the solid angle that is subtended by the detector area $d\mathscr{A}_{det}$ at a *point* on the source ring. This is determined from Eq. (4.10) with $d\mathscr{A}' = d\mathscr{A}_{det} \cos\theta$, the projected detector area; $d\Omega = d\mathscr{A}_{det} \cos\theta/\rho^2$. The flux is, therefore

$$d^2\Phi = \pi Ld\mathscr{A}_{det} \frac{\cos^2\theta}{\rho^2} d\rho^2$$

but

$$\cos^2\theta = \frac{R^2}{\rho^2}$$

Thus the total flux reaching $d\mathscr{A}_{det}$ is

$$d\Phi = \pi Ld\mathscr{A}_{det} R^2 \int_{R^2}^{R^2+a^2} \frac{d\rho^2}{\rho^4} = \pi Ld\mathscr{A}_{det} R^2 \left(\frac{1}{R^2} - \frac{1}{R^2+a^2}\right)$$

The irradiance on the detector is

$$E = \frac{d\Phi}{d\mathscr{A}_{det}} = \pi L \left(1 - \frac{1}{1 + \dfrac{a^2}{R^2}}\right) = \pi L \sin^2\theta_{max}$$

$$\left(\sin\theta_{max} = \frac{a}{\sqrt{a^2+R^2}}\right) \tag{4.30}$$

Note that as $a \to \infty$ for fixed R, E tends to πL, the analog of Eq. (4.20). Note also that for fixed a and large R, E becomes

$$E \approx \frac{\pi La^2}{R^2} = \frac{I}{R^2} \tag{4.31}$$

where $I = L \cdot \pi a^2$ is the intensity (flux/sr) of the source when emitting in a direction normal to its plane. The last result holds quite generally a large distance from a small source because then we can write

$$d\Phi = I \, d\Omega, \quad \text{with } d\Omega = \frac{d\mathscr{A}_{det}}{R^2}$$

Hence

$$d\Phi = \frac{I}{R^2} d\mathscr{A}_{det}$$

or

$$E = \frac{d\Phi}{d\mathscr{A}_{det}} = \frac{I}{R^2} \tag{4.32}$$

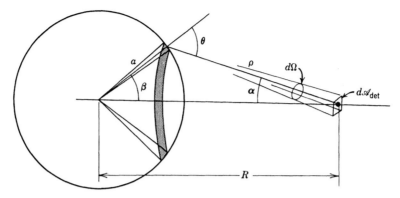

Fig. 4.21 Geometry associated with spherical Lambert source.

b. Sphere. The situation for a sphere is shown in Fig. 4.21. In this case we identify an annular ring as the differential element of the source. The flux on the detector will be

$$d^2\Phi = L d\Omega \, d\mathcal{A}_s \cos\theta$$

where

$$d\mathcal{A}_s = 2\pi(a \sin\beta)a \, d\beta$$

and

$$d\Omega = d\mathcal{A}_{det}(\cos\alpha)/\rho^2$$

Assume now that $R \gg a$. Then $\alpha \simeq 0$, $\rho \simeq R$, and $\theta \simeq \beta$. This leads to

$$d^2\Phi \simeq L\frac{2\pi a^2}{R^2} d\mathcal{A}_{det} \sin\beta \cos\beta \, d\beta$$

the total flux reaching $d\mathcal{A}_{det}$ is

$$d\Phi = L\frac{2\pi a^2}{R^2} d\mathcal{A}_{det} \int_0^{\pi/2} \sin\beta \cos\beta \, d\beta = L\frac{\pi a^2}{R^2} d\mathcal{A}_{det}$$

and

$$E = \frac{d\Phi}{d\mathcal{A}_{det}} = \frac{L}{R^2}\pi a^2 \tag{4.33}$$

This is the same result as for Eq. (4.31) for the disk of radius a. Thus a spherical Lambert emitter will appear to be a uniform disk to a distant observer.

2. Image Brightness. We are now in a position to compute the radiance at various places inside an optical system, particularly at locations of final or intermediate images.

a. Conservation of Radiance. At points that are conjugate with the source, the radiance obeys a fundamental conservation law. This idea has already been

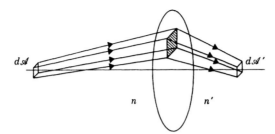

Fig. 4.22 Elementary beam passing through an optical system.

mentioned in the paraxial optics section in the form of the Lagrange relation, Eq. (3.31). There we found $nx\theta = n'x'\theta'$. Large linear magnification can only be obtained at the expense of large demagnification of the ray angles. The generalization of this concept to larger angles outside of the paraxial limit leads to what is known as the *sine condition*

$$n'x' \sin \theta' = nx \sin \theta \tag{4.34}$$

This will be justified in section 4.3A4.

Here we wish to apply the sine condition to an *elementary beam* that originates on a small part of the source, passes through a portion of the optical system, and produces a portion of the image (Fig. 4.22). To understand the elementary beam, consider an elementary pencil of rays that originate on a point of the source and that subtend an infinitesimal solid angle $d\Omega = \sin \theta \, d\theta \, d\varphi$. The geometry is pictured in Fig. 4.23, where the pencil makes the angle θ with the optical axis of the focusing system. The azimuthal angle φ will remain constant for the pencil as long as it passes through optical elements that are symmetrical about the main axis. An elementary beam consists of all such pencils passing through the area element $d\mathscr{A}$ that is located perpendicular to the axis of the system. The flux in the elementary beam is given by Eq. (4.12) as

$$d^2\Phi = L \cos \theta \, d\mathscr{A} \, d\Omega = L \, d\mathscr{A} \sin \theta \cos \theta \, d\theta \, d\varphi$$
$$= \tfrac{1}{2} L \, d\mathscr{A} \, d\varphi \, d(\sin^2 \theta)$$

We follow the elementary beam until it forms an image as in Fig. 4.22. If no

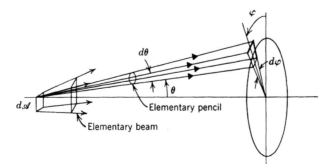

Fig. 4.23 Elementary pencil of rays—part of an elementary beam.

reflection or other losses are allowed, then the beam will have the same flux $d^2\Phi' = d^2\Phi$. This means

$$L' \, d\mathscr{A}' \, d\varphi' \, d(\sin^2 \theta') = L \, d\mathscr{A} \, d\varphi \, d(\sin^2 \theta) \tag{4.35}$$

where the primes refer to the image.

Because azimuthal angles are preserved, we must have $d\varphi' = d\varphi$, thus

$$\frac{L'}{L} = \frac{d\mathscr{A}}{d\mathscr{A}'} \frac{d(\sin^2 \theta)}{d(\sin^2 \theta')} \tag{4.36}$$

We now return to the sine condition that, when squared, becomes

$$n'^2 x'^2 \sin^2 \theta' = n^2 x^2 \sin^2 \theta$$

Let the elementary source and image areas be x^2 and x'^2, respectively, so that the sine condition leads to

$$n'^2 \, d\mathscr{A}' \sin^2 \theta' = n^2 \, d\mathscr{A} \sin^2 \theta$$

We need to consider a range of angles θ in the elementary beam, so we take the differential of both sides

$$n'^2 \, d\mathscr{A}' d(\sin^2 \theta') = n^2 \, d\mathscr{A} d(\sin^2 \theta) \tag{4.37}$$

Equation (4.37) can be combined with Eq. (4.36) to yield

$$\frac{L'}{L} = \frac{n'^2}{n^2} \tag{4.38}$$

If the two indices of refraction are the same, then we obtain the resulting conservation law

$$L' = L \tag{4.39}$$

We have shown that no amount of focusing can increase the radiance of each elementary beam from object to image. A reduced image will have more irradiance (flux per unit area), but the light rays will subtend a greater solid angle at the image than at the object.

b. Irradiance. The radiant or luminous power per unit area at the image will be the integral over the contributions of each elementary beam. A conical beam of half-angle θ' will contribute

$$L' \cos \theta' \, d\Omega' = 2\pi L' \cos \theta' \sin \theta' \, d\theta'$$

$$= \pi L' d (\sin^2 \theta')$$

to the flux per unit area at the image. We integrate this expression from $\theta' = 0$ to $\theta' = \alpha'$, where α' is the half-angle subtended by the exit pupil at the image (Fig. 4.24). Using $L' = (n'/n)^2 L$ from Eq. (4.38) and assuming L to be independent of θ (Lambert source), we obtain

$$E' = \left(\frac{n'}{n}\right)^2 L\pi \sin^2 \alpha' \tag{4.40}$$

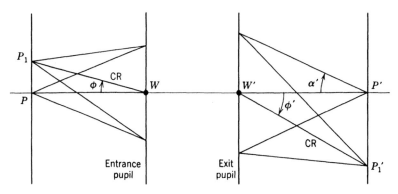

Fig. 4.24

The total solid angle subtended by the exit pupil at the image can be written

$$\Omega' = 2\pi \int_0^{\alpha'} \sin\theta \, d\theta = 2\pi(1 - \cos\alpha') = 4\pi \sin^2\frac{\alpha'}{2} \qquad (4.41)$$

For small angles α' we can approximate $\sin^2\alpha' \approx 4\sin^2(\alpha'/2)$. Then Eq. (4.40) becomes

$$E' \approx \left(\frac{n'}{n}\right)^2 L\Omega' \qquad (4.42)$$

    c. Off-Axis Behavior. An off-axis point such as P_1 in Fig. 4.24 will be imaged at P_1'. If there is no vignetting, the limiting solid angle of rays reaching P_1' will be determined by the exit pupil and will be approximately

$$\frac{\sigma_\perp'}{|\overline{W'P_1'}|^2} = \frac{\sigma'\cos\phi'}{|\overline{W'P_1'}|^2} = \frac{\sigma'\cos^3\phi'}{|\overline{W'P'}|^2} \approx \Omega'\cos^3\phi'$$

where σ' is the area of the exit pupil, $\sigma_\perp'$ is its projection perpendicular to $\overline{W'P_1'}$, and $\Omega' = 4\pi\sin^2(\alpha'/2)$ is the solid angle as seen from P'. In this case of no vignetting, as long as ϕ is less than a maximum value determined by the field stop (not shown in Fig. 4.24), all elementary beams leaving P_1 will reach P_1'. Furthermore, it can be shown that for these beams, Eq. (4.38) still holds if the system obeys the sine condition. Thus we can still use $(n'/n)^2 L$ in the off-axis version of Eq. (4.42).

    If we want the flux density at P_1' falling on a screen perpendicular to the axis, we must insert another factor of $\cos\phi'$ to account for the orientation of the screen with respect to the chief ray. Thus for the flux density at the off-axis point P_1' we obtain

$$E' \text{ (off-axis)} \cong \left(\frac{n'}{n}\right)^2 L\Omega'\cos^4\phi' \qquad (4.43)$$

This represents a fairly rapid decrease with ϕ'. The flux density is even less if vignetting is present.

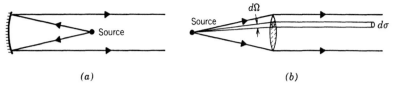

Fig. 4.25 Searchlight optics: (*a*) reflection collimating; (*b*) refraction collimating.

3. Instruments. Illumination is important in any optical instrument. However, in some, the efficient transmission of light is the essence of the device.

a. Searchlight. A searchlight is not the simple optical device that it appears to be at first sight. As shown in Fig. 4.25, a searchlight consists essentially of a small intense source placed at the focal point of a reflecting mirror or a lens. If the source were truly a point, then a parabolic mirror or a corrected lens would give a perfectly collimated parallel beam having a flux density that would be uniform, at least for small angular apertures. To see this, it is only necessary to realize that a pencil of rays from the source making a cone of solid angle $d\Omega = d\alpha/f^2$ can be found to illuminate area $d\sigma$, giving a constant flux density

$$E = E_0 = \frac{I}{f^2} \tag{4.44}$$

This result no longer holds with a source of finite size. For simplicity we let the source be a disk-shaped Lambert emitter of radius r_s. This disk is imaged by the lens at infinity to the left of the lens, where it subtends a half-angle $\delta = r_s/f$ (Fig. 4.26). Pick an observation point P to the right of the lens. The line from the center of the image at infinity to P will remain parallel to the axis as P moves, but the rays from the image to P must pass through the lens. If P is off-axis, they may not all be able to do so (Fig. 4.27).

It is useful to construct cones from P to the lens and from P to the image and project them onto a plane at a distance f to the left of P. The intersection of the cone to the image gives a disk of radius r_s, and the intersection of the cone to the lens

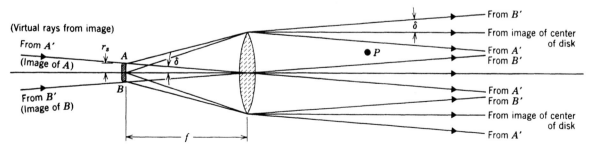

Fig. 4.26 Marginal rays from a disk-shaped Lambert source in the searchlight.

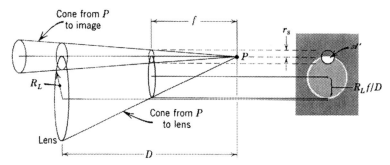

Fig. 4.27 Demonstration of the mechanism whereby a finite size source leads to uneven illumination.

gives a slightly elliptical disk of radius $R_L f/D$. The intersection of these two disks represents the image as seen from P. The area in the intersection equals $\mathscr{A}'$, the area of that part of the source that can radiate through the lens and reach P. The intersections of these cones are shown in the little figures at the bottom of Fig. 4.29.

The flux reaching area $d\sigma$ around P from area $d\mathscr{A}_s$ around a point P_s at the source is seen from Fig. 4.28 to be

$$d^2\Phi = L\, d\mathscr{A}_s\, d\Omega = \frac{L\, d\mathscr{A}_s\, d\sigma}{f^2}$$

where L is the radiance of the source. Here we assume the normal to $d\sigma$ to be along the axis and neglect the distinction between $d\sigma$ and its perpendicular projection $d\sigma_\perp$. By integrating over the area $\mathscr{A}'$ of the source that can be "seen" from P, we may calculate the total flux through $d\sigma$:

$$d\Phi = \frac{L\mathscr{A}'\, d\sigma}{f^2}$$

Then the flux density $d\Phi/d\sigma$ is

$$E = \frac{L\mathscr{A}'}{f^2} \tag{4.45}$$

Within the bright cone to the left of d in Fig. 4.29 the image subtends a smaller

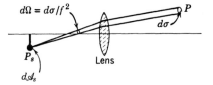

Fig. 4.28 Geometry associated with the flux from P_s to d around the observation point P.

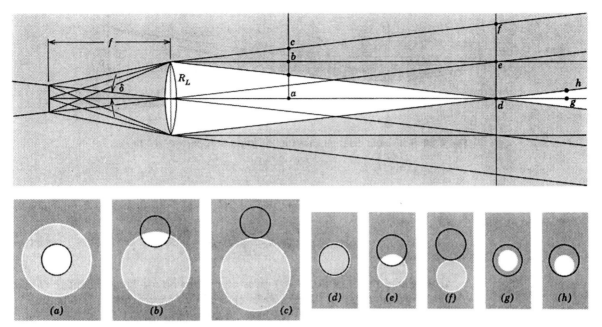

Fig. 4.29 Analysis of the illumination produced by the searchlight. The important consideration is the projection of the cones subtended by the lens and the image.

angle than does the lens, so that $r_s < R_L f / D$. The projected disk from the image is then completely inside the projected disk from the lens, so that $\mathscr{A}' = \mathscr{A}_s$. This will be the case at point a. Then

$$E = E_0 = \frac{L\mathscr{A}_s}{f^2} = \frac{\pi L r_s^2}{f^2} \tag{4.46}$$

which is a constant. Such a constant flux density is expected in a searchlight beam with an infinitely small source. The length of the shaded cone from the lens to point d is $R_L f / r_s$, that is, inversely proportional to the size of the source.

When P moves laterally out of the shaded cone from a to b or c, $\mathscr{A}'$ becomes less than $\mathscr{A}_s$ and E drops from the constant value of E_0 of Eq. (4.26).

At the point d the projections of lens and image coincide, giving $R_L f / D = r_s$. When P is to the right of d, the projection of the lens is smaller than that of the image. When P remains in the bright cone to the right of d, the lens projection is entirely inside the image projection, and the common area is

$$\mathscr{A}' = \mathscr{A} \text{ (lens projection)} = \pi \left(\frac{R_L f}{D} \right)^2$$

We may summarize these results by saying that when P is anywhere in the bright cone of Fig. 4.27, the flux density is given by

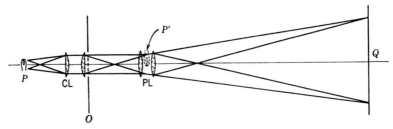

Fig. 4.30 Schematic diagram showing the important parts of the projector.

$$E = E_0 = \frac{L\pi r_s^2}{f^2}, \quad D < \frac{R_L f}{r_s}$$

$$E = \frac{LA'}{f^2} = \frac{L\pi R_L^2}{D^2}, \quad D \geq \frac{R_L f}{r_s} \tag{4.47}$$

The second result in Eq. (4.47) gives an expression for E equivalent to that obtained by assuming that the lens were replaced by a source disk with radiance L and area πR_L^2. This disk would then have an intensity

$$I = L\pi R_L^2 \tag{4.48}$$

This is what is meant by the "intensity of a searchlight." It is usually expressed in candles (or "candle power," which means the same thing).

b. The Projector. A slide projector or similar instrument has two lenses and performs two simultaneous image-formation processes. The projection lens PL images the semitransparent object O onto the screen Q (Fig. 4.30). A properly used condenser lens CL will assure that the object is uniformly illuminated with an intense beam. The condenser forms an image P' of the source P at the projection lens PL. Except for transmission losses, this image will have the same radiance as the projection lamp and will uniformly illuminate those parts of the screen (except for the $\cos^4 \phi'$ factor) that correspond to the transparent parts of the object O.

4.3 Lens Aberrations

The ray-tracing equations used in the theory of paraxial optics were correct to first order in the inclination angles of the rays and normals to refracting or reflecting surfaces. When higher-order approximations are used for the trigonometric functions of the angles, departures from the predictions of paraxial optics will be found. No longer will it be generally true that all the rays leaving a point object will exactly meet to form a point image or that the magnification in a given transverse plane is a constant. In addition, the properties of the system may be wavelength-dependent. Such deviations from ideal paraxial behavior are known as *lens aberrations*.

Monochromatic aberrations occur with light of a fixed wavelength. They may be treated mathematically in lowest order by carrying out the ray-tracing calculations to third order in the angles or, equivalently, by carrying out the calculation of the optical path difference to fourth order. The resulting "third-order theory" is itself valid only for small angles, and for many real systems calculations must be carried out to still higher order, say fifth or seventh. For a centered system with rotational symmetry, only an odd combination of the powers of the angles will appear in ray-tracing formulas, and only an even combination of powers will appear in the expression for the optical path difference. Because of their great complexity, these higher-order aberrations are usually treated numerically.

Most compound-lens systems contain enough degrees of freedom in their design so that if the third-order theory were exact, all aberrations could be eliminated. For real systems the residual higher-order aberrations would still be present, and there are not enough design parameters to eliminate all of them as well. An experienced optical designer strives for a balance between third-order and higher aberrations to give optimum performance with a given system. The performance must be judged according to the intended use of the instrument. The criteria for a telescope objective and for a camera lens for close-ups are quite different.

To go into the subject of lens aberrations in any detail would take us too far into the technicalities of optical design. Instead, we will merely describe the characteristic features of each type of aberration using the form of the third-order theory and insert occasional comments about the methods of correcting these aberrations.

A. Monochromatic Aberrations

We first consider those deviations from paraxial theory that can be identified when the source or object gives rise to light with a single wavelength. The refractive characteristics of the optical system are then contained in the geometry of the surfaces and the appropriate indices of refraction. These are constants. A given ray from a point source will have a unique path through the system to the object. If the system is to form an image, then other rays that start out at different angles at the point source must arrive at the same image point, having traveled identical optical path lengths. If the lens surfaces are spherical sections, this will not be possible. We will characterize the monochromatic aberrations by examining the deviation of the optical path length from what would be expected in the paraxial theory. This will allow us to quantify the influence that aberrations have on the form and location of the image.

This will be done for a point source. It should be understood that an extended object can be considered as a collection of points with different intensities. If the various positions on the object are incoherent, then the resulting image will be the sum of the individual images of the component point sources. If the source is coherent, then diffraction theory is a more appropriate mechanism for the discussion of lens performance. This will be discussed in Chapter 8.

1. Single-Refracting Surface. To simplify our discussion of lenses, we consider first the aberrations of a single spherical refracting interface between media of indices n and n' on the incident and the transmitting sides of the interface, respectively.

a. Optical Path Length. Figure 4.31 describes the situation. This is similar to Fig. 3.12, only now the object point is on the optical axis. We will move away from the optical axis later. Point P'_0 is the paraxial image of point P; P_S is the intercept of a general ray with the spherical refracting surface. We investigate the optical path lengths (OPL's) along $PP_SP'_0$ and compare it to the paraxial path PVP'_0. In so doing, we define the optical path length difference

$$I = (n\overline{PP_S} + n'\overline{P_SP'_0}) - (n\overline{PV} + n'\overline{VP'_0}) \tag{4.49}$$

From the geometry of Fig. 4.31, we identify

$$\overline{PP_S} = [(S + \Delta z)^2 + \rho^2]^{1/2} \tag{4.50a}$$

$$\overline{P_SP'_0} = [(S' - \Delta z)^2 + \rho^2]^{1/2} \tag{4.50b}$$

$$\overline{PV} = S \tag{4.50c}$$

$$\overline{VP'_0} = S' \tag{4.50d}$$

In Chapter 3, when we made the paraxial approximation, we assumed that P_s was located in the same plane (perpendicular to the optical axis) as is the vertex V. This is equivalent to the assumption that Δz is zero. Here we keep the size of Δz small but nonzero. By necessity then, ρ will also be small, although not as small as Δz. These restrictions still permit considerable simplification of Eqs. (4.50a and b). We require,

$$\Delta z \lll R, S, S'$$

and

$$\rho \ll R, S, S' \tag{4.51}$$

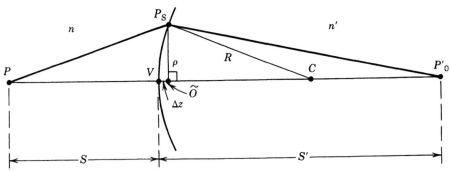

Fig. 4.31 Refraction from a point source on the optical axis through a spherical interface. Rays stay in the plane of the drawing in this example.

We proceed by expanding Eqs. (4.50a and b), neglecting the term with the highest power of Δz.

$$\overline{PP_S} \cong [S^2 + 2S \Delta z + \rho^2]^{1/2} \tag{4.52a}$$

$$\overline{P_S P_0'} \cong [S'^2 - 2S' \Delta z + \rho^2]^{1/2} \tag{4.52b}$$

We can eliminate ρ from this by using the triangle $P_S C \tilde{O}$, from which we get

$$R^2 = \rho^2 + (R - \Delta z)^2$$

Again, neglecting the term with the highest power of Δz leads to

$$\Delta z \cong \frac{\rho^2}{2R} \tag{4.53}$$

With this result in Eqs. (4.52)

$$\overline{PP_S} \cong \left[S^2 + S \frac{\rho^2}{R} + \rho^2 \right]^{1/2}$$

$$\overline{P_S P_0'} \cong \left[S'^2 - S' \frac{\rho^2}{R} + \rho^2 \right]^{1/2}$$

or

$$\overline{PP_S} \cong S \left[1 + \frac{\rho^2}{S} \left(\frac{1}{S} + \frac{1}{R} \right) \right]^{1/2} \tag{4.54a}$$

$$\overline{P_S P_0'} \cong S' \left[1 + \frac{\rho^2}{S'} \left(\frac{1}{S'} - \frac{1}{R} \right) \right]^{1/2} \tag{4.54b}$$

To simplify the square root, we use the Taylor series expansion of $(1 + \varepsilon)^{1/2}$,

$$(1 + \varepsilon)^{1/2} = 1 + \frac{\varepsilon}{2} - \frac{\varepsilon^2}{8} + \cdots$$

When $\varepsilon \ll 1$, the higher-order terms may be neglected. Applying this to Eqs. (4.54) yields

$$\overline{PP_S} \cong S \left[1 + \frac{\rho^2}{2S} \left(\frac{1}{S} + \frac{1}{R} \right) - \frac{\rho^4}{8S^2} \left(\frac{1}{S} + \frac{1}{R} \right)^2 + \cdots \right]$$

or

$$\overline{PP_S} \cong S + \frac{\rho^2}{2} \left(\frac{1}{S} + \frac{1}{R} \right) - \frac{\rho^4}{8S} \left(\frac{1}{S} + \frac{1}{R} \right)^2 + \cdots \tag{4.55a}$$

and

$$\overline{P_S P_0'} \cong S' + \frac{\rho^2}{2} \left(\frac{1}{S'} - \frac{1}{R} \right) - \frac{\rho^4}{8S'} \left(\frac{1}{S'} - \frac{1}{R} \right)^2 + \cdots \tag{4.55b}$$

Equations (4.55) are now functions of ρ. The OPL difference (I) will also be a function of ρ. When we combine Eqs. (4.55) and Eqs. (4.50c and d) with Eq. (4.49) for the OPL difference we get

$$I(\rho) \cong \frac{\rho^2}{2}\left[\frac{n}{S} + \frac{n'}{S'} - \left(\frac{n'-n}{R}\right)\right]$$

$$-\frac{\rho^4}{8}\left\{\left[n\left(\frac{1}{S} + \frac{1}{R}\right)\right]^2\frac{1}{nS} + \left[n'\left(\frac{1}{S} - \frac{1}{R}\right)\right]^2\frac{1}{n'S'}\right\} + \cdots \qquad (4.56)$$

The first observation we can make from this analysis is that the OPL difference is a second-order function of ρ. This is to be expected on the basis of Fermat's principle applied to path PVP'_0. Since ρ measures the deviation from that path, by Eq. (1.5) the OPL difference must vary according to ρ^2. Path $PP_SP'_0$ is not an arbitrary virtual path, however. We recognize this by recalling what we know about paraxial conjugate points P and P'_0. According to Eq. (3.26) the distances S and S' are related by

$$\frac{n}{S} + \frac{n'}{S'} = \frac{n'-n}{R}$$

We now can see that this condition makes the term proportional to ρ^2 in Eq. (4.56) vanish. OPL $PP_SP'_0$ is more nearly equal to OPL PVP'_0 than would be required by an arbitrary virtual path obeying Fermat's principle. This is true because this system satisfies the conditions for *approximate image formation*. The characterization of the deviation from exact image formation forms the basis of our study of aberrations.

The lowest-order nonzero contribution to the OPL difference for an on-axis object point is the term proportional to ρ^4 in Eq. (4.56). The *on-axis aberration function* is therefore

$$I(\rho) = -\frac{C\rho^4}{4} + \cdots \qquad (4.57a)$$

where

$$C \equiv \frac{1}{2}\left\{\left[n\left(\frac{1}{S} + \frac{1}{R}\right)\right]^2\frac{1}{nS} + \left[n'\left(\frac{1}{S'} - \frac{1}{R}\right)\right]^2\frac{1}{n'S'}\right\} \qquad (4.57b)$$

b. Reference Sphere. A useful way of looking at aberrations is by comparing the actual wave surface with an ideal reference surface. Consider Fig. 4.32a. Here σ represents a spherical wave surface that originated at point source P. If perfect imaging occurred, then, on the other side of the interface, we would need to have another spherical wave surface that converged on the ideal image point P'_0. We identify this ideal reference surface as σ_R. Note that the reference surface intersects the optical axis at O.

Because the refracting surface is a spherical section (with its center at C), the true wave surface will not coincide with σ_R. This situation is shown in Fig. 4.32b. Here the true wave surface that meets the optical axis at O is identified as σ'. The

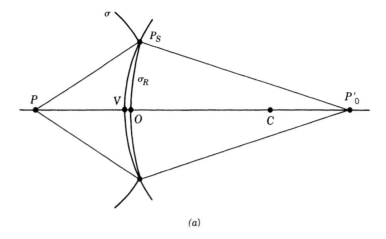

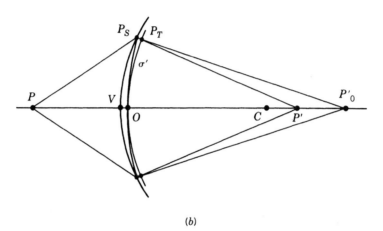

Fig. 4.32 Concept of the reference surface. (*a*) Ideal spherical reference surface that converges to the ideal image point P'_0. (*b*) The true wavefront is not spherical. A normal at P_T intersects the optical axis at P'.

rays that are perpendicular to σ' will not meet P'_0. Provided the system is cylindrically symmetrical (as in this illustration), there will be a cone of equal OPL rays that coincide at, for instance, P'. The distance between P'_0 and P' will be related to the difference between the true wave surface and the reference surface. As can be seen from the figure, this difference depends on the distance away from the optical axis.

The directed distance $\overrightarrow{P_S P_T}$ is closely related to the aberration function. To see how, reconsider the on-axis aberration function for Fig. 4.32.

$$I(\rho) = \text{OPL}(PP_S P_T P'_0) - \text{OPL}(PVOP'_0)$$

$$= \text{OPL}(PP_S P_T) + \text{OPL}(P_T P'_0) - \text{OPL}(PVO) - \text{OPL}(OP'_0)$$

Because P_T and O are on the same constant phase wave surface, the optical path lengths from P to P_T and from P to O must be equal. The first and third terms above cancel. This leaves

$$I(\rho) = \text{OPL}(P_T P_0') - \text{OPL}(O P_0')$$

$$= n'[\overline{P_T P_0'} - \overline{O P_0'}]$$

$$= n'[\overline{P_S P_0'} - \overrightarrow{P_S P_T} - \overline{O P_0'}]$$

Now because both P_s and O are on the same spherical reference surface, their distances from P_0' must be equal. The on-axis aberration function is then

$$I(\rho) = -n'\overrightarrow{P_S P_T} = -\frac{C\rho^4}{4} \tag{4.58}$$

We will use the directed distance $\overrightarrow{P_S P_T}$, the separation of the true wave surface from the spherical reference surface at the refracting interface, to examine the effect of aberrations in the vicinity of the paraxial image.

c. Off-Axis Object. If the object point lies off the optical axis, new types of aberrations appear in addition to that described by Eq. (4.58). Consider Fig. 4.33a. The optical axis is the line through V and C. We assume that the refracting surface is limited in extent such that the surface is symmetrical about the optical axis.

The point object is at P, and the paraxial image is at P_0'. Both these points are still in the plane of the figure. If there were no limitation to the refracting surface, then ray $PUCP_0'$ through the center of curvature at C would be the natural optical axis. Paraxial imaging would occur for a small pencil of rays centered about this undeviated ray (UR).

In Fig. 4.33b we identify a general ray that meets the refracting surface at P_S. If P_S lies in the same plane as P and the optical axis, then it is in the *meridional plane*. All of our previous work was restricted to the analysis of meridional rays. Here we relax this restriction and allow P_S to be away from the meridional plane. A ray that follows this type of general path is called a *skew ray* (SR).

In describing the aberrations associated with the skew ray, we could follow our established method of identifying a spherical reference surface centered on P_0' and meeting the refracting surface at P_S. Deviations of the true wavefront from the reference surface would be described by the aberration function

$$I(\rho_u) = -\frac{C\rho_u^4}{4}$$

where ρ_u, as in Eq. (4.57), is the distance of P_S from the undeviated ray (see Fig. 4.33c).

With the limitation provided by a stop, here represented by the limited extent of the refracting surface, it is more natural to refer all rays to the chief ray (CR), whose extension passes through the center of the exit pupil (Fig. 4.33d), and which meets the refracting surface at P_C. Introduction of the exit-pupil plane gives us the

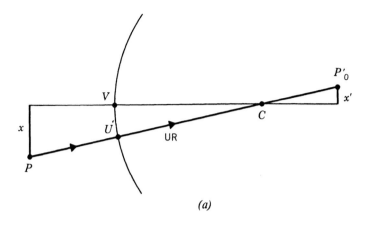

(a)

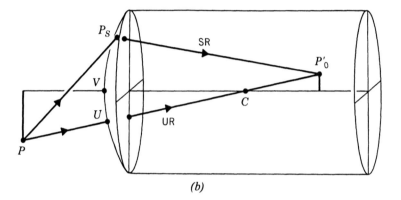

(b)

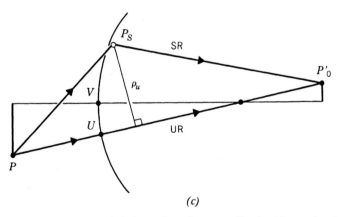

(c)

Fig. 4.33 Geometry for aberrations from an off-axis object point. (*a*) The undeviated ray. (*b*) A skew ray. (*c*) Radial parameter associated with the aberration function. The skew ray is actually not in the plane of the drawing, although it appears that way here for clarity.

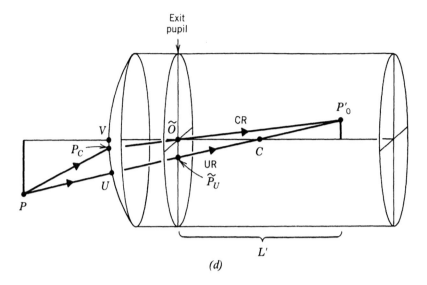

(d)

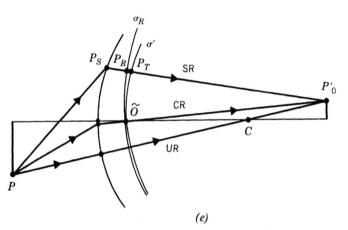

(e)

Fig. 4.33 *(continued)* (*d*) Exit pupil that is the plane for a polar coordinate system useful in the specification of rays (Fig. 4.34). (*e*) Wavefront differences for a skew ray near the exit pupil plane.

freedom to describe aberrations of general optical systems. For a single-refracting surface limited in extent by an aperture at the surface or for a simple thin lens limited by its own diameter, the exit pupil nearly coincides with the refracting element and the length L' is the same as the image distance S'.

We now must learn to describe the aberrations associated with the skew ray by referring to the chief ray. The chief ray itself will contribute to aberrations at P'_0, because, unlike in the on-axis object situation, the chief ray is not the same as the undeviated ray. In the present example the quantity of interest is the off-axis aberration function

$$W = I_{\text{skew}} - I_{\text{chief}} = \text{OPL}(PP_S P_0') - \text{OPL}(PP_C P_0') \tag{4.59}$$

To demonstrate this, consider a point P_T on the true wave surface σ' in the vicinity of P_S. The situation is shown in Fig. 4.33e (for a meridional ray). Let this point lie on the particular surface, which also passes through $\tilde{O}$, the center of the exit pupil. We must have $\text{OPL}(PP_T) = \text{OPL}(P\tilde{O})$. This leads to the following expression in place of Eq. (4.59):

$$W = \text{OPL}(P_T P_0') - \text{OPL}(\tilde{O}P_0')$$

Now construct a spherical reference surface σ_R centered at P_0' having radius $\overline{\tilde{O}P_0'}$. Let P_R be the intercept of the line $P_T P_0'$ with the reference surface. Because $\overline{\tilde{O}P_0'} = \overline{P_R P_0'}$ we have

$$W = \text{OPL}(P_T P_0') - \text{OPL}(P_R P_0') = -\text{OPL}(P_R P_T)$$

or

$$W = -n'\overrightarrow{P_R P_T} \tag{4.60}$$

which is similar to Eq. (4.58). Equation (4.60) reduces to Eq. (4.58) if the object and image are placed on the optical axis and the aperture stop coincides with the refracting surface. Then $\tilde{O}$ is nearly the same as V and P_R is the same as what we called P_S before.

To obtain a useful expression for W, we go back to Eq. (4.58) where the off-axis aberration function is expressed in terms of the aberration functions that refer to the undeviated ray. Instead of using the perpendicular distances ρ_u as shown in Fig. 4.33c, it is more convenient to identify the intersection of the rays with the exit-pupil plane. This will not seriously modify the result, because the angles that the chief ray and the skew ray make with the optical axis and the undeviated ray are still small. Figure 4.34 shows the plane of the exit pupil looking along the optical axis from the source to the image with the undeviated ray at $\tilde{P}_u$ and the skew ray at $\tilde{P}$. These points are also shown in Figs. 4.33b. The ray of interest is also identified here with the polar coordinates $\tilde{r}$ and $\tilde{\phi}$. Within our approximation the polar coordinates of $\tilde{P}$ are the same as those for P_R or P_T.

The off-axis aberration function can then be written as

$$W = I(\rho_{u\tilde{P}}) - I(\rho_{u\tilde{O}}) = -\frac{C}{4}(\rho_{u\tilde{P}}^4 - \rho_{u\tilde{O}}^4) \tag{4.61}$$

We use the law of cosines on the triangle $\tilde{O}\tilde{P}\tilde{P}_u$ to eliminate $\rho_{u\tilde{P}}$,

$$\rho_{u\tilde{P}}^2 = \rho_{u\tilde{O}}^2 + \tilde{r}^2 + 2\rho_{u\tilde{O}}\tilde{r}\cos\tilde{\phi} \tag{4.62a}$$

In the case where the aperture stop is the refracting surface, the exit pupil coincides with the lens and $L' \cong S'$. In the spirit of the small-angle approximation, we may

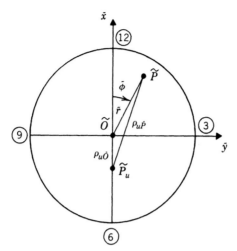

Fig. 4.34 Polar coordinates that identify ray intercepts with the exit-pupil plane.

use Fig. 4.35 to eliminate $\rho_{u\tilde{O}}$ in favor of measurable quantities,

$$\rho_{u\tilde{O}} \cong \frac{R}{(S' - R)} x' \equiv bx' \qquad (4.62b)$$

Using Eqs. (4.62) and (4.63) in Eq. (4.61) leads to our final result for the off-axis aberration function of the single-refracting surface:

$$W = -\frac{C}{4}[\tilde{r}^4 + 4bx'\tilde{r}^3 \cos \tilde{\phi} + 4b^3x'^3\tilde{r} \cos \tilde{\phi} + 2b^2x'^2\tilde{r}^2(2 \cos^2 \tilde{\phi} + 1)] \qquad (4.63)$$

Each term in this equation is associated with one of the primary aberrations.

d. General Image-Forming System. Real optical devices almost always have more than one refracting or reflecting surface. The image produced by the first surface acts as an object for the second surface. The image it produces in turn acts as an object for the next surface. And so on. At each refraction, aberrations will be introduced that affect the final image. These aberrations are additive. The final

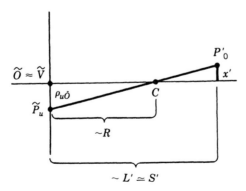

Fig. 4.35 When the aperture stop is the refracting surface, this approximate diagram can be used to eliminate $\rho_{u\tilde{O}}$ from the aberration function.

aberration function is the sum of the aberration functions associated with each refracting surface.

Often it is not desired to know the aberration function at the final refracting surface itself. We have already seen that it is more natural to refer to the aberration function in the plane of the exit pupil in the case of the single-refracting surface. For a more general system of lenses, the exit pupil may not coincide with a physical refracting surface. However, rays can be treated as if they follow straight lines from the plane of the exit pupil. With the aberration function known at the exit pupil, it becomes easy to analyze the influence of aberrations on the image.

The general aberration function will have the form of Eq. (4.63), but with different coefficients, and can be written

$$W = {}_0C_{40}\tilde{r}^4 + {}_1C_{31}x'\tilde{r}^3 \cos \tilde{\phi} + {}_2C_{22}x'^2\tilde{r}^2 \cos^2 \tilde{\phi}$$
$$+ {}_2C_{20}x'^2\tilde{r}^2 + {}_3C_{11}x'^3\tilde{r} \cos \tilde{\phi} \qquad (4.64)$$

The subscripts on the C coefficients refer to the powers of x', $\tilde{r}$, and $\cos \tilde{\phi}$, respectively, that appear in each term. These coefficients give the strength of the primary aberrations displayed by the system. They contain the details of the particular system. We will discuss the coefficients later.

2. Characteristics of Monochromatic Aberrations. In this section we assume that the C coefficients have been determined and that the aberration function can be specified in the plane of the exit pupil. The rays follow straight lines from the exit pupil to the image. These will be perpendicular to the true wave surface at the exit pupil. By contrast, the lines that are perpendicular to the spherical reference surface will intersect at the paraxial image point P'_0. Under the influence of each individual term in Eq. (4.64), the true rays will miss the paraxial image point and will intercept the paraxial image plane at P'.

Consider Fig. 4.36, which is greatly exaggerated for clarity. It illustrates the situation for rays in the meridional plane. Figure 4.36a shows the case when the object is on the optical axis, and Fig. 4.36b illustrates the off-axis situation. The true wave surface σ' and the spherical reference surface σ_R are shown in the plane of the exit pupil, here assumed to be located a distance L' from the paraxial image plane. Following our previous practice, we identify the coordinate of P_T in the meridional plane at the exit pupil as $\tilde{x}$ (Fig. 4.34). (We ignore differences in the transverse coordinate of P_R, P_T, and $\tilde{P}$.)

We will use the directed distance $\overrightarrow{P_R P_T} = l(P_T)$, which is related to the aberration function through Eq. (4.60)

$$l(P_T) = -W/n' \qquad (4.65)$$

The true ray at P_T is $P_T P'$, which is perpendicular to the wave surface σ'. On the other hand, the line $P_R P_T P'_0$ is perpendicular to the reference sphere σ_R. The true ray will miss the paraxial image point by the distance $\Delta x'$. We have the approximate result that

$$-\Delta x' = \gamma L' \qquad (4.66)$$

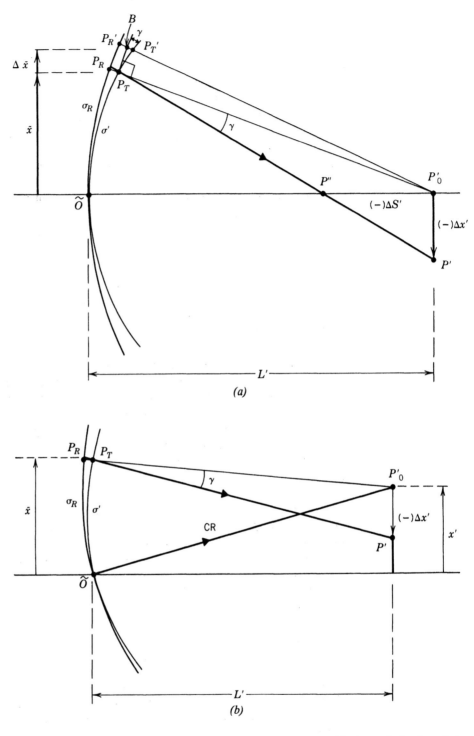

Fig. 4.36 Geometry that is used to find the true ray intercepts with the optical axis and the paraxial image plane. (*a*) On-axis paraxial image point. (*b*) Off-axis paraxial image point.

as the deviation from the paraxial image point. Here γ is the angle between the rays normal to σ_R and σ'. At our level of approximation, this relation can be used for the off-axis image as well. However, it should be clear from Figs. 4.36a and b that this equivalence will not be maintained with a higher-order theory.

To find γ in terms of the aberration function, examine a second pair of points P'_R and P'_T as shown. In our approximation these would be identified by an additional distance from the optical axis $\Delta\tilde{x}$. The directed distance between P'_R and P'_T is

$$\overrightarrow{P'_R P'_T} = l(P'_T) = l(P_T) + \Delta l$$

If we construct a line through P_T perpendicular to $P_R P'_0$ it will intersect $P'_R P'_T$ at B. If P'_T is close enough to P_T, then BP'_T will be very nearly equal to Δl, and γ will be approximated by $\Delta l / \Delta\tilde{x}$. In the limit as P'_T approaches P_T we have

$$\gamma \approx \frac{\partial l}{\partial \tilde{x}} = -\frac{1}{n'}\frac{\partial W}{\partial \tilde{x}} \tag{4.67}$$

Hence the deviation in the x'-direction will be

$$\Delta x' = \frac{L'}{n'}\frac{\partial W}{\partial \tilde{x}} \tag{4.68a}$$

We could repeat this argument for the deviation along the y'-direction. This yields

$$\Delta y' = \frac{L'}{n'}\frac{\partial W}{\partial \tilde{y}} \tag{4.68b}$$

(Remember here that the object and the paraxial image points are still located in the meridional plane, $y = 0$ and $y' = 0$, respectively.)

We now know how the aberration function will modify the image. Next we investigate each term in the aberration function separately, assuming that only each, in turn, is present.

a. Spherical Aberration. As a result of the first term in Eq. (4.64)

$$W = {}_0C_{40}\tilde{r}^4 = {}_0C_{40}(\tilde{x}^2 + \tilde{y}^2)^2 \tag{4.69}$$

spherical aberration occurs. Because W due to spherical aberration is independent of x', the error it produces is the same for all image points in the field of view. This is the only primary monochromatic aberration that exists if the object is on the optical axis.

Following Eqs. (4.68), we find that the coordinates of the deviation that describe the *transverse spherical aberration* are

$$\Delta x' = 4\frac{L'}{n'}{}_0C_{40}\tilde{r}^2\tilde{x}$$

and

$$\Delta y' = 4 \frac{L'}{n'} {}_0 C_{40} \tilde{r}^2 \tilde{y}$$

or

$$\overline{P_0' P'} = \Delta r' \equiv \{(\Delta x')^2 + (\Delta y')^2\}^{1/2} = 4 \frac{L'}{n'} {}_0 C_{40} \tilde{r}^3 \qquad (4.70a)$$

If we introduce a screen perpendicular to the optical axis at P_0', a circle of radius $\Delta r'$ will be observed instead of a point image. By moving the screen closer to the exit pupil, the image circle is reduced. As shown in Fig. 4.37, however, there will be a position E at which the circle is smallest. This is called the "circle of least confusion." Moving still closer to the exit pupil, we come to the position on the optical axis where the marginal rays meet. This identifies P'' in Fig. 4.36a. Provided the deviation is small (not as in the exaggeration of Fig. 4.36a), we can identify the *longitudinal spherical aberration* as

$$\overline{P_0' P''} = \Delta S' \approx \frac{\Delta r'}{\tilde{r}} L' = \frac{4L'^2}{n'} {}_0 C_{40} \tilde{r}^2 \qquad (4.70b)$$

b. Coma. For off-axis object points, the other primary aberrations come into play in addition to spherical aberration. The most important of these is coma. *Ignoring all other aberrations,* we have

$$W = {}_1 C_{31} x' \tilde{r}^3 \cos \tilde{\phi}$$

or

$$W = {}_1 C_{31} x' (\tilde{x}^2 + \tilde{y}^2) \tilde{x} \qquad (4.71)$$

where $\tilde{x} = \tilde{r} \cos \tilde{\phi}$ and $\tilde{y} = \tilde{r} \sin \tilde{\phi}$ (Fig. 4.34).

The paraxial image should be at P_0', a distance x' from the optical axis. The intersections of the true rays with the paraxial image plane will occur at points that deviate from the ideal image point by

$$\Delta x' = \frac{L'}{n'} \frac{\partial W}{\partial \tilde{x}} = {}_1 C_{31} \frac{x' L'}{n'} (3\tilde{x}^2 + \tilde{y}^2) = {}_1 C_{31} \frac{x' L' \tilde{r}^2}{n'} (2 + \cos 2\tilde{\phi}) \quad (4.72a)$$

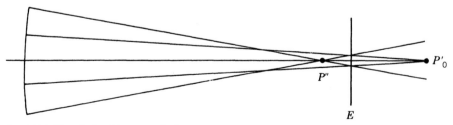

Fig. 4.37 The circle of least confusion.

and

$$\Delta y' = \frac{L'}{n'}\frac{\partial W}{\partial \tilde{x}} = {}_1C_{31}\frac{x'L'}{n'}2\tilde{x}\tilde{y} = {}_1C_{31}\frac{x'L'\tilde{r}^2}{n'}\sin 2\tilde{\phi} \tag{4.72b}$$

Note that rays from points 180° apart, that is, with the same $\tilde{r}$ but with azimuths $\tilde{\phi}$ and 180°-$\tilde{\phi}$, have the same values of $\Delta x'$ and $\Delta y'$. Thus, referring to Fig. 4.34, rays from the edge of the exit pupil at 12 o'clock and 6 o'clock, identified by $\tilde{\phi} = 0°$ and 180°, will meet at the same point in the image plane. Also rays from 3 o'clock and 9 o'clock, identified by $\tilde{\phi} = 90°$ and 270°, meet at the same point in the image plane. As we move around the edge of the exit pupil 360°, the associated rays intersect the image plane so as to define a circle that is traced out twice. The radius of this circle is

$$a = \frac{{}_1C_{31}x'L'\tilde{r}^2}{n'} \tag{4.73}$$

The center of the circle is at $\Delta x' = 2a$, $\Delta y' = 0$. Such a circle is shown in Fig. 4.38a for a negative value of the C coefficient [as would be the case for a single-refracting surface—Eq. (4.63)]. This circle represents the locus of image "points" coming from an annular ring of radius $\tilde{r}$ at the exit pupil. The superposition of such rings for all values of $\tilde{r}$ from zero to a maximum value yields a comet-shaped flare pattern with its apex at the paraxial image point P'_0 (Fig. 4.38b). Because a is proportional to x', we see that the size of this flare pattern is directly proportional to the amount of off-axis displacement of the image.

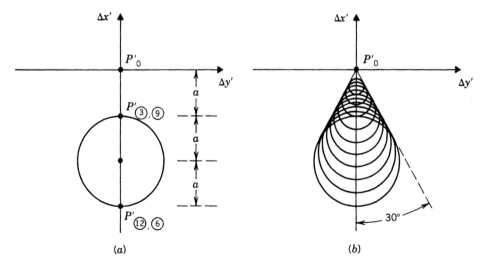

Fig. 4.38 Aberration pattern for coma. (a) Double-valued circle corresponding to rays that pass through an annular ring at the exit pupil. The image point P'_0 is the paraxial location of the off-axis ideal image. (b) For an exit pupil that is completely open, the circles merge into a flare pattern that converges on the paraxial image point.

c. Astigmatism and Curvature of Field. These two aberrations are closely related and are usually discussed together. The appropriate aberration function is

$$W = (_2C_{22} \cos^2 \tilde{\phi} + _2C_{20})x'^2\tilde{r}^2 = (_2C_{22} + _2C_{20})x'^2\tilde{x}^2 + _2C_{20}x'^2\tilde{y}^2 \quad (4.74)$$

In the plane of the paraxial image this leads to deviations

$$\Delta x' = \frac{L'}{n'}\frac{\partial W}{\partial \tilde{x}} = (_2C_{22} + _2C_{20})x'^2\left[\frac{2L'\tilde{x}}{n'}\right] \quad (4.75a)$$

and

$$\Delta y' = \frac{L'}{n'}\frac{\partial W}{\partial \tilde{y}} = _2C_{20}x'^2\left[\frac{2L'}{n'}\tilde{y}\right] \quad (4.75b)$$

To understand these aberrations more completely we must move away from this plane and investigate the rays as they intersect other transverse planes. Figure 4.39 shows how this can be accomplished. Here $\Delta S'$ identifies the location of the test plane with respect to the paraxial image plane. The ray from the exit pupil at $\tilde{P}$ will intersect this plane at P''. This will identify a deviation, $\Delta x''$, from the intersection of the chief ray. If all the angles remain small, and provided $\Delta S'$ is also small, we find

$$\Delta x'' = \Delta x' - \Delta S'\frac{\tilde{x}}{L'} = b_x\tilde{x} \quad (4.76a)$$

Likewise we find

$$\Delta y'' = \Delta y' - \Delta S'\frac{\tilde{y}}{L'} = b_y\tilde{y} \quad (4.76b)$$

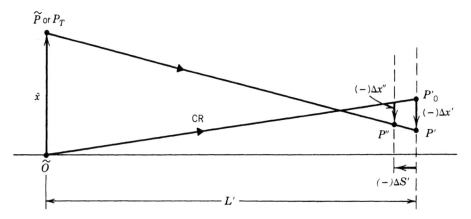

Fig. 4.39 Considerations of the intersection of a true ray with a plane that is longitudinally displaced with respect to the paraxial image plane. This identifies point P''.

where

$$b_x = ({}_2C_{22} + {}_2C_{20})x'^2 \frac{2L'}{n'} - \frac{\Delta S'}{L'} \qquad (4.77a)$$

and

$$b_y = {}_2C_{20}x'^2 \frac{2L'}{n'} - \frac{\Delta S'}{L'} \qquad (4.77b)$$

Note that

$$b_x = b_y + {}_2C_{22}x'^2 \frac{2L'}{n'} \qquad (4.78)$$

The rays that pass through an annular ring at the rim of the exit pupil are defined by $\tilde{x}^2 + \tilde{y}^2 = \tilde{r}^2 = $ constant, the equation of a circle. The resulting image will follow

$$\frac{(\Delta x'')^2}{b_x^2 \tilde{r}^2} + \frac{(\Delta y'')^2}{b_y^2 \tilde{r}^2} = 1 \qquad (4.79)$$

the equation of an ellipse having semiaxes of length

$$b_x \tilde{r} \quad \text{and} \quad b_y \tilde{r}$$

When either b_x or b_y is zero, the ellipse degenerates into a straight line. The case for $b_y = 0$ occurs at P'_s, where

$$\Delta S'_s = {}_2C_{20}x'^2 \left[\frac{2L'^2}{n'} \right] \qquad (4.80)$$

Then we have $\Delta y'' = 0$, and $\Delta x''$ in the form of a line between the limits

$$\Delta x'' = \pm \, {}_2C_{22}x'^2 \left[\frac{2L'}{n'} \tilde{r} \right] \qquad (4.81)$$

This line extends in the plane containing the optical axis and P'_0. This is called the *sagittal* or *radial* image. Rigorously, the line image is perpendicular to the chief ray. This is shown in Fig. 4.40. Here our derivation assumes that the line would be parallel to the $\tilde{x}$ axis for the small-angle approximation.

When $b_x = 0$ the image plane must be at P'_t, given by

$$\Delta S'_t = ({}_2C_{22} + {}_2C_{20})x'^2 \left[\frac{2L'^2}{n'} \right] \qquad (4.82)$$

Then we have $\Delta x'' = 0$, and $\Delta y''$ extending between

$$\Delta y'' = \pm \, {}_2C_{22}x'^2 \left[\frac{2L'}{n'} \tilde{r} \right] \qquad (4.83)$$

This line is parallel to the $\tilde{y}$ direction and is called the *tangential* image.

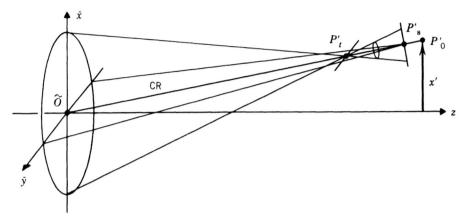

Fig. 4.40 The sagittal and tangential images along the chief ray.

When $\Delta S'$ is halfway between the two values just defined, it is easy to show that the image is a circle, because in that case $b_x = -b_y$. The radius of the circle is

$$\Delta = {}_2C_{22}x'^2\left[\frac{L'}{n'}\tilde{r}\right] \qquad (4.84)$$

which represents the best compromise between the two line images at P'_t and P'_s in Fig. 4.40.

The longitudinal displacements identified in Eqs. (4.80) and (4.82) are both proportional to x'^2. As the off-axis distance x' varies, the loci of these two image points sweep out two paraboloids of revolution σ_s and σ_t as shown in Fig. 4.41. When ${}_2C_{22} = 0$, there is no astigmatism, and σ_s and σ_t coincide to form a single curved surface called the *Petzval surface*. This yields a curved image field from a plane object field.

Astigmatism, as is commonly found among eye defects, is not the same

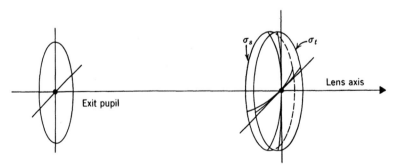

Fig. 4.41 Sagittal and tangential surfaces created by varying x' and rotating the image point about the optical axis.

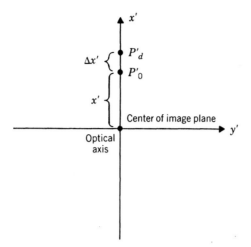

Fig. 4.42 Coordinates for the calculation of distortion.

phenomenon as described here. That effect is due to a distortion of the eye lens tending toward a cylindrical contour rather than a sphere. What we have described here is a primary aberration present in spherical refraction for an off-axis object.

d. Distortion. With a system that shows only primary distortion, we obtain a perfect point image from an off-axis point object, but the image is not at the paraxial image point P'_0. Instead, it is at P'_d, a transverse distance $\Delta x'$ away in the plane containing the optical axis and P'_0. This is shown in the image plane of Fig. 4.42. The aberration function is

$$W = {}_3C_{11}x'^3\tilde{r}\cos\tilde{\phi} = {}_3C_{11}x'^3\tilde{x} \tag{4.85}$$

and thus the deviation from P'_0 is

$$\Delta x' = {}_3C_{11}\frac{L'}{n'}x'^3 \tag{4.86}$$

Because $\Delta x'$ is independent of $\tilde{r}$ and $\tilde{\phi}$, all rays normal to the emerging wave surface will form the image at P'_d.

Now consider the situation when the object point is no longer in the meridional plane. The paraxial image point is shown in Fig. 4.43 at P'_0. This is most easily handled by substituting r' for x' and $\Delta r'$ for $\Delta x'$ in Eq. (4.86). This is equivalent to a rotation of the problem about the optical axis by an angle $\tilde{\phi}$. By this technique we can identify

$$\Delta r' = {}_3C_{11}\frac{L'}{n'}r'^3$$

Once this has been done, we can return to the original coordinate system orientation to find the coordinates of an image. These will be at

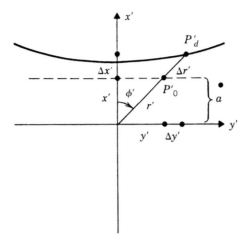

Fig. 4.43 Distortion for points not in the meridional plane.

$$\Delta x' = \Delta r' \cos \phi' = {}_3C_{11} \frac{L'}{n'} (x'^2 + y'^2)^{3/2} \frac{x'}{(x'^2 + y'^2)^{1/2}}$$

$$= {}_3C_{11} \frac{L'}{n'} (x'^3 + x'y'^2) \tag{4.87a}$$

and

$$\Delta y' = \Delta r' \sin \phi' = {}_3C_{11} \frac{L'}{n'} (x'^2 y' + y'^3) \tag{4.87b}$$

where the paraxial image would have been at (x', y').

Now consider a line object that yields a paraxial image at $x' = a$. Instead of this, when distortion is present, we find image points that deviate

$$\Delta x' = {}_3C_{11} \frac{L'}{n'} (a^3 + ay'^2)$$

$$\Delta y' = {}_3C_{11} \frac{L'}{n'} (ay' + y'^3)$$

For positive ${}_3C_{11}$ the distorted image will bend away from the paraxial line image as shown in Fig. 4.43. If the object plane contains a grid of lines on a checkerboard pattern, the image plane will appear as in Fig. 4.44a. This is called *pincushion distortion*. *Barrel distortion* occurs when ${}_3C_{11}$ is negative (Fig. 4.44b).

3. Thin Lens. In the foregoing discussion, the C coefficients were left general and unspecified so that the geometrical influence of each aberration could be brought out. Here we wish to illustrate the primary monochromatic aberrations' strengths by identifying the form of the C coefficients in the case of a thin lens in air that acts

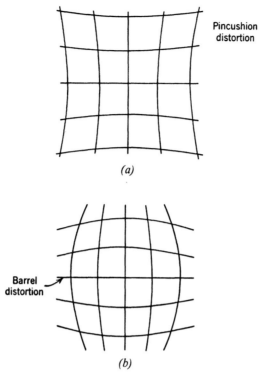

Pincushion distortion

(a)

Barrel distortion

(b)

Fig. 4.44

as its own aperture stop. We will not derive the results here but merely present the third-order formulas and discuss their implications (see Born and Wolf for details).

For a thin lens with focal length f and refractive index n, the spherical aberration is proportional to

$$_0C_{40} = -\frac{1}{32f^3n(n-1)}$$

$$\times \left[\frac{n+2}{n-1}q^2 + 4(n+1)pq + (3n+2)(n-1)p^2 + \frac{n^3}{n-1}\right] \quad (4.88)$$

Here q is the so-called "shape factor" determined from the radii of curvatures of the two surfaces

$$q = \frac{R' + R}{R' - R} \quad (4.89)$$

and p is the so-called "position factor"

$$p = 1 - \frac{2f}{S'} \quad (4.90)$$

If we normalize all lengths to the focal length f, we obtain from Eq. (4.70b) the convenient dimensionless formula for the longitudinal spherical aberration (with $L' = S'$)

$$\frac{\Delta S'}{f} = 4_0 C_{40} f^3 \left[\left(\frac{\tilde{r}}{f} \right) \left(\frac{S'}{f} \right) \right]^2 \tag{4.91}$$

From the functional form of Eq. (4.88) we see that for fixed position factor p the longitudinal aberration will give a parabola when plotted as a function of the shape factor q, and vice versa. Note further the proportionality to f^{-3}; when f changes sign, so will the spherical aberration.

In Fig. 4.45 we have plotted the coefficient $4_0 C_{40} f^3$ as a function of q for three values of p, for $n = 1.5$. Sketches of the lens shapes are also included. The minimum spherical aberration occurs when the two surfaces of the lens deviate the off-axis rays in the light beam by equal amounts. The minimum will not be zero for real objects and images (p between -1 and $+1$). The $p = 0$ curve applies when $S' = 2f = S$; the lens is used symmetrically and gives a magnification of (-1). The symmetrical equiconvex lens ($q = 0$) then has the smallest spherical aberration. The $p = -1$ curve corresponds to $S' = f$. The minimum value of spherical aberration occurs when q is near the value of $+1$.

Primary spherical aberration can be eliminated by combining two thin lenses with focal lengths of opposite sign. We then exploit the fact that $\Delta S'$ is proportional to f^3, so that it is negative for a positive lens and positive for a negative lens. For a given position factor we choose a shape factor q so that the negative primary

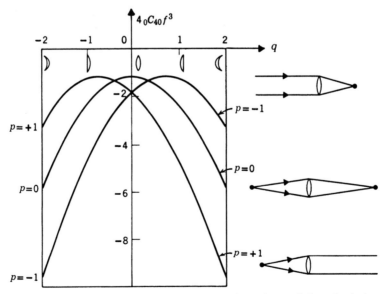

Fig. 4.45 Lens shape and imaging geometry dependence of the spherical aberration parameter for a thin lens in air.

spherical aberration of the positive component has nearly its minimum magnitude. The shape factor of the negative component can then be chosen to give a relatively large amount of positive spherical aberration for its small negative power. The net power of the combination will be positive, the primary spherical aberration is zero. Such a lens is frequently manufactured with the adjoining surfaces cemented together.

For our example, the third-order theory for the coma coefficient produces the result

$$_1C_{31} = \frac{1}{4nf^2S'}\left[(2n + 1)p + \frac{(n + 1)}{(n - 1)}q\right] \tag{4.92}$$

where p and q are defined in Eqs. (4.89) and (4.90). According to Eq. (4.73), the radius of the comatic circle is given by

$$\frac{a}{f} = \frac{(\tilde{r}/f)^2(x'/f)}{4n}\left[(2n + 1)p + \frac{(n + 1)}{n - 1}q\right] \tag{4.93}$$

This is linear in both p and q. It is relatively easy to find reasonable values of p and q that will make this third-order coma zero for a thin lens.

At this level of approximation the thin lens is free from distortion, $_3C_{11} = 0$. However, astigmatism and curvature of field will always be present. The third-order theory gives

$$_2C_{22} = -\frac{1}{2fS'^2} \tag{4.94}$$

and

$$_2C_{20} = -\left(\frac{n + 1}{4n}\right)\frac{1}{fS'^2} \tag{4.95}$$

These aberrations can be reduced or eliminated in multilens systems.

4. Sine Condition. We have seen that primary coma produces a comet-shaped "image" with a size that varies linearly with the off-axis distance x'. Unless this effect is small, the image of an extended object will be considerably blurred and distorted, even though the image of the on-axis point might be of higher quality. A similar effect can be produced by aberrations of higher order, namely, those with aberration functions W that depend on high powers of $\tilde{r}$, but are linear in x'. There is a general requirement, known as the *sine condition*, that must be met by an image-forming system if there is to be no such linear scaling of the size of an aberrant off-axis image with x'.

If the system in Fig. 4.46 has no spherical aberration for point P'_0 on the symmetry axis, the imaging is perfect, and the system is "Cartesian" for the conjugate points P_0 and P'_0. Now consider the point P_1 located off axis from P_0 by the small distance x. Let P'_1 be its paraxial image. Except for a plane mirror, no practical optical system that produces perfect imaging of P_0 onto P'_0 can also

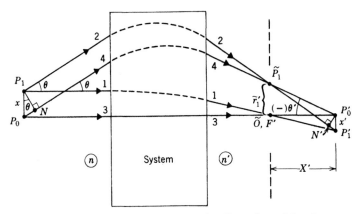

Fig. 4.46 Geometrical relationships in the discussion of the sine condition.

produce perfect imaging of P_1 onto P'_1. That is, all optical path lengths from P_1 to P'_1 cannot be made exactly equal. It is possible, however, to make the optical path lengths equal to first order, in the off-axis image distance x'. In other words, the system can be designed so that the aberration function W for light coming from P_1 will contain no terms proportional to x'. This will be true if the sine condition holds.

It is convenient, but not necessary, to put the exit pupil of the system in Fig. 4.46 in the second focal plane through F'. Then F' may be identified with $\tilde{O}$ in Fig. 4.34. Then the independent variable $\tilde{r}_1$ for the aberration function W represents the distance of a ray from the axis when it passes through the focal plane. Rays leaving the object point P_1 parallel to the axis pass through F' and have $\tilde{r}_1 = 0$. Ray 1 is such a ray. Ray 2 leaves P_1 at a finite angle θ and intersects the exit pupil at $\tilde{P}_1$ a distance $\tilde{r}_1$ from the axis while making a different angle with the axis. The optical path difference for these rays is given by the aberration function

$$\text{OPL}(2) - \text{OPL}(1) = W \tag{4.96}$$

and our requirement is that this be zero to first order in x'.

We first show that the parallel rays 1 and 3 have equal optical path lengths to first order in x'. Note that the OPL's from P_1 to F' and P_0 to F' are exactly equal. Thus

$$\text{OPL}(1) - \text{OPL}(3) = n'\overline{P'_1 F'} - \overline{P'_0 F'}$$

$$= n'(\sqrt{x'^2 + X'^2} - X') = n'\frac{x'^2}{2X'} + \dots \tag{4.97}$$

The last equality results from an expansion of the square root in a power series in $(x'/X')^2$.

Keep in mind that according to Eq. (4.96) we ultimately want to compare the path lengths of rays 1 and 2. Rays 1 and 3 are equal to first order in x'. Rays 3 and 4 have exactly equal path lengths, because P_0' is a perfect image of P_0. Thus

$$\text{OPL}(3) = \text{OPL}(4) \tag{4.98}$$

Together with Eq. (4.97) this gives

$$\text{OPL}(4) - \text{OPL}(1) = \mathcal{O}(x'^2) \tag{4.99}$$

To obtain Eq. (4.96) we need to compare path lengths of rays 2 and 4. Before any refracting surfaces they are parallel, both making an angle θ with the axis. The OPL's from P_1 to $\tilde{P}_1$ and from N to $\tilde{P}_1$ are exactly equal, because $P_1 N$ is perpendicular to the ray direction. The OPL's from $\tilde{P}_1$ to P_0' and $\tilde{P}_1$ to N' are equal to first order in x'. Specifically,

$$\overline{\tilde{P}_1 P_0'} = \sqrt{(\overline{\tilde{P}_1 N'})^2 + (\overline{N'P_0'})^2} = \overline{\tilde{P}_1 N'}\left(1 + \frac{(\overline{N'P_0'})^2}{(\overline{\tilde{P}_1 N'})^2}\right)^{1/2}$$

$$= \overline{\tilde{P}_1 N'}\left(1 + \frac{1}{2}\frac{(\overline{N'P_0'})^2}{(\overline{\tilde{P}_1 N'})^2} + \dots\right) = \overline{\tilde{P}_1 N'} + \mathcal{O}(x'^2)$$

Thus, to first order in x' the part from P_1 to N' of ray 2 has the same OPL as the part from N to P_0' of ray 1. This leaves the contribution from $\overline{P_0 N}$ and $\overline{N'P_1'}$:

$$\text{OPL}(2) - \text{OPL}(4) = n'\overline{N'P_1'} - n\overline{P_0 N} + \mathcal{O}(x'^2) \tag{4.100}$$

When we add Eqs. (4.99) and (4.100), we obtain the desired comparison of the OPL's of rays 1 and 2:

$$\text{OPL}(2) - \text{OPL}(1) = n'\overline{N'P_1'} - n\overline{P_0 N} + \mathcal{O}(x'^2) \tag{4.101}$$

This can be further simplified by use of the triangles $P_0 N P_1$ and $P_0' N' P_1'$:

$$\overline{P_0 N} = x \sin \theta, \quad \overline{N'P_1'} = x' \sin \theta'$$

Thus

$$W = \text{OPL}(2) - \text{OPL}(1) = [n'x' \sin \theta' - nx \sin \theta] + \mathcal{O}(x'^2) \tag{4.102}$$

We will have our desired result that W be independent of x' to first order if and only if the *sine condition*

$$n'x' \sin \theta' = nx \sin \theta \tag{4.103}$$

is satisfied. This represents a generalization of the Lagrange equation [Eq. (3.31)]

$$n'x'\theta' = nx\theta$$

of paraxial optics to large angles.

A system that (a) is Cartesian for a pair of points P and P', and (b) satisfies the sine condition is called *aplanatic*.

Some implications of the sine condition will be explored in the problems. It can be rearranged to say that the ray angle magnification condition takes the form

$$\frac{\sin \theta'}{\sin \theta} = \frac{nx}{n'x'} = \frac{n}{n'm_x} = \text{const.} \qquad (4.104)$$

Here $m_x = x'/x$ is the transverse magnification. Thus, if a series of rays leaves P at the angles $\theta_1, \theta_2, \ldots$ and arrives at P' at the angles $\theta'_1, \theta'_2, \ldots$, we must have

$$\frac{\sin \theta'_1}{\sin \theta_1} = \frac{\sin \theta'_2}{\sin \theta_2} = \cdots \qquad (4.105)$$

Equation (4.105) can be used to test for an aplanatic system, that is, one with perfect imaging of slightly off-axis points. Interestingly, this test involves only the behavior of rays from an on-axis point.

The sine condition has already been used to derive Eq. (4.39), the law of conservation of luminance or radiance in an elementary beam. It is possible to derive this conservation law from the laws of thermodynamics and then derive the sine condition from the conservation law.

Real optical systems can be made aplanatic or nearly aplanatic by a variety of devices such as the use of meniscus lenses. They will be useful only for small off-axis displacements and only for a single object plane. They can be used in microscopes and certain telescopes where the field of view is small and the object plane is fixed. For wider fields of view the aberrations that are quadratic or higher in x' become important, and one cannot reduce them in general without giving up the sine condition and accepting some coma and related effects.

B. Chromatic Aberrations

Because most optical systems are used with light of varying wavelengths, the dispersions of the refracting media must be taken into account. The optical designer must consider the variation of refractive indices with wavelength so that the resulting performance of the system is more or less wavelength independent.

We can distinguish two general types of chromatic aberrations: (1) chromatic variations in the paraxial image-forming properties of a system, and (2) wavelength dependence of the monochromatic aberrations. The second type is evaluated by treating the refractive indices as variable in section A of this chapter. We will concentrate here primarily on the first type. With that in mind, we return to the paraxial imaging conditions.

In paraxial optics the properties of an image-forming system depend solely on the locations of the principal planes and the focal planes. Chromatic aberrations result when the position of any of these planes is wavelength dependent. We can discuss the effects of monochromatic aberrations in terms of a variation in image distance along the axis and a variation in transverse magnification. Because a single-refracting surface shows behavior that can be observed in more complex systems, we will start with it.

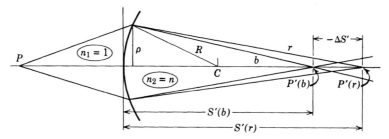

Fig. 4.47 When chromatic aberration is present, the paraxial image point depends on the wavelength of the light. Blue light (*b*) and red light (*r*) create different images that are separated by $\Delta S'$.

1. Single-Refracting Surface. Consider first an object point P on the axis of the system. With refractive indices as shown in Fig. 4.47, the image formation condition is

$$\frac{1}{S} + \frac{n}{S'} = \frac{n-1}{R} \tag{4.106}$$

Let us denote two different wavelengths by r and b (for red and blue, typically) and assume that the indices have the property $n(b) > n(r)$. The change in index $\Delta n = n(b) - n(r)$ will be small compared with unity for transparent optical media, and we may use approximations to first order in Δn. Then we let $\Delta S' = S'(b) - S'(r)$ represent the longitudinal chromatic aberration, and by differentiating Eq. (4.106) we obtain

$$\Delta\left(\frac{1}{S'}\right) = -\frac{\Delta S'}{S'^2} = \frac{\Delta n}{n}\left(\frac{1}{R} - \frac{1}{S'}\right) \tag{4.107}$$

Here n and S' can be evaluated at any convenient wavelength, but best accuracy is obtained with a wavelength between r and b.

At $P'(b)$, the focus for b light, the rays of r light will form a cone converging toward $P'(r)$. The cone will have a half-angle of approximately ρ/S', and it will intersect the b image plane in a disk of radius

$$\Delta x' = -\Delta S'\frac{\rho}{S'} = S'\rho\frac{\Delta n}{n}\left(\frac{1}{R} - \frac{1}{S'}\right) \tag{4.108}$$

A disk of approximately the same radius is formed by b rays in the r image plane. This is shown in Fig. 4.48.

An off-axis object point will produce images in r and b light at different heights from the axis. This is shown in Fig. 4.49. The object point P has chromatic paraxial point images at $P'(b)$ and $P'(r)$ that lie along the undeviated ray (UR) that passes through the center of curvature at C. This leads to a wavelength-dependent transverse magnification.

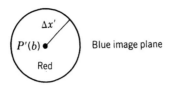

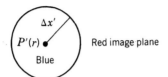

Red image plane

Fig. 4.48 Appearance of the "focus" in the vicinity of the paraxial image points for blue and red light.

The fractional change in magnification

$$\frac{\Delta m_x}{m_x} = \frac{x'(b) - x'(r)}{x'} \tag{4.109}$$

where x' is the average of $x'(b)$ and $x'(r)$, can be found through a consideration of the similar triangles shown in the inset of Fig. 4.49. We have

$$\frac{x'}{(S' - R)} = \frac{-[x'(b) - x'(r)]}{\Delta S'}$$

so that

$$\frac{\Delta S'}{(S' - R)} = \frac{x'(b) - x'(r)}{x'} = \frac{\Delta m_x}{m_x} \tag{4.110}$$

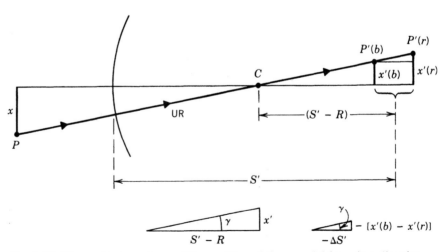

Fig. 4.49 Chromatic dependence of the position of the paraxial focus in a direction perpendicular to the optical axis.

This can be modified by rewriting Eq. (4.107) as

$$\frac{\Delta S'}{S'^2} = -\frac{\Delta n}{n}\frac{(S'-R)}{RS'}$$

leading to

$$\frac{\Delta S'}{(S'-R)} = -\frac{\Delta n}{n}\frac{S'}{R} \tag{4.111}$$

Comparing this to Eq. (4.110) yields the final result

$$\frac{\Delta m_x}{m_x} = -\frac{\Delta n}{n}\frac{S'}{R} \tag{4.112}$$

which is independent of x'.

In complex systems with several refracting surfaces we can often remove the longitudinal chromatic aberration so that the blue and red images coincide for an on-axis object point at a particular distance. Unless the different colored off-axis rays themselves coincide, in general, there will still be chromatic aberration for off-axis object points. Thus the wavelength dependent transverse magnification is harder to correct.

2. Lenses in Contact. The cemented doublet is one of the simplest and most common of the "achromatic" designs. Given two thin lenses with essentially zero separation d, we can write the power of the combination from Eq. (3.79) as

$$\frac{1}{f} = \mathscr{P} = \mathscr{P}_1 + \mathscr{P}_2$$

with

$$\mathscr{P}_1 = \frac{1}{f_1} = (n_1 - 1)\left(\frac{1}{R_1} - \frac{1}{R_1'}\right) \equiv (n_1 - 1)K_1 \tag{4.113a}$$

and

$$\mathscr{P}_2 = \frac{1}{f_2} = (n_2 - 1)\left(\frac{1}{R_2} - \frac{1}{R_2'}\right) \equiv (n_2 - 1)K_2 \tag{4.113b}$$

The quantities K_1, K_2 are called the *total curvature* of each lens. The change in total power caused by a change in wavelength is

$$\Delta\mathscr{P} = \Delta n_1 K_1 + \Delta n_2 K_2$$

$$= \left[\frac{\Delta n_1}{n_1 - 1}\right](n_1 - 1)K_1 + \left[\frac{\Delta n_2}{n_2 - 1}\right](n_2 - 1)K_2 = \frac{\mathscr{P}_1}{V_1} + \frac{\mathscr{P}_2}{V_2} \tag{4.114}$$

We used Eq. (4.113) in the second equality and have introduced the *dispersion constants* of the glasses in the two lenses, which are defined as follows:

$$V_1 \equiv \frac{n_1 - 1}{\Delta n_1}, \quad V_2 \equiv \frac{n_2 - 1}{\Delta n_2} \tag{4.115}$$

The wavelengths at which the n's and Δn's in Eq. (4.115) are determined are taken conventionally to be the following:

$$\Delta n = n(F) - n(C)$$

$$n = n(D)$$

where F represents the blue line of hydrogen at 486.1 nm, C the red line of hydrogen at 656.3 nm, and D the yellow lines of sodium at 589.3 nm. The D line is close to the wavelength of maximum response of the human eye. The F and C lines gave a convenient spread across the visible spectrum. Data on indices of refraction and dispersion constants are supplied by the glass manufacturers often as a six-digit number; for example, 517645 means $n(D) = 1.517$, $V = 64.5$.

Some black-and-white film has peak sensitivity in the blue region of the spectrum; we would then want to select a camera lens achromatized for shorter wavelengths than the F and C lines.

We return now to Eq. (4.114). For achromatization at the two wavelengths used to define Δn, we set $\Delta \mathscr{P} = 0$ and through use of Eq. (4.115) obtain

$$\frac{\mathscr{P}_2}{\mathscr{P}_1} = -\frac{V_2}{V_1} \tag{4.116}$$

or

$$f_1 V_1 + f_2 V_2 = 0 \tag{4.117}$$

By writing

$$\mathscr{P} = \mathscr{P}_1 + \mathscr{P}_2 = \mathscr{P}_2 - \left(\frac{V_1}{V_2}\right)\mathscr{P}_2$$

we obtain

$$\mathscr{P}_2 = \frac{V_2}{V_2 - V_1}\mathscr{P}$$

and

$$\mathscr{P}_1 = \frac{-V_1}{V_2 - V_1}\mathscr{P} \tag{4.118}$$

Equations (4.118) give the necessary condition for our thin lens to be achromatized for the two wavelengths used to define Δn. The chromatic aberration that remains at other wavelengths is called the *secondary spectrum* and is often small enough to be neglected.

Note that $\mathscr{P}_1$ and $\mathscr{P}_2$ must have opposite signs. If $\mathscr{P}$ is to be positive, we need $\mathscr{P}_1 > -\mathscr{P}_2$, assuming that $\mathscr{P}_1$ is positive. Then by Eq. (4.116) we want

$$\frac{V_1}{V_2} = \frac{\mathscr{P}_1}{\mathscr{P}_2} > 1$$

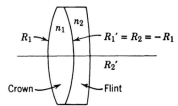

Fig. 4.50 Achromatic lens.

Thus the positive component must have the larger dispersion constant (and hence the smaller dispersion Δn).

To control the secondary spectrum and to minimize the monochromatic aberrations, we want to keep the magnitudes of $\mathscr{P}_1$ and $\mathscr{P}_2$ small. This can be done if $|V_2 - V_1|$ is large.

The shape factors of the individual lenses do not influence Eqs. (4.118). These can be varied to reduce spherical aberration and coma. For inexpensive achromatic doublets, the shape factors are more likely to be chosen for ease of manufacture. For instance, the positive element is often equiconvex; the negative element then becomes almost planoconcave for typical glasses used in a cemented doublet.

As an example, consider a cemented doublet made of borosilicate crown glass No. 517645, $n(D) = 1.517$, $V = 64.5$, as the positive element and dense flint glass No. 617366 as the negative element. Then if the focal length is to be 10 cm = 0.1 m, the power $\mathscr{P}$ will be 10 m^{-1} (diopter). Equations (4.118) then give $\mathscr{P}_2 = -13.118$ diopters and $\mathscr{P}_1 = 23.118$ diopters. If the positive element is equiconvex, the radii of curvature then are $R_1 = -R' = 4.47$ cm. Because the elements are in contact, we must have $R_2 = -4.47$ cm. Then we can calculate R_2' to be 91.2 cm. The lens is illustrated in Fig. 4.50.

3. Separated Doublet. The system consisting of two thin lenses separated by a distance d in air is a model for several important optical instruments. If the components have powers of $\mathscr{P}_1$ and $\mathscr{P}_2$ then, from Eq. (3.79), the total power will be

$$\frac{1}{f} = \mathscr{P} = \mathscr{P}_1 + \mathscr{P}_2 - \mathscr{P}_1\mathscr{P}_2 d$$

Then the change in $\mathscr{P}$ at two different wavelengths is given by

$$\Delta\mathscr{P} = \Delta\mathscr{P}_1 + \Delta\mathscr{P}_2 - d(\mathscr{P}_2\,\Delta\mathscr{P}_1 + \mathscr{P}_1\,\Delta\mathscr{P}_2) \qquad (4.119)$$

Proceeding as before, we write

$$\mathscr{P}_1 = (n_1 - 1)K_1 = (n_1 - 1)\left(\frac{1}{R_1} - \frac{1}{R_1'}\right)$$

$$\mathscr{P}_2 = (n_2 - 1)K_1 = (n_2 - 1)\left(\frac{1}{R_2} - \frac{1}{R_2'}\right)$$

and obtain the results

$$\Delta \mathscr{P}_1 = \mathscr{P}_1/V_1$$

and

$$\Delta \mathscr{P}_2 = \mathscr{P}_2/V_2$$

used in Eq. (4.114).

To determine the achromatic configuration set $\Delta \mathscr{P} = 0$ and obtain

$$0 = \frac{\mathscr{P}_1}{V_1} + \frac{\mathscr{P}_2}{V_2} - d\mathscr{P}_1\mathscr{P}_2\left(\frac{1}{V_1} + \frac{1}{V_2}\right)$$

If the lenses are made from glass of the same index, we have $V_1 = V_2$, and we obtain

$$\mathscr{P}_1 + \mathscr{P}_2 - 2d\mathscr{P}_1\mathscr{P}_2 = 0$$

or

$$2d = \frac{1}{\mathscr{P}_1} + \frac{1}{\mathscr{P}_2} = f_1 + f_2 \tag{4.120}$$

Equation (4.120) is the condition that the doublet have a focal length independent of wavelength (to first order in Δn). In general, when it is satisfied we have

$$f = \frac{f_1 f_2}{d} \tag{4.121}$$

Doublets are often used as low-cost eyepieces in telescopes and microscopes. The function of the eyepiece is to magnify the intermediate image formed by the objective lens. The final virtual-image distance S' is much greater than the focal length of the eyepiece, and the object distance is very close to the focal length f. Under these conditions it is the angular size $\alpha_E = x/f$ of the image that is important as in Eq. (3.103) or, alternatively, its linear size at the conventional viewing distance of 250 mm. Thus the magnification of the eyepiece will be achromatized if f is. We say that the eyepiece is corrected for "lateral color."

Now as an example of a doublet corrected for lateral color, consider the *Huygens eyepiece* of Fig. 4.51. For simplicity it is convenient to assume that the final virtual image is at infinity. Then the intermediate image formed by the objective lens of the telescope or microscope must be in the first focal plane through F of the eyepiece. It acts as a virtual object for the eyepiece. The rays from the object are actually brought to a focus at F_2, the first focal plane of L_2. This is also the second principal plane for the eyepiece. This is where the stop that limits the field of view (field stop) should be placed. Here also should be put the cross hairs or reticles, so that they can be imaged by L_2 in the same plane as the final image. Unfortunately, this imaging is uncorrected, since L_2 alone is used; thus the Huygens eyepiece is not suitable for use with a reticle.

The eyepiece can be corrected for coma by proper selection of n and of f_2/f_1, but it has a relatively large amount of spherical aberration.

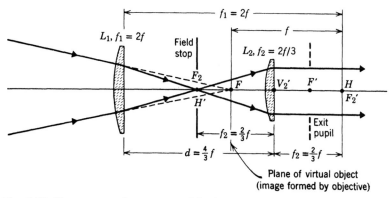

Fig. 4.51 Huygens eyepiece, corrected for lateral color.

The objective lens of a telescope or microscope is the aperture stop. Its distance from the eyepiece is so much larger than the focal length of the eyepiece that the image of it formed by the eyepiece is almost in its second focal plane, through F' in Fig. 4.51. This image, of course, is the exit pupil, and it should coincide approximately with the pupil of the observer's eye. The distance $\overline{L_2 F'}$ from the second vertex of the second lens is called the *eye relief*. For the lens shown in Fig. 4.51 this distance is $f/3$, if we may ignore the thickness of L_2.

Now let us pick $f_1 = f_2$. Then to correct for lateral color, Eq. (4.120) yields $d = f_1$ and Eq. (4.121) gives $f = f_1$. The resulting eyepiece is shown in Fig. 4.52. Here L_1 acts as a field lens, imaging the objective onto L_2 and preventing vignetting. The intermediate image is now coincident with L_1 and is inaccessible for placement of cross hairs or reticles. Another disadvantage comes with the scratches or dust on L_1 that is in focus along with the image. A third disadvantage lies in the zero eye relief, because the exit pupil is at L_2.

The disadvantages of the lens of Fig. 4.52 can be overcome if we sacrifice some correction for lateral color and move the lenses closer together. The case where $d = 2f_1/3$ is illustrated in Fig. 4.53. This is called a *Ramsden eyepiece* and will have $f = 3f_1/4$. The lenses are oriented as shown to reduce spherical aberration and eliminate coma. The eye relief is $f/3$. There is now some transverse chromatic aberration. The *Kellner eyepiece* Fig. 4.54) is a Ramsden eyepiece with the eye lens L_2 achromatized to reduce this lateral color.

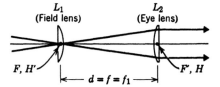

Fig. 4.52

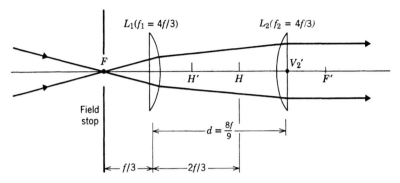

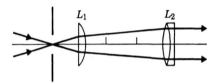

Fig. 4.53 Ramsden eyepiece.

Fig. 4.54 Kellner eyepiece, a Ramsden design with an achromatized eye lens.

With these practical examples of eyepieces, we end our discussion of geometrical optics.

REFERENCES

Born, Max, and Emil Wolf. *Principles of Optics*. Pergamon Press, Oxford, 1980.

Cagnet, Michel, Maurice Francon, and Jean Claude Thrierr. *Atlas of Optical Phenomena*. Springer-Verlag, Berlin, 1962.

Conrady, A. E. *Applied Optics and Optical Design*. Dover, New York, 1957.

Cox, Arthur. *Photographic Optics*. Focal Press, London, 1966.

Driscoll, Walter G., and William Vaughan. *Handbook of Optics*. McGraw-Hill, New York, 1978.

Hardy, A. C. and F. H. Perrin. *The Principles of Optics*. McGraw-Hill, New York, 1932.

Herzberger, Max. *Modern Geometrical Optics*. Interscience Publishers, New York, 1958.

Hopkins, H. H. *Wave Theory of Aberrations*. Clarendon Press, Oxford, 1950.

Kingslake, Rudolf. *Lens Design Fundamentals*. Academic Press, New York, 1978.

Levi, Leo. *Applied Optics*. Wiley, New York, 1968.

Martin, L. C. *The Theory of the Microscope*. Elsevier, New York, 1965.

Nicodemus, Fred E., "Radiometry" in Robert R. Shannon, and James C. Wyant. eds., *Applied Optics and Optical Engineering*, vol. 9, p. 13. Academic Press, New York, 1983.

Palmer, J. M. *Lens Aberration Data*. Elsevier, New York, 1971.

Smith, Warren J. *Modern Optical Engineering*. McGraw-Hill, New York, 1966.

Stavroudis, O. N. *The Optics of Rays, Wavefronts and Caustics*. Academic Press, New York, 1972.

Walsh, J. W. T. *Photometry*. Dover, New York, 1965.

Welford, W. T. *Aberrations of the Symmetrical Optical System*, Academic Press, New York, 1974.

PROBLEMS

Section 4.1   Stops and Apertures

1. A thick lens as shown in Fig. 4.55 is used in air. The first and second radii of curvature are $R_1 > 0$ and $R_2 < 0$, the index is $n > 1$, and the thickness is d. What will be the aperture stop for this lens for an axial object at a general distance S_1 to the left of V_1? Is the aperture stop always the same? (No calculation is necessary to solve this problem.)

2. An object is located a distance $3R/2$ from the surface of a spherical marble ($n = 3/2$) that has a diameter of $2R$. An aperture stop (diameter R) is positioned at the front vertex. Find the locations of the principal planes and the image, the location and size of the exit pupil, and trace a chief ray and a marginal ray on a diagram demonstrating your solutions.

3. A thin positive lens with a focal length of 2 cm is placed behind a diaphragm at a distance of 4 cm. The diaphragm is symmetrically positioned about the optical axis of the lens and contains an opening of 3 cm. An object is 2 cm in front of the diaphragm. The object extends from the optical axis 1.5 cm vertically. *Graphically* determine the location of the exit pupil, the marginal rays, and a chief ray from the tip of the object. From this diagram, locate the position and size of the image.

4. A circular diaphragm of radius 2 cm is placed 1 cm behind a thin lens that has a focal length of 3 cm. An object that is 2 cm tall is situated with its base on the optical axis a distance of 6 cm in front of the lens. *Graphically* determine the sizes and locations of the entrance and exit pupils, the chief ray from the top of the object, the marginal rays. From this diagram, locate the position and size of the image.

5. A thin positive lens with diameter of 1 in. and focal length 2 in. covers the end of a 1-in.-diameter tube that is 5 in. long. The closed end of the tube is pointed toward an object. Find an expression for the location of the aperture stop as a function of the object distance.

6. A thin positive lens that has a radius a and a focal length f is placed midway between an object and the eye of an observer. If the object is planar and located a distance f from the eye, find the extent of the object that can be observed through the lens.

7. The Coddington magnifier is cut as shown in Fig. 4.56 from a single sphere.

    (a) Where must an object be held so that a virtual image is formed 250 mm in front of the lens?

    (b) Under the conditions of (a), find the entrance and exit pupils.

8. For the optical system given in Fig. 4.57, locate (a) the image, (b) the position and size of the entrance window.

9. A thin lens L_1 with a 5.0-cm diameter aperture and focal length $+4.0$ cm is placed 4.0 cm to the left of another lens L_2 4.0 cm in diameter with a focal length

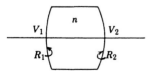

Fig. 4.55

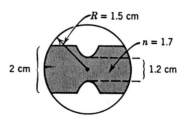

Fig. 4.56

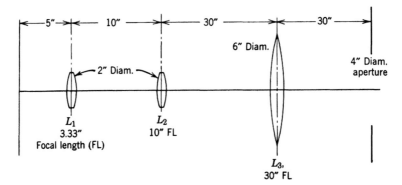

Fig. 4.57

of $+10.0$ cm. A 2.0-cm-high object is located with its center on the axis 5 cm in front of L_1. There is a 3.0-cm-diameter stop centered halfway between L_1 and L_2. Find analytically the position and size of the entrance pupil, the exit pupil, the image. Make a brief sketch to scale.

10. In Fig. 4.58 the following dimensions are given: $f_1 = 12$ cm, $R_1 = 1.0$ cm, $R = 0.9$ cm, $f_2 = 12$ cm, $R_2 = 1.0$ cm. Locate the aperture stop for the system for an object at A as a function of the distance S, where S ranges from infinity to zero. For each value of S, find the field stop. (Assume the lenses are thin.)

11. Standard "35-mm" cameras have a 24×37 mm aperture just in front of the film that acts as a field stop. Consider a camera with a 50-mm focal length lens.

What is the size of the field of view for an object distance of 1 m? 30 m?

12. The f/# of a lens is defined as the ratio of the focal length to the diameter of the entrance pupil with the iris diaphragm (if any) wide open and with a distant object. Consider an f/2.8 lens, that is, one with an f/# of 2.8, having a focal length of 45 mm. What will be the half-angle of the cone of light converging from the exit pupil toward the image of an object at a distance of 0.50 m? 5.0 m? 50 m?

13. A field stop 4 mm in diameter is placed at the position of the intermediate image in the compound microscope of Fig. 3.38. What is the size of the field of view? The exit pupil is 4 mm in diameter. What is the half-angle of the cone of rays from an object point, if the cone just fills the entrance pupil?

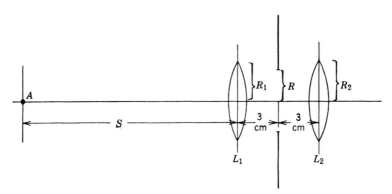

Fig. 4.58

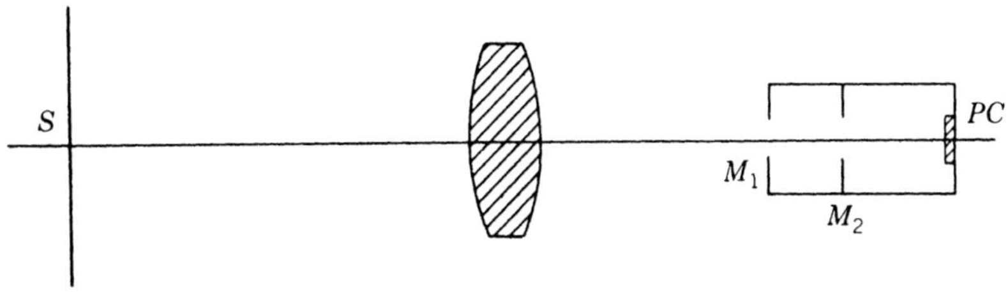

Fig. 4.59

Section 4.2   Radiometry and Photometry

14. Assume the sun obeys Lambert's law. (It does not, because of a phenomenon known as "darkening of the limb").

(a) Find its luminance.

(b) Find the intensity of light from the sun (in lumens/steradian).
An image of the sun is formed on the earth by a lens of focal length 10 m and radius 1 m.

(c) Find the illuminance in that image (neglecting losses in the lens).

(d) Find the total luminous flux hitting the screen at the image. Assume that the sun is overhead and that its radiation reaches the earth with an illuminance of 10^5 lux. The radius of the sun is 7×10^8 m. The earth-sun distance is 1.5×10^{11} m.

15. Using the information given in Problem 14, calculate the radiant intensity of the sun's image formed by a simple lens of diameter 4 cm and focal length 25 cm. Compare this to the case with no lens.

16. A light source that is square and 2 mm on a side emits 2 W of radiation. Find the radiance of the source assuming it behaves as a Lambert emitter. Find the radiant flux collected by a lens that is 20 mm in diameter situated 80 mm from the source.

17. A small lamp with a radiant intensity of 10 W/sr is 2 m above a floor. Find the irradiance on the floor directly below the lamp and on the floor 1 m away from the spot directly below the lamp.

18. Repeat the calculations of Problem 17, only now assume that the source is a disk-shaped Lambert emitter of radius 10 cm with a radiance of 1000 W/sr-m^2.

19. A Lambert source S with radiance 10^6 W/sr-m^2,

1 cm in diameter is located 45 cm from a thin lens of $+30$ cm focal length and 10 cm diameter. A detector is located 120 cm to the right of the lens. It consists of two masks, M_1 and M_2, each with 0.1 mm diameter pinholes on axis 1 cm apart, followed closely by a large photocell PC with a sensitivity of 10^3 A/W. Neglect losses in the lens and find the value of the photocurrent coming from the photocell (see Fig. 4.59).

20. The moon is 2.5×10^5 miles from the earth and is 2200 miles in diameter. Assuming that the moon acts like a uniform disk as observed on earth and with the knowledge that the moonlight amounts to 0.2 lm/m^2 on earth, calculate the luminance of the moon.

21. Refer to Fig. 3.8. This shows a coordinate system at the center of which we now assume there is an oscillating dipole $\mathbf{p}$ oriented along the $+x$-axis. The maximum amplitude of the dipole is 10^{-30} Cm. The frequency of oscillation is such that the radiation has a wavelength of 500 nm. The intensity of the electric field a radial distance r from the origin is

$$\frac{\pi}{\varepsilon_0} \frac{p}{\lambda^2 r} e^{i\omega(t-r/c)} \sin \theta_x$$

From this information calculate the irradiance at a distance of 0.1 m along the directions $\hat{\mathbf{s}} = \hat{\mathbf{z}}$, and $\hat{\mathbf{s}} = (1/\sqrt{2})(\hat{\mathbf{x}} + \hat{\mathbf{z}})$, calculate the radiant intensity along these two directions, and determine the radiant flux emitted by the dipole.

22. Let the object in Fig. 4.60 be a *large* uniform Lambert source of radiance 10 W/cm^2/sr. Assume 10% loss in each lens and calculate the total power transmitted by the system.

23. A projector uses a flat filament lamp with a luminance of 2000 cd/cm^2. The condenser lens completely

Object

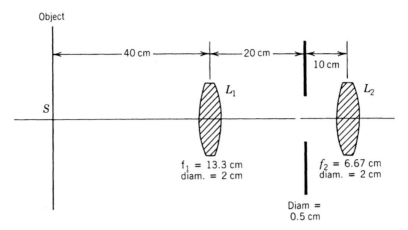

Fig. 4.60

fills the 1-in.-diameter aperture of the 2-in. focal length projection lens with the image of the lamp filament. What is the illuminance on a screen 30 ft from the projection lens if each lens transmits 90% of the light incident onto it?

24. What is the maximum luminous flux density (illuminance) at a distance of 1 km from a 100-million candlepower searchlight if, at this distance, the inequality

$$D > \frac{R_L f}{r_s}$$

is assumed to hold? If the reflector has a diameter of 1 m, what is the required luminance of the arc lamp used as a source? Does the size of the arc effect the candlepower of the searchlight? If not, what effect does the arc size have?

Section 4.3 Lens Aberrations

25. A uniform parallel beam 3 mm in radius from a helium-neon laser is brought to a focus by a 1-m focal length planocovex lens of 1-cm diameter made with glass of index of refraction 1.5. The lens is oriented to give the best performance. Calculate the transverse spherical aberration as a multiple of the wavelength, 633 nm. Repeat with $f = 20$ mm.

26. Consider the symmetrical biconvex lens of radius a when $S = 2f$. If the index of refraction of the glass is n, derive a simplification for the expression giving longitudinal spherical aberration.

27. For a single-refracting surface with $_0C_{40} = -C$ and radius r and for a paraxial image on axis and located at S', analytically determine the location and size of the circle of least confusion.

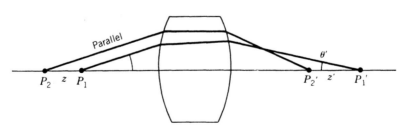

Fig. 4.61

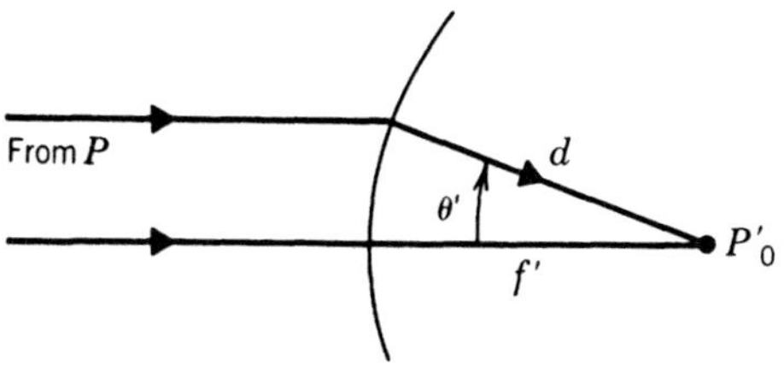

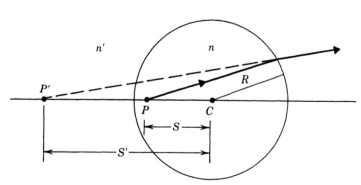

From P

θ'

d

P'_0

f'

Fig. 4.62

28. Consider a lens in air that displays only coma (hypothetical). If the object and image (paraxial) distances are equal at 20 cm and if the object (a point) is 5 cm from the optical axis, sketch the appearance of the image. Suppose that the lens has a radius of 2 cm and that $_1C_{31} = -(1/400)$ cm^{-3}. The sketch must be accurately drawn to scale.

29. Locate the conjugate *aplanatic* points of a spherical glass refracting surface of radius 6 cm if the index of refraction of the glass is 1.60.

30. Find the relationship for the image distance in terms of the object distance and the parameter R under the conditions that the primary coma for a thin lens is to be zero. The radii of curvature for the lens are $R_1 = R$ and $R_2 = -2R$, and its index is 1.5.

31. A given optical system has at most one pair of aplanatic points. In fact, the system gives perfect image formation only for the aplanatic pair of points and for points displaced laterally a small distance x. It will not give perfect image formation for points displaced axially a small distance z. Show that the conditions for

perfect image formation of P_1 onto P'_1 and of P_2 onto P'_2 gives this expression to first order in z:

$$n'z' \sin^2 \frac{\theta'}{2} = nz \sin^2 \frac{\theta}{2}$$

(*Hint:* Use the rays shown in Fig. 4.61).
Except for a plane mirror, it is incompatible with the sine condition. Why?

32. Show that to have an aplanatic pair of points with the object at infinity we must have $d = x'/\sin \theta' = $ const $= f'$. (See Fig. 4.62.)

33. The spherical system in Fig. 4.63 was already considered in Problem 9 of Chapter 3, where you were asked to show that P' is a perfect image of P provided that $n^2S = n'^2S' = nn'R$. Now show that P and P' form an aplanatic pair; that is, show that the sine condition is obeyed.

34. One practical method of obtaining an object at P in Problem 33 is to use a meniscus lens and to put the object at the center of curvature C_1 of the first surface (Fig. 4.64). Then the first surface forms a perfect image of C_1 at C_1. The meniscus lens has a natural axis through the two centers of curvature C_1 and C_2. Show that the imaging by the first surface of an object O a short distance x above this axis from C_1 will satisfy the sine condition, and that the image I lies along the line from C_1 to O. (The imaging of the second surface was shown to be aplanatic in Problem 33. The final image will be along the line C_2I.) Hence the overall imaging is aplanatic. This property of meniscus lenses is often used in microscope objectives.

n' n

P'

R

P C

S

S'

Fig. 4.63

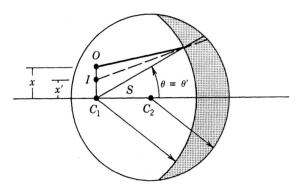

Fig. 4.64

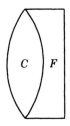

Fig. 4.65

35. Consider a thin lens with $n = 1.5$, $f = 10$ cm, and $r = 2$ cm in air that displays only astigmatism and curvature of field. For a point object at 20 cm in front of the lens located 2 cm off the optical axis, find the locations of the sagittal focus, the circle of least confusion, and the tangential focus in terms of deviations from the paraxial focus along the chief ray.

36. If the tangential focal point is 30.1 cm and the radial focal point is 31.4 cm from the simple lens, find the maximum size of a lens that demonstrates only astigmatism and curvature of field that will have a circle of confusion less than 1 mm in diameter.

37. A commercial spectrophotometer uses a 1/2-meter focal-length mirror in an $f/10$ system (see Problem 12 for a definition of $f/\#$) with an exit pupil 1/4 meter from the mirror. Parallel light incident on the mirror gives a paraxial point image 1 cm off axis. Discuss the size and shape of the true "image" resulting from primary spherical aberration, coma, and astigmatism.

38. Design an achromatic doublet with the contour indicated in Fig. 4.65. Use glass 510635 as the crown component and glass 620364 as the flint component. Treat the lenses as thin lenses with zero separation and find a combination of radii of curvature that will produce a lens having a focal length of 100 cm.

39. The Huygens eyepiece of Fig. 4.51 made with glass 500650 is adjusted to give an image of an achromatic object (virtual) that is at infinity for blue light. Where is the image for red light?

40. Calculate the lateral chromatic aberration, expressed as a fractional change in magnification, for the Ramsden eyepiece of Fig. 4.53 if the glass used is the same as that of Problem 39.

5 Interference

We now begin our study of physical optics where the details of the wavelike properties of light become more important. This will be the case where the effects of superposition of the optical electric fields of several waves play a major role in the problem. *Linear superposition* is the process whereby, at a specific point in space and at a specific time, the electric fields from several sources are vector-combined to yield a resultant electric field:

$$\mathbf{E} = \mathbf{E}_1 + \mathbf{E}_2 + \mathbf{E}_3 + \dots \tag{5.1}$$

This concept is also important in geometrical optics. The idea of superposition, including relative phase differences among the various components of the "optical disturbance," was brought out in our discussions supporting Huygen's and Fermat's principles. We also used the superposition principle in our treatment of reflection and refraction. We will see how the laws of image formation can be derived from the perspective of physical optics as well.

As a special case of physical optics, geometrical optics can be described by a formalism that does not explicitly keep track of the phases of the optical electric fields. The superposition principle is at work behind the scenes, however, in nearly every classical optics phenomenon. The reason for this is that the electric and magnetic fields that describe light also must obey Maxwell's equations. Maxwell's formalism leads to a linear second-order partial differential wave equation,

$$\nabla^2 \mathbf{E} - \frac{1}{v^2} \frac{\partial^2 \mathbf{E}}{\partial t^2} = 0 \tag{5.2}$$

which restricts the type of vector functions that can be used to describe the optical electric field. An important characteristic of the wave equation is that any two vector functions that satisfy Eq. (5.2) may be vectorially combined to yield a third vector function that also will satisfy Eq. (5.2). Thus the optical superposition principle can be directly traced to the electromagnetic character of light.

This is rigorously true for light in a vacuum. When a medium is introduced into the problem, the superposition principle holds provided that the response of the medium to the optical field can be approximated as linear. Within the approximation of linear, isotropic media, as discussed in Chapter 2, the induced dipole moment per unit volume, $\mathbf{P}_L$, is related to the electric field in the medium by

$$\mathbf{P}_L = \varepsilon_0 \chi_L \mathbf{E} \tag{5.3}$$

where χ_L is the linear susceptibility. Nonlinear effects, by contrast, will involve an additional contribution to $\mathbf{P}$ of $\mathbf{P}_{NL}$, whose components are

$$(P_{NL})_i = \sum_{k=1}^{3} \sum_{j=1}^{3} d_{ijk} E_j E_k \quad (i = 1, 2, 3) \tag{5.4}$$

In addition to possessing anisotropic character through the nonlinear susceptibility, d_{ijk}, the nonlinear polarizability bears a quadratic dependence on the optical fields. When $\mathbf{P}_{NL}$ can be neglected relative to $\mathbf{P}_L$, we refer to the medium as "linear." This is always the case for sufficiently small E^2 (i.e., for sufficiently weak energy density).

5.1 Two-Beam Interference

A. General Considerations

A wide variety of quite different-appearing interference experiments involve the superposition of the fields in two separate light beams, which in practically all cases are derived from the same source. If the two beams travel separate paths having different optical path lengths, there will be a difference in the phases of the fields at the point of recombination. If the source is monochromatic, the phase difference will be independent of time. We detect the time average of the power density that is proportional to the time average of the energy density in the local optical field (see section 1.C2)

$$\langle U \rangle = \varepsilon \langle \mathbf{E} \cdot \mathbf{E} \rangle \tag{5.5}$$

Here the electric field, $\mathbf{E}$, is the resultant of the vector sum of the electric fields in each of the two beams at the point of recombination, $\mathbf{E} = \mathbf{E}_1 + \mathbf{E}_2$. The explicit introduction of the two contributions leads to

$$\langle U \rangle = \varepsilon \langle \mathbf{E}_1 \cdot \mathbf{E}_1 + \mathbf{E}_2 \cdot \mathbf{E}_2 + 2\mathbf{E}_1 \cdot \mathbf{E}_2 \rangle$$

or

$$\langle U \rangle = \langle U_1 \rangle + \langle U_2 \rangle + 2\varepsilon \langle \mathbf{E}_1 \cdot \mathbf{E}_2 \rangle \tag{5.6}$$

The first two terms in Eq. (5.6) are the average energy densities of the two beams considered independently. The last term is caused by interference. The influence of interference may be constructive or destructive depending on whether the sign of the interference term is positive or negative, respectively.

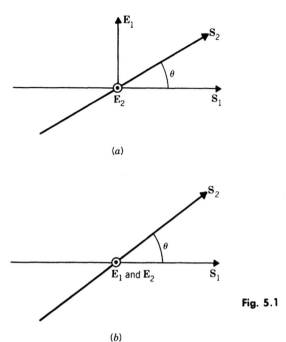

(a)

(b)

Fig. 5.1

The vector character of this term cannot, in general, be ignored. If the fields are perpendicular, that is, if the polarization directions are orthogonal, the interference term vanishes. This situation is shown in Fig. 5.1a. When the fields are parallel, as in Fig. 5.1b, interference may take place. However, unless the angle is small, the interference fringes will occur on a length scale so small that they may be unobservable, unless special techniques, such as those of holography, are used. To avoid the difficulties associated with Fig. 5.1, we will in most cases, restrict our attention to cases in which two beams are superimposed at a point where their directions of propagation are nearly the same and where their electric fields are colinear (linearly polarized in the same direction). The latter limitation does not restrict us from using unpolarized light for the following reason. It is possible to treat unpolarized light as composed of two perpendicularly oriented fields that are out of phase with respect to each other in a random and rapidly varying way. Each of these components can be considered independently. Therefore, when unpolarized monochromatic light is used as the source for an interference experiment that depends on variation of optical path lengths, the outcome will be the same as if linearly polarized light had been used.

We will suppress the vector nature of the fields from now on by considering

$$\langle U \rangle = \varepsilon \langle E^2 \rangle$$

or equivalently

$$\langle S \rangle = v \langle U \rangle = v \varepsilon \langle E^2 \rangle \tag{5.7}$$

The electric fields in each of the light beams must satisfy Eq. (5.2). Appropriate solutions have the form of the familiar sinusoidal travelling wave:

$$E = Ae^{i\phi} \tag{5.8}$$

where the phase is given by

$$\phi = 2\pi\left(\frac{t}{T} - \frac{R}{\lambda}\right) + \varphi = 2\pi v\left(t - \frac{R}{v}\right) + \varphi = (\omega t - kR) + \varphi \tag{5.9}$$

and where φ is a characteristic of the source. Here, as in Chapter 1, the relationships among the frequency v, angular frequency ω, wavenumber k, wavelength λ, period T, and phase velocity v of the wave are:

$$\omega = 2\pi v \tag{5.10a}$$

$$T = \frac{1}{v} \tag{5.10b}$$

$$k = \frac{2\pi}{\lambda} \tag{5.10c}$$

$$v = v\lambda \tag{5.10d}$$

It is also useful to recall that the phase velocity of the wave propagating through a nonabsorbing medium of index n will be

$$v = \frac{c}{n} \tag{5.11}$$

If the source is a geometrical point, then R in Eq. (5.9) is the radial distance from the source and the amplitude A in Eq. (5.8) is inversely proportional to R, $A \to A/R$. For spherical waves the source strength A must have the dimensions of (electric field × length). The advancing wavefronts, as identified by the surfaces of constant phase, are spheres. If the source is a geometrical line, then R is still measured from the source but the amplitude becomes inversely proportional to the square root of R, (provided R is much larger than λ) $A \to A/\sqrt{R}$. For the cylindrical wave the source strength A must have the dimensions of (electric field × $\sqrt{\text{length}}$). The surfaces of constant phase are then cylinders. Far from the sources in either of these cases, the R dependence loses its significance in the amplitude factor when compared to the rapidly varying dependence on R resulting from the phase factor. By assuming that A in Eq. (5.8) behaves as a constant, we approximate the surfaces of constant phase as planes. In this case, R appearing in the phase ϕ is the distance along the direction of propagation of the light beam, measured perpendicular to the planes of constant phase.

It should be understood that the exponential notation is a convenient representation of the actual field, which, of course, cannot be complex. We use this form because of the straightforward way in which the phases of the fields can be manipulated. At the end, before inserting the resultant field into Eq. (5.7), we must take the real part as the physically significant quantity.

B. Phasor Addition

Consider the fields in our two beams as

$$E_1 = A_1 e^{i\phi_1} \quad \text{and} \quad E_2 = A_2 e^{i\phi_2} \tag{5.12}$$

which are represented as phasors in the complex plane in Fig. 5.2a. If in the phases

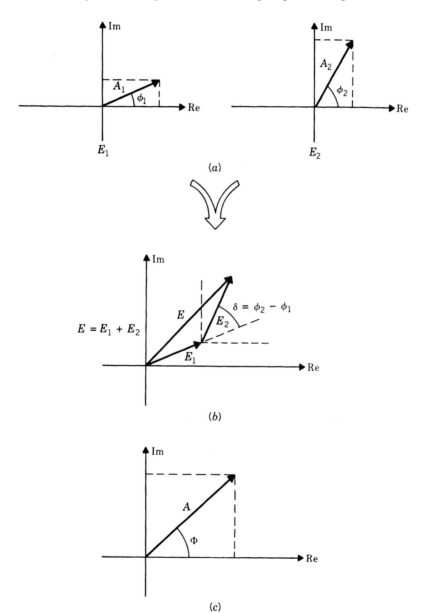

Fig. 5.2 Phasor construction showing complex addition of E_1 and E_2. (a) Component phasors. (b) Vector addition of phasors to yield E. (c) Representation of the sum E.

of these two fields [as in Eq. (5.9)] the coefficients of the terms proportional to time are the same, the interference conditions do not change as functions of time. The two phasors in Fig. 5.2a then rotate about the origin at the same rate with the phase angle difference, $\delta = \phi_2 - \phi_1$, remaining the same. We can see from this argument that the resultant field will have the same frequency as the components but will differ from each of the components with regard to its amplitude and phase. The two components are combined to form the resultant

$$E = E_1 + E_2 = Ae^{i\Phi} = A(\cos \Phi + i \sin \Phi) \tag{5.13}$$

This operation is represented in the complex plane as vector addition, as shown in Fig. 5.2b. The magnitude and phase of the resultant, as shown in Fig. 5.2c, can be expressed in terms of the real and imaginary components of the phasor:

$$A = [(\operatorname{Re} E)^2 + (\operatorname{Im} E)^2]^{1/2} = [EE^*]^{1/2} \tag{5.14}$$

$$\Phi = \tan^{-1}(\operatorname{Im} E/\operatorname{Re} E) \tag{5.15a}$$

$$= \cos^{-1}(\operatorname{Re} E/A) \tag{5.15b}$$

Where, because of the vectoral character of the phasor construction, we have,

$$\operatorname{Re} E = A_1 \cos \phi_1 + A_2 \cos \phi_2 \tag{5.16a}$$

$$\operatorname{Im} E = A_1 \sin \phi_1 + A_2 \sin \phi_2 \tag{5.16b}$$

Substituting Eqs. (5.16) into Eq. (5.14) for the amplitude of the resultant field leads to

$$A = [A_1^2 \cos^2 \phi_1 + A_2^2 \cos^2 \phi_2 + 2A_1A_2 \cos \phi_1 \cos \phi_2$$
$$+ A_1^2 \sin^2 \phi_1 + A_2^2 \sin^2 \phi_2 + 2A_1A_2 \sin \phi_1 \sin \phi_2]^{1/2}$$

or

$$A = [A_1^2(\cos^2 \phi_1 + \sin^2 \phi_1) + A_2^2(\cos^2 \phi_2 + \sin^2 \phi_2)$$
$$+ 2A_1A_2(\cos \phi_1 \cos \phi_2 + \sin \phi_1 \sin \phi_2)]^{1/2}$$

By the use of two common trigonometric identities, this becomes

$$A = [A_1^2 + A_2^2 + 2A_1A_2 \cos(\phi_2 - \phi_1)]^{1/2}$$
$$= [A_1^2 + A_2^2 + 2A_1A_2 \cos \delta]^{1/2} \tag{5.17}$$

For an analytical expression that gives rise to the phase of the resultant, we start with Eq. (5.15b) using $\phi_1 = \omega t$ and $\phi_2 = \omega t + \delta$.

$$\cos \Phi = [\operatorname{Re} E/A] = [(A_1 \cos \omega t + A_2 \cos(\omega t + \delta))/A] \tag{5.18}$$

This can be rewritten as

$$\cos \Phi = \frac{A_1}{A} \cos \omega t + \frac{A_2}{A} \cos \omega t \cos \delta - \frac{A_2}{A} \sin \omega t \sin \delta$$
$$= \cos \omega t \left[\frac{A_1}{A} + \frac{A_2}{A} \cos \delta \right] - \sin \omega t \left[\frac{A_2}{A} \sin \delta \right]$$

Let us now define the time-independent phase angle Φ', where

$$\cos \Phi' = \frac{A_1}{A} + \frac{A_2}{A} \cos \delta \qquad (5.19a)$$

and

$$\sin \Phi' = \frac{A_2}{A} \sin \delta \qquad (5.19b)$$

It can easily be demonstrated that $\cos^2 \Phi' + \sin^2 \Phi' = 1$ by direct substitution, thus justifying our assignment. This enables us to write

$$\cos \Phi = \cos \omega t \cos \Phi' - \sin \omega t \sin \Phi'$$

$$= \cos (\omega t + \Phi')$$

Therefore the phase of the resultant is

$$\Phi = \omega t + \Phi'$$

where Φ' is given by Eqs. (5.19).

These concepts are clarified in Fig. 5.3, where the instant in time corresponding to the conditions of Fig. 5.2 is illustrated.

C. Time-Averaged Power Density

Care must be exercised in the calculation of the power density of the resultant. The complex notation needs to be replaced with the sinusoidal formalism for this development because we must evaluate the time average of the field squared. From Eq. (5.7)

$$\langle S \rangle = v\varepsilon \langle E^2 \rangle$$

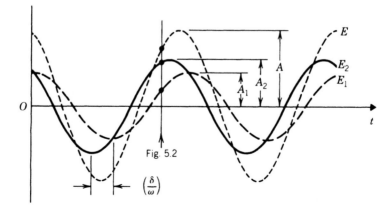

Fig. 5.3 Algebraic superposition of $E_1(t)$ and $E_2(t)$ to yield $E(t)$. The conditions at the vertical line are the same as those that give rise to the phasor construction of Fig. 5.2.

where

$$E^2 = A^2 \cos^2(\omega t + \Phi')$$

$$= A^2[\cos \omega t \cos \Phi' - \sin \omega t \sin \Phi']^2$$

$$= A^2[\cos^2 \omega t \cos^2 \Phi' + \sin^2 \omega t \sin^2 \Phi'$$

$$- 2 \cos \omega t \sin \omega t \cos \Phi' \sin \Phi'] \qquad (5.20)$$

The time average of Eq. (5.20) involves

$$\langle \cos^2 \omega t \rangle = \langle \sin^2 \omega t \rangle = \frac{1}{2} \qquad (5.21a)$$

and

$$\langle \cos \omega t \sin \omega t \rangle = 0 \qquad (5.21b)$$

So we are left with

$$\langle E^2 \rangle = A^2 \left[\frac{1}{2} (\cos^2 \Phi' + \sin^2 \Phi') \right]$$

$$= \frac{A^2}{2} \qquad (5.22)$$

This leads to

$$\langle S \rangle = \frac{v\varepsilon}{2} A^2 \qquad (5.23)$$

and using Eq. (5.17) we end up with,

$$\langle S \rangle = \frac{v\varepsilon}{2} [A_1^2 + A_2^2 + 2A_1 A_2 \cos \delta]$$

This can be rewritten in terms of the component beam power densities as

$$\langle S \rangle = \langle S_1 \rangle + \langle S_2 \rangle + 2\sqrt{\langle S_1 \rangle \langle S_2 \rangle} \cos \delta \qquad (5.24)$$

From now on in this chapter we will drop the time average symbolism $\langle \ldots \rangle$ with it being understood that the power density S is always the time average $\langle S \rangle$.

An important special case occurs when $A_1 = A_2$. Then S is given by

$$S = 2S_1[1 + \cos \delta] \qquad (5.25a)$$

or

$$S = 4S_1 \cos^2(\delta/2) \qquad (5.25b)$$

This equation is plotted in Fig. 5.4. Note that, although the peak in the time-average power density is twice as large as the sum of the source power densities, energy conservation is still maintained. The spatial average of Eq. (5.25) over an integral number of oscillations is $2S_1$ as it must be. The interference piles up the energy in some regions of space while stealing it from other regions of space.

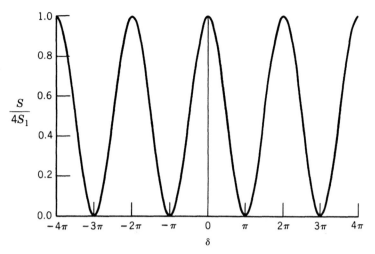

Fig. 5.4 Power density distribution in the two-beam interference pattern.

D. Interference Conditions

1. Polarization. We have already seen how the polarization of the electric fields of the component beams must be colinear at the time and location of superposition if those beams are to interfere with each other. In addition to this, restrictions exist on the phases of the component beams.

2. Frequency. Assume *for the time being* that two light beams are superimposed after having traveled the *same optical paths* as measured from their sources. Then the phase difference is

$$\delta = \phi_2 - \phi_1 = (\omega_2 - \omega_1)t + (\varphi_2 - \varphi_1) \tag{5.26}$$

As shown in the phasor presentation in the previous section, if interference is to be observable, this quantity must not change during the time, T, of the measurement. This requires that the two sources have the same frequency. Then the first term vanishes.

3. Coherence. The remaining term in the phase difference is the contribution $\varphi_2 - \varphi_1$ coming from the sources themselves. Any true source of light possesses random phase fluctuations that are slowly varying compared to the time required for one period of oscillation of the optical field.

Let us define a mutual coherence time τ_c as the time necessary for the mean square value of $\varphi_2 - \varphi_1$ to change by, say, 1 rad. (We can make this more precise later.) Then if $\tau_c \gg T$, we observe interference effects, whereas if $\tau_c < T$, we do not. In the former case we say the two sources are *mutually coherent*; in the latter case they are *mutually incoherent*. Note that the concept of coherence is a quantitative one, because it depends on the relative sizes of the mutual coherence time and the response time of the measuring apparatus.

Consider the case of two radio transmitters as an example. Each can be made quite stable, but unless they are actually locked into phase by some kind of control unit, they will drift out of phase with respect to one another. It is the same with *lasers*. Unless special synchronization schemes are used, two lasers will drift out of phase in a small fraction of a second ($\tau_c \simeq 10^{-3}$ sec). Ordinary gas-discharge sources of "monochromatic" radiation such as neon lamps or mercury lamps will have coherence times shorter than 10^{-8} sec. With fast photography (exposure times of 10^{-3} sec or less) we might be able to photograph the interference pattern from two separate unsynchronized lasers, but for ordinary sources this would be virtually impossible. We will return to the subject of coherence in Chapter 8.

The traditional method of obtaining two sources that are mutually coherent is to have two sources related ultimately to the same single source. Two famous examples are provided by Young's experiment and by the Michelson interferometer. The Young's geometry involves a division of the wavefront whereas the Michelson device exploits a division of the amplitude of the optical wave. We will investigate both cases in detail.

E. Young's Configuration

In 1807, Thomas Young performed one of the earliest experiments involving interference. Sunlight from a single pinhole in a screen was arranged so as to illuminate a second screen containing two pinholes. These behaved as mutually coherent sources. Modern duplication of this experiment usually involves slits instead of pinholes to simplify the analysis and to improve the throughput. The basic elements are shown in Figure 5.5a. Light from the source L impinges on two very narrow slits at L_1 and L_2 in a barrier screen. The cylindrical secondary Huygens wavelets there act as if they originated at L_1 and L_2. The spreading of the light beyond the slits will be pronounced if the slits are narrow. This is a manifestation of the phenomenon of diffraction, which we will discuss in much detail later. For now we simply assume L_1 and L_2 to be coherent line sources with well-defined amplitudes and the same phase. This phase difference will depend only on the difference in optical paths ($n_1\overline{LL_1} - n_2\overline{LL_2}$) and hence remains constant if the mechanical mounts are sturdy and free from vibration.

We measure the flux density on an observation plane parallel to and a distance D' from the barrier plane. At L', as identified by coordinate x' in the observation plane, the two component beams are superimposed after having traveled optical paths that differ by ($R'_1n_1 - R'_2n_2$). If both beams pass through the same medium (here assumed to have $n = 1$) then the optical path length (OPL) difference is

$$R'_1 - R'_2 = \overrightarrow{L_1M'}$$

If L' is close to the z-axis ($x' \ll D'$) and if the line sources are close together ($a \ll D'$), then we can exploit the following approximation

$$\overrightarrow{L_1M'} \cong a \sin \theta' \qquad (5.27)$$

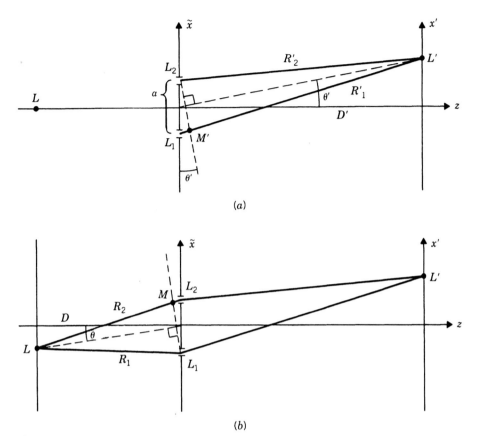

Fig. 5.5 Young's geometry in which parallel slits are illuminated by the radiation from a single source slit. In (a) the source slit is on the z-axis. In (b) the source is displaced away from the axis.

But because $\sin \theta' \cong x'/D' \cong \theta'$, the OPL difference is

$$R'_1 - R'_2 \cong \frac{ax'}{D'} \cong a\theta' \tag{5.28}$$

The amplitudes of the optical fields at L' bear the characteristic $R^{-1/2}$ dependence of cylindrical waves. We can ignore this compared with the effects of differences in phase between the two beams because in our approximation $R'_1 \cong R'_2 \cong D'$.

The phase difference from Eq. 5.9 will be

$$\delta = \phi_2 - \phi_1 = -k(R'_2 - R'_1) = k(R'_1 - R'_2)$$

which, using Eq. (5.28), becomes

$$\delta \cong kax'/D' = \frac{2\pi}{\lambda}\frac{ax'}{D'} \tag{5.29}$$

If we also assume that L_1 and L_2 have equal brightness, then the time-averaged power density at L' is given by Eq. 5.25

$$S = 4S_1 \cos^2\left(\frac{\pi}{\lambda}\frac{ax'}{D'}\right) \tag{5.30}$$

As L' moves in the observation plane toward or away from the z-axis, the power density varies in a pattern of fringes illustrated in Fig. 5.4, where the abscissa may be interpreted as proportional to observation position through Eq. (5.29). The separation $\Delta x'$ between two successive maxima (or minima) is obtained by setting $\Delta \delta = 2\pi$ and solving for $\Delta x'$:

$$\Delta x' = \frac{\lambda D'}{a}$$

If the source is not on the z-axis as shown in Fig. 5.5b, then an additional OPL difference is introduced on the incident side of the barrier

$$R_2 - R_1 = \overrightarrow{ML_2} \cong a \sin \theta$$

The total OPL difference from L to L' between beams passing through L_1 and L_2 is

$$(OPL)_2 - (OPL)_1 = (R_2 + R'_2) - (R_1 + R'_1)$$
$$= (R_2 - R_1) - (R'_1 - R'_2)$$
$$\cong a(\sin \theta - \sin \theta')$$

The total phase difference is then a function of both θ and θ'.

$$\delta = \phi_2 - \phi_1 = -k[(R_2 + R'_2) - (R_1 + R'_1)]$$
$$\cong ka(\sin \theta' - \sin \theta)$$
$$= 2\pi \frac{a}{\lambda}(\sin \theta' - \sin \theta) \tag{5.31}$$

F. Other Split-Source Configurations

There are other arrangements for producing interference by splitting the source. Some are more efficient than Young's geometry and lead to patterns whose maxima have larger power density. *Lloyd's mirror* consists of a line source and a single mirror used at almost grazing incidence. The interference occurs between the source and the reflection, as shown in Fig. 5.6. Two virtual line sources can be produced from a single line source by refraction through a pair of thin prisms whose bases are cemented together as shown in Fig. 5.7. This configuration is known as *Fresnel's biprism*. In both of these examples the two beams overlap without the aid of diffraction.

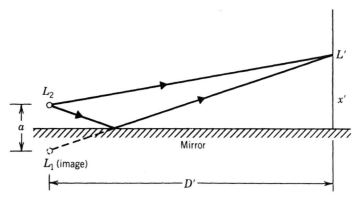

Fig. 5.6 Lloyd's mirror.

5.2 Multiple-Beam Interference

A. Phasor Addition

The procedure followed in Section 5.1A can be generalized to the case of several coherent light beams. We wish to find the time-average power density at the point of superposition of N beams. This will follow from Eq. (5.23) with the amplitude of the resultant given by Eq. (5.14). For the case of multiple-beam interference, however, we must replace Eq. (5.13) and (5.16) with

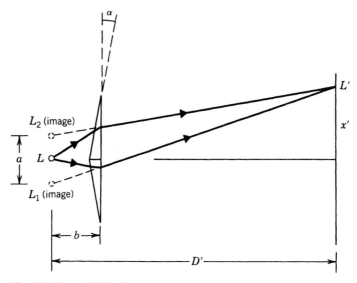

Fig. 5.7 Fresnel's biprism.

$$E = E_1 + E_2 + \cdots + E_N = \sum_{n=1}^{N} E_n = Ae^{i\Phi} = A(\cos\Phi + i\sin\Phi) \quad (5.32)$$

$$\text{Re } E = A_1\cos\phi_1 + A_2\cos\phi_2 + \cdots + A_N\cos\phi_N = \sum_{n=1}^{N} A_n\cos\phi_n \quad (5.33a)$$

$$\text{Im } E = A_1\sin\phi_1 + A_2\sin\phi_2 + \cdots + A_N\sin\phi_N = \sum_{n=1}^{N} A_n\sin\phi_n \quad (5.33b)$$

When Eqs. (5.33) are substituted into Eq. (5.17) for the amplitude, A, of the resultant, we will find two types of terms under the square root. One type has the form $A_n^2(\cos^2\phi_n + \sin^2\phi_n) = A_n^2$, and the other type has the form

$$A_n A_m(\cos\phi_n\cos\phi_m + \sin\phi_n\sin\phi_m),$$

where n and m are not equal. When A^2, determined in this way, is substituted into Eq. (5.23), we see that terms of the first type lead to the independent beam power densities, whereas those of the second type provide the interference

$$S = \sum_{n=1}^{N} S_n + \sum_{\substack{n=1 \\ n\neq m}}^{N} \sum_{m=1}^{N} \sqrt{S_n S_m} \cdot \cos(\phi_m - \phi_n) \quad (5.34)$$

B. Mathematical Solution

It is convenient to use the exponential notation to determine the resultant in the multiple-beam situation.

$$E = A_1 e^{i\phi_1} + A_2 e^{i\phi_2} + \cdots + A_N e^{i\phi_N} = \sum_{n=1}^{N} A_n e^{i\phi_n}$$

$$= A e^{i\Phi} \quad (5.35)$$

When all the amplitudes are equal and if $\phi_n = \omega t + \delta(n-1)$, as in Fig. 5.8, we have

$$E(\delta) = A_1 e^{i\omega t}(1 + e^{i\delta} + e^{i2\delta} + \cdots + e^{i(N-1)\delta}) = A_1 e^{i\omega t}\sum_{n=0}^{N-1} e^{in\delta} \quad (5.36)$$

The resultant as expressed by Eq. (5.36) takes the form of a geometric progression with $e^{i\delta} \equiv r$ as the common ratio. To evaluate the sum consider

$$\sigma_N \equiv \sum_{n=0}^{N} r^n = (1 + r + r^2 + \cdots + r^N)$$

$$= \sigma_{N-1} + r^N \quad (5.37a)$$

We also can form another expression by subtracting one from the sum.

$$\sigma_N - 1 = r\sigma_{N-1} \quad (5.37b)$$

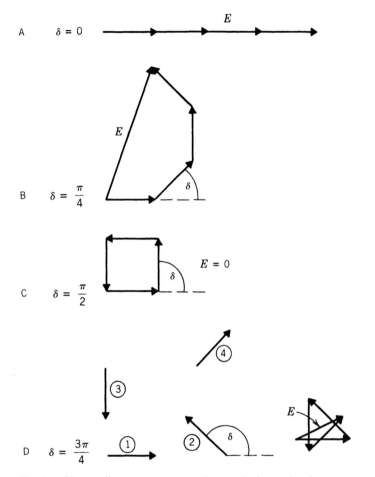

Fig. 5.8 Phasor diagrams corresponding to the lettered points on the power-density plot of Fig. 5.10a.

Now eliminate σ_N from Eqs. (5.37a) and (5.37b) to yield

$$\sigma_{N-1} = \left(\frac{1-r^N}{1-r}\right) = \sum_{n=0}^{N-1} r^n \tag{5.37c}$$

By comparison we can see that the sum in Eq. (5.36) is exactly of this form. Therefore,

$$E(\delta) = A_1 \, e^{i\omega t}\left(\frac{1-e^{iN\delta}}{1-e^{i\delta}}\right) = A_1 \, e^{i\omega t} \frac{e^{iN\delta/2}}{e^{i\delta/2}}\frac{(e^{-iN\delta/2}-e^{iN\delta/2})}{(e^{-i\delta/2}-e^{i\delta/2})}$$

$$= A_1 \exp[i(\omega t + (N-1)\delta/2)]\frac{\sin(N\delta/2)}{\sin(\delta/2)} \tag{5.38}$$

To find the power density according to Eq. (5.23) we must evaluate the square of the amplitude of the resultant, $A^2 = EE^*$. If we associate S_1 with $v\varepsilon A_1^2/2$, the power density due to one of the beams alone, then the power density of the resultant is

$$S(\delta) = v\varepsilon \frac{A^2}{2} = S_1 \left(\frac{\sin(N\delta/2)}{\sin(\delta/2)} \right)^2 \qquad (5.39)$$

We can determine the value of this function at $\delta \to 0$ through an application of L'Hospital's rule.

$$\lim_{\delta \to 0} \left(\frac{\sin(N\delta/2)}{\sin(\delta/2)} \right) = N \lim_{\delta \to 0} \left(\frac{\cos(N\delta/2)}{\cos(\delta/2)} \right) = N$$

This leads to

$$\lim_{\delta \to 0} S(\delta) = S_1 N^2 \equiv S(0) \qquad (5.40)$$

Under this condition all the fields add constructively. This will also occur when

$$\delta = m2\pi \qquad (5.41)$$

where m is an integer. Equation (5.41) identifies the conditions for the *principal maxima*. These occur under the condition that the OPL difference between adjacent beams is a multiple of λ. The integer m is the *order* of the maximum and is the number of wavelengths by which adjacent beam OPL's differ.

The numerator in Eq. (5.39) can be zero without the denominator being zero. This occurs for

$$\delta = \frac{2\pi}{N}, \frac{4\pi}{N}, \frac{6\pi}{N}, \ldots, \frac{N-1}{N} 2\pi \qquad (5.42)$$

There are $N - 1$ of these minima between adjacent principal maxima. Between these minima there are $N - 2$ *minor maxima*. These are identified by the condition

$$\frac{d}{d\delta} \left(\frac{\sin(N\delta/2)}{\sin(\delta/2)} \right) = 0$$

or

$$\frac{\tan(N\delta/2)}{N \tan(\delta/2)} = 1 \qquad (5.43)$$

The case for $N = 4$ is shown in Fig. 5.8. Complex plane diagrams of the phasors are also given for various values of the phase shift. Here we have used Eq. (5.40) to normalize the power density with respect to its value at the principal maxima.

$$S(\delta) = S(0) \left(\frac{\sin(N\delta/2)}{N \sin(\delta/2)} \right)^2 \qquad (5.44)$$

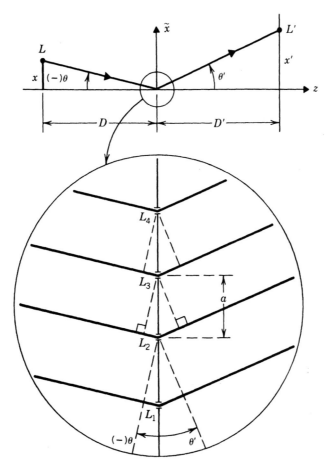

Fig. 5.9 Geometry of the transmission grating showing the detail of the OPL differences between adjacent contributions from slits separated by a. Angles are measured positive counterclockwise from the z-axis.

A practical geometry in which the four slit interference pattern may be observed is pictured in Fig. 5.9.

If $N = 2$, then Eq. (5.44) becomes

$$S(\delta) = S(0)\left(\frac{\sin(\delta)}{2\sin(\delta/2)}\right)^2$$

And using $\sin\delta = 2\sin(\delta/2)\cos(\delta/2)$ we have

$$S(\delta) = S(0)\cos^2(\delta/2) \tag{5.45}$$

which is the same as Eq. (5.25), which we derived for the case of Young's experiment.

C. Gratings

For large N, the power density in the interference pattern will be concentrated near $\delta = 2\pi m$, the locations of the principal maxima. The width of the principal maxima [as identified by the separation between the first minima on either side of a principal maximum, from Eq. (5.42), this is $\Delta\delta = 4\pi/N$] becomes very narrow.

We call this type of interference device a *grating*. Because diffraction is required for the contribution from each of the individual slits to overlap in the plane of superposition, these are commonly called *diffraction gratings*. However, the primary use of the grating is as a product of multiple-beam interference, as is being dealt with here. Later we will introduce the features of the pattern that result from the size and shape of the slits. For now, we continue to restrict our treatment to the case of infinitely long, very narrow slits.

Not all of the power density is delivered to the principal maxima, however. The minor maxima are attenuated so much that for large N they can be ignored. To see this more clearly consider Eq. (5.44) at the values of δ given by Eq. (5.43). These will be very *close* to $\delta = 3\pi/N, 5\pi/N, \ldots, (2N-3)\pi/N$. If the first minor maximum is identified with $\ell = 1$, the second with $\ell = 2$, and so forth up to $\ell = N - 2$, we can index the minor maxima with ℓ and express δ as

$$\delta = (2\ell + 1)\pi/N; \quad \ell = 1, 2, \ldots, (N - 2) \tag{5.46}$$

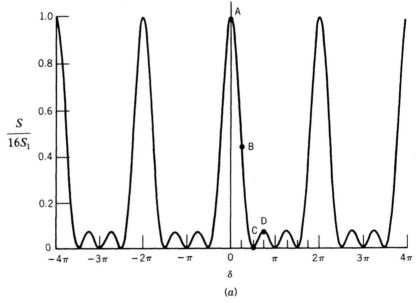

Fig. 5.10 (*a*) Power density distribution in the four-source multiple-beam interference pattern. (*b*) Location of the first minimum ◯ and the first minor maximum ◇ on either side of a principal maximum at $\delta = 0$. This shows that as N, the number of slits, increases the pattern becomes concentrated at $\delta = 0$. (*c*) Normalized relative power density for the first minor maximum ◇ and at the midpoint between principal maxima ▢. This shows how, as N increases, virtually no power appears in the middle of the pattern.

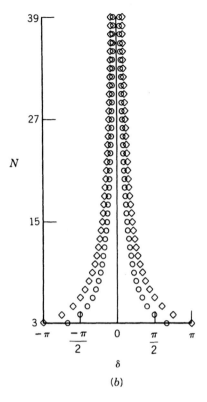

(b)

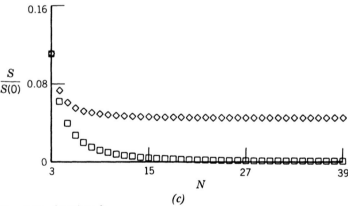

(c)

Fig. 5.10 (continued)

An approximation for the power density at these minor maxima follows directly.

$$S(\prime) = \frac{S(0)}{N^2}\left[\sin^2\left(\frac{(2\prime + 1)\pi}{2N}\right)\right]^{-1} \qquad (5.47)$$

If we have an odd number of slits, then one of the minor maxima will be halfway between the principal maxima, at $\ell = (N-1)/2$. Here the power density is $S(0)N^{-2}$. Figure 5.10 shows these characteristics as functions of N.

Gratings have been extremely useful in spectroscopy because of their ability to separate polychromatic light into its monochromatic components. This process is called *dispersion*. The condition for an interference maximum can be written in the form of the familiar grating equation by combining Eqs. (5.31) and (5.41).

$$a(\sin \theta' - \sin \theta) = m\lambda \tag{5.48}$$

The dispersion implied here can be seen from the wavelength dependence of the angle of observation, θ', for a fixed incident angle, θ, and nonzero-order m. Differentiating, we obtain

$$a \cos \theta' \, d\theta' = m \, d\lambda$$

$$\frac{d\theta'}{d\lambda} = \frac{m}{a \cos \theta'} = \frac{m}{a_\perp} \tag{5.49}$$

where $a_\perp = a \cos \theta'$ is the projected slit separation.

Suppose that light of two wavelengths λ_i and λ_j is present in the incident beam propagating in direction θ. The zeroth-order interference occurs in the same direction. For λ_i the first-order interference occurs at $(\sin \theta' - \sin \theta) = \lambda_i/a$, for λ_j at

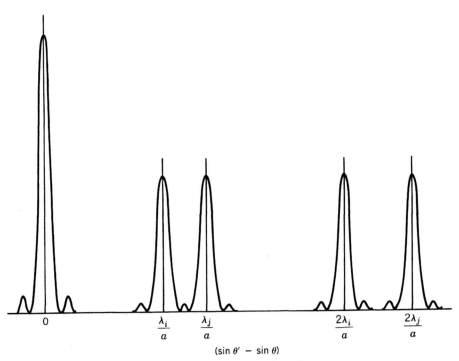

$(\sin \theta' - \sin \theta)$

Fig. 5.11 Grating interference pattern when two wavelengths are present.

λ_j/a. The second-order interference for λ_i is at $2\lambda_i/a$, that for λ_j at $2\lambda_j/a$, and so on as indicated in Fig. 5.11. Note that the separation of the two peaks in the mth order is

$$\Delta(\sin \theta') = \frac{m\,\Delta\lambda}{a} \tag{5.50}$$

We now ask: How close together may λ_i and λ_j be so that they are just barely distinguishable as separate peaks? If the two lines are equally intense, the criterion usually applied, the *Rayleigh criterion*, is to say that they are just "resolved" in mth order if the mth-order maximum of λ_j occurs at the first zero just beyond the mth maximum of λ_i (Fig. 5.12). From maximum to first minimum involves a change in δ of $2\pi/N$, and hence the change in $\sin \theta'$ is $\lambda/(Na)$, where λ could be either λ_i or λ_j, as they are almost equal. Equating the expression for $\Delta \sin \theta'$ with that in Eq. (5.50), we obtain $\lambda/(Na) = m\,\Delta\lambda/a$. We define the *resolving power* as

$$\mathscr{R} \equiv \frac{\lambda}{\Delta\lambda} \tag{5.51}$$

which in this case becomes

$$\mathscr{R} = Nm \tag{5.52}$$

This number can be quite large. Moderately-priced grating monochromators are available that use gratings having typically 1200 lines/mm with a useful area 100 mm long. (The best gratings for most spectroscopic purposes are those produced holographically. They tend to be freer of irregularities than ruled gratings.) Thus,

$$N = 100\ \text{mm} \times 1200\ \text{lines/mm} = 120{,}000\ \text{lines}$$

This would give a theoretical resolving power in first-order of 120,000. If $\lambda = 600$ nm, $\Delta\lambda = 1/200$ nm.

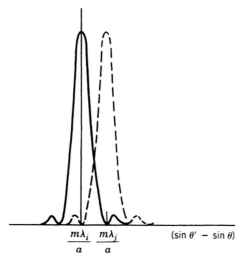

Fig. 5.12 Interference in mth order at the limit of resolution.

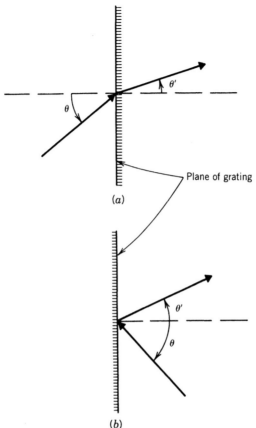

Fig. 5.13 Sign conventions for grating interference in (a) transmission and (b) reflection.

In actual use we find more gratings of the reflection type than of the transmission type. The simplest reflection grating is an exact analog of the simplest transmission grating—where the latter has an opening, the former has a reflecting surface, and where the latter is opaque, the former is nonreflecting, as indicated in Fig. 5.13.

Zero-order interference, where $\theta' = \theta$, corresponds now to specular reflection from the plane of the grating. The mth-order interference maxima still occur at $\delta = 2\pi m$, which gives the same grating equation

$$a(\sin \theta' - \sin \theta) = m\lambda$$

5.3 Two-Beam Interference: Parallel Interfaces

A. Optical Path Length Difference in a Dielectric Layer

Another method of effectively obtaining two coherent sources from the same physical source is to use the double-reflection process illustrated in Fig. 5.14.

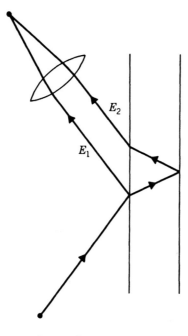

Fig. 5.14 Two-beam interference produced by reflection from two parallel interfaces.

The pertinent physical and geometrical relationships can be derived with the help of Fig. 5.15. The parallel sided slab of index n_2 is imbedded in a medium with refractive index n_1. We consider here only those materials that lead to low reflectivities at each interface, such as a glass slab in air. In section 5.4 we will treat

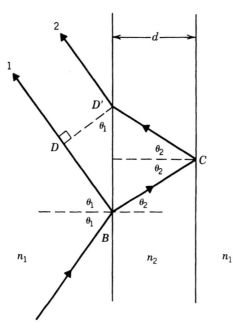

Fig. 5.15 Detail of the parallel interface geometry.

the case of highly reflecting interfaces. Our current approximations ensure that the dominant effects can be described by the two-beam interference formalism. E_1 and E_2 are the optical fields of the component beams that must be brought together with a lens (not shown) to demonstrate interference.

At the point of superposition we must evaluate the relative amplitudes and phases of the component beams. This can be done by comparing them at the line DD', because beyond this line both beams will be subjected to the same optical processing. We consider variations measured with respect to point B, outside the slab, because both beams here are one in the same incident beam. Let R at B be zero so that the incident optical field is $Ae^{+i\omega t}$. The amplitude reflection and transmission coefficients have been derived in Chapter 2. If the beam is incident onto the interface from medium 1, we label these ρ and τ, respectively. If the beam is incident from medium 2, they will be ρ' and τ'. The amplitudes of the pertinent fields are listed in Table 5.1. The arrows show how beam 2 is changed in becoming the transmitted beam at D'. Here the amplitude factors $\tau'\rho'\tau$ can be rewritten with the aid of Eqs. (2.71) and (2.72).

$$\tau'\rho'\tau = -\rho(1 - \rho^2)$$

Since ρ is small, ρ^2 will be even smaller—for an air-glass system $(1 - \rho^2) \cong 0.96$. The amplitudes of the component beams are then very nearly

$$A_1 = \rho A$$
$$A_2 = -\rho A \tag{5.53}$$

The minus sign comes from the phase shift of 180° that occurs on reflection from an interface where the incident medium is optically less dense than the transmitting medium. Rather than retaining the minus sign in the amplitude expression for A_2, we choose to maintain this effect by including an additional term in the phase of E_2. The magnitudes of the two components within our approximation are equal. We may now use Eq. (5.25) to find the time-averaged power-density as a function of the phase difference introduced by the nonequal optical path lengths and the reflection at B.

Table 5.1. Demonstration of the contributions to the amplitude factors for the fields in two-beam parallel interface interference

	Incident	Reflected	Transmitted
B	A	$\boxed{\rho A}$	τA
C	τA	$\rho'\tau A$	
D'	$\rho'\tau A$		$\boxed{\tau'\rho\tau A}$

In general the phase is given by

$$\phi = \omega t - \frac{2\pi R}{\lambda} = \omega t - \frac{2\pi n R}{\lambda_0}$$

where λ_0 is the vacuum wavelength and n is the refractive index. The phase difference between E_1 at D and E_2 at D' will be

$$\phi_2 - \phi_1 = \frac{-2\pi}{\lambda_0}[n_2(\overline{BC} + \overline{CD'}) - n_1\overline{BD}] + \pi \tag{5.54}$$

From Fig. 5.15 we see that

$$\overline{BC} = \overline{CD'} = \frac{d}{\cos\theta_2} \tag{5.55}$$

This also shows that $\overline{BD} = \overline{BD'}\sin\theta_1$. We want to express this in terms of the film thickness. This is done by recognizing that $\overline{BD'} = 2d\tan\theta_2$. Therefore,

$$\overline{BD} = 2d\tan\theta_2\sin\theta_1 \tag{5.56}$$

Combining Eqs. (5.55) and (5.56) for the physical path lengths with Eq. (5.54), we arrive at the expression for the phase difference

$$\phi_2 - \phi_1 = \frac{-2\pi}{\lambda_0}\left[2d\left(\frac{n_2}{\cos\theta_2} - n_1\tan\theta_2\sin\theta_1\right)\right] + \pi \tag{5.57}$$

Use of Snell's law helps to reduce this to a simpler form.

$$n_1\tan\theta_2\sin\theta_1 = n_1\frac{\sin\theta_2}{\cos\theta_2}\sin\theta_1 = n_2\frac{\sin^2\theta_2}{\cos\theta_2}$$

so

$$\phi_2 - \phi_1 = \frac{-2\pi}{\lambda_0}\left[\frac{2dn_2}{\cos\theta_2}(1 - \sin^2\theta_2)\right] + \pi$$

$$= \frac{-4\pi n_2 d}{\lambda_0}\cos\theta_2 + \pi \tag{5.58}$$

Eq. (5.25) then leads to

$$S(\delta) = 2S_1[1 + \cos(-\delta + \pi)]$$

$$= 2S_1[1 - \cos(\delta)]$$

$$= 4S_1\sin^2(\delta/2) \tag{5.59}$$

where

$$\delta = \frac{4\pi n_2 d}{\lambda_0}\cos\theta_2 \tag{5.60}$$

This is illustrated in Fig. 5.16.

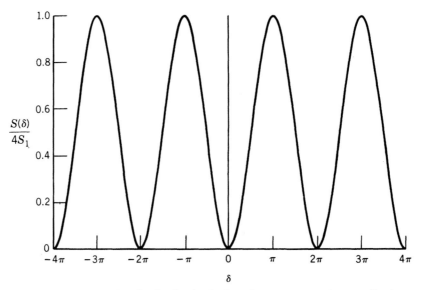

Fig. 5.16 Power density distribution in the interference pattern due to reflection from two low-reflectance parallel interfaces.

B. Haidinger Interference Fringes

Haidinger fringes are the type of interference pattern that results with an extended source where partial reflections occur from a plane-parallel dielectric slab. The interference considered in the previous section then occurs for light coming from each separate part of the source.

Figure 5.17 shows a typical experimental arrangement for the observations of Haidinger fringes. The extended source must be essentially monochromatic. The beam splitter is a partially transmitting plate that allows the incident angle θ_1 to be small. The lens could be in a telescope or in the eye of an observer. The fringes result from interference occurring in the focal plane of the lens.

The ray of light leaving the source at P_1 will be split into two beams by partial reflections off the slab. The two partial rays will leave at an angle θ_1 and will have a relative phase shift given by $-\delta + \pi$, where δ is given by Eq. (5.60). The angle θ_2 can be determined from the angle θ_1 by application of Snell's law.

The two beams will enter the lens making an angle θ_1 with its axis and hence will be brought together in the focal plane at an off-axis distance $x = \theta_1 f$ (in the small-angle approximation), where f is the focal length of the lens.

Another ray of light parallel to the first one leaving the source at a different point P_{II} will also be split into two partial rays. Because they have the same value of θ_1, and hence of θ_2 as the first ray (from P_1), there will be the same relative phase shift of $-\delta + \pi$ as before.

Because of rotational symmetry about the axis of the lens, the phase shift will be a constant on a circle in the focal plane having the lens axis as its center. The interference between the partial beams will be the same on this circle. From Eq.

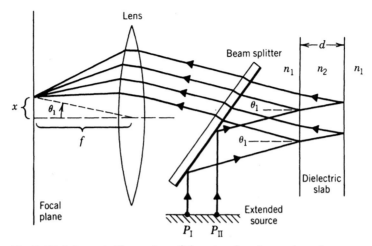

Fig. 5.17 Schematic illustration of the setup for observation of Haidinger fringes.

(5.59) we see that constructive interference, which produces a bright ring, will result when δ is an odd integral multiple of π and destructive interference, which produces a dark ring, will result when δ is zero or an even integral multiple of π. The interference pattern in the focal plane of the lens then consists of a series of concentric rings.

The total number of multiples of 2π in the phase shift at a certain point in the pattern is the order m of the interference pattern. We can use m to index the pattern by writing $\delta = (2m - 1)$ (m need not be an integer), then

$$m = \frac{2n_2 d}{\lambda_0} \cos \theta_2 + \frac{1}{2} \tag{5.61}$$

For the case under consideration, an integral value of m corresponds to a bright band, a half-integral value to a dark band. Note that the highest order in the interference pattern occurs at the center, where $\theta_2 = 0$ and $\cos \theta_2 = 1$.

This maximum-order number is thus given by

$$m_{\text{max}} = \frac{2n_2 d}{\lambda_0} + \frac{1}{2} \tag{5.62}$$

In terms of it, Eq. (5.61) becomes

$$m = \left(m_{\text{max}} - \frac{1}{2} \right) \cos \theta_2 + \frac{1}{2} \approx m_{\text{max}} \left(1 - \frac{1}{2} \theta_2^2 \right) \tag{5.63}$$

if the angle θ_2 is small and m_{max} is large. For small θ_2, Snell's law becomes $n_1 \theta_1 \approx n_2 \theta_2$, so that we can write

$$m = m_{\text{max}} \left[1 - \frac{1}{2} \left(\frac{n_1}{n_2} \right)^2 \theta_1^2 \right] \tag{5.64}$$

We can use the relation $\theta_1 = x/f$ to obtain the following expression relating the order of the fringe to its radius in the focal plane in the lens:

$$m(x) = m_{max}\left[1 - \frac{1}{2}\left(\frac{n_1 x}{n_2 f}\right)^2\right] \qquad (5.65)$$

This can be solved for x as a function of m_{max}:

$$x = \frac{n_2 f}{n_1}\left[\frac{2(m_{max} - m)}{m_{max}}\right]^{1/2} \qquad (5.66)$$

The number of visible bright fringes between $x = 0$ and x will be given by $p \equiv m_{max} - m$. If d is known, we can determine m_{max} from Eq. (5.62). This yields the following expression for the radius of the pth visible fringe:

$$x_p = \frac{f}{n_1}\left(\frac{p n_2 \lambda_0}{d}\right)^{1/2} \qquad (5.67)$$

From Eq. (5.65) we see that equal increments in m correspond to equal decrements in x^2, hence the area $\pi\Delta(x^2)$ enclosed between two successive rings is a constant for fixed d. In fact, we can write, from Eq. (5.65),

$$1 = \Delta m = -\frac{m_{max}}{2\pi}\left(\frac{n_1}{n_2 f}\right)^2 \Delta(\pi x^2)$$

Thus the area between the pth ring and the $(p + 1)$th ring is independent of p:

$$\Delta(\pi x^2) = 2\pi\left(\frac{n_2 f}{n_1}\right)^2 \frac{1}{m_{max}} \approx \pi\left(\frac{f}{n_1}\right)^2 \frac{n_2 \lambda_0}{d} \qquad (5.68)$$

As d increases, the area between fringes decreases, and the fringes get closer and closer together.

C. Michelson Interferometer

The Michelson interferometer is often used in such a way that it exhibits Haidinger fringes. This is shown in Fig. 5.18 where the similarities between the two-beam interference from a dielectric slab and the Michelson interferometer are emphasized. In the former, the OPL difference is provided by partial reflections at interfaces I_1 and I_2. Because these mirrors are made of the same material, the reflection coefficients that must be applied to the optical fields of beams 1 and 2 as they reflect off the mirrors will be equal. In contrast to this, the reflection off the interfaces of the dielectric slab introduce a phase difference between the two beams of 180°. There is added asymmetry in the Michelson interferometer, however, because beam 2 goes through the beam splitter three times and the reflection for beam 2 at the beam splitter is off a dielectric–metal interface. However, the reflection for beam 1 at the beam splitter is off an air–metal interface. The asymmetry in the beam-splitter reflections produces a phase difference of $\varphi_2 - \varphi_1$. By introducing a compensator plate in beam 1, we can eliminate the effect of the thickness of the

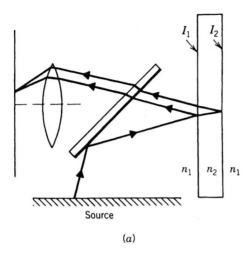

(a)

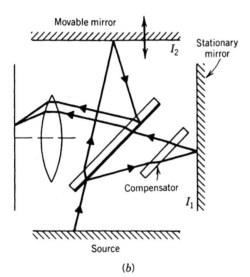

(b)

Fig. 5.18 Demonstration of the similarity between (a) the Haidinger geometry and (b) the Michelson interferometer.

beam splitter. The physical length difference between the two branches then will be the only OPL difference in the problem. The total phase difference will be given by

$$\phi_2 - \phi_1 = -\delta + (\varphi_2 - \varphi_1)$$

where

$$\delta = \frac{4\pi n d}{\lambda_0} \cos \theta \tag{5.69}$$

Here n is the index of refraction of the air, λ_0 the vacuum wavelength of light, d the separation of the mirrors caused by movement of I_2, and θ is the angle with which the light rays are inclined with respect to the axis of the interferometer. The phase δ can be varied over a large range by simply moving one of the mirrors and thus changing d.

A few changes are necessary to carry over the expressions of the previous section to the Michelson interferometer. This is particularly straightforward if, as is usually the case, $\varphi_2 - \varphi_1 = 0$. Then S is given by Eq. (5.25) as

$$S(\delta) = 4S_1 \cos^2(\delta/2)$$

This pattern is the complement of that described in Eq. (5.59). If we index the pattern with m, as before, and set $\delta = 2\pi m$, then

$$m = \frac{2nd}{\lambda_0} \cos\theta \qquad (5.70)$$

At the center of the pattern $m_{max} = 2nd/\lambda_0$, so that $m = m_{max} \cos\theta$. Again if θ is small, but unlike before with no restriction on m_{max}, $m \approx m_{max}(1 - \theta^2/2)$. Equations (5.65) through (5.68) carry over directly with the substitution $n_1 = n_2 = n$.

Figure 5.19 shows the irradiance distribution in the Michelson pattern as a

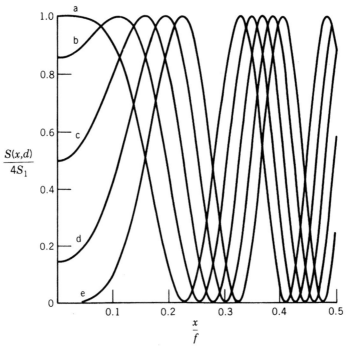

Fig. 5.19 Power density distribution in the Michelson interferometer pattern of fringes as a function of the difference in path length between the two beams. (a) $d = 10\lambda$, (b) $d = 10\lambda + \lambda/8$, (c) $d = 10\lambda + \lambda/4$, (d) $d = 10\lambda + 3\lambda/8$, (e) $d = 10\lambda + \lambda/2$. This shows how the $m = 20$ fringe moves from the center of the pattern to larger x as d increases.

function of distance from the center of the pattern for several values of d. Note how the fringes form at $x = 0$ and move out as d increases.

D. Fizeau Interference

Consider monochromatic light falling on a thin dielectric film (Fig. 5.20). The existence of interference effects and their localization in space depends critically on the way the reflected (or transmitted) light is observed. Suppose that an observer's eye is focused at a point P', or is looking through a microscope focused at P', and let the film be illuminated by an extended source. Consider a small element of the source P_1. The rays of interest then are those from P_1 to P' reflected from the two surfaces of the film. If the film is very thin compared with the distances $\overline{P_1 A}$ and $\overline{AP'}$ and if the surfaces are almost parallel, our earlier results for the phase difference for the two rays will still hold, namely,

$$\phi_2 - \phi_1 = \frac{-4\pi n_2 d}{\lambda_0} \cos \theta_2 + \pi$$

From another part P_{II} of the source there will be another pair of rays going to P'. For these rays d and θ_2 will generally differ enough from those for the rays from P_1 so that the two different values of the phase shift will not be equal. The interference effects will not combine constructively, and as other parts of the source are also considered, we generally obtain no net interference of any significance. We

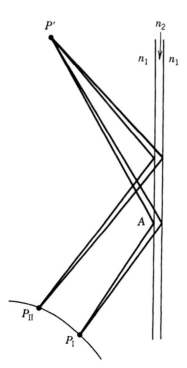

Fig. 5.20 Thin-film interference from an extended source.

must find places on which to focus for which the phase difference is constant for all observable parts of the source.

However, if P' is close to the surface of the film, the pairs of rays from both P_I and P_{II} will have practically the same values of d. They will have different values for the phase shift for a thin film only if their $\cos \theta_2$ values differ appreciably, provided that d is not very large. The variation in $\cos \theta_2$ can be reduced by either or both of the following devices: (1) If the eye or the optical instrument focused on the film has a small entrance pupil, all the rays that can pass from the source of P' off the film through the pupil will have just about the same value of θ_2. (2) If the fringes are observed at nearly normal incidence $\theta_2 = 0$, $\cos \theta_2$ will be almost unity even if there is a moderate spread in θ_2 values about $\theta_2 = 0$. Because one must be focused on (or very near) the film to see these *Fizeau fringes*, they are said to be "localized in the film."

As the eye, or observing microscope, moves over the surface of the film, $\cos \theta_2$ does not change rapidly, particularly near normal incidence, and the order m of the fringes will depend mainly on d. The fringes can be considered to be the loci of constant optical film thickness. We observe a "contour map" of the cross section of the film.

Figure 5.21 illustrates this effect for an air wedge between two plates. The upper plate is an optical flat, and the lower plate may be a sample to be tested. If both pieces are very flat, a series of equidistant straight bands is observed. Between one fringe and the next, the wedge thickness must change by $\lambda/2$ (for a nearly perpendicular observation direction). Hence, if α is the wedge angle, we have the following relation for the transverse separation w between the fringes:

$$\alpha w = \frac{\lambda}{2} \tag{5.71}$$

Here $\lambda = \lambda_0/n_2$ represents the wavelength in the medium of the wedge. If air, $n_2 = 1$.

Region of depression

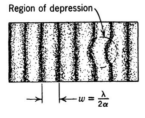

$w = \frac{\lambda}{2\alpha}$

Fig. 5.21 Fizeau fringes in an air wedge.

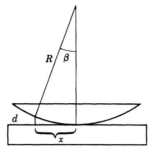

Fig. 5.22 Setup for the observation of Newton's rings.

In the region of a slight depression in the lower plate, the gap thickness will increase. The fringes will be convex toward the point of contact between the plates, as shown in Fig. 5.21.

A famous example of Fizeau fringes occurs when a concave spherical surface is placed on an optical flat. When we view from above with monochromatic illumination, we see a series of concentric rings (Newton's rings) around a central dark spot (Fig. 5.22). For the mth dark ring from the center the gap has a thickness of

$$d_m = m \frac{\lambda}{2}$$

But d may be approximated by analyzing the right triangle.

$$R^2 = x^2 + (R - d)^2 = x^2 + R^2 - 2Rd + d^2$$

$$0 = x^2 - 2Rd + d^2 \approx x^2 - 2Rd$$

so

$$d \approx \frac{x^2}{2R}$$

Thus, $d_m = x_m^2/2R$, and the radius of the mth dark ring will be given by

$$x_m = \sqrt{m\lambda R} \tag{5.72}$$

5.4 Multiple-Beam Interference: Parallel Interfaces

A. Matrix Formalism

We now wish to treat in some detail the multiple reflections neglected in section 5.3. They will be important in practice when the reflectivities of the interfaces are large. The effect can be studied by partially "silvering" the surfaces of a plane parallel dielectric slab or the mirrors of a *Fabry-Perot etalon* (an interferometer wherein the

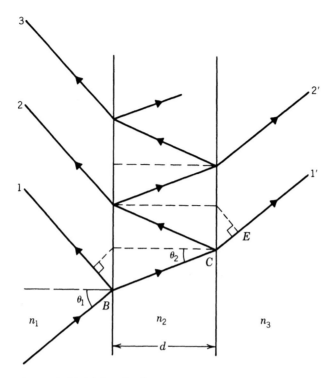

Fig. 5.23 Multiple-reflection geometry.

"slab" is really an air gap). The "silvering" can be a metallic layer, or it can be multilayer dielectric coating.

It is possible to follow the path of each ray explicitly when multiple reflections occur if the slab is a single layer as shown in Fig. 5.23. The resultant is the superposition of all the components, each of which is phase-shifted from the others by multiples of the same factor and each of which is attenuated by the appropriate product of reflection and transmission coefficients. This approach soon becomes cumbersome when more than one layer is present. We develop here a more general formalism that can be used to specify the reflectance and transmittance of a stack having any number of layers. The technique is not restricted to dielectrics at normal incidence, although later we will demonstrate several concepts under these circumstances.

Consider the multilayer configuration shown in Fig. 5.24. The light is incident on the system from left to right. Within the stack, two arbitrary adjacent layers are i and j. In any layer, including the bounding media, the total optical field, in general, consists of two components, E_{rj} traveling to the right, and $E_{\ell j}$ traveling to the left. Each layer has two sides. We will distinguish between the fields on the left side (unprimed notation) and the right side (primed notation). Thus E_{rj} is the field at the left side of layer j that is associated with the wave that is traveling to the right in layer j and E'_{ri} is the field for the wave on the right side of layer i.

In crossing a given interface we employ the reflection and transmission

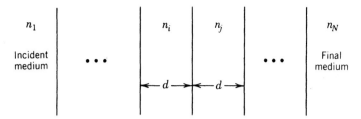

Fig. 5.24 Notation for the matrix method of multiple-reflection, multiple-medium optics.

coefficients ρ_{ij} and τ_{ij}. Here the first subscript refers to the beginning medium, the second to the ending medium. By examining Fig. 5.25 we can write down relationships among the fields on either side of the interface. For instance, E_{rj} is the sum of the transmitted portion of E'_{ri} and the reflected portion of $E_{\ell j}$.

$$E_{rj} = E'_{ri}\tau_{ij} + E_{\ell j}\rho_{ji} \tag{5.73a}$$

In a similar manner we find

$$E'_{\ell i} = E_{\ell j}\tau_{ji} + E'_{ri}\rho_{ij} \tag{5.73b}$$

We use Eq. (5.73a) to eliminate E'_{ri} from Eq. (5.73b)

$$E'_{\ell i} = E_{\ell j}\left[\frac{\tau_{ji}\tau_{ij} - \rho_{ji}\rho_{ij}}{\tau_{ij}}\right] + E_{rj}\frac{\rho_{ij}}{\tau_{ij}} \tag{5.74}$$

The first term in this expression can be simplified through an application of the symmetry relationships derived from the Fresnel equations in Chapter 2 [Eq. (2.71) $\rho_{ij} = -\rho_{ji}$ and Eq. (2.72) $\tau_{ji}\tau_{ij} + (\rho_{ij})^2 = 1$]. Equations (5.73a) and (5.74) then become

$$E'_{ri} = \left(\frac{\rho_{ij}}{\tau_{ij}}\right)E_{\ell j} + \left(\frac{1}{\tau_{ij}}\right)E_{rj} \tag{5.75a}$$

$$E'_{\ell i} = \left(\frac{1}{\tau_{ij}}\right)E_{\ell j} + \left(\frac{\rho_{ij}}{\tau_{ij}}\right)E_{rj} \tag{5.75b}$$

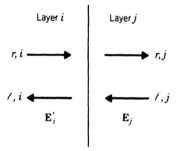

Fig. 5.25 Notation for the waves traveling to the right and to the left on either side of an interface between layers i and j.

Equations (5.75) are coupled linear equations that tell us what the fields on the left side of the interface between layers i and j are in terms of the fields on the right side of the interface.

It is convenient to define a matrix formalism that represents Eqs. (5.75). We will find that propagation across a layer can also be handled within the matrix technique. The advantage of this approach is that through matrix multiplication we can model the optical effect of a multilayer stack with relative ease.

First we must identify column matrices that contain the field components at a particular point in the stack. Let

$$\mathbf{E}_j \equiv \begin{pmatrix} E_{\ell j} \\ E_{rj} \end{pmatrix} \tag{5.76a}$$

be associated with the fields on the left side of layer j and

$$\mathbf{E}'_j \equiv \begin{pmatrix} E'_{\ell j} \\ E'_{rj} \end{pmatrix} \tag{5.76b}$$

represent the fields on the right side of layer j. We now define the interface transition matrix

$$\mathbf{H}_{ij} \equiv \frac{1}{\tau_{ij}} \begin{pmatrix} 1 & \rho_{ij} \\ \rho_{ij} & 1 \end{pmatrix} \tag{5.77}$$

With this collection of matrices, Eqs. (5.75) become

$$\mathbf{E}'_i = \mathbf{H}_{ij} \mathbf{E}_j \tag{5.78}$$

In crossing a given layer from the left side to the right side a phase factor $e^{-i\beta_j}$ is introduced where

$$\beta_j \equiv \frac{2\pi}{\lambda_0} n_j d_j \cos \theta_j \tag{5.79}$$

This results from the difference in length between the optical paths within layer j and the incident or final medium. For the case of one layer (Fig. 5.23)

$$\beta = \frac{2\pi}{\lambda_0} (n_2 \overline{BC} - n_1 \overline{BD}) = \frac{2\pi}{\lambda_0} (n_2 \overline{BC} - n_3 \overline{CE}) \tag{5.80}$$

This can be expressed as

$$E'_{rj} = E_{rj} e^{-i\beta_j} \tag{5.81a}$$

Likewise, for propagation from the right side of a layer to the left side of that same layer

$$E_{\ell j} = E'_{\ell j} e^{-i\beta_j} \tag{5.81b}$$

If we define a layer propagation matrix

$$\mathbf{L}_j \equiv \begin{pmatrix} e^{-i\beta_j} & 0 \\ 0 & e^{i\beta_j} \end{pmatrix} \tag{5.82}$$

then Eqs. (5.81) can be written in matrix notation as

$$\mathbf{E}_j = \mathbf{L}_j \mathbf{E}'_j \tag{5.83}$$

This tells us what the fields on the left side of a layer area are in terms of those on the right side of the layer.

Finally, we identify an important boundary condition. Within the final medium there can be no light propagation to the left. Therefore,

$$\mathbf{E}_N = \begin{pmatrix} 0 \\ E_{rN} \end{pmatrix} \tag{5.84}$$

Starting with the field matrix in the final medium, we can apply the interface transition matrix to determine the field matrix at the incident side of the last interface,

$$\mathbf{E}'_{N-1} = \mathbf{H}_{N-1,N} \mathbf{E}_N$$

At the left side of medium $N-1$ we will have

$$\mathbf{E}_{N-1} = \mathbf{L}_{N-1} \mathbf{E}'_{N-1} \tag{5.85}$$

where Eq. (5.83) has been used to propagate the fields across the layer. To transform the example into medium $N-2$ we use the interface transition matrix again.

$$\begin{aligned} \mathbf{E}'_{N-2} &= \mathbf{H}_{N-2,N-1} \mathbf{E}_{N-1} \\ &= \mathbf{H}_{N-2,N-1} \mathbf{L}_{N-1} \mathbf{H}_{N-1,N} \mathbf{E}_N \end{aligned} \tag{5.86}$$

In a similar manner we continue until we reach the incident medium. The combination of matrices that represent the entire multilayer stack is

$$\mathbf{H}_{12} \mathbf{L}_2 \dots \mathbf{L}_{N-1} \mathbf{H}_{N-1,N} \equiv \mathbf{S}_{1N} = \begin{pmatrix} S_{11} & S_{12} \\ S_{21} & S_{22} \end{pmatrix} \tag{5.87}$$

The relationships among fields in the incident medium (1) and the final medium (N) are then expressed by,

$$\mathbf{E}'_1 = \mathbf{S}_{1N} \mathbf{E}_N \tag{5.88}$$

The stack matrix $\mathbf{S}_{1N}$ *contains the combined effect of all the layers including multiple reflections.* We can always construct $\mathbf{S}_{1N}$ for any number of layers if we know the individual β_j, τ_{ij} and ρ_{ij}. If the algebra gets out of hand, we can program a computer to do the matrix multiplications required in Eq. (5.87). Once the stack matrix is determined, we can easily find the relationship among fields on either side of the stack.

The matrix method that is developed in this chapter is used to determine field intensity ratios. Ray trajectories are handled through the matrix formalism presented in Chapter 3. The two techniques are closely related. Both rely on sequential processing of optical information through encounters with a collection of elements. However, the two techniques would not be suited for use on the same problem. It is important to recognize that the multilayer stack matrix involves the manipulation of the phases of optical fields within parallel-sided layers, whereas the matrices of Chapter 3 deal with the geometry of rays and spherical interfaces. We require here that coherence be maintained through the overall thickness of the multilayer stack. It is best therefore to restrict our stack method to the analysis of thin-film situations.

The reflection and transmission coefficients for the entire stack derived by using the boundary conditions of Eq. (5.84) in Eq. (5.88)

$$\begin{pmatrix} E'_{\ell 1} \\ E'_{r1} \end{pmatrix} = \begin{pmatrix} S_{11} & S_{12} \\ S_{21} & S_{22} \end{pmatrix} \begin{pmatrix} 0 \\ E_{rN} \end{pmatrix}$$

$$= \begin{pmatrix} S_{12} E_{rN} \\ S_{22} E_{rN} \end{pmatrix}$$

Therefore,

$$\rho \equiv \frac{E'_{\ell 1}}{E'_{r1}} = \frac{S_{12}}{S_{22}} \tag{5.89}$$

and

$$\tau = \frac{E_{rN}}{E'_{r1}} = \frac{1}{S_{22}} \tag{5.90}$$

The reflectance R and transmittance T are formed from these as in Chapter 2.

$$R = |\rho|^2 \tag{5.91}$$

$$T = \frac{n_N \cos \theta_N}{n_1 \cos \theta_1} |\tau|^2 \tag{5.92}$$

B. Single Dielectric Slab

Returning to the situation of Fig. 5.23, one layer of index n_2 sandwiched between media of index n_1, n_1 and n_2 both real, we recognize that the stack matrix will be

$$\mathbf{S} = \mathbf{H}_{12} \mathbf{L}_2 \mathbf{H}_{21} \tag{5.93}$$

From Eq. (5.77), the interface transition matrices are

$$\mathbf{H}_{12} = \frac{1}{\tau_{12}} \begin{pmatrix} 1 & \rho_{12} \\ \rho_{12} & 1 \end{pmatrix}, \quad \mathbf{H}_{21} = \frac{1}{\tau_{21}} \begin{pmatrix} 1 & \rho_{21} \\ \rho_{21} & 1 \end{pmatrix}$$

and from Eq. (5.82), the layer propagation matrix is

$$\mathbf{L}_2 = \begin{pmatrix} e^{-i\beta} & 0 \\ 0 & e^{i\beta} \end{pmatrix}$$

where

$$\beta = \frac{2\pi}{\lambda_0} n_2 d \cos \theta_2$$

Substituting these into Eq. (5.93) yields

$$\mathbf{S} = \frac{1}{\tau_{12}\tau_{12}} \begin{pmatrix} e^{-i\beta} + e^{i\beta}\rho_{12}\rho_{21} & e^{-i\beta}\rho_{21} + e^{i\beta}\rho_{12} \\ e^{-i\beta}\rho_{12} + e^{i\beta}\rho_{21} & e^{-i\beta}\rho_{12}\rho_{21} + e^{i\beta} \end{pmatrix}$$

from which the reflection and transmission coefficients directly follow.

$$\rho = \frac{S_{12}}{S_{22}} = \frac{\rho_{12}(e^{i\beta} - e^{-i\beta})\,e^{-i\beta}}{1 - \rho_{12}^2\,e^{-i2\beta}} \tag{5.94}$$

and

$$\tau = \frac{1}{S_{22}} = \frac{\tau_{12}\tau_{21}\,e^{-i\beta}}{1 - \rho_{12}^2\,e^{-i2\beta}} \tag{5.95}$$

The reflectance and transmittance for the single <u>dielectric</u> layer are then

$$R = |\rho|^2 = \frac{4|\rho_{12}|^2\sin^2\beta}{|1 - \rho_{12}^2\,e^{-i2\beta}|^2} \tag{5.96}$$

and

$$T = |\tau|^2 = \frac{|\tau_{12}\tau_{21}|^2}{|1 - \rho_{12}^2\,e^{-i2\beta}|^2} \tag{5.97}$$

If the slab is coated with a thin metal film on both sides to enhance the multiple reflections, then we really should treat these films as two additional layers. We can take a shortcut, however, that will preserve the form of Eqs. 5.96 and 5.97. We recognize that the influence of the metal coating will be to introduce some absorption and additional phase changes. We therefore write

$$\rho_{12} = \rho_0\,e^{i\gamma}$$

where ρ_0 is real and positive and where γ is a real angle that represents the change in phase of the wave on reflection at the 1-2 interface. The denominator in Eq. (5.97) then becomes

$$|1 - \rho_{12}^2\,e^{-i2\beta}|^2 = (1 - \rho_0^2\,e^{-i(2\beta - 2\gamma)})(1 - \rho_0^2\,e^{i(2\beta - 2\gamma)})$$

$$= 1 + \rho_0^4 - 2\rho_0^2\cos\delta \tag{5.98}$$

Here $\delta \equiv 2\beta - 2\gamma$ represents the total phase difference between successive beams in the sequence of multiple reflections. Part of this change results from propagation (2β), part from reflection (2γ).

It is convenient to introduce the reflectance and transmittance for a single interaction with interface 1-2:

$$R_1 \equiv |\rho_{12}|^2 = \rho_0^2 \tag{5.99}$$

and

$$T_1 \equiv |\tau_{12}\tau_{21}| \tag{5.100}$$

In general, energy conservation requires

$$R_1 + T_1 = 1 - A_1 \tag{5.101}$$

where A_1 is the absorptance of the interface. This is required because of the possibility of absorption in the metallic coating. Using Eqs. (5.98) through (5.101) in Eq. 5.97 leads to a simple result.

$$T = \frac{T_1^2}{1 + R_1^2 - 2R_1 \cos \delta} \tag{5.102}$$

Maximum transmission occurs where $\cos \delta = 1$. The transmittance is then

$$T_{max} = \frac{T_1^2}{(1 - R_1)^2} = \left[1 - \frac{A_1}{1 - R_1} \right]^2 \tag{5.103}$$

For dielectric interfaces $A_1 = 0$, so $T_{max} = 1$. However, under these conditions R_1 would be small. When this is the case, the minimum transmittance is still rather large. The minimum occurs when $\cos \delta = -1$.

$$T_{min} = T_{max} \frac{(1 - R_1)^2}{(1 + R_1)^2} \tag{5.104}$$

To reduce T_{min} and thus to achieve high contrast in the interference pattern we require $R_1 \approx 1$. In this event T stays small unless δ is close to $2\pi m$, where m is an integer. We can show this easiest after some rewriting:

$$T = T_{max} \frac{(1 - R_1)^2}{1 - 2R_1 \cos \delta + R_1^2} = \frac{T_{max}}{\dfrac{1 + R_1^2 - 2R_1 + 2R_1 - 2R_1 \cos \delta}{1 + R_1^2 - 2R_1}}$$

$$= \frac{T_{max}}{1 + \dfrac{2R_1(1 - \cos \delta)}{(1 - R_1)^2}}$$

Now use $1 - \cos \delta = 2 \sin^2 \delta/2$. Then we can write

$$T = T_{max} \Bigg/ \left[1 + \frac{4R_1 \sin^2(\delta/2)}{(1 - R_1)^2} \right] = \frac{T_{max}}{1 + F \sin^2(\delta/2)} \tag{5.105}$$

where

$$F = \frac{4R_1}{(1 - R_1)^2} \tag{5.106}$$

is called the *contrast*. Equation (5.105) is known as the *Airy formula*.

When $\delta = \pi + 2\pi m$, $\sin(\delta/2)$ equals ± 1, and T equals T_{min}. The minima are halfway between the maxima. Thus we have

$$T_{min} = \frac{T_{max}}{1 + F}$$

or

$$\frac{T_{max}}{T_{min}} = 1 + F \tag{5.107}$$

This ratio can be quite large when F is large, which in turn will be the case if R_1 is close to 1. For example, take $R_1 = 0.9$. Then $F = 3.6(1 = 0.9)^2 = 3.6/0.01 = 360$. If $F \gg 1$, we can derive approximate expressions for T when δ is close to $2\pi m$ where m is an integer, that is, near a maximum in T. We can write

$$\sin^2\left(\frac{\delta}{2}\right) = \sin^2\left(\frac{\delta}{2} - m\pi\right) \approx \left(\frac{\delta}{2} - m\pi\right)^2 \qquad (5.108)$$

Then we have

$$T = \frac{T_{\max}}{1 + F\sin^2(\delta/2)} \approx \frac{T_{\max}}{1 + F(\delta/2 - m\pi)^2} = T_{\max}\frac{1/F}{1/F + (\delta/2 - m\pi)^2} \quad (5.109)$$

This approximation holds as long as

$$\left|\frac{\delta}{2} - m\pi\right| \ll 1$$

The functional dependence on δ in the last equality of Eq. (5.109) is called *Lorentzian*.

The full-width at half-maximum W is twice the value of $\delta - 2\pi m$ at which $T = T_{\max}/2$. This will happen when

$$\left(\frac{\delta}{2} - m\pi\right)^2 = \frac{1}{F} \qquad (5.110)$$

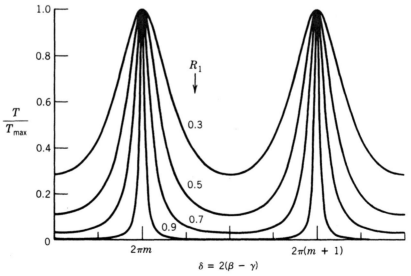

Fig. 5.26 Airy function characteristic of the transmittance of a high-reflectance slab.

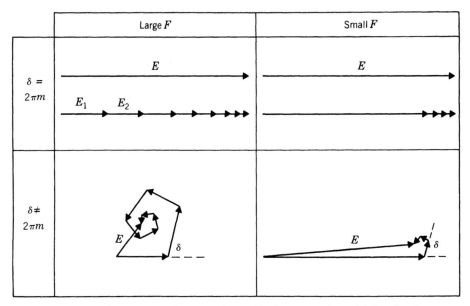

Fig. 5.27 Phasor diagrams showing the origin of high contrast in multiple-reflection interference when F is large.

Thus

$$W = \frac{4}{\sqrt{F}} = \frac{2(1 - R_1)}{\sqrt{R_1}} \; (\approx 2(1 - R_1) \text{ for } R_1 \text{ near } 1) \tag{5.111}$$

Figure 5.26 shows what the transmission curve looks like for high and low F.

The reason there is much more contrast when R_1 is close to unity than for small R_1 can be seen from the phasor diagrams in Fig. 5.27. They show graphically the complex addition treated analytically by the matrix method. When $R_1 \approx 1$, many small components add to form the resultant field.

When they are in phase, the phasors all line up, giving a large resultant. When δ, the relative phase difference, is not equal to an integral multiple of 2π, the individual contributions lie on a polygon that almost closes, giving a small resultant. (We will deal with similar phasor diagrams when we discuss diffraction.) When R_1 is small, the component phasors decrease rapidly in size, and the resultant field varies only slightly with δ.

The effect of multiple-beam interference here is qualitatively the same as in the case of the grating. The pattern becomes concentrated near $\delta = 2\pi m$. Equation (5.25) describes the two-beam interference pattern for both the irradiance in Young's slit geometry and the transmittance of the low reflectivity slab. As more beams enter in the determination of the interference patterns, the results are described by Eq. (5.44) for the multiple-slit geometry and by Eq. (5.105) for the high reflectivity slab.

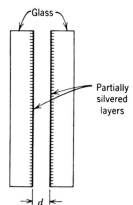

Fig. 5.28 Fabry-Perot interferometer.

C. Fabry-Perot Interferometer

The basic elements of a Fabry-Perot interferometer are indicated in Fig. 5.28. We often use it in visual operation by looking through it at a diffuse source. This produces fringes quite similar to the circular Haidinger fringes discussed earlier, but now the transmission T in the fringes is not a sinusoidal function of δ but shows the sharp maxima with broad minima of Fig. 5.26. The phase difference between contributions from successive multiple reflections is

$$\delta = \frac{4\pi nd}{\lambda_0} \cos\theta - 2\gamma \qquad (5.112)$$

where n, d, and θ are measured in the gap between the plates. The condition for an interference maximum leads to

$$4\pi nd \cos\theta = 2\pi m\lambda_0 \qquad (5.113)$$

where γ in Eq. (5.112) has been neglected because it is, in most cases, very much smaller than d/λ_0.

The wavelength selectivity of the Fabry-Perot interferometer can be seen from an examination of Eq. (5.113). For a given order m longer wavelength contributions give rise to ring maxima that are found at smaller values of θ.

If two wavelengths, λ_i and $\lambda_j = \lambda_i + \Delta\lambda$, are present we see a superposition of the transmission patterns for each wavelength separately. Provided $\Delta\lambda$ is small, the two rings are separated in the mth order by

$$\Delta(\delta\lambda) = \Delta(4\pi nd \cos\theta) = 2\pi m \, \Delta\lambda \qquad (5.114)$$

As in our treatment of the grating, we are interested in the minimum size of $\Delta\lambda$ that can be detected. Here we adopt as our resolution criterion the following: If two components λ_i and λ_j are of equal irradiance, then we define them to be just resolvable if their peak values are separated by an amount equal to full width at half maximum of either peak. The peak separation corresponds to a phase change

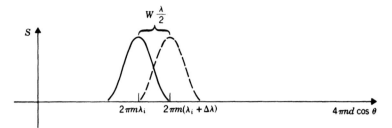

Fig. 5.29 Illustration of our criterion for resolving power determination.

of W. This is illustrated in Fig. 5.29. As a function of $4\pi nd \cos\theta$, the peak separation is

$$\Delta(\delta\lambda) = W\lambda \tag{5.115}$$

where λ is either λ_i or λ_j. By use of Eq. (5.111) with Eqs. (5.115) and (5.114), the resolving power is found to be

$$\mathcal{R} \equiv \frac{\lambda}{\Delta\lambda} = \frac{\pi\sqrt{F}}{2} m \tag{5.116}$$

This is to be compared with the resolving power of the grating, Eq. (5.52), which was shown to be typically around 120,000. With a reflectivity, R_1, of 0.95 the factor in front of m in Eq. (5.116) is about 63. The big difference between Eq. (5.52) and (5.116) is that the order m which is relevant for the Fabry-Perot instrument, is very large. For $\lambda = 600$ nm and a plate separation of 3 mm the resolving power of the Fabry-Perot interferometer is close to 6.1×10^5.

Complications now arise. As illustrated in Fig. 5.30, it may be difficult to tell if maxima A and B are two peaks in the mth order or if A belongs to the mth order and B to the $(m + 1)$th order. If λ_j exceeds λ_i by a certain amount, $\Delta\lambda_{fsr}$, called the free spectral range, the mth-order fringe for $\lambda_j(A)$ will fall on top of the $(m + 1)$th-order fringe for $\lambda_i(B)$.

In this case we have

$$(m + 1)\lambda_i = m\lambda_j$$

Now let $\lambda_j = \lambda_i + \Delta\lambda$ and solve for the free special range $\Delta\lambda_{fsr}$:

$$(m + 1)\lambda_i = m(\lambda_i + \Delta\lambda_{fsr}); \quad \lambda_i = m\,\Delta\lambda_{fsr}; \quad \Delta\lambda_{fsr} = \frac{\lambda_i}{m}$$

Thus, if we are working in high order (large m), which usually means that d the plate separation is large, we must use wavelengths that are quite close together to avoid confusion of orders. Actually, to avoid all confusion, we should have $(\lambda_j - \lambda_i) < \Delta\lambda_{fsr}/2$. Preliminary dispersion with a prism or grating monochromator can be used to eliminate unwanted wavelengths.

The ratio $\Delta\lambda_{fsr}/\Delta\lambda$ of the free-spectral range (or fringe separation) to the

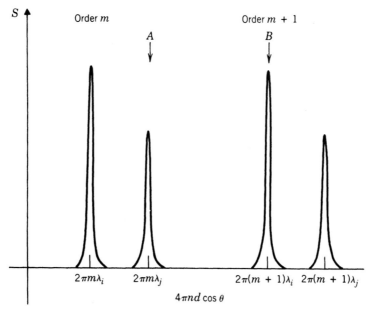

Fig. 5.30 Fabry-Perot transmission pattern when two nearly equal wavelengths are present.

minimum resolvable wavelength separation is called the *finesse* $\mathscr{F}$. Since $\Delta\lambda_{\text{fsr}} = \lambda/m$, we obtain

$$\mathscr{F} = \frac{\pi\sqrt{F}}{2} \approx \frac{\pi}{1 - R_1} \tag{5.117}$$

where the last approximation is valid when R_1 is close to unity.

5.5 Applications of Interference

Many powerful techniques have been developed using the principles of interference that we have now covered. Several examples are discussed in this section. However, for a complete understanding of many of the advanced applications of interference in physical optics, a working knowledge of Fourier analysis and coherence theory is required. These will be presented in Chapters 6 and 8.

A. Interferometry

Devices can be constructed to measure small changes in distance, refractive index, or wavelength by comparing the relative phase differences between two or more beams through changes in an interference pattern. We have already seen how multiple-beam interference leads to dispersive features in the grating [Eq. (5.49)]

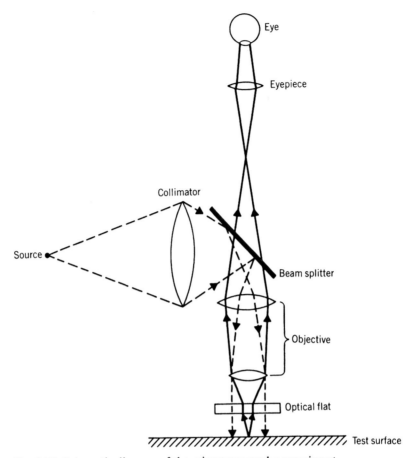

Fig. 5.31 Schematic diagram of the microtopography experiment.

and the Fabry-Perot interferometer [Eq. (5.114)]. We have also discussed how Fizeau interference can be used to monitor thickness changes in a film [Eq. (5.71)].

1. Surface Microtopography. Surface topography can be evaluated using Fizeau interference in a configuration similar to Fig. 5.21. The surface under investigation is the bottom element of a two-plate wedge. The top element must be an optical flat. For maximum use, optical elements must be provided that focus the test surface in the same plane as the interference fringes. If the features occur on a small scale, then a microscope is called for. A schematic design is illustrated in Fig. 5.31.

Usually the contrast is increased by coating both the optical flat and the test surface with partially transmitting metallic layers. A common application is the measurement of the thickness of an evaporated film. Figure 5.32a shows the configuration of a flat sample and the resulting interference pattern. If half of the

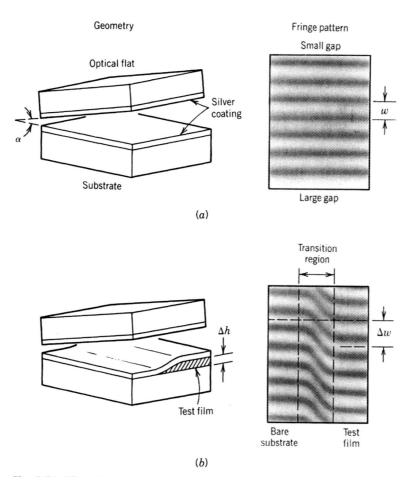

Fig. 5.32 Film thickness determination by Fizeau interference.

sample consists of a test film that effectively decreases the gap between plates, then in the image of the sample the fringes corresponding to a given order of interference will shift across the transition region to indicate the contours of constant gap thickness. This is shown in Fig. 5.32b.

Equation (5.71) tells us that the change in gap thickness is associated with one fringe separation (that is, the vertical height between contour lines)

$$h = \alpha w = \frac{\lambda}{2} m \tag{5.118}$$

The gap decreases by

$$\Delta h = \alpha \, \Delta w \tag{5.119}$$

as one crosses the transition region up the step to the test film. In combining these two equations, we arrive at an internally calibrated expression for the thickness of the test film,

$$\Delta h = \left(\frac{\Delta w}{w}\right)\frac{\lambda}{2} \tag{5.120}$$

In its most refined application, surface microtopography is capable of identifying steps only a few atomic layers thick.

2. Optical-Element Testing. While the Fizeau method is useful for measuring surface irregularities, a more appropriate evaluation for lenses and prisms should involve a test of the optical path through the element. The Michelson interferometer can be modified for use with collimated light to perform such a test. The resulting instrument, the *Twyman-Green* interferometer, is shown in Fig. 5.33. The

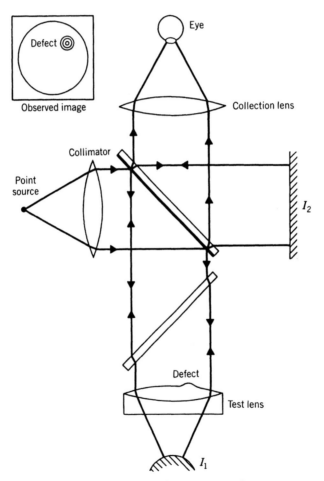

Fig. 5.33 Schematic diagram of the Twyman-Green interferometer in use as a lens-testing instrument.

reference branch that reflects off I_2 provides a plane wave that interferes with the wavefront coming from the test branch. The test beam must pass through the test element to a mirror whose focal length matches that of the test element (in testing a prism this becomes a plane mirror). If the element is perfect, the returning beam will again be a plane wave after having passed through the test element a second time.

The interfering beams are collected by a lens and evaluated by eye, electronically, or photographically. A uniformly illuminated field is observed in the image of a perfect test element. If an imperfection is present, the optical path length of rays passing through that region of the element will cause localized fringes in the image that represents constant optical path length contours.

To determine if the contours surround an optically thick or an optically thin region, the spherical mirror in the test branch can be moved slightly so as to increase the optical path length in that branch. The contours will move away from regions in the image associated with added retardation by the test element. If the contours open up, then the enclosed region is more optically dense than are the surrounding regions.

This scheme is particularly convenient during the final grinding of an optical surface. Identifying paint can be applied to the test element in precisely the location requiring further work by observing the image of the paint applicator superimposed on the interference pattern and the image of the lens. A computer-interpreted topography map can be produced if the interference pattern is photometrically recorded and digitized.

3. Gas-Density Measurement. To gain flexibility over the Michelson configuration, the return paths from I_1 and I_2 can be separated, as shown in Fig. 5.34. In this, the *Mach-Zehnder* interferometer, an additional beam splitter is required to combine the beams. By slightly rotating I_1, the two interfering beams can be made to appear to cross at a point along one of the branches. Because the wavefronts under these circumstances are inclined with respect to one another, the interference pattern is a series of linear fringes. The pattern is localized in the region where the

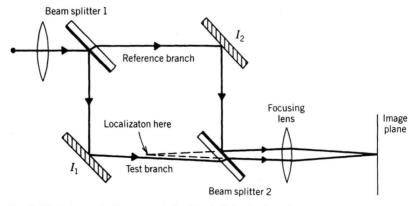

Fig. 5.34 Schematic diagram of the Mach-Zehnder interferometer.

beams appear to cross. If a lens is set up so as to produce an image of that region on a screen or a detector, then the interference pattern will also be in focus in the same plane as the image.

The Mach-Zehnder interferometer is commonly employed to study flow around obstacles in wind tunnels and in other similar applications. This is possible because the index of refraction under many circumstances is proportional to the gas density. The test chamber is placed in the branch of the interferometer in the region where the fringes are localized. A reference chamber is used in the other branch to equalize the optical paths. The interference fringes represent the total difference in optical path length between the two branches. When the gas density changes as a result of localized compression, the fringes are modified so as to follow contours of constant optical path length around the test object.

B. Spectroscopy

The dispersive characteristics of the diffraction grating are most commonly exploited to produce monochromatic light for analytical experiments. The principal elements of such a device are illustrated in Fig. 5.35. They consist of an entrance slit onto which the polychromatic source is focused, a collimating mirror that creates a parallel beam, the grating, a focusing mirror that forms a monochromatic image of the entrance slit on the exit slit plane, and the exit slit that allows only a portion of the dispersed spectrum to pass. The amount of light that the monochromator is able to deliver depends on the size of the slits and the solid angle subtended by the grating at the slits. In addition, most gratings are *blazed* to improve the efficiency in a given order of the interference pattern. The blazing is produced by controlling the contour of the individual grating groove profiles. We will study this in more detail after the concepts of diffraction have been presented (Chapter 6).

1. Plane Gratings. In addition to the angular dispersion and the resolution, which we have already discussed, the linear dispersion is an important measure of performance of a grating monochromator. Equation (5.49) gives the angular dispersion that is related to the linear dispersion by $dx'/d\theta' = f$, where x' is linear displacement in the exit slit plane and f is the focal length of the focusing mirror. This leads to

$$\frac{dx'}{d\lambda} = \frac{dx'}{d\theta'}\frac{d\theta'}{d\lambda} = \frac{fm}{a_\perp} \tag{5.121}$$

This demonstrates the advantage of monochromators with long focal lengths. However, a high resolution device with a large focal length is usually associated with low light transmission because the solid angle subtended by the grating at the slits is small.

Most mounting techniques employ equal path lengths for the input and the output branches. A popular system is pictured in Fig. 5.35, the *Czerny-Turner* configuration. The collimating and focusing mirrors are both used off-axis. The

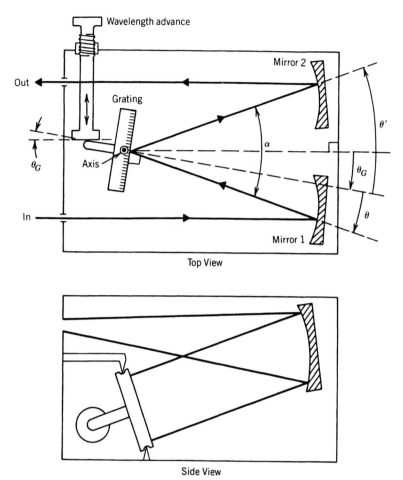

Fig. 5.35 Czerny-Turner monochromator in the off-axis configuration.
$\lambda = (\text{const}) \times \sin \theta_G$.

angles on these mirrors can be arranged so that the coma resulting from the two reflections cancels at the exit slit. Spherical aberration and astigmatism are present, however. These defects can be partially compensated by the use of curved slits or off-axis parabolic mirrors instead of spherical mirrors.

Usually the slits are fixed and the grating rotates about an axis parallel to the slits. We can write $\theta' = \alpha/2 + \theta_G$ and $\theta = \alpha/2 - \theta_G$, as shown in Fig. 5.35. When these are used in Eq. (5.48), we find the relationship between the grating angle (θ_G) and the selected wavelength,

$$\lambda = \frac{a}{m} 2 \cos \frac{\alpha}{2} \sin \theta_G \qquad (5.122)$$

This is convenient, because it is easy to design a mechanical device to change $\sin \theta_G$ in a linear fashion.

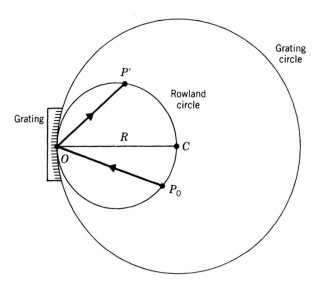

Fig. 5.36 Geometry of the concave grating. The entrance slit is at P_0, the dispersed slit images appear on the Rowland circle, that is, at P'.

2. Concave Gratings. Each of the optical elements in a monochromator introduces some small attenuation of the radiation throughput of the instrument. This can be very significant, particularly in the far ultraviolet where reflectances are small. In addition to this, if the elements have dust particles on them, the resulting scattered radiation gives rise to light within the monochromator that propagates in the wrong direction. This results in light of the wrong wavelength reaching the exit slit.

To reduce these problems the monochromator can be constructed with a concave grating that is designed to image the entrance slit on the exit slit while at the same time dispersing the light into its component wavelengths. This eliminates several optical elements and greatly improves the efficiency of the monochromator.

A common mounting technique for concave gratings is based on the *Rowland circle*. The principle is illustrated in Fig. 5.36. Here the radius of curvature of the grating is R with the grating grooves perpendicular to the plane of the figure. The Rowland circle identifies the correct locations of the entrance slit (P_0) and the exit slit (P'). If P_0 is anywhere on the circle whose diameter is equal to R and which contacts the center of the grating at O as shown, then the specular beam and the dispersed beams in all orders will be focused at other points on the same circle.

To prove this we consider Fig. 5.37, which concentrates on the essential geometry. We will show that P_0 and P' satisfy the grating equation if they lie on the Rowland circle.

The coordinate origin is at $O(0, 0)$ so that $P_0(x_0, z_0)$ and $P'(x', z')$ are identified by

$$x_0 = -r \sin \theta \qquad z_0 = r \cos \theta$$
$$x' = r' \sin \theta' \qquad z' = r' \cos \theta' \tag{5.123}$$

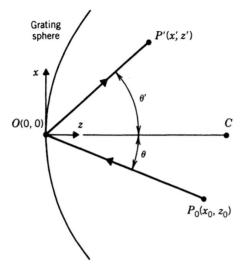

Fig. 5.37 Concave grating geometry for O at the center of the grating.

Along the grating we also identify position $P(x, z)$ shown in Fig. 5.38. The arc OP is required to contain an integral number of grooves that is (if the arc length is small compared to R) approximately x/a, where a is the groove spacing. If P' is to be at an interference maximum, then each groove contributes a beam whose optical path length as measured from P_0 differs from contributions due to adjacent grooves by a factor of $m\lambda$ (for dispersion into order m). The total optical path length difference between contributions from O and P is

$$\overline{P_0OP'} - \overline{P_0PP'} = m\lambda x/a \tag{5.124}$$

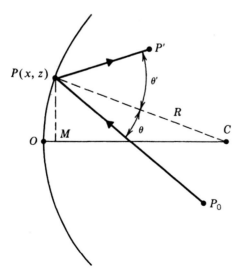

Fig. 5.38 P a distance ma from O.

where

$$\overline{P_0OP'} = r + r'$$

and

$$\overline{P_0PP'} = [(x_0 - x)^2 + (z_0 - z)^2]^{1/2} + [(x' - x)^2 + (z' - z)^2]^{1/2} \quad (5.125)$$

Now z is very small compared to the other dimensions in the example, so Eq. (5.125) can be approximated by

$$\overline{P_0PP'} \approx [x_0^2 + z_0^2 - 2(xx_0 + zz_0) + x^2]^{1/2} + [x'^2 + z'^2 - 2(xx' + zz') + x^2]$$

This can be simplified even further by using our standard approximation for the curvature of a spherical surface [see discussion prior to Eq. (5.72)],

$$z \approx x^2/2R \qquad (5.126)$$

Substituting this into Eq. (5.126) along with Eq. (5.123) leads to

$$\overline{P_0PP'} \approx \left\{ r^2 - 2\left[x(-r\sin\theta) + \frac{x^2}{2R}(r\cos\theta) \right] + x^2 \right\}^{1/2}$$
$$+ \left\{ r'^2 - 2\left[x(r'\sin\theta') + \frac{x'^2}{2R}(r'\cos\theta') \right] + x^2 \right\}^{1/2} \qquad (5.127)$$

We now use the fact that x/r and x/r' are small to simplify the square root operation. Because $(1 + \varepsilon)^{1/2} \approx 1 + \varepsilon/2 - \varepsilon^2/8 + \dots$ where in this case, for instance

$$\varepsilon = \left(\frac{x}{r} \right)^2 + \frac{2x\sin\theta}{r} - \frac{x^2\cos\theta}{Rr}$$

the first term on the right side of Eq. (5.127) is approximately

$$\overline{P_0P} \approx r\left\{ 1 + \frac{x^2}{2r^2} + \frac{x\sin\theta}{r} - \frac{x^2\cos\theta}{2Rr} - \frac{x^2\sin^2\theta}{2r^2} \right\}$$
$$\approx r + x\sin\theta + \frac{x^2}{2r}\cos^2\theta - \frac{x^2\cos\theta}{2R}$$

We have retained only those terms that are of second or lower order in x. A similar step leads to a simplification of the second term on the right side of Eq. (5.127). We are left with

$$\overline{P_0PP'} \simeq (r + r') + x(\sin\theta - \sin\theta')$$
$$+ \frac{x^2}{2}\left[\left(\frac{\cos^2\theta}{r} - \frac{\cos\theta}{R} \right) + \left(\frac{\cos^2\theta'}{r'} - \frac{\cos\theta'}{R} \right) \right] \qquad (5.128)$$

This can now be substituted into Eq. (5.124), which expresses the conditions for an interference maximum at P' given the source at P_0:

$$\overline{P_0OP'} - \overline{P_0PP'} \simeq x(\sin\theta' - \sin\theta) + \frac{x^2}{2}\left[\left(\frac{\cos^2\theta}{r} - \frac{\cos\theta}{R}\right)\right.$$

$$\left.+ \left(\frac{\cos^2\theta'}{r'} - \frac{\cos\theta'}{R}\right)\right] = m\lambda x/a \qquad (5.129)$$

The first term on the left of Eq. (5.129) with the right side is just the familiar form of the grating equation. The second term will be zero (at this level of approximation) provided

$$r = R\cos\theta \qquad (5.130a)$$

and

$$r' = R\cos\theta' \qquad (5.130b)$$

But this implies that OP_0C and $OP'C$ define right triangles, and because $\overline{OC}$ is common to both right triangles, then points P_0 and P' must lie on a *common circle* whose diameter is $\overline{OC}$. This is the Rowland circle.

We recognize that P' is the image of P_0 in order m and wavelength λ because all paths P_0PP' are equivalent within an integral number of wavelengths. Thus the contributions at P' that originate at P_0 are all in phase provided the angles θ and θ' satisfy

$$a(\sin\theta' - \sin\theta) = m\lambda$$

3. Fabry-Perot Spectroscopy. For resolving power larger than that provided by a grating monochromator, the Fabry-Perot *etalon* can be used. This is essentially a Fabry-Perot interferometer having a fixed separation d between plates. It is conventionally employed with incident light covering a continuous range of values of the angle θ. The spectrum is observed photographically in the focal plane of a camera lens. The equations of section 5.3B apply to this case as modified in Section 5.3C for the Michelson interferometer. The order of a fringe is given by Eq. (5.70) as $m = 2nd\cos\theta/\lambda_0$. The maximum order is found at the center of the pattern $m_{max} = 2nd/\lambda_0$, which does not have to be exactly an integer. The radius of the pth bright fringe from the center, where $p = m_{max} - m$, is given by the adaptation of Eq. (5.67):

$$x_p = f\left[\frac{p\lambda_0}{nd}\right]^{1/2} \qquad (5.131)$$

Here also, p need not be an integer.

Suppose only monochromatic green light is incident on the entrance slit of a grating monochromator. Then a green image of the entrance slit will be formed for the optimized grating order at the place on the photographic plate labeled "green" in Fig. 5.39. With the etalon inserted in the indicated position, the image of the slit will not be continuous but will consist of sections of the interference pattern produced by the etalon as shown in the inset. If the green light is not truly

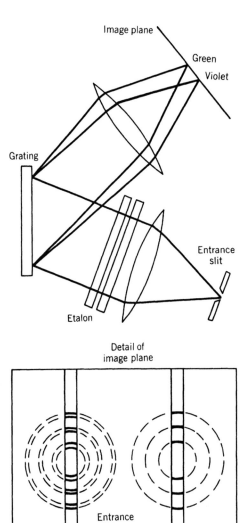

Fig. 5.39 Use of a Fabry-Perot etalon with a grating. The image consists of small sections of arcs. The green line in this example is a doublet.

monochromatic but contains several closely spaced wavelengths, then the interference pattern will show the superimposed patterns of each component wavelength as required by Eq. (5.131). The function of the grating is to separate the green group from other groups that might be present in the spectrum under analysis.

Another way of using the Fabry-Perot interferometer in spectroscopy is the *center-spot scanning* technique shown in Fig. 5.40. It is applicable when only a few wavelength groups are present, so that preliminary separation can be accomplished with a filter or premonochromator. We either shine light from a diffuse source directly into the interferometer or use collimated light from a broad source. Then,

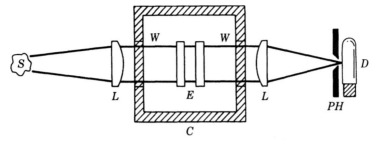

Fig. 5.40 Center-spot scanning with a Fabry-Perot etalon: *S*, source; *L*, lenses; *C* vacuum chamber for pressure scanning; *W*, windows; *E* etalon; *PH* pinhole; *D* detector.

at the focal plane of a lens, we observe the circular pattern. A pinhole is placed at the center of the pattern through which light passes to a detector. We are therefore measuring the flux density of the "center spot" at $\theta = 0$. From Eq. (5.70) $\lambda = 2nd/m$. To scan wavelength and thus to record the spectral distribution of flux density, we must change n or d. Piezoelectric crystals can be used to change d, or with d fixed n can be changed by varying the pressure of the gas between the plates.

C. Optical Coatings

The optical properties of thin films (as described by the formalism in Section 5.4A) can be exploited to enhance or reduce the reflectance from an optical element. This is useful in the design of antireflection coatings for lenses and in the design of dielectric mirrors. These systems are usually meant to be employed at normal incidence.

1. Antireflection Coating. The idea in antireflection coating is to produce destructive interference between the beams reflected from the front face of the film and the back face of the film. This is achieved by (1) having both reflections at interfaces characterized by an incident medium that is less optically dense, and (2) having a 180° phase shift resulting from propagation in the second medium. Thus we choose the layer thickness d so that

$$\beta = 2\pi n_2 d/\lambda_0 = \pi/2$$

or

$$d = \frac{1}{4}\frac{\lambda_0}{n_2} = \frac{\lambda_2}{4}$$

(Hence the name "quarter-wave layer.") Often the wavelength λ_0 is chosen to be near 550 nm, at the maximum of the visual response. Then we choose the materials so that $n_1 < n_2 < n_3$. Figure 5.41 illustrates the phase shift between the first two contributions from the multiple reflections. In practice these are the dominant

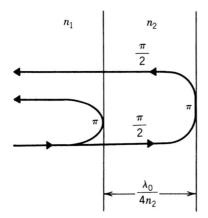

n_1 n_2 n_1

$\dfrac{\pi}{2}$

π

π

π $\dfrac{\pi}{2}$

$\dfrac{\lambda_0}{4n_2}$

Fig. 5.41 Illustration of phase changes on reflection and transmission in the quarter-wave film used at normal incidence.

fields, although our theory includes multiple reflections. The stack matrix is given by Eq. (5.93) with

$$\mathbf{H}_{12} = \frac{1}{\tau_{12}} \begin{pmatrix} 1 & \rho_{12} \\ \rho_{12} & 1 \end{pmatrix}, \quad \mathbf{L}_2 = \begin{pmatrix} e^{-i\beta} & 0 \\ 0 & e^{i\beta} \end{pmatrix}, \quad \text{and } \mathbf{H}_{23} = \frac{1}{\tau_{23}} \begin{pmatrix} 1 & \rho_{23} \\ \rho_{23} & 1 \end{pmatrix}$$

The reflection coefficient follows from the elements of **S**,

$$S_{12} = \frac{1}{\tau_{12}\tau_{23}} (e^{-i\beta}\rho_{23} + e^{i\beta}\rho_{12})$$

and

$$S_{22} = \frac{1}{\tau_{12}\tau_{23}} (e^{-i\beta}\rho_{12}\rho_{23} + e^{i\beta})$$

We have from Eq. (5.94)

$$\rho = \frac{S_{12}}{S_{22}} = \frac{\rho_{23}\, e^{-i\beta} + \rho_{12}\, e^{i\beta}}{e^{i\beta} + \rho_{12}\rho_{23}\, e^{-i\beta}} \tag{5.132}$$

But since β for the design wavelength is $\pi/2$, this leads to the requirement that $\rho_{12} = \rho_{23}$. Using Eq. (2.62a) for the reflection coefficient of a dielectric layer at normal incidence

$$\rho_{ij} = \frac{n_i - n_j}{n_i + n_j} \tag{5.133}$$

we find that the requirement $\rho_{12} = \rho_{23}$ is equivalent to

$$n_2 = \sqrt{n_1 n_3} \tag{5.134}$$

Introducing this information into Eq. (5.132) and assuming that the dominant

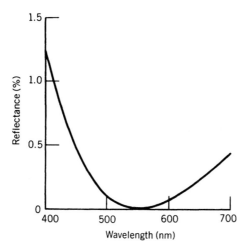

Fig. 5.42 Performance of antireflection coating for the case of $n_1 = 1$, $n_2 = 1.22$, and $n_3 = 1.5$.

changes will be in the phase factors, we can estimate what the reflection coefficient will be for wavelengths in a range close to the design wavelength:

$$\rho = \frac{\rho_{12}(e^{-i\beta} + e^{i\beta})}{e^{i\beta} + \rho_{12}^2 e^{-i\beta}} = \frac{2\rho_{12} \cos\beta \, e^{-i\beta}}{1 + \rho_{12}^2 e^{-i2\beta}}$$

The reflectance follows from this (assuming that the film is a dielectric)

$$R = |\rho|^2 = \frac{4R_1 \cos^2\beta}{1 + 2R_1 \cos(2\beta) + R_1^2} \tag{5.135}$$

where $R_1 = |\rho_{12}|^2$. Because $\beta \approx \pi/2$, this is approximately

$$R \approx \frac{4R_1 \cos^2\beta}{(1 - R_1)^2}$$

Figure 5.42 shows how R remains small over a large wavelength region about the design wavelength, 550 nm. For the system chosen, the reflectance of the interface between the first and third media, without the antireflection coating, would be 4%.

2. Reflectance-Enhancing Coating. We can construct an all-dielectric multilayer reflection-enhancing coating by alternating layers of high (H) and low (L) index materials evaporated onto a substrate as shown in Fig. 5.43. If the thickness of the layers are controlled so that

$$n_H d_H = n_L d_L = \frac{\lambda_0}{4}$$

then the phases of the beams contributing to the resultant optical field will be as shown. The partially reflected contributions all interfere constructively, giving a large net reflection coefficient.

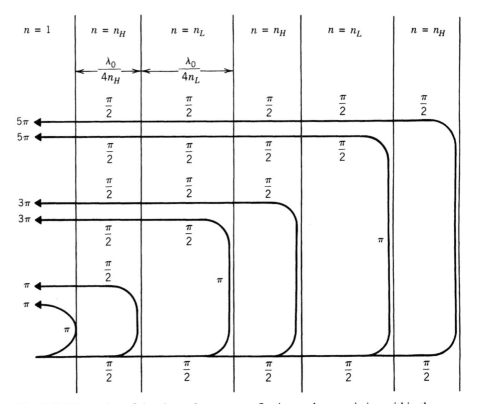

Fig. 5.43 Illustration of the phase changes on reflection and transmission within the multiple-periodic reflectance-enhancing coating.

Using the theory of Section 5.4A, we can identify a repeating unit consisting of a pair of high and low index films. The stack matrix for this unit is

$$\mathbf{S}_{\text{pair}} = \mathbf{H}_{LH}\mathbf{L}_H\mathbf{H}_{HL}\mathbf{L}_L$$

where

$$\mathbf{H}_{HL} = \frac{1}{\tau_{HL}}\begin{pmatrix} 1 & \rho_{HL} \\ \rho_{HL} & 1 \end{pmatrix}, \quad \mathbf{H}_{LH} = \frac{1}{\tau_{LH}}\begin{pmatrix} 1 & \rho_{LH} \\ \rho_{LH} & 1 \end{pmatrix}$$

and

$$\mathbf{L}_L = \mathbf{L}_H = \begin{pmatrix} -i & 0 \\ 0 & i \end{pmatrix}$$

so that

$$\mathbf{S}_{\text{pair}} = \frac{-1}{\tau_{LH}\tau_{HL}}\begin{pmatrix} 1 + \rho_{LH}^2 & 2\rho_{LH} \\ 2\rho_{LH} & 1 + \rho_{LH}^2 \end{pmatrix} \qquad (5.136)$$

If we add a second pair, the stack matrix becomes

$$\mathbf{S}_{\text{pair}}\,\mathbf{S}_{\text{pair}} = \left(\mathbf{S}_{\text{pair}}\right)^2$$

$$= \left(\frac{-1}{\tau_{LH}\tau_{HL}}\right)^2 \begin{pmatrix} 1 + 6\rho_{LH}^2 + \rho_{LH}^4 & 4\rho_{LH} + 4\rho_{LH}^3 \\ 4\rho_{LH} + 4\rho_{LH}^3 & 1 + 6\rho_{LH}^2 + \rho_{LH}^4 \end{pmatrix} \quad (5.137)$$

At this point we can simplify the process by making an interesting observation. The elements of the double-pair stack matrix are alternating terms in the expansion

$$(1 + \rho_{LH})^4 = 1 + 4\rho_{LH} + 6\rho_{LH}^2 + 4\rho_{LH}^3 + \rho_{LH}^4 \quad (5.138)$$

Using the fact that

$$(1 - \rho_{LH})^4 = 1 - 4\rho_{LH} + 6\rho_{LH}^2 - 4\rho_{LH}^3 + \rho_{LH}^4 \quad (5.139)$$

Eq. (5.137) can be rewritten as

$$(\mathbf{S}_{\text{pair}})^2 = \frac{1}{2}\left(\frac{-1}{\tau_{LH}\tau_{HL}}\right)^2 \begin{pmatrix} (1 + \rho_{LH})^4 + (1 - \rho_{LH})^4 & (1 + \rho_{LH})^4 - (1 - \rho_{LH})^4 \\ (1 + \rho_{LH})^4 - (1 - \rho_{LH})^4 & (1 + \rho_{LH})^4 + (1 - \rho_{LH})^4 \end{pmatrix}$$

$$(5.140)$$

It is a straightforward exercise to demonstrate that the general stack matrix for N pairs of layers will be

$$(\mathbf{S}_{\text{pair}})^N = \frac{1}{2}\left(\frac{-1}{\tau_{LH}\tau_{HL}}\right)^N \begin{pmatrix} (1 + \rho_{LH})^{2N} + (1 - \rho_{LH})^{2N} & (1 + \rho_{LH})^{2N} - (1 - \rho_{LH})^{2N} \\ (1 + \rho_{LH})^{2N} - (1 - \rho_{LH})^{2N} & (1 + \rho_{LH})^{2N} + (1 - \rho_{LH})^{2N} \end{pmatrix}$$

$$(5.141)$$

But because by Eq. (5.133)

$$1 \pm \rho_{LH} = \frac{(n_L \pm n_L) + (n_H \mp n_H)}{n_L + n_H}$$

the general stack matrix becomes

$$(\mathbf{S}_{\text{pair}})^N = \frac{1}{2}\left(\frac{-4}{\tau_{LH}\tau_{HL}(n_L + n_H)^2}\right)^N \begin{pmatrix} n_L^{2N} + n_H^{2N} & n_L^{2N} - n_H^{2N} \\ n_L^{2N} - n_H^{2N} & n_L^{2N} + n_H^{2N} \end{pmatrix} \quad (5.142)$$

by Eq. (5.89) the reflection coefficient due to the N-pairs of layers is

$$\rho = \frac{n_L^{2N} - n_H^{2N}}{n_L^{2N} + n_H^{2N}}$$

This formalism does not account for the fact that the incident medium is air with a refractive index slightly greater than unity. However, the concept is sufficiently demonstrated with this development. The reflectance of the stack of $2N$ layers is

$$R = \left(\frac{1 - \left(\dfrac{n_H}{n_L}\right)^{2N}}{1 + \left(\dfrac{n_H}{n_L}\right)^{2N}}\right)^2 \quad (5.143)$$

As N increases, the reflectance approaches unity. The convergence improves as the ratio n_H/n_L increases. One type of reflectance-enhancing coating is that used to construct mirrors for helium–neon laser cavities. It consists of 13 layers of zinc sulfide ($n_H = 2.32$) and magnesium fluoride ($n_L = 1.38$) with a peak reflectance of 98.9% at 633 nm.

D. Optical Cavities and Waveguides

1. Standing Waves. An important type of interference between two optical beams of the same frequency moving in opposite directions will produce a *standing wave*. For example, consider the superposition of plane waves

$$E_1 = A_1 \, e^{i2\pi((t/T)-(z/\lambda))}$$

and

$$E_2 = A_2 \, e^{i2\pi((t/T)+(z/\lambda))} \qquad (5.144)$$

If the amplitudes of the two components are equal, then the total optical field becomes

$$E = A_1 \, e^{i2\pi(t/T)}\left(e^{i2\pi(z/\lambda)} + e^{-i2\pi(z/\lambda)}\right) = 2A_1 \cos\left(\frac{2\pi z}{\lambda}\right) e^{i2\pi(t/T)} \qquad (5.145)$$

This shows that the amplitude of the resultant is a sinusoidal function of position along the direction of propagation with a spatial period equal to the wavelength of the light.

A standing wave of this kind can be established between two infinite, parallel mirrors a distance L apart, as shown in Fig. 5.44. If there are no reflection losses, the maximum amplitude will not change with time. There will be a net change in phase in one round trip between mirrors of

$$\delta = -\frac{2\pi}{\lambda} 2L + 2\gamma = -\frac{4\pi}{c} \nu L + 2\gamma$$

where γ is the phase change that occurs at reflection off each mirror. In this cycle the two oppositely directed beams are created. However, if the optical field is to reproduce itself after one round trip, the change in phase must be an integral

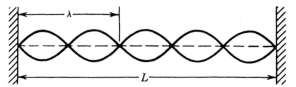

Fig. 5.44 Standing-wave amplitude versus distance along a cavity that is $5\lambda/2$ long.

multiple of 2π; that is, with m = integer the frequency must satisfy

$$\delta = v\frac{4\pi L}{c} - 2\gamma = 2\pi m$$

$$v = \frac{mc}{2L} + \frac{\gamma c}{2\pi L} \tag{5.146}$$

2. Laser Cavities. One of the most important applications of standing-wave interference is the laser cavity. If an active medium is placed between the mirrors of the cavity in the previous section, then each time the light travels through the cavity it is amplified by the process of stimulated emission. In general, there will be a range of frequencies that can be amplified in a given active medium. However, only light having frequencies satisfying Eq. (5.146) will constructively interfere within the cavity. Through the amplification processes, these frequencies rapidly dominate. These are the longitudinal modes of the cavity. If one of the end mirrors is partially transmitting, the amplification process still dominates, but some of the radiation is allowed to leak out, thus creating the laser output.

The end mirrors in this example represent an infinite Fabry-Perot etalon. Equation (5.146) is the condition for peak transmission through such an etalon. We recall from our discussion of such a system that for large energy reflection coefficients the transmission through the etalon can be very strongly peaked as a function of the frequency. We see from Eq. (5.146) that from one peak to the next there will be a separation Δv given by

$$\Delta v = \frac{c}{2L} \tag{5.147}$$

Furthermore, each peak will be approximately Lorentzian with a full phase width at half maximum given from Eq. (5.111) by

$$W = \frac{4}{\sqrt{F}}$$

This is equal to the phase width $d\delta$ associated with a frequency full width dv obeying

$$d\delta = \frac{4\pi L}{c}(dv) = W$$

or

$$dv = \frac{c}{\pi L\sqrt{F}} \tag{5.148}$$

The mirrors in the He-Ne laser have values of R_1 of around 99%. Using Eq. (5.106) and (5.148) this leads to a longitudinal mode peak width of 4.8×10^5 Hz for a 1-m long cavity. The frequency difference between modes is, from Eq. (5.147), $1.5 \times$

10^8 Hz. These are very small fractions of the mean frequency of the output radiation at 632.8 nm, 4.74×10^{14} Hz. The origin of the laser output is schematically illustrated in Fig. 5.45.

Measures of the sharpness of the cavity resonance are related to the efficiency of the Fabry-Perot interferometer. Thus the finesse of a laser cavity is defined by

$$\mathscr{F} \equiv \frac{\Delta \nu}{d\nu} = \frac{\pi \sqrt{F}}{2} \tag{5.149}$$

An alternative measure of quality is the cavity Q, which can be defined as

$$Q \equiv \frac{\nu}{d\nu} \approx \frac{2L\mathscr{F}}{\lambda} \tag{5.150}$$

In addition to the longitudinal modes just discussed, the standing fields in a laser cavity also exist in transverse modes. These result when the infinite plane mirrors are replaced by finite-sized mirrors. Further discussion of the transverse character of the laser radiation will be delayed until Chapter 7 where near field diffraction theory is presented.

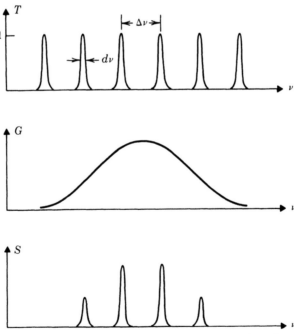

Fig. 5.45 Representation of the frequency distribution of laser output power in terms of the transmission characteristics of the laser cavity T, and the spectral character of the emission line G, which contributes to gain.

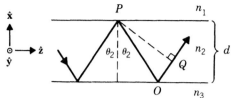

Fig. 5.46 Geometry of the thin-film optical waveguide.

3. Optical Waveguides. *Integrated optics* is that branch of electronics wherein signals are carried by light beams that are trapped by internal reflection within thin film waveguides. Interference plays an important role in these devices because only modes that are constructively reinforced are able to propagate within the film.

Figure 5.46 shows a cross section of a planar waveguide that we consider to be infinite in extent in the *y-z* plane. The internal angle θ_2 must satisfy

$$\sin \theta_2 \geq n_{\text{out}}/n_2 \tag{5.151}$$

where n_{out} is the larger of n_1 or n_3. If this condition is not met, then the mode is *radiative* and light is lost at each internal reflection.

The path of the ray shown in Fig. 5.46 is only one of a continuum of possible rays that would exist within this waveguide at θ_2. The others are generated by translating points P and O along the *z*-direction. Normal to each of these rays will be planes of constant phase of the optical electric field. The dashed line PQ represents the surface on which one of these planes might lie. This surface belongs to the plane associated with the beam immediately *before* reflection at P and also with the beam at Q. The phase difference between the optical fields associated with these points is

$$\phi_Q - \phi_P = \frac{-2\pi\nu}{c}[n_2(\overline{PO} + \overline{OQ})] + \gamma_{12} + \gamma_{23} \tag{5.152}$$

where γ_{12} and γ_{23} are associated with reflection from the interfaces. We can evaluate $\overline{PO} + \overline{OQ}$ by noting that $\overline{PO} = d/\cos \theta_2$ and $\overline{OQ} = \overline{PO} \cos(2\theta_2)$. Combining these relations and using $\cos 2\theta = 2\cos^2\theta - 1$ we find

$$\overline{PO} + \overline{OQ} = 2d \cos \theta_2 \tag{5.153}$$

For constructive interference, the phase difference must be an integral multiple of 2π. Thus

$$\frac{2\pi\nu}{c}(2n_2 d \cos \theta_2) - 2\pi \, \Delta m = 2\pi m \tag{5.154}$$

where $\Delta m = (\gamma_{12} + \gamma_{23})/2\pi$ accounts for the phase changes in the reflections. This can be rewritten to identify the allowed angles at which radiation will propagate. At other angles, destructive interference dominates.

$$\cos \theta_{2,m} = \frac{(m + \Delta m)c}{2n_2 d\nu} \quad m = 0, 1, 2 \dots \tag{5.155}$$

For a given angle, there will be a characteristic mode that is associated with a single value of m.

Equation (5.151) must also be satisfied. This places restrictions on the frequency for a given waveguide of thickness d and index n_2 and bounded by n_{out},

$$v = \frac{c(m + \Delta m)}{2n_2 d \cos \theta_{2,m}}$$

But

$$\cos \theta_{2,m} = (1 - \sin^2 \theta_{2,m})^{1/2}$$

and from Eq. (5.151)

$$\cos \theta_{2,m} \leq \left(1 - \frac{n_{out}}{n_2}\right)^{1/2}$$

Therefore

$$v \geq \frac{c(m + \Delta m)}{2d(n_2^2 - n_{out}^2)^{1/2}} \tag{5.156}$$

Propagation within the waveguide is cut off if the frequency is lower than the critical value when $m = 0$; Δm is a characteristic of the waveguide structure and, for a given system, is fixed.

REFERENCES

Arfken, G. *Mathematical Methods for Physicists*. Academic Press, New York, 1970.

Aspnes, David E. "Analysis of Modulation Spectra of Stratified Media." *Journal of the Optical Society of America*, 63:1380, 1973.

Born, Max, and Emil Wolf. *Principles of Optics*. Pergamon Press, Oxford, 1980.

Cagnet, Michel, Maurice Françon, Jean Claude Thrierr. *Atlas of Optical Phenomena*. Springer-Verlag, Berlin, 1962.

Dainty, J. C., ed. *Laser Speckle and Related Phenomena*. Springer-Verlag, Berlin, 1984.

Driscoll, Walter, G., and William Vaughan. *Handbook of Optics*. McGraw-Hill, New York, 1978.

Fowles, Grant R. *Introduction to Modern Optics*. Holt, Rinehart and Winston, New York, 1968.

Françon, Maurice. *Optical Image Formation and Processing*. Academic Press, New York, 1979.

Heavens, O. S. *Optical Properties of Thin Solid Films*. Dover, New York, 1955.

James, J. F., and R. S. Sternberg. *The Design of Optical Spectrometers*. Chapman and Hall, Ltd., London, 1969.

Kline, Morris, and Irwin W. Kay. *Electromagnetic Theory and Geometrical Optics*. Interscience, New York, 1965.

Lengyel, Bela A. *Introduction to Laser Physics*. Wiley, New York, 1966.

Marcuse, D. *Theory of Dielectric Optical Waveguides*. Academic Press, New York, 1974.

Okoshi, Takanori. *Optical Fibers*. Academic Press, London, 1982.

Smith, William V., and Peter P. Sorokin. *The Laser*. McGraw-Hill, New York, 1966.

Steel, W. H. *Interferometry*. Cambridge University Press, Cambridge, 1967.

Stone, J. M. *Radiation and Optics*. McGraw-Hill, New York, 1963.

Tolansky, S. *Microstructures of Surfaces Using Interferometry*. Elsevier, New York, 1968.

Tolansky, S. *Surface Microtopography*. Interscience, New York, 1960.

PROBLEMS

Section 5.1   Two-Beam Interference

1. Two monochromatic planewaves are propagating in *nearly* the same direction. One is inclined with respect to the other by 20 mrad. As a function of the average propagation direction, identify the location of the interference maxima (their spacing). The common wavelength of the light is 532 nm. Assume perfect coherence for each beam.

2. Suppose that a reflecting surface has a complex amplitude reflection coefficient $\rho = \rho_0 \, e^{i\gamma}$. This means that if the incident electric field is given by

$$E = A \cos\left[2\pi\nu\left(t - \frac{z}{v}\right) + \varphi\right]$$

the reflected field will be given by

$$E'' = \rho_0 A \cos\left[2\pi\nu\left(t + \frac{z}{v}\right) + \varphi + \gamma\right]$$

Suppose that $\rho_0 \neq 1$. Show that the combined field has a part consisting of a standing wave and a part consisting of a propagating wave.

3. Plot the phasors that represent the optical electric field $E = A \exp(i(\omega t - kr))$. Here $A = 100$ V/cm, $\omega = 3.77 \times 10^{15}$ sec^{-1}, $k = 1.25 \times 10^7$ m^{-1}. Determine the phasor under the following conditions: $t = 0.14 \times 10^{-15}$ sec, $r = 0$; $t = 10^{-15}$ sec, $r = 0$; $t = 2.22 \times 10^{-15}$ sec, $r = 0$; $t = 0$, $r = 2 \times 10^{-7}$ m; $t = 10^{-15}$ sec, $r = 10^{-7}$ m; $t = 10^{-15}$ sec, $r = 8.75 \times 10^{-7}$ m.

4. Graphically and analytically determine the result of the phasor addition of $e^{i\pi/6}$ and $3\,e^{i\pi/3}$.

5. Analytically determine the phasor sum of $E_1 = A \exp\{i[\omega t - k(0.3r)]\}$ and $E_2 = 4A \exp\{i[\omega t - k(0.6r)]\}$.

6. Prove that the spatially integrated flux density on the observation plane in Young's experiment satisfies the requirement of energy conservation as stated following Eq. 5.25.

7. Plot the functional form of the flux density in Young's configuration if a neutral density filter that cuts the transmitted *power* by a factor of 3 is placed over one slit.

8. The fringes in Young's experiment are not actually constant in separation as suggested by Eq. (5.30). That equation depends on $x' \ll D'$. Improve on this result by carrying the approximations to the next higher order and in this way calculate the deviation from constant fringe width as a function of the position in the observation plane.

9. If a dielectric slab of unknown maximum thickness is inserted over one of the slits in Young's configuration, the central maximum shifts 3.4 fringes to one side. (This can be monitored by introducing the slab in the form of a wedge so that one fringe can be observed as the thickness changes.) If the index of refraction of the material is 1.467, find the maximum thickness of the slab as a multiple of λ.

10. It is possible to displace the central maximum in Young's configuration by inserting a transparent material in front of one of the slits. If, instead of monochromatic light, the slits are illuminated with white light, then this "central" maximum represents the only position in the pattern that will be a maximum for all the wavelengths. Provided the transparent material is non-dispersive (n not a function of wavelength), then this situation is maintained even with the transparent material present. If the material is dispersive, then it is not the phase velocity but the group velocity that must be used in determining optical path length differences. (See the development of this concept in Problem 2.55). In this case the maximum that is the same for all wavelengths may be different from that which we would identify as the "central" maximum for any one wavelength. Using this information and the equations given in Problem 2.55, determine the location of this fringe of achromatic character, with a dielectric plate thickness d and index n inserted over one slit.

11. Consider Young's geometry in which the source emits light of two wavelengths λ_i and λ_j. Derive a relationship for the position on the observation screen at which the fringes vanish due to complementary superposition of two interference patterns.

12. In Fig. 5.5 consider the original Young's experiment in which the openings in the screen were pinholes instead of slits. As a function of the coordinates x', y' in the observation plane, derive an expression for the shape of the fringes; that is, determine the locus of points in the observation plane that satisfy constructive interference.

13. In a Lloyd's mirror experiment with light of wavelength 500 nm it is found that the bright fringes on a screen 1 m away from the source are 1 mm apart. Calculate the (perpendicular) height of the source slit from the mirror.

14. In an experiment with Lloyd's mirror the source is 2 mm above the plane of the mirror. The mirror is halfway between the source and the screen and is 40 cm long. The source is 1.5 m from the screen and emits radiation with a wavelength of 500 nm. Calculate the fringe separation and the maximum and minimum values of x' (the vertical coordinate on the observation screen) for which fringes are seen. How many fringes are observed?

15. Consider a Fresnel biprism with index of refraction n, where the prism angle α is small compared with unity (in radians). Use the small-angle prism approximation described in Problem 2.35. Now suppose that $a \ll b$ and suppose that the image L_1 and L_2 of L formed by the two halves of the biprism are a distance $a/2$ from L. What then is the necessary value of b? Under these circumstances, and for a general value of $D' > 0$, calculate the fringe separation on the screen. How many fringes are observed? (See Fig. 5.7.)

Section 5.2 Multiple-Beam Interference

16. Work out the details for the multiple-beam interference algebra so as to derive Eq. (5.34) starting with Eq. (5.32).

17. Instead of the asymmetrical phasor addition expression in Eq. (5.35), consider

$$E = \sum_{n=-(N-1/2)}^{(N-1/2)} A_n \, e^{i\phi_n}$$

where N is odd and $\phi_n = \omega t + n\delta$. This is a symmetrical form in which the central member of the sum ($n = 0$) is the reference member (that is, $\phi_0 = \omega t$). Starting from this expression derive a formula similar to Eq. (5.38) for the total electric field due to the sum of N components. Now compute the power density. How does this result compare with Eq. (5.39), which was derived from the asymmetrical situation?

18. A plane wave is normally incident on a screen containing five *very* narrow long slits separated from each other by equal distances.

(a) Plot the flux density in the interference pattern as a function of δ (which is proportional to position in the interference plane) from $\delta = -2\pi$ to $\delta = 2\pi$.

(b) Draw phasor diagrams illustrating the optical electric field in the interference pattern for $\delta = 0$, $\delta = \pi/5$, $\delta = 2\pi/5$, $\delta = \pi$.

19. Use a graphical or computerized technique to solve the transcendental equation (5.43) to find the locations of the minor maxima in the case of five slits. Determine the error in your calculation and compare your result with the approximation $\delta = 3\pi/5$, π, $7\pi/5$.

Section 5.3 Two-Beam Interference: Parallel Interfaces

20. Following arguments similar to those that precede Eq. (5.58), derive an expression for the difference in phases between the first two transmitted beams in the final medium beyond the dielectric slab.

21. Consider Haidinger fringes observed at normal incidence with a slab of thickness d and index n. Let D be the diameter of the lens and f its focal length. Monochromatic light of wavelength λ is used. For d much larger than λ, the requirements on uniformity of n and on the flatness and parallelism of the surfaces of the slab are quite severe if the ideal Haidinger fringes described in the text are to be observed. Explain why this should be so, and estimate the tolerances on d and n if no observable departures from ideal behavior are to be noticed.

22. Haidinger fringes are observed with a 2.00-mm-thick slab of index $n = 1.60000$ having accurately flat surfaces. The light used has a wavelength of 500.0 nm. Calculate the maximum order at the center of the circular fringe pattern. How many bright fringes are observed within an observation angle of $1/30$ rad.

23. A Michelson interferometer can be used to measure the wavelength of light incident on it from a monochromatic source. However, first the mirror position must be found where both arms have identical optical path lengths. How could this be accomplished? Assuming that that task has been performed, how many fringes would emerge from the center of the pattern if the mirror moved 0.1 mm under illumination by light from a source having wavelength 492 nm? At what angle does the zero-order fringe appear after this mirror movement?

24. A Michelson interferometer is used with light having a wavelength of 600 nm, and observations are made at two values of the effective plate separation d, namely 0.5 mm and 5 mm. What is the angular separation between the fringes at the two values of d? If the interferometer mirrors are circular and 50 mm in diameter and observed with the naked eye at a distance of 400 mm, about how many fringes appear across the field of the mirror for the two values of d?

25. The derivation of the flux density at the center of the pattern in a Michelson interferometer with a moving mirror can be developed in the time domain by considering the light reflected from the moving mirror to be *Doppler shifted*. Follow this reasoning to determine the rate at which fringes appear at the center of the pattern when the mirror is moving at velocity v.

26. Describe quantitatively the fringes observed when a cylindrical glass surface is placed in contact at one end with an optical flat as shown in Fig. 5.47 and observed from above with light of wavelength 450 nm.

27. An oil film ($n = 1.2$) on water ($n = 1.33$) is viewed from directly above with light of wavelength 600 nm in

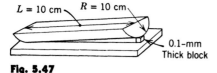

$L = 10$ cm $R = 10$ cm

0.1-mm
Thick block

Fig. 5.47

air. The film appears circular and is known to have a center thickness of 1 μm, decreasing to zero thickness at the edge. Tell whether the edge will appear dark or bright and why. How many dark rings appear in the interference pattern?

28. A circular oil film has a thickness profile that follows $d = d_0 \exp(-r^2/r_0^2)$ where $d_0 = 2 \mu m$, $r_0 = 2$ cm, and the index of refraction for the oil is 1.44. Quantitatively describe the pattern of Fizeau fringes if viewed from above with

(a) Light from a Na lamp (589.3 nm wavelength).

(b) White light in a band from 450 nm (blue) to 650 nm (red).

Section 5.4 Multiple-Beam Interference: Parallel Interfaces

29. Derive Eq. (5.95) for the multiple-reflection case by explicit development of the individual beams. The first stages of this method are contained in Table 5.1. In this way you can arrive at an expression that leads to a geometrical series. By evaluating the series you arrive at Eq. (5.95).

30. Calculate the ratio of full width at half-maximum to the separation between maxima (as a function of the phase difference δ) for a Fabry-Perot etalon with $R_1 = 0.5, 0.8, 0.9, 0.98$.

31. A Fabry-Perot interferometer is used with yellow light of two wavelengths of equal intensity at 599.4 nm and 600.0 nm. At what approximate value of the total order number m will they be just resolvable according to our criterion if $R_1 = 0.85$?

32. Calculate the reflectance and the transmittance of a stack with the configuration shown in Fig. 5.48. In terms of the incident field, calculate the field in the third layer at X. Assume that the measurement will be made at normal incidence with light having wavelength 500 nm.

33. A Fabry-Perot etalon is used with monochromatic light as described in the text. Let the first bright fringe from the center of the pattern occur at $p = q$ (not necessarily an integer), the first beyond that at $p = q + 1$, the second beyond that at $p = q + 2$, and so forth up to the Nth beyond the central fringe at $p = q + N$. Let x_N be the radius of the Nth bright fringe beyond the first. Show that q is given by

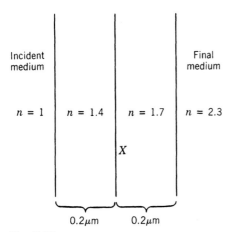

Fig. 5.48

$$q = \frac{x_N^2}{\Delta x^2} - N$$

where Δx^2 is the difference in x^2 between fringes, which does not depend on N. All the quantities on the right of this equation can be measured. Assuming n and λ to be known, find the plate separation in terms of Δx^2 and f, the focal length of the lens.

34. Determine the reflectance of the following configuration using the matrix method: Substrate glass with

$n = 1.517$, 50-nm thick layer of silver with $n = 0.12 - i3.63$, 5 nm overlayer of amorphous quartz $n = 1.458$. You will need to calculate the matrices first and then perform the matrix multiplication involving complex numbers. This problem and others like it are ideally suited for computerization. Assume that the measurement is made at normal incidence with light having a wavelength of 589.2 nm.

35. How does the development of the transmittance of the single-slab multiple reflection change if the material from which the slab is constructed demonstrates gain? By this is meant that the layer propagation matrix takes the form

$$\mathbf{L} = \begin{pmatrix} e^{-i\beta + \beta'} & 0 \\ 0 & e^{i\beta + \beta'} \end{pmatrix}$$

where the added term in the exponential is due to the nonlinear character of the medium. In practice, nonlinear amplification of this type is always limited by relaxation phenomena. How does your answer change if β' is negative?

Section 5.5 Applications of Interference

36. What would the quantitative characteristics of the fringe pattern be in the case of a wedge made from two

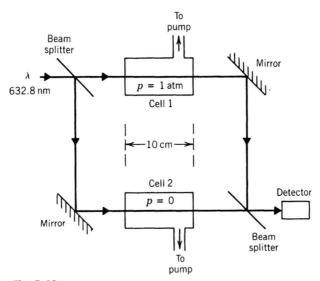

Fig. 5.49

optical flats 5-cm long and held apart at the open end by a 2-μm shim with illumination by 589.3-nm light? How does this change if a small rectangular raised area is introduced in the middle of the wedge whose dimensions are 2-cm long (along the wedge axis) and 450-nm thick? Assume that the raised area is not as wide as the optical flats.

37. The optical paths through a Mach-Zehnder interferometer are shown in Fig. 5.49. The two arms pass through identical gas cells that can be independently pressurized or evacuated. Assume that the following relationship between pressure p and index of refraction n for a test gas is valid: $n = 1 + Ap$. Initially, both cells are filled with an unknown gas at 1 atm of pressure. As one cell is evacuated, 100 minima are observed in the output of the detector. Find the constant of proportionality, A, in the $n(p)$ equation.

38. In the Czerny-Turner grating mount the collimating mirrors have a focal length of 1/4 m in a common example. If the grating has a groove density of 800/mm, the fixed angle α is 20°, and the grating is used in first-order, find the size of the linear spread on the exit slit of a band of light 0.1 nm wide centered around 400 nm. See Fig. 5.35.

39. Compare the resolving power of a grating that is 2-cm square having rulings of 600 lines/mm with a Fabry-Perot interferometer that has a plate spacing of 1 mm and a reflectivity per plate of 0.92 when both are used to resolve a doublet at 516.9 nm and 516.7 nm.

40. Derive a resolution criterion based on the assumption that two peaks can be separated if the flux density in the interference pattern is "flat" between the values of $\delta\lambda$ for the maxima of the two peaks (see Fig. 5.50). Use Eq. (5.51) to express your answer in a form that can be compared with the Rayleigh criterion. Perform your calculation for both the interference grating and for the Fabry-Perot interferometer.

41. In the center-spot scanning technique, the question of the size of the pinhole in front of the detector often arises. If it is too small, we lose illuminance unnecessarily; if too large, resolution falls off. Let us assume that a good compromise is reached when the pinhole radius ρ corresponds to a change in phase δ of amount $W/2$, half the full width at half-maximum. Find an expression for ρ in terms of n, d, λ, F, and f, the focal length of the final lens.

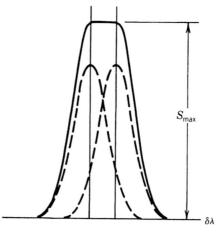

Fig. 5.50

42. A camera lens has an antireflection coating made of MgF_2 ($n = 1.38$). If the lens is to operate best at 520 nm and the front element of the lens is made of a glass with an index of refraction of 1.562, find the thickness of the required coating and the percentage improvement in transmittance through the first interface of the lens when the coating is present compared with the non-coated lens.

43. Consider a reflectance reducing coating made with MgF_2 ($n = 1.38$) on a glass of index of refraction 1.7. The system is observed at nearly normal incidence. How thick should the coating be if the wavelength of the light is 550 nm? Calculate the net reflectance at this wavelength as well as at 400 nm and 750 nm.

(a) Perform these calculations considering only the first two reflected beams.

(b) Include multiple reflections in your calculation.

44. Find the maximum reflectance of a dielectric layer mirror constructed from alternating layers (five pairs) of materials with indices of refraction 1.45 and 1.72. What thickness would each member of a pair need to be if these indices are given for 550-nm radiation?

45. An interference filter can be constructed from the concept behind the Fabry-Perot interferometer. If the spacing between the plates is small enough, then there will be only one allowed principal maximum within the spectral range of visible light (400 nm to 700 nm). Using this idea, design an interference filter that will have a peak transmittance of 50% at 514 nm and a full-width at half-maximum of 5 nm. Assume that you have a dielectric material from which extremely uniform thin films can be made and that the index of refraction of the material is 1.421 at 514 nm. You will need to specify the film thickness and the reflectance of the coating that will sandwich the film. How much will the filter need to be tilted so as to shift the central wavelength of the maximum to 488 nm?

46. What is the frequency difference between longitudinal modes of a 1/4-m laser cavity? Ignore the phase change on reflection at the end mirrors. How many nodes exist within the cavity if it is used to produce radiation at 632.80 nm? If the mirrors at the ends have reflectance of 0.98, what is the finesse of this cavity?

47. A thin-film waveguide is 0.8 μm thick and is constructed of a material whose index of refraction is 1.49, covered with an overlayer of index 1.35. If for this structure Δm is 0.7, find the longest wavelength radiation that will propagate within the waveguide without undergoing destructive interference.

48. The insertion of an uncompensated dielectric slab in one arm of a Michelson interferometer changes the nature of the white-light fringes, which occur when the optical path difference in the two arms is very close to zero. In a compensated interferometer the white-light fringes have very large radii, when the plates are accurately parallel. In this problem you are asked to find expressions for the radii of the white-light fringes in the uncompensated case. The setup is illustrated in Fig. 5.51a. Assume the beam splitter to be a self-supporting semireflecting film. Mirror I_1 has a parallel dielectric slab in front of it (an antireflection coating for the mirror, for example). Mirror I'_2 is the image of the mirror I_2 formed by the beam splitter. Ray 0 from the source splits into rays 1 and 2 at the beam splitter. Ray 2', the image of ray 2, is reflected by I'_2. The situation may be analyzed as shown in Fig. 5.51b. It is useful to refer the phases of rays 1 and 2' to the reference ray that would be reflected from the surface of the slab if there were no antireflecting coating there. Discuss the order and radii of the fringes as a function of the equivalent separation d'. Where are the white-light fringes located? What are their radii? You will need an auxiliary lens of focal length f to form images of the fringes.

49. When a slit is used with a monochromator, as shown in Fig. 5.52, we obtain a series of collimated beams whose directions (parallel to OP) have a projec-

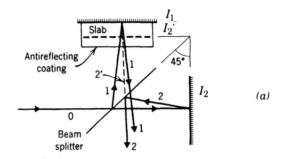

(a)

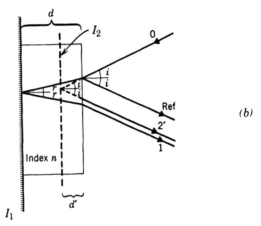

(b)

Fig. 5.51

tion in the x-z plane along the line Oz' through the center of the slit and the center of the lens. The direction OP may be specified by the angles ϕ_x and ϕ_y. Alternatively, we may take the angle ϕ_{x0} that the projection of OP in the x-z plane makes with the z axis. This angle ϕ_{x0} will be a constant if the slit and the lens are used as shown, but ϕ_x will not be constant. Show that the interrelationship is

$$\sin \phi_x = \sin \phi_{x0} \cos \phi_y$$

A transmission diffraction grating with spacing d and rulings parallel to the y axis lies in the x-y plane. It will diffract monochromatic light moving parallel to OP into a beam in mth order moving parallel to OP', which will have a projection OP'_c onto the xz plane and direction $\phi'_y = \langle (P'OP'_c)$ and $\phi'_{x0} = \langle (P'_cOz)$. Find expressions for ϕ'_y and ϕ'_x in terms of ϕ_y and ϕ_{x0}, and show that for small ϕ_y, ϕ'_{x0} changes quadratically with ϕ_y.

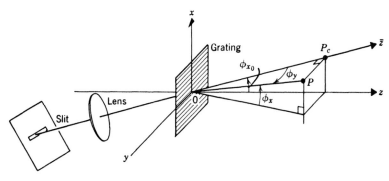

Fig. 5.52

Another lens is used to the right of the grating to bring this *m*th order beam to a focus. Show that the resulting monochromatic image of the slit will be curved, and derive a formula for the *sagitta*, that is, the deviation of the curve from a straight line. Is this an important effect? Compared with what? Discuss, using realistic numbers for the focal lengths, grating constants, wavelength, and slit height.

50. Show that the equation for the transmittance of the Fabry-Perot etalon can be written in the form $T = C(S_{in}/S)$, where C is a constant of the particular system and S and S_{in} are the incident and the internal power densities, respectively. Assume now that the index of refraction within the region between the plates (internal) depends on the power density there, $n = n_0 + gS_{in}$. This might arise through a nonlinear phenomenon that might be subject to external control. Use this in the equation for the transmission Eq. (5.105) that we derived earlier in the chapter. These two equations can be simultaneously solved to eliminate S_{in}. Show that the result for $T(S)$ demonstrates more than one solution for high enough values of S. This is called the *dispersive Fabry-Perot etalon* and is a model system for optical bistability. Because the transmittance can take two forms, depending on the system past history, this can be used as an optical switch and, therefore, can form the basis for an optical computer.

6 Diffraction I

6.1 General Concepts of Diffraction

The concept of interference involves superposition of a finite number of component fields. Each field may have its own amplitude and phase.

In Chapter 5 these characteristics were related to the different optical paths through which the light beams associated with the component fields traveled. In diffraction we generalize the same phenomenon to describe the superposition of infinitesimal contributions. Each contribution is associated with its own phase and amplitude. When superimposed, these infinitesimal components represent the effect of a continuous variation in path experienced by light coming from an extended region in space.

In the classic example of diffraction, a point source illuminates a barrier in which there is an opening (Fig. 6.1). The solution to the problem involves the

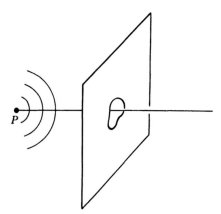

Fig. 6.1 Diffraction is concerned with the description of the behavior of light as it interacts with apertures and barriers. Here a point source creates spherical waves that encounter an opening in a planar barrier.

337

description of the optical field distribution beyond the barrier. From this result the spatial distribution of the time-averaged flux density can be determined. Diffraction is related to the performance of lenses and optical instruments. It is essential in the description of laser beam optics. We use it to tailor the performance of interference gratings (thus leading to the more common term "diffraction grating") and to create three-dimensional images through holography. The role of diffraction in image formation can be exploited to improve picture quality. It can also be used in analog devices capable of pattern recognition.

Although a rigorous solution of the wave equation for each situation would be desirable, many diffraction phenomena can be quantitatively described by a mathematical application of Huygens' principle. Huygens' principle can be shown to be an excellent approximation to the behavior of the solutions to the electromagnetic wave equation.

In this chapter we develop the theory and then apply it to the case where the source and observation points are far from the barrier. The near-observation situation is dealt with in the next chapter. These concepts are justified by a rigorous derivation of the Fresnel-Kirchhoff formalism for diffraction in the Appendix.

Huygens' principle says that each part of the optical wave surface in space acts as a source for a spherical wavelet. All the wavelets combine to yield the new wave surface at a distant point. In geometrical optics we only worry about the envelope of the secondary wavelets. Here we treat the secondary wavelets explicitly by keeping track of their respective phases and amplitudes at the observation point.

A. Diffraction Integral

The interference effects studied in Chapter 5 dealt with the superposition of a limited number of beams. For interference to occur, these were required to have the same polarization and frequency. The sources of the beams also needed to be coherent. Once this was established, we developed the theory using phasors to represent the scalar part of the optical field. We will follow the same approach here. Now the source for the interfering wavelets will be a monochromatic primary wavefront from the original point source, which ensures that all the secondary wavelets will have the same frequency and that they will be coherent.

This situation is similar to Young's geometry for two-beam interference. Insight about diffraction starts with a thorough understanding of this example. In section 5.2E (Fig 5.5) we simplified the problem by dealing with parallel, infinitely long slits. That led to one-dimensional variation of the interference pattern in the observation plane. Here we are interested in two-dimensional patterns that are more appropriately linked to the original Young's geometry involving pinholes.

Recall the form of a spherical wave.

$$\tilde{E} = \frac{A}{R} e^{i\phi} \tag{6.1}$$

with

$$\phi = \omega t - kR \tag{6.1a}$$

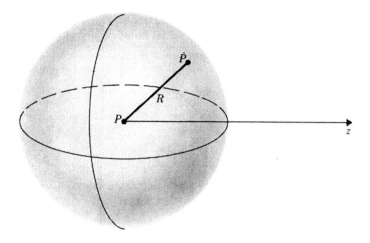

This is depicted in Fig. 6.2. Eq. (6.1) is a valid solution to the wave equation. It describes the magnitude of the optical field at $\tilde{P}$ a distance R from a point source at P.

Now introduce a barrier perpendicular to and a distance D along a line leading from P. Let this line represent the z axis. Assume there is a single pinhole at $\tilde{P}$ with coordinates in the barrier plane $(\tilde{x}, \tilde{y})$ as shown in Fig. 6.3. The pinhole serves to isolate a single Huygens' wavelet as shown in Fig. 6.4. The optical field associated with the wavelet at R' will be proportional to

$$\frac{e^{i\phi'}}{R'}$$

where $\phi' = \omega t - kR'$. It will also be proportional to the field on the incident side of

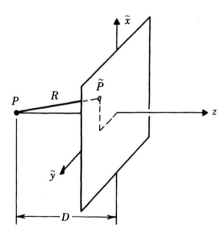

Fig. 6.3 The z axis is the line perpendicular to the barrier leading to P. Here a single pinhole is found at $\tilde{P}(\tilde{x}, \tilde{y})$.

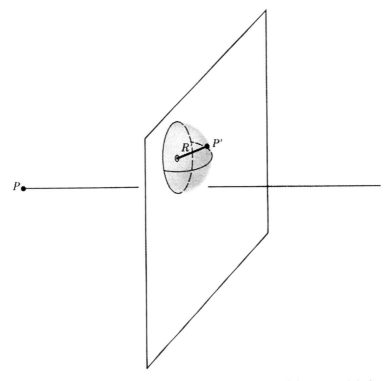

Fig. 6.4 A single Huygens' wavelet propagates outward from the pinhole.

the barrier at $\tilde{P}$, and to the area of the pinhole at $\tilde{P}$, $\Delta\tilde{\sigma}$. Because the time dependence already appears in ϕ', we must multiply $\tilde{E}$ by $e^{-i\omega t}$ to isolate the time-independent part of the incident field. Thus at P' we have

$$E' = C(\tilde{E} e^{-i\omega t}) \Delta\tilde{\sigma} \frac{e^{i\phi'}}{R'}$$

and

$$E' = C\tilde{E} \frac{e^{-ikR'}}{R'} \Delta\tilde{\sigma} \tag{6.2}$$

In this case, where $\tilde{E}$ is the result of a point source at P, we can write

$$E' = CAe^{i\omega t} \frac{e^{-ik(R+R')}}{RR'} \Delta\tilde{\sigma} \tag{6.3}$$

The area element $\Delta\tilde{\sigma}$ is a reasonable ingredient because the amplitude of the spherical wavelet should depend on the size of the available source field. We must require, however, that the hole remain small enough so that it behaves as a point source. The factor C in Eq. (6.2) has the dimensions of inverse length and will be found to be complex. We will determine C later. However, it is worth pointing out

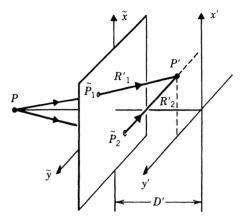

Fig. 6.5 The Huygens' wavelets from two pinholes are superimposed at P'. This is similar to the original geometry used by Young.

here that its complex character is a reflection of the fact that, in order to reproduce an unobstructed wave, the secondary Huygens' wavelets at the barrier must be out of phase with respect to the incident wavefront at the barrier. The C factor turns out to be a slowly varying function of the angles that describe the directions of propagation of the incident wave and the secondary wavelet at the pinhole. In most cases these angles are small enough that we can ignore this dependence and treat C as being independent of the geometry. This is certainly true in the far-field diffraction limit.

With two identical pinholes at $\tilde{P}_1$ and $\tilde{P}_2$ shown in Fig. 6.5, the configuration becomes the same as Young's original experiment. We examine an observation point at P' identified with coordinates (x', y') in a plane perpendicular to the z-axis a distance D' from the barrier. Superposition leads to the resultant field

$$E' = C\left[\tilde{E}_1 \frac{e^{-ikR_1'}}{R_1'} + \tilde{E}_2 \frac{e^{-ikR_2'}}{R_2'}\right]\Delta\tilde{\sigma} \tag{6.4}$$

The flux density at P' will be

$$S' = \frac{v\varepsilon}{2}|E'|^2 \tag{6.5}$$

For N pinholes (Fig. 6.6) each area $\Delta\tilde{\sigma}_n$ located at $\tilde{P}_n$ $(n = 1, 2, \ldots N)$ this result generalizes to

$$E' = C \sum_{n=1}^{N} \tilde{E}_n \frac{e^{-ikR_n'}}{R_n'} \Delta\tilde{\sigma}_n \tag{6.6}$$

Each $\tilde{E}_n$ and R_n' are functions of the position of the nth pinhole in the barrier.

Now let the number of pinholes become so large and their positions so close together that, collectively, they identify a continuous opening, an aperture $\tilde{\sigma}_0$ as shown in Fig. 6.7. One of the pinholes is explicitly drawn in this picture. In the limit that the number of pinholes tends to infinity and the size of each pinhole tends to

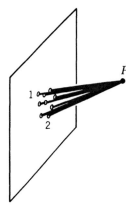

Fig. 6.6 A large number of pinholes. Each contributes a Huygens' wavelet to the field at P'.

zero, while keeping the size of $\tilde{\sigma}_0$ fixed, the sum in Eq. (6.6) becomes an integral

$$E' = C \iint\limits_{\tilde{\sigma}_0} \tilde{E} \frac{e^{-ikR'}}{R'} d\tilde{\sigma} \tag{6.7}$$

The resultant field at P' can be seen to be a superposition of the contributions from secondary wavelets, each associated with one of the imaginary pinholes comprising $\tilde{\sigma}_0$. This is the essence of Huygens' principle.

Huygen's principle, as we have employed it here, still retains an inconsistency. There is nothing to prevent a secondary wavelet from propagating backward toward the source. We need to introduce a factor that will eliminate this unphysical situation. When the Fresnel–Kirchhoff theory is solved we find that this factor is present automatically. Here we treat the added influence as a correction (which can frequently be ignored). Figure 6.8 shows an increment $d\tilde{\sigma}$ of the aperture $\tilde{\sigma}_0$ with normal unit vector $\hat{\eta}$ identified as positive when directed away from the source.

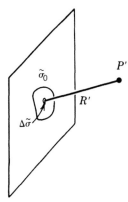

Fig. 6.7 A continuous opening can be considered to be a combination of infinitesimal pinholes.

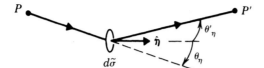

Fig. 6.8 Geometry associated with the inclination factor.

Angles θ_η and θ'_η are the angles that the incident and diffracted ray would make with $\hat{\eta}$ at $d\tilde{\sigma}$. The *inclination factor* is

$$Q \equiv \tfrac{1}{2}(\cos\theta_\eta + \cos\theta'_\eta) \tag{6.8}$$

and is used under the integral in Eq. (6.7) as an additional factor. Q is nearly zero when P' is on the same side of $d\tilde{\sigma}$ as P and vanishes in the exact backward direction when $\theta'_\eta = \theta_\eta - \pi$. Usually both θ_η and θ'_η are near zero, under which conditions $Q \simeq 1$. Because Q is a slowly varying function of θ_η and θ'_η, we can frequently treat Q as a constant with respect to variation in $\tilde{\sigma}_0$ even in cases where Q is different from unity. In that situation, Q depends only on the location of P and P' through the average angles θ_η and θ'_η at the aperture.

B. Discussion of Diffraction Integral

The diffraction integral in Eq. (6.7) has been applied to a wide variety of diffraction phenomena with results in a great many cases that are surprisingly good, considering that it is not a rigorous deduction from the wave equation, but only an approximation. It works best when both source and observation points are a very large number of wavelengths away from the aperture and when the size of the opening is somewhat larger than a wavelength. It does not take any polarization effects into account. These can be important in some experiments.

1. Choice of the Surface. We have assumed in our derivation that the aperture is a plane. The extension of the plane into the opening to form the integration surface $\tilde{\sigma}_0$ is the most convenient choice. However, it does not follow that this is necessarily the best choice. After all, the opening represents the absence of a physical surface. How do we decide what $\tilde{\sigma}_0$ will be? In principle, we can choose any reasonably smooth surface. Sometimes we may choose to make $\tilde{E}$ or R' constant. This is illustrated in Fig. 6.9, where the incident light is from the point source at P. The

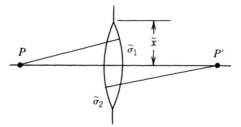

Fig. 6.9 Two choices of the surface of integration in the calculation of the diffraction integral.

surface $\tilde{\sigma}_1$ would make $\tilde{E}$ constant, whereas $\tilde{\sigma}_2$ would make R' constant. In the "parabolic" approximation, in which calculations are carried out to second order in $(\tilde{x}, \tilde{y})$, the results with $\tilde{\sigma}_1$ or $\tilde{\sigma}_2$ or any intermediate surface will all be the same. We are then allowed to choose the most convenient surface for $\tilde{\sigma}_0$.

2. Transmission Function. If the opening is covered by a partly transmitting film with amplitude transmission factor $\tilde{\tau}(\tilde{x}, \tilde{y})$, we should replace $\tilde{E}$ in Eq. (6.7) with $\tilde{\tau}\tilde{E}$ to represent the transmitted light at the aperture.

The use of such a *transmission function* can be generalized to describe the location of a clear opening. Let the open part of the barrier surface be designated with $\tilde{\sigma}_0$, and the closed part $\tilde{\sigma}_c$. Now define $\tilde{\tau}$ as a step function

$$\tilde{\tau}(\tilde{x}, \tilde{y}) = \begin{cases} 1 & \text{if } (\tilde{x}, \tilde{y}) \text{ within } \tilde{\sigma}_0 \\ 0 & \text{if } (\tilde{x}, \tilde{y}) \text{ within } \tilde{\sigma}_c \end{cases}$$

Then Eq. (6.7) can be written as

$$E' = C \iint\limits_{\tilde{\sigma}} \tilde{\tau}\tilde{E}\, \frac{e^{-ikR'}}{R'}\, d\tilde{\sigma} \tag{6.9}$$

More generally, any function $\tilde{\tau}(\tilde{x}, \tilde{y})$ that describes the modification of the incident field at the barrier can be used in Eq. (6.9). This applies also to functions that change the phase, that is, functions of the form

$$\tilde{\tau} = |\tilde{\tau}|\, e^{-i\delta}$$

with nonzero phase angle of δ that may depend on $\tilde{x}$ and $\tilde{y}$. This can be produced by a transparent plate with varying thickness, as shown in Fig. 6.10. For nearly

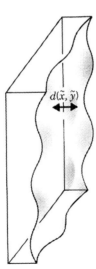

Fig. 6.10 A transparent plate with varying thickness might serve as the modulating element in the diffraction integral.

normal incidence and small changes in the plate thickness, d, the phase change produced by a plate referred to the case with no plate is

$$\delta = \frac{2\pi}{\lambda_0}(n-1)d$$

where n is the index of refraction and λ_0 is the vacuum wavelength.

The effect of this plate in a diffracting aperture would be approximately described by the transmission function

$$\tilde{\tau} = |\tilde{\tau}| \exp\left[-i\frac{2\pi}{\lambda_0}(n-1)d\right] \tag{6.10}$$

with $|\tilde{\tau}|$ and d both being functions of $\tilde{x}$, and $\tilde{y}$. Losses caused by reflection and absorption would be described by a value of $|\tilde{\tau}|$ less than unity.

3. Babinet's Principle. An important and useful conclusion can be drawn from the fact that the diffraction formula Eq. (6.7) expresses the field E' as an integral over the open aperture $\tilde{\sigma}_0$. If $\tilde{\sigma}_0$ is split into two parts, $\tilde{\sigma}_1$ and $\tilde{\sigma}_2$, so that $\tilde{\sigma}_1 + \tilde{\sigma}_2 = \tilde{\sigma}_0$ (Fig. 6.11), the field at the point P' will be the sum of integrals over $\tilde{\sigma}_1$ and over $\tilde{\sigma}_2$. But the integral over $\tilde{\sigma}_1$ represents the field at P' if $\tilde{\sigma}_1$ is the only opening, and the integral over σ_2 represents the field at P' if σ_2 is the only opening. Thus, if we denote these fields by

$$E_1' = C \iint_{\tilde{\sigma}_1} (\ldots)\, d\tilde{\sigma}$$

and

$$E_2' = C \iint_{\tilde{\sigma}_2} (\ldots)\, d\tilde{\sigma}$$

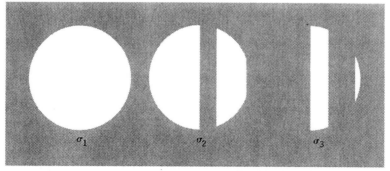

Fig. 6.11 Babinet's principle. The last two apertures are complementary.

where (...) denotes the integrand in Eq. (6.7), we obtain

$$E'_0 = E'_1 + E'_2 \tag{6.11}$$

This equation expresses a principle due to Babinet. The apertures $\tilde{\sigma}_1$ and $\tilde{\sigma}_2$ are said to be *complementary*. The fields observed with complementary apertures add to give the field observed with the combined open surface of both.

Equation (6.11) can be very helpful when, say, $\tilde{\sigma}_1$ is a relatively complicated aperture composed of the difference $\tilde{\sigma}_0 - \tilde{\sigma}_2$ of two simple apertures. Then we calculate the field at P' resulting from $\tilde{\sigma}_0$ and $\tilde{\sigma}_2$ separately and subtract to obtain the field resulting from $\tilde{\sigma}_1$.

Babinet's principle is particularly useful when the field at P' is zero (or very small) with the aperture $\tilde{\sigma}_0$. This occurs because of cancellation effects with the component contributions from the different parts of the opening. In this case we obtain

$$E'_1 = -E'_2 \tag{6.12}$$

The flux density will be proportional to the square of the absolute value of these fields. We have thus shown that, in this case

$$S'_1 = S'_2 \tag{6.13}$$

Therefore, a point in the diffraction pattern of $\tilde{\sigma}_0$ that would be in the dark has the same nonzero flux density when the opening is changed to either of the complementary apertures $\tilde{\sigma}_1$ and $\tilde{\sigma}_2$, where $\tilde{\sigma}_1 + \tilde{\sigma}_2 = \tilde{\sigma}_0$.

6.2 Far-Field Diffraction

Fraunhofer or far-field diffraction is a special case of the general situation that occurs when the optical path from points in the diffraction aperture to the observation point depends at most linearly on the coordinates of the aperture point. This can be accomplished by arranging to have the source and observation points very far from the aperture. It can also be achieved by the proper use of the lenses. We will learn more about Fraunhofer diffraction by lenses later in this chapter and in Chapter 7. Here we restrict our attention to the case where the source and observation points are in the far-field region.

A. Linear Approximation

Figure 6.12a defines the geometry for our calculation. The aperture is the $\tilde{x}\tilde{y}$ plane that is perpendicular to the z axis at $\tilde{O}$. A point source at P is in the xy plane that is parallel to and a distance D from the aperture plane. The rectangular coordinates of P are $(x, y, -D)$. Here position $\tilde{P}$ with coordinates $(\tilde{x}, \tilde{y}, 0)$ is a typical point in the aperture. The observation plane is parallel to and a distance D' from the aperture plane. An observation point is P' with coordinates (x', y', D').

From the origin of the barrier plane we can identify the directions in which the

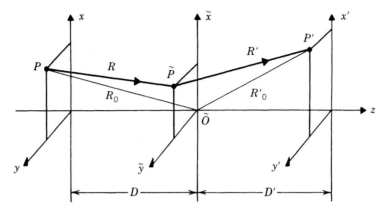

Fig. 6.12a Standard geometry for diffraction mathematics. $P(x, y)$ is the source point; $\tilde{P}(\tilde{x}, \tilde{y})$ is a point in the aperture plane; $P'(x', y')$ is the observation point. The perspective is compressed in the z direction.

incident and the diffracted waves are traveling. This is shown in Fig. 6.12b. Angles θ_x, θ_y, and θ_z are those that the incident propagation vector at $\tilde{O}$ makes with the $\tilde{x}$, $\tilde{y}$, and $\tilde{z}$ axes. The direction cosines are also useful

$$\alpha = \cos \theta_x = \frac{-x}{R_0} = \sin \theta$$

$$\beta = \cos \theta_y = \frac{-y}{R_0} \qquad (6.14a)$$

$$\gamma = \cos \theta_z = \sqrt{1 - \alpha^2 - \beta^2} = \frac{D}{R_0}$$

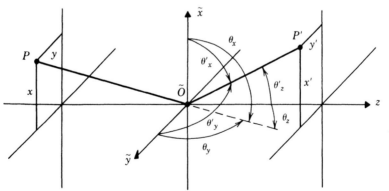

Fig. 6.12b Important angles that define the directions from the source and to the observation point as measured with respect to the origin of the aperture plane. $\theta = \pi/2 - \theta_x$, and $\theta' = \pi/2 - \theta'_x$ are used in §5.1, 5.2, and 6.4.

Likewise, we identify

$$\alpha' = \cos\theta'_x = \frac{x'}{R'_0} = \sin\theta'$$

$$\beta' = \cos\theta'_y = \frac{y'}{R'_0} \tag{6.14b}$$

$$\gamma' = \cos\theta'_z = \sqrt{1 - \alpha'^2 - \beta'^2} = \frac{D'}{R'_0}$$

These definitions follow the previously established convention for directions at an interface. The perspective of Fig. 6.12 is distorted. In actual practice D and D' would have to be much larger than they are shown here. We have reduced the size of the drawing along the z direction so as to emphasize the transverse coordinates.

Our starting point is the diffraction integral in the form of Eq. (6.9). If we use Eq. (6.1) to express the explicit form of $\tilde{E}$, the optical field at the aperture, we arrive at the result for the field at P':

$$E' = CAe^{i\omega t}\int_{\tilde{\sigma}}\tilde{\tau}\frac{e^{-ik(R+R')}}{RR'}\,d\tilde{\sigma} \tag{6.15}$$

Here A is the amplitude coefficient of the source. The integration surface $\tilde{\sigma}$ is the entire $\tilde{x}\tilde{y}$ plane. But $\tilde{\tau}(\tilde{x}, \tilde{y})$ is the transmission function that describes the characteristics of the aperture and that effectively limits the integration. Both R and R' are functions of $\tilde{x}$ and $\tilde{y}$ but, for a given aperture, E' is only dependent on the locations of P and P'.

In Eq. (6.14) the propagation distances R and R' are

$$R = [(\tilde{x} - x)^2 + (\tilde{y} - y)^2 + D^2]^{1/2} \tag{6.16a}$$

and

$$R' = [(x' - \tilde{x})^2 + (y' - \tilde{y})^2 + D'^2]^{1/2} \tag{6.16b}$$

In the far-field approximation, these distances are considered to be linear in $\tilde{x}$ and $\tilde{y}$. To develop this form for Eq. (6.15) we will expand R and R' about R_0 and R'_0, shown in Fig. 6.12a and given by

$$R_0 = (x^2 + y^2 + D^2)^{1/2}$$

and

$$R'_0 = (x'^2 + y'^2 + D'^2)^{1/2}$$

These distances to the origin of the aperture plane are independent of $\tilde{x}$ and $\tilde{y}$. Following this procedure for the incident distance, we have

$$R = [R_0^2 - 2(x\tilde{x} + y\tilde{y}) + \tilde{x}^2 + \tilde{y}^2]^{1/2} = R_0\left[1 - \frac{2}{R_0^2}(x\tilde{x} + y\tilde{y}) + \frac{\tilde{x}^2 + \tilde{y}^2}{R_0^2}\right]^{1/2}$$

Now, provided the second and third terms are small, we can use the expansion

$$R = R_0\sqrt{1 + \varepsilon} \simeq R_0(1 + \tfrac{1}{2}\varepsilon - \tfrac{1}{8}\varepsilon^2)$$

with

$$\varepsilon = -\frac{2}{R_0^2}(x\tilde{x} + y\tilde{y}) + \frac{\tilde{x}^2 + \tilde{y}^2}{R_0^2}$$

and calculate to second order in $\tilde{x}$ and $\tilde{y}$,

$$R \simeq R_0\left[1 - \frac{x\tilde{x} + y\tilde{y}}{R_0^2} + \frac{\tilde{x}^2 + \tilde{y}^2}{2R_0^2} - \frac{(x\tilde{x} + y\tilde{y})^2}{2R_0^4}\right] \tag{6.17}$$

If this expansion is to remain valid, then $\tilde{x}$ and $\tilde{y}$ must be small compared with R. That will be the case in the "far-field" limit where R is large and the aperture small. The third term will also be much smaller than the second term because it depends on a higher power of $\tilde{x}$ and $\tilde{y}$. These conditions will all be met provided the last term is small compared with the optical wavelength. Thus we must have

$$|\tilde{x}| \ll \sqrt{R_0\lambda} \tag{6.18a}$$

$$|\tilde{y}| \ll \sqrt{R_0\lambda} \tag{6.18b}$$

We are left with

$$R \simeq R_0 - \left(\frac{x}{R_0}\right)\tilde{x} - \left(\frac{y}{R_0}\right)\tilde{y} \tag{6.19a}$$

as the linearized form for the incident distance. A similar derivation leads to the linearized form for the distance from the aperture point to the observation point.

$$R' \simeq R_0' - \left(\frac{x'}{R_0'}\right)\tilde{x} - \left(\frac{y'}{R_0'}\right)\tilde{y} \tag{6.19b}$$

Before substituting Eqs. (6.19) into the diffraction integral, Eq. (6.15), we will make one additional approximation. The aperture position information carried by the total optical path $R + R'$ is essential in the determination of the phase factor, $e^{-ik(R + R')}$. However, the amplitude factor $(RR')^{-1}$ is less sensitive to small changes. To simplify the integration, and with no serious limitation, we then assume

$$(RR')^{-1} \simeq (R_0 R_0')^{-1} \tag{6.20}$$

We also introduce the variable changes

$$u \equiv -\left(\frac{x}{R_0} + \frac{x'}{R_0'}\right)\frac{1}{\lambda} = \frac{(\alpha - \alpha')}{\lambda} = \frac{(\cos\theta_x - \cos\theta_x')}{\lambda} \tag{6.21a}$$

$$v \equiv -\left(\frac{y}{R_0} + \frac{y'}{R_0'}\right)\frac{1}{\lambda} = \frac{(\beta - \beta')}{\lambda} = \frac{(\cos\theta_y - \cos\theta_y')}{\lambda} \tag{6.21b}$$

so that

$$R + R' = (R_0 + R'_0) + (u\tilde{x} + v\tilde{y})\lambda \tag{6.22}$$

The phase factors in the diffraction integral are

$$e^{i\omega t} e^{-ik(R + R')} = e^{i\phi_0} e^{-i2\pi(u\tilde{x} + v\tilde{y})}$$

where

$$\phi_0 = \omega t - k(R_0 + R'_0) \tag{6.23}$$

is independent of $\tilde{x}$ and $\tilde{y}$. The coordinates of the source (x, y) and of the observation point (x', y') are now contained within (u, v).

The observed optical electric field at P' due to a point source at P and an aperture characterized by $\tilde{\tau}(\tilde{x}, \tilde{y})$ will be

$$E'(u, v) = \frac{CA\, e^{i\phi_0}}{R_0 R'_0} \int\int\limits_{-\infty}^{\infty} \tilde{\tau}(\tilde{x}, \tilde{y})\, e^{-i2\pi(u\tilde{x} + v\tilde{y})}\, d\tilde{x}\, d\tilde{y} \tag{6.24}$$

The integral in Eq. (6.24) is the *Fourier transform* of $\tilde{\tau}(\tilde{x}, \tilde{y})$.

$$T(u, v) \equiv \int\int\limits_{-\infty}^{\infty} \tilde{\tau}(\tilde{x}, \tilde{y})\, e^{-i2\pi(u\tilde{x} + v\tilde{y})}\, d\tilde{x}\, d\tilde{y} \tag{6.25}$$

We see that: *The Fraunhofer diffraction pattern for the electric field is proportional to the Fourier transform of the transmission function.* We will study the characteristics of Fourier transforms a bit later. For now, we may treat Eq. (6.25) as the definition of an operation that is to be performed on $\tilde{\tau}(\tilde{x}, \tilde{y})$ in the process of determining the diffracted field.

If, as in Fig. 6.12c, P, $\tilde{O}$, and P' are situated on a straight line, then $(u, v) = (0, 0)$ and

$$E'(0, 0) = \frac{CA\, e^{i\phi_{00}}}{R_{00} R'_{00}} T(0, 0) \tag{6.26}$$

where the double subscript refers to the "straight-through" orientation of P, $\tilde{O}$, and P'. We may use Eq. (6.26) to normalize the diffracted field at any general point P' in terms of the field for P' associated with the straight-through geometry,

$$E'(u, v) = E'(0, 0)\left[\frac{R_{00} R'_{00}}{R_0 R'_0}\right] \frac{T(u, v)}{T(0, 0)} e^{i(\phi_0 - \phi_{00})} \tag{6.27}$$

The time-averaged optical energy flux density at P' is proportional to the square of the absolute value of the electric field. Thus, from Eq. (6.27) we find

$$S'(u, v) = S'(0, 0)\left[\frac{R_{00} R'_{00}}{R_0 R'_0}\right]^2 \frac{|T(u, v)|^2}{|T(0, 0)|^2} \tag{6.28}$$

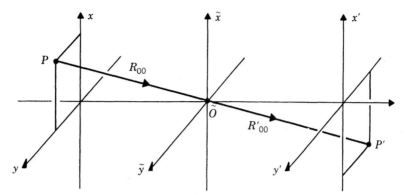

Fig. 6.12c The straight-through geometry.

The ratio $(R_{00}R'_{00})/(R_0 R'_0)$ is an implicit function of the direction cosines α, β, α', and β' and thus will vary *weakly* with u and v. In many cases, however, this ratio is sufficiently close to unity that it may be ignored. This will be the situation whenever we are investigating the diffraction pattern relatively near to the straight-through geometry. Most of the time R_{00} will equal R_0.

B. Rectangular Aperture

An important special case of far-field diffraction is that for which $\tilde{\tau}(\tilde{x}, \tilde{y})$ describes a rectangle. Because this is an ideal prototype, we will study it in detail.

1. Analytical Result. Let the transmission function be

$$\tilde{\tau}(\tilde{x}, \tilde{y}) = \begin{cases} 1 & \text{if } |\tilde{x}| < \tilde{x}_0 \text{ and } |\tilde{y}| < \tilde{y}_0 \\ 0 & \text{Otherwise} \end{cases} \tag{6.29}$$

This defines an aperture shown in Fig. 6.13 of area $4\tilde{x}_0\tilde{y}_0 = T(0, 0)$. We must evaluate the Fourier transform of $\tilde{\tau}$ according to Eq. (6.25). This is an example of a case in which $T(u, v)$ may be factored into two integrals.

$$T(u, v) = T(u)T(v) = \int_{-\tilde{x}_0}^{\tilde{x}_0} e^{-i2\pi u\tilde{x}} \, d\tilde{x} \int_{-\tilde{y}_0}^{\tilde{y}_0} e^{-i2\pi v\tilde{y}} \, d\tilde{y} \tag{6.30}$$

This factorization corresponds to the integrating over strips. The contribution from the identified strip in Fig. 6.13 is

$$e^{-i2\pi u\tilde{x}} \, d\tilde{x} \int_{-\tilde{y}_0}^{\tilde{y}_0} e^{-i2\pi v\tilde{y}} \, d\tilde{y}$$

The integrations are readily evaluated, for instance,

$$T(v) = \frac{1}{-i2\pi v} (e^{-i2\pi v\tilde{y}_0} - e^{+i2\pi v\tilde{y}_0}) = \frac{\sin(2\pi v\tilde{y}_0)}{\pi v} = 2\tilde{y}_0 \frac{\sin(2\pi v\tilde{y}_0)}{2\pi v\tilde{y}_0} \tag{6.31}$$

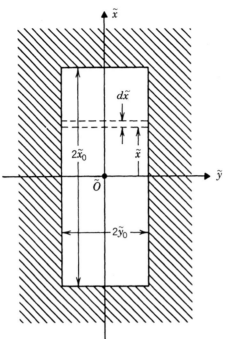

Fig. 6.13 Coordinates in the aperture plane for a rectangular opening.

The last form is so common that we identify a special function to describe it

$$\text{sinc}(w) \equiv \frac{\sin(w)}{w} \tag{6.32}$$

The function sinc (w) is plotted in Fig. 6.14 along with $\text{sinc}^2(w)$, which we will also need. Using L'Hospital's rule, we find

$$\lim_{w \to 0} \left(\frac{\sin w}{w} \right) = \lim_{w \to 0} \left(\frac{\cos w}{1} \right) = 1$$

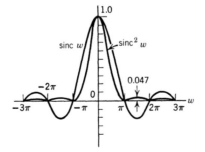

Fig. 6.14

This identifies the central maximum. Zeros occur when w is an integral multiple of π. That is for

$$w = \pm \pi, \pm 2\pi, \ldots \tag{6.33}$$

Additional extrema are located at values of w for which

$$\frac{d\left(\dfrac{\sin w}{w}\right)}{dw} = 0$$

or for

$$\tan w = w$$

From graphical or numerical solution of this equation, we find $w = \pm 1.4303\pi$, $\pm 2.4590\pi$ as the next two extrema with $(\sin w/w) = -0.2172$ and $+0.1284$ at these locations, respectively.

Equation (6.31) becomes

$$T(v) = 2\tilde{y}_0 \, \mathrm{sinc}(2\pi v \tilde{y}_0) \tag{6.34a}$$

Likewise

$$T(u) = 2\tilde{x}_0 \, \mathrm{sinc}(2\pi u \tilde{x}_0) \tag{6.34b}$$

At P' the field is given by Eq. (6.27)

$$E'(u, v) = E'(0,0) \frac{2\tilde{x}_0 \, \mathrm{sinc}(2\pi u \tilde{x}_0) 2\tilde{y}_0 \, \mathrm{sinc}(2\pi v \tilde{y}_0)}{4\tilde{x}_0 \tilde{y}_0} \left[\frac{R_{00} R'_{00}}{R_0 R'_0}\right] e^{i(\phi_0 - \phi_{00})}$$

$$= E'(0,0) \, \mathrm{sinc}(2\pi u \tilde{x}_0) \, \mathrm{sinc}(2\pi v \tilde{y}_0) \left[\frac{R_{00} R'_{00}}{R_0 R'_0}\right] e^{i(\phi_0 - \phi_{00})}$$

and the flux density by Eq. (6.28)

$$S'(u, v) = S'(0, 0) \, \mathrm{sinc}^2(2\pi u \tilde{x}_0) \, \mathrm{sinc}^2(2\pi v \tilde{y}_0) \left[\frac{R_{00} R'_{00}}{R_0 R'_0}\right]^2 \tag{6.35}$$

2. Phasor Representation. The integral for $T(u)$ or $T(v)$ in Eq. (6.30) can be represented as a curve in the complex plane. Such a representation is also useful for other diffraction problems. This is an extension of the discrete phasor addition method used in the previous chapter. Here we write

$$T(u) = \int_{-\tilde{x}_0}^{\tilde{x}_0} dT(u)$$

where

$$dT(u) = e^{-i2\pi u \tilde{x}} \, d\tilde{x}$$

can be represented as a phasor in the complex plane of length $d\tilde{x}$ and phase angle

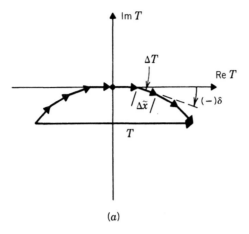

(a)

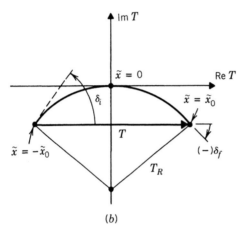

(b)

Fig. 6.15 Phasor diagrams that show the graphical representation of the diffraction integral in the far field limit. (a) Discrete approximation. (b) Continuous phasor diagram. As we rotate clockwise from the left end, we represent the contributions to the integral starting from the lower edge of the slit through the center of the slit to the top edge.

$\delta = -2\pi u\tilde{x}$. The integral represents the phasor sum of all the infintesimal phasors $dT(u)$. We can approximate the dT by the finite, but small, phasor ΔT. Then T is approximated by the sum $\sum \Delta\tilde{x}\, e^{i\delta}$. The resultant is the phasor that connects the origin of the initial ΔT with the end of the final ΔT, as shown in Fig. 6.15. As $\Delta T \to 0$, the polygon becomes a smooth curve that is a segment of a circle. The integration proceeds from $\tilde{x} = -\tilde{x}_0$ to $\tilde{x} = \tilde{x}_0$. These points are identified in Fig. 6.15b. The center of the phasor curve representing the symmetrical rectangular opening will be at the origin where $\tilde{x} = 0$. The resultant is the chord of the circular segment and is equal to $T(u)$.

The total arc length of the curve is given by

$$T(0) = \int dT(0) = 2\tilde{x}_0$$

The radius of curvature is given by

$$T_R = \frac{d\tilde{x}}{d\delta} \tag{6.36}$$

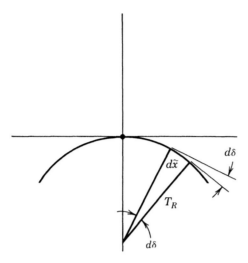

Fig. 6.16 The curvature of the phasor diagram depends on the value of u that is a function of the geometry associated with the location of the source and the observation point.

(see Fig. 6.16). This yields

$$T_R = \frac{d\tilde{x}}{-2\pi u \, d\tilde{x}} = -\frac{1}{2\pi u}$$

The minus sign means that the curve is concave downward. The initial and final phase angles δ_i and δ_f are given by

$$\delta_i = 2\pi u \tilde{x}_0 \quad \text{and} \quad \delta_f = -2\pi u \tilde{x}_0 \tag{6.37}$$

Some of these curves are shown in Fig. 6.17.

3. Pattern of Flux Density. Far-field diffraction patterns for two rectangular apertures are shown in Fig. 6.18. The distribution of flux density follows Eq. (6.35). From Eqs. (6.21) we recover the spatial dependence in the observation plane. Assume here that the source is on the z axis so that $\alpha = \beta = 0$ and $x = y = 0$. Then the coordinates in the x', y' plane map into u, v as

$$x' = -R_0' \lambda u$$

and

$$y' = -R_0' \lambda v$$

The width of the central maximum may be defined to be

$$|\Delta x'| = R_0' \lambda \, \Delta u$$

where Δu is measured between the first zeros adjacent to the central maximum. From Eq. (6.33) $\Delta(2\pi u \tilde{x}_0) = 2\pi$, so

$$\Delta u = \frac{1}{\tilde{x}_0} \tag{6.38}$$

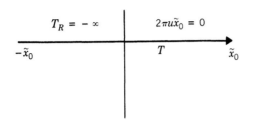
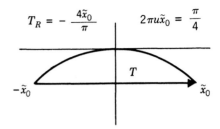
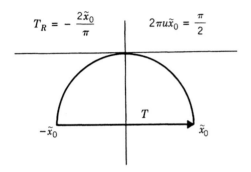
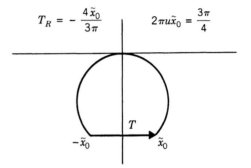
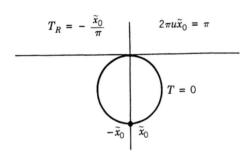
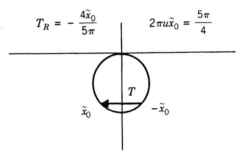

Fig. 6.17 Phasor diagrams for the diffraction integral versus u.

and

$$|\Delta x'| = \frac{R_0' \lambda}{\tilde{x}_0} \tag{6.39a}$$

Likewise

$$|\Delta y'| = \frac{R_0' \lambda}{\tilde{y}_0} \tag{6.39b}$$

Alternatively, we may express the width of the central maximum in terms of the direction cosines

$$|\Delta \alpha'| = \frac{\lambda}{\tilde{x}_0} \tag{6.39c}$$

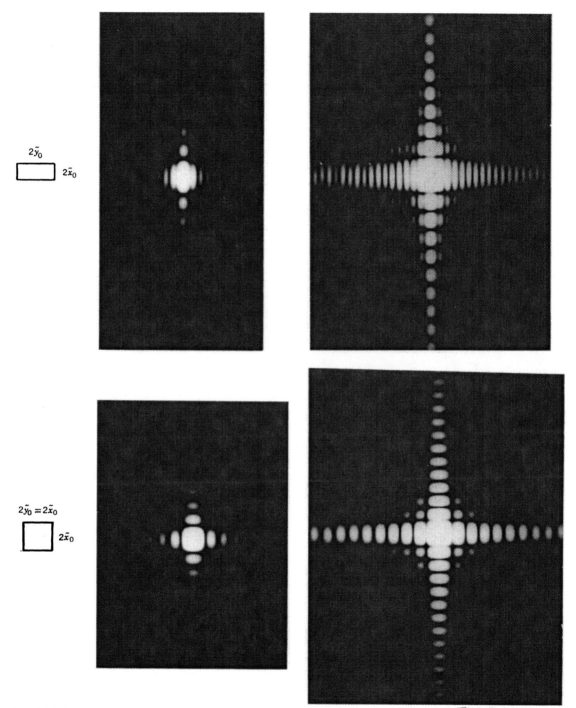

Fig. 6.18 Rectangular and square apertures and their far-field diffraction patterns at two different exposures. Longer exposure is needed to bring out the weak parts of the pattern.

and

$$|\Delta\beta'| = \frac{\lambda}{\tilde{y}_0} \qquad (6.39d)$$

Note the reciprocal relationship between the width of the pattern and the widths of the slit. This, as we will see, is a general characteristic of diffraction that, in the far-field case, is elegantly represented by the properties of Fourier transforms.

If we multiply Eqs. (6.39) by $1/\tilde{x}_0$ and $1/\tilde{y}_0$, we can generate

$$\left|\frac{\Delta x'}{\tilde{x}_0}\right| = \frac{R_0'\lambda}{\tilde{x}_0^2} \gg 1 \qquad (6.40a)$$

and

$$\left|\frac{\Delta y'}{\tilde{y}_0}\right| = \frac{R_0'\lambda}{\tilde{y}_0^2} \gg 1 \qquad (6.40b)$$

These inequalities follow from Eqs. (6.18), which are true in the far-field limit. We see from Eqs. (6.40) that the dimensions of the Fraunhofer diffraction pattern are much greater than those of the diffracting aperture. This result holds only for diffraction occurring at great distances without the use of a lens.

If the aperture is very long in the $\tilde{y}$ direction (with the source still remaining at a point), then the width of the pattern in the $\tilde{y}$ direction will be very narrow. When $\tilde{y}_0 \gg \lambda$ we can ignore diffraction in this direction. Then $v = 0$, thus implying that $y' = 0$. This would be demonstrated at very large R' or in the focal plane of a lens used to bring the pattern in closer to the aperture plane. Under these circumstances, the pattern is a function of u alone

$$u = -\left(\frac{x}{R_0} + \frac{x'}{R_0'}\right)\frac{1}{\lambda} = \frac{\alpha - \alpha'}{\lambda} = \frac{\cos\theta_x - \cos\theta_x'}{\lambda}$$

If, in addition, $\beta = \beta' = 0$, then the problem can be solved in the x–z plane, and $\cos\theta_x = \sin\theta$, $\cos\theta_x' = \sin\theta'$, thus in this case

$$u = \frac{\sin\theta - \sin\theta'}{\lambda} \qquad (6.41)$$

Note that, if the source became an infinite line parallel to the y axis at coordinates $(x, z) = (x, -D)$ and the slit was also infinitely long, we would need to replace the amplitude factor.

$$\left[\frac{R_{00}R_{00}'}{R_0 R_0'}\right]^2 \quad \text{with} \quad \left[\frac{R_{00}R_{00}'}{R_0 R_0'}\right].$$

This comes from the characteristics of cylindrical waves. In practice, the spherical wave approximation is better, even for a source in the shape of a finite line segment. In the far-field case, the distances R_{00} and R_{00}' are large enough that any cylindrical wave features are minimal. The wavefronts are essentially planes.

C. Circular Aperture

Circular apertures occur in many optical instruments, and the Fraunhofer diffraction they produce can play an important role in the performance of these instruments.

Figure 6.19 shows the aperture plane with the coordinates $\tilde{r}$ and $\tilde{\phi}$ identified. The opening is characterized by the transmission function

$$\tilde{\tau}(\tilde{r}, \tilde{\phi}) = \begin{cases} 1 & \tilde{r} < \tilde{r}_0 \\ 0 & \text{otherwise} \end{cases} \tag{6.42}$$

where

$$\tilde{x} = \tilde{r} \cos \tilde{\phi}$$

$$\tilde{y} = \tilde{r} \sin \tilde{\phi}$$

It is also useful to define circular polar coordinates that are related to u and v by

$$u = \rho \cos \phi'$$

$$v = \rho \sin \phi'$$

Then

$$\rho = \sqrt{u^2 + v^2} = \frac{1}{\lambda} \sqrt{\left(\frac{x}{R_0} + \frac{x'}{R_0'}\right)^2 + \left(\frac{y}{R_0} + \frac{y'}{R_0'}\right)^2}$$

In most cases we will assume that the source is on the optical axis so that $x = y = 0$, then

$$\rho = \frac{1}{\lambda} \frac{r'}{R_0'} = \frac{\sin \theta_z'}{\lambda} \tag{6.43}$$

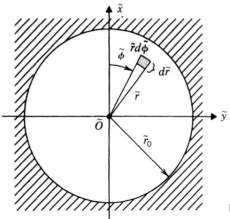

Fig. 6.19 Coordinates in the aperture plane for a circular opening.

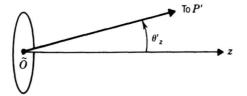

Fig. 6.20 Definition of the angular coordinate with which the circular opening diffraction pattern is annotated.

where

$$r' = \sqrt{x'^2 + y'^2}$$

is the radial distance in the observation plane from the z-axis to the observation point and θ'_z is the angle that identifies the observation direction from the center of the aperture with respect to the z-axis (see Fig. 6.20).

We must compute the Fourier transform of $\tilde{\tau}$ according to Eq. (6.25). The differential area element $d\tilde{\sigma}$ becomes

$$d\tilde{\sigma} = \tilde{r}\, d\tilde{\phi}\, d\tilde{r}$$

This is identified in Fig. 6.19. The exponential factor in the integrand becomes

$$\exp\left[-i2\pi(u\tilde{x} + v\tilde{y})\right] = \exp\left[-i2\pi\rho\tilde{r}(\cos\phi'\cos\tilde{\phi} + \sin\phi'\sin\tilde{\phi})\right]$$
$$= \exp\left[-i2\pi\rho\tilde{r}\cos(\tilde{\phi} - \phi')\right]$$

Eq. (6.25) is then reexpressed in circular polar coordinates

$$T(\rho, \phi') \cong \int_0^{\tilde{r}_0}\tilde{r}\int_0^{2\pi} \exp\left[-i2\pi\rho\tilde{r}\cos(\tilde{\phi} - \phi')\right] d\tilde{r}\, d\tilde{\phi} \qquad (6.44)$$

Eq. (6.44) may be evaluated with the help of the formula

$$J_0(w) = \frac{1}{2\pi}\int_0^{2\pi} \exp\left[-iw\cos(\tilde{\phi} - \phi')\right] d\tilde{\phi} \qquad (6.45)$$

where $J_0(w)$ is a *Bessel function* of zero order. Bessel functions turn up in many areas of science and engineering as solutions to differential equations, particularly in problems where the boundary conditions bear circular symmetry. The numerical values for Bessel functions are tabulated in mathematical reference books. There are also higher-order Bessel functions $J_n(w)$ given by

$$w\frac{dJ_n}{dw} + nJ_n = wJ_{n-1} \qquad (6.46)$$

or alternatively by

$$J_n = -w^{n-1}\frac{d}{dw}\left(w^{-(n-1)}J_{n-1}\right) \qquad (6.47)$$

Some useful values of Bessel functions with $n = 1$ and 2 are given in Table 6.1.

Table 6.1 Useful Bessel Function Values

w	$J_1(w)$	$J_2(w)$	$[2J_1(w)/w]^2$
3.83171	0		0
5.13562	-0.33967	0	0.0175
7.01559	0		0
8.41724	$+0.27138$	0	0.00416
10.17347	0		0
11.61984	-0.23244	0	0.00160
13.32369	0		0
14.79595	$+0.20654$	0	0.00078

With Eq. (6.45) in Eq. (6.44), the Fourier transform of the circular hole function, Eq. (6.41), becomes

$$T(\rho) = 2\pi \int_0^{\tilde{r}_0} \tilde{r} J_0(2\pi\rho\tilde{r}) \, d\tilde{r} \tag{6.48}$$

which is independent of the azimuthal angle ϕ'. This is a result of the angular independence of the transmission function of Eq. (6.42). Change the integration variable to $w' = 2\pi\rho\tilde{r}$ to get

$$T(\rho) = \frac{1}{2\pi\rho^2} \int_0^{2\pi\rho\tilde{r}_0} w' J_0(w') \, dw' \tag{6.49}$$

This may be evaulated by manipulating Eq. (6.46) with $n = 1$.

$$w \frac{dJ_1}{dw} + J_1 = wJ_0$$

or

$$\frac{d(wJ_1)}{dw} = wJ_0 \tag{6.50}$$

The last relation is easily integrated to give

$$wJ_1(w) = \int_0^w w' J_0(w') \, dw' \tag{6.51}$$

This is the same form as Eq. (6.49), thus

$$T(\rho) = \frac{1}{2\pi\rho^2} 2\pi\rho\tilde{r}_0 J_1(2\pi\rho\tilde{r}_0)$$

or

$$T(\rho) = 2\pi\tilde{r}_0^2 \frac{J_1(2\pi\rho\tilde{r}_0)}{2\pi\rho\tilde{r}_0} \tag{6.52}$$

At the center of the diffraction pattern $\rho = 0$. From the properties of Bessel functions we find

$$\lim_{w \to 0} \left(\frac{J_1(w)}{w} \right) = \frac{1}{2}$$

So

$$T(0) = \pi \tilde{r}_0^2$$

Therefore, from Eq. (6.27), the electric field in the diffraction pattern is

$$E'(\rho) = E'(0) \left[\frac{2J_1(2\pi\rho\tilde{r}_0)}{2\pi\rho\tilde{r}_0} \right] \left[\frac{D'}{R'_0} \right] e^{i(\phi_0 - \phi_{00})} \tag{6.53}$$

and the time-averaged energy flux density will be

$$S'(\rho) = S'(0) \left[\frac{2J_1(2\pi\rho\tilde{r}_0)}{2\pi\rho\tilde{r}_0} \right]^2 \left[\frac{D'}{R'_0} \right]^2 \tag{6.54}$$

where

$$\rho = (\sin \theta'_z)/\lambda = \frac{r'}{(R'_0 \lambda)}$$

The Bessel function factors in Eqs. (6.48) and (6.49) are plotted in Fig. 6.21. The general behavior is similar to that of sinc and (sinc)^2, but the "feet" beyond the first

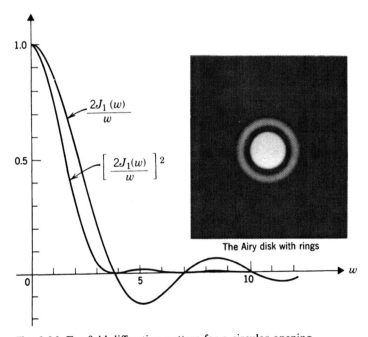

The Airy disk with rings

Fig. 6.21 Far-field diffraction pattern for a circular opening.

zero are weaker. Eighty-four percent of the total area under $(2J_1(w)/w)^2$ is in the first maximum. The other maxima are considerably weaker. The main maximum gives a circular pattern called the *Airy disk*. Its radius corresponds to the value of w at the first zero of $J_1(w)$, which is

$$w = 3.83171 = \frac{2\pi \tilde{r}_0 r'_{\text{disk}}}{R'_0 \lambda}$$

The radius of the disk is

$$r'_{\text{disk}} = \frac{0.61\lambda R'_0}{\tilde{r}_0} \tag{6.55}$$

Notice again the inverse relationship between the size of the aperture and the size of the diffraction pattern. The angular measure of the central maximum is

$$\Delta\theta'_z \simeq \Delta(\sin \theta'_z) = \frac{0.61\lambda}{\tilde{r}_0} \tag{6.56}$$

The other zeros can be evaluated with the help of Table 6.1. To calculate the radii of the maxima we need

$$\frac{d}{dw}\left(\frac{J_1(w)}{w}\right) = 0$$

$$\frac{d}{dw}\left(\frac{J_1(w)}{w}\right) = -\left(\frac{J_2(w)}{w}\right) \tag{6.57}$$

Therefore, $J_1(w)$ will have an extrema where $J_2(w)$ is zero. These values of w are also found in Table 6.1.

6.3 Fourier Analysis

The mathematics of far-field or Fraunhofer diffraction is most conveniently handled within the framework of Fourier analysis. Previously we had merely defined the two-dimensional Fourier transform as the integral that must be evaluated when we apply the linear approximation to the phase term in the integrand. We saw in section 6.2B2 how the Fourier transform of a unit amplitude "box" function can be viewed as the resultant of a phasor curve. This concept can be generalized to describe any transmission function by a phasor sum of incremental pieces, each with the appropriate phase angle *and* incremental length.

This is the essence of Fourier analysis. We view the transform as a weighted sum. Each component that represents the contribution associated with the phase factor $e^{-i2\pi(u\tilde{x}+v\tilde{y})}$ is weighted by the amplitude factor $\tilde{\tau}(\tilde{x}, \tilde{y})$ (which may also shift the phase if $\tilde{\tau}$ is complex).

In Eq. (6.25) the phase factors represent differences among plane waves. To see this write

$$\exp[-i2\pi(u\tilde{x} + v\tilde{y})] = \exp\{-ik[(\alpha - \alpha')\tilde{x} + (\beta - \beta')\tilde{y}]\}$$

But $k\alpha\tilde{x} + k\beta\tilde{y} = \mathbf{k} \cdot \tilde{\mathbf{r}}$ and $k\alpha'\tilde{x} + k\beta'\tilde{y} = \mathbf{k}' \cdot \tilde{\mathbf{r}}$, where $\tilde{\mathbf{r}}$ is a position vector in the $\tilde{x}\,\tilde{y}$ plane. Therefore, the phase in the integrand is

$$e^{i(\mathbf{k}' - \mathbf{k}) \cdot \tilde{\mathbf{r}}} = e^{i\Delta\mathbf{k} \cdot \tilde{\mathbf{r}}}$$

Relative to the center of the aperture ($\tilde{\mathbf{r}} = 0$) we find that the phase of the diffracted wave (in the direction of $\mathbf{k}'$) at the point $\tilde{\mathbf{r}}$ in the aperture differs from the phase of the incident wave (in the direction of $\mathbf{k}$) by $\Delta\mathbf{k} \cdot \tilde{\mathbf{r}}$.

Because of its importance in this and in later chapters, we here present some of the formal features of Fourier analysis. With the mathematical details collected in one place, it will be easier, from now on, to quote results from this section. We have explicitly derived the form of the Fourier transform for two simple cases. Now we want to generalize the procedure.

A. Basic Definitions in Fourier Analysis

In this section we employ the form $f(x, y)$ to emphasize the generality of Fourier analysis. When applied to far-field diffraction the appropriate function is $\tilde{\tau}(\tilde{x}, \tilde{y})$.

1. Delta Function. One of the most important functions whose Fourier transformation characteristics we will need is that which represents a single point, say $x = x_0$, $y = y_0$, defined within a very small (infinitesimal) area $\sigma_0(x_0, y_0)$

$$f(x, y) = \begin{cases} 1, & \text{if } x, y \text{ is within } \sigma_0(x_0, y_0) \\ 0, & \text{otherwise} \end{cases} \tag{6.58}$$

The two-dimensional Fourier transform of this is

$$F(u, v) = \iint f(x, y)\, e^{-i2\pi(ux + vy)}\, dx\, dy = \sigma_0\, e^{-i2\pi(ux_0 + vy_0)} \tag{6.59}$$

Although it is not a true function in the rigorous sense of the word, we define what is called a *delta function* to describe this behavior by setting

$$f(x, y) = \sigma_0\, \delta(x - x_0)\, \delta(y - y_0) \tag{6.60}$$

We write $\delta(x - x_0)$ and $\delta(y - y_0)$ as the delta functions for one dimension. They represent an "impulse" located where the argument equals zero. The delta function has some special properties that we must present now.

a. Filter Characteristics. The delta function acts like a filter that selects a single value among those possible in the range of integration. Strictly speaking, the delta function has no mathematical meaning unless it is used in conjunction with an integration. Provided the range of the integration includes x_0, we have

$$\int f(x)\, \delta(x - x_0)\, dx = f(x_0) \tag{6.61}$$

If x_0 is outside of the integration range, then the integral is zero.

b. Scaling Relationship. For real values of the constant b we can prove that

$$\delta(bx) = \frac{\delta(x)}{|b|} \tag{6.62}$$

This follows directly from the definition in Eq. (6.61)

$$\int_{x_1}^{x_2} f(x)\delta(bx)\,dx = \frac{1}{b}\int_{x_1/b}^{x_2/b} f\!\left(\frac{x'}{b}\right)\delta(x')\,dx' = \frac{f(0)}{|b|}$$

A special case involves $b = -1$. Then $\delta(-x) = \delta(x)$.

c. Representations. Because the delta function is not a true function, we must be careful when writing down an analytical form that might replace it in calculations. We can identify families of functions that approach the behavior of a delta function in some limit. Such a family of functions is called a *representation* of the delta function. These must be sharply peaked, tending to infinite height and zero width as the limit is approached. They must also integrate to unity, because from Eq. (6.61)

$$\int \delta(x)\,dx = 1$$

An example of this is a sequence of Gaussian functions (Fig. 6.22)

$$\delta(x) = \lim_{b \to \infty} \frac{b}{\sqrt{\pi}}\exp(-b^2 x^2) \tag{6.63}$$

As $b \to \infty$, this function is zero everywhere except near $x = 0$. There it approaches ∞. We can also show

$$\int \delta(x)\,dx = \frac{b}{\sqrt{\pi}}\int \exp(-b^2 x^2)\,dx = 1 \tag{6.64}$$

if the integration includes $x = 0$.

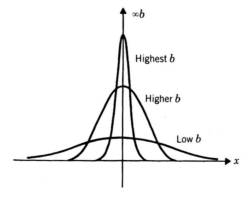

Fig. 6.22 Representation of the delta function by a sequence of Gaussian profiles with increasing height and decreasing width.

Another example that we have already encountered [Eq. (6.31)] is

$$\delta(u) = \lim_{b \to \infty} 2b \, \mathrm{sinc}(2\pi u b) = \lim_{b \to \infty} \int_{-b}^{b} e^{-i2\pi u x} \, dx \qquad (6.65)$$

The oscillations, with period $1/b$, become rapidly attenuated, losing their significance as $b \to \infty$.

It will also be convenient to have a representation that is expressed as an integral over a positive variable

$$\delta(x) = \lim_{b \to \infty} 2 \int_{0}^{b} \cos(2\pi u x) \, du = \lim_{b \to \infty} 2b \, \mathrm{sinc}(2\pi b x) \qquad (6.66)$$

2. Fourier Integrals. Armed with the representations of the delta function, we are now ready to examine the Fourier transform more closely.

a. Complex Fourier Integral. This is the form that we have already defined in Eq. (6.25). Let us formalize that definition.

Consider a function $f(x)$, which may be complex, with the following properties:

1. It is "square integrable," that is, the limit of the integral

$$\int_{-x_0}^{x_0} |f(x)|^2 \, dx \qquad (6.67)$$

exists as $x_0 \to \infty$.

2. The function is continuous, and df/dx exists at all but a finite number of points in any finite range of the variable x.

In practice, these conditions are met by any function that represents a physical quantity. Frequently, however, we use unphysical functions because of their simplicity. Examples of this are the delta function, the infinite sine and cosine functions, and a constant that exists at all values of x. In these cases, as we did for the delta function representation, we can define a family of functions that approach the desired function as a limit is reached. If the definition is chosen so that each member of the family satisfies conditions (1) and (2), then the limit can be taken *after* the Fourier integral has been evaluated.

Under these conditions, the generalized Fourier transform

$$F(u) = \int_{-\infty}^{\infty} f(x) \, e^{-i2\pi u x} \, dx \qquad (6.68)$$

exists and is continuous at all but a finite number of points in any finite range of the *real* variable u. At points of discontinuity, the integral converges to the average of the right-hand and left hand limits.

To determine $f(x)$ from $F(u)$, multiply both sides of Eq. (6.68) by $e^{i2\pi u x'}$ and integrate both sides with respect to u.

$$\int_{-\infty}^{\infty} F(u) \, e^{i2\pi u x'} \, du = \int_{-\infty}^{\infty} \int_{-\infty}^{\infty} f(x) \, e^{i2\pi u x'} \, e^{-i2\pi u x} \, dx \, du$$

Interchanging the order of the integrations on the right side leads to

$$\int_{-\infty}^{\infty} F(u)\, e^{i2\pi ux'}\, du = \int_{-\infty}^{\infty} f(x) \left[\int_{-\infty}^{\infty} e^{i2\pi u(x'-x)}\, du \right] dx$$

From Eq. (6.65) we see that the factor within brackets is a representation of $\delta(x' - x)$. Thus the right side becomes $f(x')$, and (by an elimination of the prime) we have

$$f(x) = \int_{-\infty}^{\infty} F(u)\, e^{i2\pi ux}\, du \tag{6.69}$$

The variables x and u are called *Fourier variable pairs*. They must be in units that are inverses of each other. Note that if $f(x)$ is unitless, then $F(u)$ must have the same units as x.

It is often convenient to adopt a notation that takes the place of the integral in Eq. (6.68) to indicate the Fourier transform operation

$$F(u) \equiv \mathscr{F}[f(x)] \equiv \int_{-\infty}^{\infty} f(x)\, e^{-i2\pi ux}\, dx \tag{6.70}$$

Likewise

$$f(x) \equiv \mathscr{F}^{-1}[F(u)] \equiv \int_{-\infty}^{\infty} F(u)\, e^{i2\pi ux}\, du \tag{6.71}$$

denotes the *inverse transform* operation.

b. Two-Dimensional Form. In diffraction we deal with two-dimensional functions. These were the conditions under which we first defined the Fourier transform in Eq. (6.25). For completeness we review here the two-dimensional forms of the transform and inverse transform equations.

$$\mathscr{F}[f(x, y)] = F(u, v) = \iint_{-\infty}^{\infty} f(x, y)\, e^{-i2\pi(ux + vy)}\, dx\, dy \tag{6.72}$$

$$\mathscr{F}^{-1}[F(u, v)] = f(x, y) = \iint_{-\infty}^{\infty} F(u, y)\, e^{i2\pi(ux + vy)}\, du\, dv \tag{6.73}$$

If $f(x, y)$ can be written as $f_1(x)f_2(y)$, we say that f is "separable." In that case, as for the rectangular transmission function with which we have already dealt, the Fourier transform will also be separable. In other words, $F(u, v)$ becomes $F_1(u)F_2(v)$. For simplicity, we will develop the characteristics of the Fourier transform in the one-dimensional case. These results can be directly applied to the separable two-dimensional problems.

In general, the form of $f(x, y)$ does not permit independence of the x and y integrations. Such was the case for the circular aperture. This led to a rather complicated result. Fortunately this is by far the most common example. Other cases must be explicitly calculated or deduced by use of Babinet's principle.

Two-dimensional analogs exist for all the Fourier transform characteristics that we will examine in detail for one dimension. These will be collected at the end in tabular form.

c. *Real Fourier Integral.* If $f(x)$ is real, then we can establish an equivalent Fourier transformation that is based on superposition of sines and cosines rather than complex exponentials. Assume

$$f^*(x) = f(x)$$

Then take the complex conjugate of Eq. (6.70)

$$F^*(u) = \int_{-\infty}^{\infty} f(x)\, e^{i2\pi ux}\, dx = F(-u) \tag{6.74}$$

Now write $F(u)$ in terms of its absolute value and phase factor:

$$F(u) = |F(u)|\, e^{i\theta(u)} \tag{6.75}$$

Equation (6.74) implies that

$$|F(-u)| = |F(u)| \tag{6.76}$$

and

$$\theta(-u) = -\theta(u) \tag{6.77}$$

Then we may write

$$f(x) = \operatorname{Re} f(x) = \operatorname{Re} \int_{-\infty}^{\infty} F(u)\, e^{i2\pi ux}\, du$$

$$= \operatorname{Re} \int_{-\infty}^{\infty} |F(u)|\, e^{i[2\pi ux + \theta(u)]}\, du$$

$$= \int_{-\infty}^{\infty} |F(u)| \cos[2\pi ux + \theta(u)]\, du$$

$$= 2 \int_{0}^{\infty} |F(u)| \cos[2\pi ux + \theta(u)]\, du \tag{6.78}$$

This may be rewritten in the form

$$f(x) = \int_{0}^{\infty} [2|F(u)|\cos\theta(u)]\cos(2\pi ux)\, du$$

$$+ \int_{0}^{\infty} [-2|F(u)|\sin\theta(u)]\sin(2\pi ux)\, du \tag{6.79}$$

Now define

$$F_c(u) \equiv 2|F(u)| \cos\theta(u) = 2\operatorname{Re} F(u) \tag{6.80a}$$

and

$$F_s(u) \equiv -2|F(u)| \sin \theta(u) = -2\mathrm{Im}\, F(u) \tag{6.80b}$$

From Eq (6.68) these are

$$F_c(u) = 2 \int_{-\infty}^{\infty} f(x) \cos(2\pi ux)\, dx \tag{6.81a}$$

and

$$F_s(u) = 2 \int_{-\infty}^{\infty} f(x) \sin(2\pi ux)\, dx \tag{6.81b}$$

These are known as the cosine and sine Fourier transforms of $f(x)$. Eq. (6.79) becomes

$$f(x) = \int_{0}^{\infty} F_c(u) \cos(2\pi ux)\, du + \int_{0}^{\infty} F_s(u) \sin(2\pi ux)\, du \tag{6.82}$$

3. Fourier Transform Characteristics. Later we will need to apply some of the fundamental properties of Fourier transforms. We collect them here for convenience.

   a. Linearity. This property follows from the concept of linear superposition on which the Fourier integrals were defined. Therefore,

$$\mathscr{F}[a_1 f_1(x) + a_2 f_2(x)] = a_1 F_1(u) + a_2 F_2(u) \tag{6.83}$$

where

$$F_1(u) = \mathscr{F}[f_1(x)] \quad \text{and} \quad F_2(u) = \mathscr{F}[f_2(x)].$$

   b. Reciprocal Relationship. This important characteristic manifests itself in many physical phenomena. It involves an interpretation of scaling operations for the Fourier variable pairs.

$$\mathscr{F}[f(ax)] = \int_{-\infty}^{\infty} f(ax)\, e^{-i2\pi ux}\, dx$$

With a change of variable in the integration to $x' = ax$ we have

$$\mathscr{F}[f(ax)] = \frac{1}{|a|} \int_{-\infty}^{\infty} f(x') \exp\left(-i2\pi \frac{u}{a} x'\right) dx'$$

which may be written as

$$\mathscr{F}[f(ax)] = \frac{1}{|a|} F\left(\frac{u}{a}\right) \tag{6.84}$$

This shows that the scale change appears in reciprocal form in the Fourier transform. If we expand the function $f(x)$ by the magnification factor a, then the

Fourier transform appears the same except that it will be reduced in amplitude and will be compressed by the factor $1/a$.

A special case is when $a = -1$. Then

$$\mathscr{F}[f(-x)] = F(-u)$$

c. Conjugation. By direct application of the complex conjugation operation we may prove that

$$\mathscr{F}[f^*(x)] = \int_{-\infty}^{\infty} f^*(x)e^{-i2\pi ux}\, dx = \left[\int_{-\infty}^{\infty} f(x)e^{i2\pi ux}\, dx\right]^* = F^*(-u) \quad (6.85)$$

d. Shifting. This very important property allows us to compute the transform of $f(x - x_0)$ or the inverse transform of $F(u - u_0)$ where x_0 and u_0 are constants. These are expressed in terms of the transform or the inverse transform of the respective unshifted function.

$$\mathscr{F}[f(x - x_0)] = \int_{-\infty}^{\infty} f(x - x_0)e^{-i2\pi ux}\, dx$$

$$= e^{-i2\pi ux_0}\int_{-\infty}^{\infty} f(x - x_0)e^{-i2\pi u(x - x_0)}\, dx$$

Now let $x' = x - x_0$ in the integral, which becomes

$$\int_{-\infty}^{\infty} f(x')e^{-i2\pi ux'}\, dx' = F(u)$$

Therefore,

$$\mathscr{F}[f(x - x_0)] = e^{-i2\pi ux_0} F(u) \quad (6.86)$$

In a similar manner we can prove that

$$\mathscr{F}^{-1}[F(u - u_0)] = e^{i2\pi u_0 x} f(x) \quad (6.87)$$

e. Repeated Application. Suppose we have calculated the Fourier transform of $f(x)$ to be $F(u)$. Now compute the Fourier transform of $F(u)$. This must bring forth a function that depends on a variable with the same units as x, say x'.

$$\mathscr{F}[F(u)] = \int_{-\infty}^{\infty} e^{-i2\pi x'u}\int_{-\infty}^{\infty} f(x)e^{-i2\pi ux}\, dx\, du$$

$$= \int_{-\infty}^{\infty} f(x)\left[\int_{-\infty}^{\infty} e^{-i2\pi(x + x')u}\, du\right] dx$$

$$= \int_{-\infty}^{\infty} f(x)\,\delta(x + x')\, dx$$

$$= f(-x') \quad (6.88)$$

On the other hand, the inverse transform operation yields the original function.

$$\mathscr{F}^{-1}[F(u)] = \int_{-\infty}^{\infty} e^{+i2\pi ux'} \int_{-\infty}^{\infty} f(x) e^{-i2\pi ux}\, dx\, du$$

$$= \int_{-\infty}^{\infty} f(x) \left[\int_{-\infty}^{\infty} e^{-i2\pi(x-x')u}\, du \right] dx$$

$$= \int_{-\infty}^{\infty} f(x)\delta(x - x')\, dx$$

$$= f(x') \tag{6.89}$$

f. Conservation. Equation (6.69) shows that $f(x)$ may be decomposed into an integral in "u-space." The coefficients $F(u)$ are the weighting factors. In the diffraction pattern, the measured quantity, the radiation power density, is proportional to $|F|^2$. The incident power density must be proportional to $|f|^2$. If we integrate these two functions over their respective variables, u for F and x for f, we should get the same result. This is required by energy conservation. Let us prove this mathematically.

$$\int_{-\infty}^{\infty} |F(u)|^2\, du = \int_{-\infty}^{\infty} \left[\int_{-\infty}^{\infty} f(x) e^{-i2\pi ux}\, dx \int_{-\infty}^{\infty} f^*(x') e^{i2ux'}\, dx' \right] du$$

$$= \int_{-\infty}^{\infty} \int_{-\infty}^{\infty} f(x) f^*(x') \left[\int_{-\infty}^{\infty} e^{i2\pi u(x'-x)}\, du \right] dx\, dx'$$

$$= \int_{-\infty}^{\infty} \int_{-\infty}^{\infty} f(x) f^*(x')\delta(x' - x)\, dx\, dx'$$

$$= \int_{-\infty}^{\infty} |f(x)|^2\, dx \tag{6.90}$$

This conservation statement is frequently called *Parseval's theorem*.

4. Convolution. Frequently in optics, and in other areas of physical science, we need to combine the influence of two functions. This may take the form, for instance, of an interference grating that consists of an array of finite width apertures. We already know how to describe the effects of multiple-beam interference in the case where the slits of infinitesimally small, and we know how to describe the diffraction pattern for a single finite-width slit. The pattern that will be produced by the realistic grating is the result of the action of the two phenomena together.

This example, as well as many others, can be treated through the concept of *convolution.* When we convolute two functions, the result is a third function that may be formally defined as

$$g(x) \equiv \int_{-\infty}^{\infty} f_1(x') f_2(x - x')\, dx' \equiv f_1 \circledast f_2 \tag{6.91}$$

This process is illustrated in Fig. 6.23. The resulting function g has characteristics of

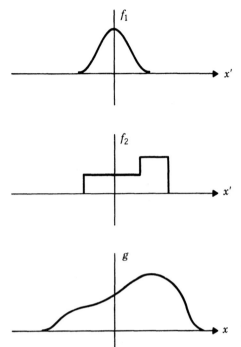

Fig. 6.23 The convolution of two functions bears a similarity to each of the component functions.

both of the input functions, f_1 and f_2. At a specified value of x, the convolution represents the area of the product of $f_1(x')$ and $f_2(x - x')$. As x changes, the relative displacement of the two input functions changes, thus modifying the product and the area of the product.

In two dimensions the convolution takes the form

$$f_1 \circledast f_2 \equiv \int\!\!\!\int_{-\infty}^{\infty} f_1(x', y')f_2(x - x', y - y')\, dx'\, dy' \tag{6.92}$$

a. Convolution Characteristics. The convolution operation for real functions displays several fundamental properties that we list here without proof. The demonstration of these statements are straightforward exercises.

If $f_1(x)$ and $f_2(x)$ are *real* valued functions and if $f_1(x) \circledast f_2(x) = g(x)$, then

(i) $$f_2(x) \circledast f_1(x) = g(x) \tag{6.93}$$

(ii) $$f_1(x - x_0) \circledast f_2(x) = g(x - x_0) \tag{6.94}$$

(iii) $$f_1(ax) \circledast f_2(ax) = \frac{1}{|a|}\, g(ax) \tag{6.95}$$

b. Convolution Theorem. The convolution operation is often complicated, even for simple component functions. However, when the Fourier transform of a convolution is examined, it turns out to be very simple.

Assume $\mathscr{F}[f_1 \circledast f_2] = \mathscr{F}[g] = G(u)$.

Then by explicit substitution we find

$$\mathscr{F}[f_1 \circledast f_2] = \int_{-\infty}^{\infty} \left[\int_{-\infty}^{\infty} f_1(x')f_2(x - x')\, dx' \right] e^{-i2\pi ux}\, dx$$

$$= \int_{-\infty}^{\infty} f_1(x') \left[\int_{-\infty}^{\infty} f_2(x - x')e^{-i2\pi ux}\, dx \right] dx'$$

Now we can apply Eq. (6.86) to this.

$$\mathscr{F}[f_1 \circledast f_2] = \int_{-\infty}^{\infty} f_1(x') \left[e^{-i2\pi ux'} F_2(u) \right] dx'$$

But $F_2(u)$ can be removed from the integral leaving

$$\mathscr{F}[f_1 \circledast f_2] = F_1(u)F_2(u) = G(u) \tag{6.96}$$

The result is merely the product of the Fourier transforms of the component functions. This turns out to be a very useful result.

A complementary relation can be derived by considering the inverse transform of the convolution of two Fourier transforms.

Define

$$G(u) \equiv \int_{-\infty}^{\infty} F_1(u')F_2(u - u')\, du' = F_1 \circledast F_2$$

such that

$$\mathscr{F}^{-1}[F_1 \circledast F_2] = \mathscr{F}^{-1}[G] = g(x),$$

then

$$\mathscr{F}^{-1}[F_1 \circledast F_2] = f_1(x)f_2(x)$$

or equivalently,

$$\mathscr{F}[f_1(x)f_2(x)] = F_1 \circledast F_2 \tag{6.97}$$

This form of the convolution theorem is equally useful.

5. Correlation. Another useful quantity may be developed to describe the degree to which two functions "match up," as they are shifted relative to each other. This is closely related to convolution but displays some important differences. For two, in general complex, functions we define the *cross correlation* as

$$f_1 \otimes f_2 \equiv \int_{-\infty}^{\infty} f_1(x')f_2^*(x' - x)\, dx' \tag{6.98a}$$

With the substitution $x'' = x' - x$, this may also be written as

$$f_1 \otimes f_2 \equiv \int_{-\infty}^{\infty} f_1(x + x'') f_2^*(x'') \, dx'' \qquad (6.98b)$$

We must pay attention to the order in which the functions are written and to which of the two functions enter the integrand in the complex conjugate form.

In two dimensions the cross-correlation integral takes the form

$$f_1 \otimes f_2 \equiv \iint_{-\infty}^{\infty} f_1(x', y') f_2^*(x' - x, y' - y) \, dx' \, dy' \qquad (6.99)$$

When f_1 and f_2 are the same function, then we have the *auto correlation* integral. This we denote by

$$\hat{\gamma} \equiv f \otimes f \qquad (6.100)$$

The autocorrelation will always be largest at $x = 0$. This represents the condition where the function is not shifted with respect to itself. Here the product in the integrand is as large as it can be for all values of the range of integration. As the relative shift increases, the autocorrelation will drop off according to the extent and shape of the original function.

The Fourier transform of the correlation can be evaluated starting from Eq. (6.98b)

$$\mathscr{F}[f_1 \otimes f_2] = \int_{-\infty}^{\infty} \int_{-\infty}^{\infty} f_1(x + x'') f_2^*(x'') \, dx'' e^{-i2\pi ux} \, dx$$

$$= \int_{-\infty}^{\infty} \left[\int_{-\infty}^{\infty} f_1(x + x'') e^{-i2\pi ux} \, dx \right] f_2^*(x'') \, dx''$$

$$= \int_{-\infty}^{\infty} e^{i2\pi ux''} F_1(u) f_2^*(x'') \, dx''$$

$$= F_1(u) \left[\int_{-\infty}^{\infty} f_2(x'') e^{-i2\pi ux''} \, dx'' \right]^*$$

$$= F_1(u) F_2^*(u) \qquad (6.101)$$

A special case of this relationship involves the autocorrelation where Eq. (6.101) becomes

$$\mathscr{F}[\hat{\gamma}] = \mathscr{F}[f \otimes f] = |F(u)|^2 \qquad (6.102)$$

This relationship is known as the *Wiener-Khintchine theorem*. The importance of the autocorrelation is readily apparent from this. We can see by comparison to Eq. (6.28) that the power density in the far-field diffraction pattern is proportional to $|F|^2$, where both F and $\hat{\gamma}$ refer to the transmission function in the aperture plane. These concepts assume an even more important role when we discuss coherence theory in Chapter 8.

6.4   Examples of Fourier Analysis in Diffraction

We have seen that the analysis of far-field diffraction is largely concerned with the specification of the transmission function at the aperture and the calculation of its Fourier transform. Here we collect the Fourier analysis results of the previous sections, provide for some useful extensions, and discuss the application of these concepts in several important diffraction problems.

A.   Fourier Summary

A summary of properties of Fourier transforms in one and two dimensions along with transform pairs for several useful functions is presented in Table 6.2.

The results of Table 6.2 may be extended using the convolution theorem to find transforms of more complicated functions. For instance, consider the convolution of two identical box functions as shown in Fig. 6.24. The result is a triangular function. We write this in dimensionless form as

$$g(x) = \frac{1}{2x_0} f(x) \circledast f(x) \tag{6.103}$$

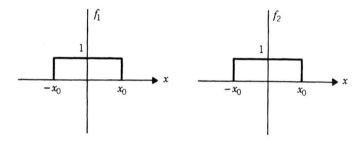

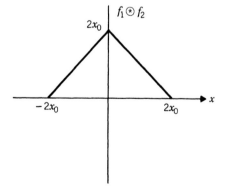

Fig. 6.24 The convolution of two identical "box" functions is a triangular function.

Table 6.2. Fourier Transform Properties

One Dimension

$$f(x) = \int_{-\infty}^{\infty} F(u)e^{i2\pi ux}du \qquad F(u) = \int_{-\infty}^{\infty} f(x)e^{-i2\pi ux}dx$$

$$f(ax) \qquad\qquad \frac{1}{|a|}F\left(\frac{u}{a}\right)$$

$$f^*(x) \qquad\qquad F^*(-u)$$

$$f(x - x_0) \qquad\qquad e^{-i2\pi ux_0}F(u)$$

$$e^{i2\pi u_0 x}f(x) \qquad\qquad F(u - u_0)$$

$$f_1 \circledast f_2 \qquad\qquad F_1 F_2$$

$$f_1 f_2 \qquad\qquad F_1 \circledast F_2$$

$$f_1 \otimes f_2 \qquad\qquad F_1 F_2^*$$

Two Dimensions

$$f(x, y) = \iint_{-\infty}^{\infty} F(u, v)e^{i2\pi(ux + vy)}du\,dy \qquad F(u, v) = \iint_{-\infty}^{\infty} f(x, y)e^{-i2\pi(ux + vy)}dx\,dy$$

$$f(ax, cy) \qquad\qquad \frac{1}{|a|}\frac{1}{|c|}F\left(\frac{u}{a}, \frac{v}{c}\right)$$

$$f^*(x, y) \qquad\qquad F^*(-u, -v)$$

$$f(x - x_0, y - y_0) \qquad\qquad e^{-i2\pi(ux_0 + vy_0)}F(u, v)$$

$$e^{i2\pi(u_0 x + v_0 y)}f(x, y) \qquad\qquad F(u - u_0, v - v_0)$$

$$f_1 \circledast f_2 \qquad\qquad F_1 F_2$$

$$f_1 f_2 \qquad\qquad F_1 \circledast F_2$$

$$f_1 \otimes f_2 \qquad\qquad F_1 F_2^*$$

One-Dimensional Examples

$$\delta(x) \qquad\qquad 1$$

$$1 \qquad\qquad \delta(u)$$

$$\cos(2\pi u_0 x) \qquad\qquad \frac{1}{2}\left[\delta(u - u_0) + \delta(u + u_0)\right]$$

$$\sin(2\pi u_0 x) \qquad\qquad \frac{1}{2i}\left[\delta(u - u_0) - \delta(u + u_0)\right]$$

$$e^{i2\pi u_0 x} \qquad\qquad \delta(u - u_0)$$

$$\exp(-\pi b^2 x^2) \qquad\qquad \frac{1}{|b|}\exp\left[-\pi\left(\frac{u}{b}\right)^2\right]$$

$$\begin{Bmatrix} 1 & |x| \le x_0 \\ 0 & |x| > x_0 \end{Bmatrix} \qquad\qquad 2x_0 \operatorname{sinc}(2\pi u x_0)$$

$$\begin{Bmatrix} e^{-xb} & x \ge 0 \\ 0 & x < 0 \end{Bmatrix} \qquad\qquad (b + i2\pi u)^{-1}$$

Table 6.2   (continued)

Two-Dimensional Examples	
$\begin{cases} 1 & \|x\| \le x_0, \|y\| \le y_0 \\ 0 & \|x\| > x_0, \|y\| > y_0 \end{cases}$	$4x_0 y_0 \,\text{sinc}(2\pi u x_0)\,\text{sinc}(2\pi v y_0)$
$r^2 = x^2 + y^2$	$\rho^2 = u^2 + v^2$
$\begin{cases} 1 & r \le r_0 \\ 0 & r > r_0 \end{cases}$	$\pi r_0^2 \dfrac{2J_1(2\pi\rho r_0)}{2\pi\rho r_0}$
$\exp[-\pi b^2 r^2]$	$\dfrac{1}{b^2}\exp\left[-\pi\left(\dfrac{\rho}{b}\right)^2\right]$

When applied to far-field diffraction, then $f \to \tilde{\tau}$, $F \to T$, and $x \to \tilde{x}$. Equation (6.103) would present the functional form of, for instance, the electric field amplitude across a long slit of width $2\tilde{x}_0$ that was covered with a variable density transmission filter. Use of the convolution theorem allows us to immediately write down the Fourier transform

$$G(u) = \frac{1}{2\tilde{x}_0}[T(u)]^2 = 2\tilde{x}_0 \,\text{sinc}^2(2\pi u \tilde{x}_0) \tag{6.104}$$

This would be related to the field intensity in the far-field diffraction pattern through Eq. (6.41) $-\sin\theta' = \lambda u - \sin\theta$. The power density follows from Eq. (6.28).

$$S(u) = S(0)\left[\frac{R_{00}R'_{00}}{R_0 R'_0}\right]^2 \text{sinc}^4(2\pi u \tilde{x}_0)$$

Another example is shown in Fig. 6.25, the truncated cosine wave. This is actually of more practical use in the Fourier analysis of the *time* dependence of optical signals. This is formed as the product of an infinite cosine wave and the box function. Here $f_1 = \cos(2\pi\tilde{x}/x_0)$. The Fourier transform of this is

$$F_1(u) = \frac{1}{2}\left[\delta\left(u - \frac{1}{x_0}\right) + \delta\left(u + \frac{1}{x_0}\right)\right].$$

The transform of a product is the convolution of the transforms thus,

$$G(u) = \mathscr{F}[f_1 f_2]$$

$$= \int_{-\infty}^{\infty}\frac{1}{2}\left[\delta\left(u' - \frac{1}{x_0}\right) + \delta\left(u' + \frac{1}{x_0}\right)\right]2x_0\text{sinc}\left[2\pi(u - u')x_0\right]du'$$

$$= x_0\left\{\text{sinc}\left[2\pi\left(u - \frac{1}{x_0}\right)\right] + \text{sinc}\left[2\pi\left(u + \frac{1}{x_0}\right)\right]\right\} \tag{6.105}$$

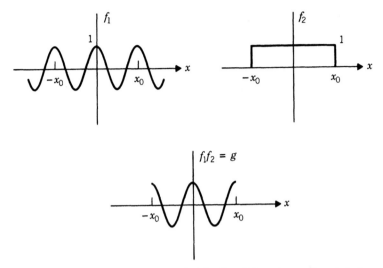

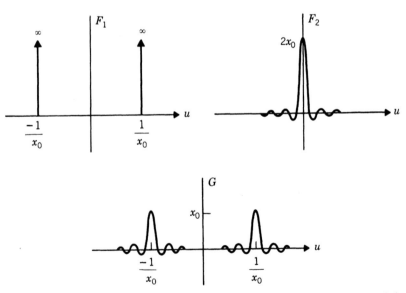

Fig. 6.25 The product of the unit height "box" function and an infinite cosine function is a truncated cosine.

We see that the result of truncation is to "smear out" the singular behavior of the delta functions, replacing them with sinc functions. This is illustrated in Fig. 6.26. This effect is a result of the fundamental uncertainty involved in specifying the period of a truncated wave.

Fig. 6.26 The Fourier transform of the three functions in Fig. 6.25. The result is the convolution of the two component transforms.

B. Array Theorem

1. General Result. One of the most useful applications of the convolution theorem is found when one of the members is a sum of delta functions. To see this write

$$g = f_1(x) \circledast \sum_m \delta(x - x_m) = \int_{-\infty}^{\infty} f_1(x') \sum_m \delta(x - x' - x_m)\, dx'$$

$$= \sum_m f_1(x - x_m) \tag{6.106}$$

This represents a linear addition of functions, each of which has the local form $f_1(x)$ but shifted to new origins that are identified by x_m. This is shown in Fig. 6.27 for $m = 1, 2, 3$.

Through an application of the convolution theorem the Fourier transform of $g(x)$ is

$$G(u) = \mathscr{F}[f_1(x)]\mathscr{F}\left[\sum_m \delta(x - x_m)\right]$$

$$= F_1(u) \sum_m e^{-i2\pi u x_m} \tag{6.107}$$

The transformed result of the spatially shifted local function is just the product of

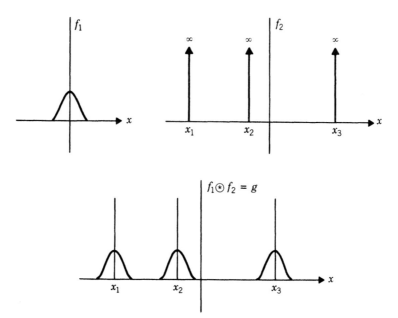

Fig. 6.27 The convolution of a local function f_1 with a collection of delta functions is a collection of local functions centered on the positions of the delta functions. This illustrates the concept behind the Array theorem.

the transform of the local function and a phase factor. This result is called the *Array theorem*. In two dimensions

$$G(u, v) = F_1(u, v) \sum_m e^{-i2\pi(ux_m + vy_m)} \tag{6.108}$$

where

$$g(x, y) = \sum_m f_1(x - x_m, y - y_m)$$

is formed by translating $f_1(x, y)$ to (x_m, y_m) for each m.

2. Young's Double Slit. The Array theorem enables us to find the Fourier transform, and thus the far-field diffraction pattern, associated with transmission functions that represent a combination of identical holes. For instance, consider

$$\tilde{\tau}(\tilde{x}) = \tilde{\tau}_1\left(\tilde{x} - \frac{a}{2}\right) + \tilde{\tau}_1\left(\tilde{x} + \frac{a}{2}\right) \tag{6.109}$$

where $\tilde{\tau}_1$ is the box function of width b,

$$\tilde{\tau}_1(\tilde{x}) = \begin{cases} 1, & |\tilde{x}| \leqslant \dfrac{b}{2} \\[2mm] 0, & |\tilde{x}| > \dfrac{b}{2} \end{cases}$$

In this example, $\tilde{\tau}(\tilde{x})$ shown in Fig. 6.28 would be the transmission function for a pair of slits that were separated by center-to-center distance a and that each had width b. This is the realistic situation in the geometry of Young's interference experiment. In Chapter 5 we ignored the finite width of the slits, assuming that only one Huygens wavelet originated at each. That would be equivalent, in the present situation, to setting $\tilde{\tau}(\tilde{x}) \propto \delta(\tilde{x})$. We are now equipped to investigate the influence of diffraction on Young's experiment.

According to Eq. (6.107)

$$T(u) = b \, \text{sinc}(\pi u b)[e^{+i\pi u a} + e^{-i\pi u a}]$$

$$T(u) = b \, \text{sinc}(\pi u b) \, 2 \cos(\pi u a) \tag{6.110}$$

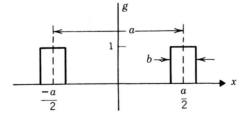

Fig. 6.28 Transmission function appropriate for Young's double-slit experiment where the width of the slits are finite.

The flux density will be

$$S'(u) = S'(0)\left[\frac{R_{00}R'_{00}}{R_0 R'_0}\right]^2 \frac{|T(u)|^2}{|T(0)|^2}$$

$$= S'(0)\left[\frac{R_{00}R'_{00}}{R_0 R'_0}\right]^2 \text{sinc}^2(\pi u b) \cos^2(\pi u a) \quad (6.111)$$

This is plotted in Fig. 6.29. We see that the $\cos^2(\pi u a)$ term is identical to Eq. (5.30) wherein the interference from infinitely narrow slits was considered. Here we have

$$\pi u a = \pi\left(\frac{-x}{R_0} + \frac{-x'}{R'_0}\right)\frac{a}{\lambda}$$

$$= \frac{\pi a}{\lambda}(\sin\theta - \sin\theta')$$

$$= \frac{-\delta}{2}$$

where δ was introduced in Chapter 5. Maxima are found for $\cos^2(\pi u a) = 1$ or at $\pi u a = \pi$ (integer), then, consistent with our established convention

$$(\sin\theta' - \sin\theta) = \frac{m\lambda}{a} \quad (6.112)$$

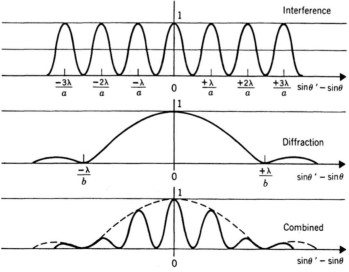

Fig. 6.29 Modulation of the interference pattern in Young's experiment in the case where the slits have finite width.

This factor is modulated by an envelope function $\text{sinc}^2(\pi ub)$ that is due to diffraction. The width of the central maximum of the envelope is inversely proportional to the width of the slits. The envelope is zero where $\text{sinc}^2(\pi ub) = 0$ or at

$$(\sin\theta' - \sin\theta) = \frac{m'\lambda}{b} \tag{6.113}$$

For small slits the central maximum is large enough that many interference fringes fit within its boundaries. This effect was implied in Chapter 5. Without diffraction in that development, there would be no way the contributions from the two slits would overlap. We now understand the origin of the assumption of overlap that was made in Chapter 5.

The width b of the slits must be less than the separation a between slits. Hence, as can be seen by setting $m' = m = 1$, the first zero in the diffraction factor must occur at a larger value of $(\sin\theta' - \sin\theta)$ than the first maximum in the interference factor. At least three interference fringes will appear under the central envelope. Zeros in the diffraction pattern can coincide with the maxima in the interference pattern. This causes the phenomenon of *missing orders*. These will be found at

$$(\sin\theta' - \sin\theta) = \frac{m\lambda}{a} = \frac{m'\lambda}{b}$$

or for

$$m = \frac{a}{b}m' \tag{6.114}$$

where both m and m' are integers.

3. Periodic Functions. An important class of functions represent common apertures that involve periodicity. This is a special case with which the Array theorem can be helpful. Let f_1 be the locally defined contour of a one-dimensional periodic function such that $f_1(x)$ is defined on interval $-a/2$ to $a/2$. Then by displacing this function through N integral multiples of a, we generate periodic behavior over the range Na.

$$f(x) = \sum_{n=1}^{N} f_1(x - x_n), \quad \text{where } x_n = (n - 1)a \tag{6.115}$$

Some examples of periodic functions are shown in Fig. 6.30. It should be obvious that $f(x) = f(x + ma)$, with m an integer.

a. Finite Periodicity. This is the practical situation, for no function that represents a physical quantity can be truly infinite. The most important example is the grating. We have already studied this in Chapter 5. However there, as was the case of Young's geometry, the width of a single slit was assumed to be zero. Here the properties of different slit contours will be explored.

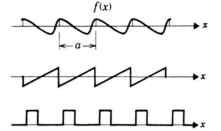

Fig. 6.30 Examples of periodic functions.

Applying the Array theorem to a transmission function such as Eq. (6.115) we find

$$T(u) = T_1(u) \sum_{n=1}^{N} e^{-i2\pi u(n-1)a} \qquad (6.116)$$

Changing the summation index and reintroducing $\delta = -2\pi u a$, this becomes

$$T(u) = T_1(u) \sum_{m=0}^{N-1} e^{im\delta}$$

This sum has already been evaluated in Eq. (5.37c), where we derived

$$\sum_{m=0}^{N-1} e^{im\delta} = e^{i(N-1)\delta/2} \left[\frac{\sin\left(\dfrac{N\delta}{2}\right)}{\sin\left(\dfrac{\delta}{2}\right)} \right] \qquad (6.117)$$

Now assume that N is an an odd integer. This will be necessary when we study the case of an infinitely periodic function. We can now shift our origin to $\tilde{x} = (N-1)a/2 \equiv \Delta\tilde{x}$. Equivalently, we can imagine that the pattern represented by $\tilde{\tau}(x)$ is displaced by $-\Delta\tilde{x}$. Now the problem is symmetrical. We accomplish this in the formal sense by the change $\tilde{\tau}(\tilde{x}) \to \tilde{\tau}(\tilde{x} + \Delta\tilde{x})$. From the Fourier transform properties in Table 6.2 we see how this change effects the Fourier transform: $T(u) \to T(u)e^{i2\pi u\,\Delta\tilde{x}} = T(u)e^{-i\delta(\Delta\tilde{x}/a)}$. With this information and Eq. (6.116) we find

$$T(u) = T_1(u)e^{i(N-1)\delta/2} \left[\frac{\sin\left(\dfrac{N\delta}{2}\right)}{\sin\left(\dfrac{\delta}{2}\right)} \right] e^{-i(N-1)\delta/2}$$

or

$$T(u) = T_1(u) \left[\frac{\sin\left(\dfrac{N\delta}{2}\right)}{\sin\left(\dfrac{\delta}{2}\right)} \right] \qquad (6.118)$$

with $\delta = -2\pi u a$.

We recognize the second factor from Chapter 5, where the interference grating was discussed. The first factor is due to diffraction and will modulate the interference pattern.

b. Infinite Periodicity. In the fictitious limiting case when the number of repeating elements in a periodic function goes to infinity, the maxima at $\delta = 2\pi m$ or $|ua| = m$ in Eq. (6.118) become infinitely high, infinitely narrow "spikes."

$$\lim_{N \to \infty} \left[\frac{\sin\left(\frac{N\delta}{2}\right)}{\sin\left(\frac{\delta}{2}\right)} \right] \to \sum_{m=-\infty}^{\infty} \delta(ua - m) \tag{6.119}$$

Equations of the type (6.118) assume the form

$$F(u) = F_1(u) \sum_{m=-\infty}^{\infty} \delta(ua - m) \tag{6.120}$$

Thus the Fourier transform of an infinitely periodic function is nonzero only for values of u equal to integral multiples of the inverse of the spatial periodicity a.

The significance of this result can be seen when we write $f(x)$ in terms of its Fourier transform and use Eq. (6.120).

$$f(x) = \int_{-\infty}^{\infty} \left[F_1(u) \sum_{m=-\infty}^{\infty} \frac{1}{|a|} \delta\left(u - \frac{m}{a}\right) \right] e^{i2\pi ux} \, du$$

$$= \sum_{m=-\infty}^{\infty} \left[\frac{1}{|a|} \int_{-\infty}^{\infty} F_1(u) \, \delta\left(u - \frac{m}{a}\right) e^{i2\pi ux} \, du \right]$$

$$= \sum_{m=-\infty}^{\infty} \left[\frac{1}{|a|} F_1\left(\frac{m}{a}\right) \exp\left(i2\pi m \frac{x}{a}\right) \right]$$

This can be written in the form of the *Fourier series*

$$f(x) = \sum_{m=-\infty}^{\infty} F_m \exp\left(i2\pi m \frac{x}{a}\right) \tag{6.121}$$

with the *Fourier coefficient* given by

$$F_m \equiv \frac{1}{|a|} F_1\left(\frac{m}{a}\right) = \frac{1}{|a|} \int_{-a/2}^{a/2} f(x) \exp\left(-i2\pi m \frac{x}{a}\right) dx \tag{6.122}$$

C. Diffraction Grating

1. General Grating We return now to Eq. (6.118) for the Fourier transform of a finite periodic function. Let this represent the transmission function for an interference grating that consists of an array of finite-width slits. The far-field diffraction pattern will be given by Eq. (6.28) with $T(u)$ determined from Eq. (6.118). Here we have

$$T(0) = T_1(0)N.$$

$$S'(u) = S'(0)\left[\frac{R_{00}R'_{00}}{R_0 R'_0}\right]^2 B(u) \left[\frac{\sin\left(\dfrac{N\delta}{2}\right)}{N\sin\left(\dfrac{\delta}{2}\right)}\right]^2 \tag{6.123a}$$

where

$$B(u) \equiv \frac{|T_1(u)|^2}{|T_1(0)|^2} \tag{6.123b}$$

is the *blaze function* that describes the influence of diffraction on the pattern. If $\tilde{\tau}_1(\tilde{x}) = \delta(\tilde{x})$, then $T_1(u) = 1$ and $B(u) = 1$. This returns us to the special case treated in Chapter 5. The behavior of Eq. (6.123) is then exactly the same as Eq. (5.44). All of the arguments concerning dispersion and resolution for gratings are associated with the behavior of this factor. Because this is retained in the general form of Eq. (6.123), we may carry those arguments directly over to the current discussion.

The full expression for the radiation flux density includes the blaze function for a single slit in the array of apertures. For a simple box function of width b, $T_1(u) = b \operatorname{sinc}(\pi u b)$ and

$$B(u) = \operatorname{sinc}^2(\pi u b) \tag{6.124}$$

Its effect on the overall pattern is indicated in Fig. 6.31, which should be compared with Fig. 6.29. Because b must be less than a, the main maximum of the pattern will cover at least three interference maxima, including the one at zero, which is useless

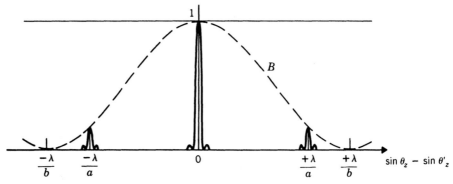

Fig. 6.31 Flux-density pattern for the interference grating modulated by diffraction due to finite-size slits.

for spectroscopy and which receives the greatest amount of diffracted energy. Note also that equal amounts of radiation energy are diffracted into both positive and negative orders, which again is a waste of energy.

This state of affairs makes this simple kind of grating inefficient for practical spectroscopy. Efficiency can be greatly improved by *blazing* the grating.

2. Blazed-Transmission Grating. The technique of blazing involves the creation of a grating with a slit transmission function that will shift the blaze so that its maximum coincides with one of the nonzero orders in the interference pattern. For simplicity we assume that the $\theta = 0$ and that θ' remains small. If the array of slits is replaced by an array of transparent dielectric wedges, then the light from each wedge will be deviated. Consider the geometrical optics limit for the wedge grating shown in Fig. 6.32. A ray is deviated according to Snell's law by θ', where $\theta' = \theta'' - \theta_B$. Here θ_B is the blaze angle representing the inclination of a facet with respect to the plane of the grating, and θ'' is the refracted angle that enters into Snell's law. We must have

$$n \sin \theta_B = \sin \theta''$$
$$= \sin \theta_B \cos \theta' + \cos \theta_B \sin \theta'$$

If θ_B is small, then $\cos \theta_B \simeq 1$. Because θ' should be small, we approximate $\cos \theta' \simeq 1$. Therefore,

$$\sin \theta' \simeq \sin \theta_B (n - 1) \tag{6.125}$$

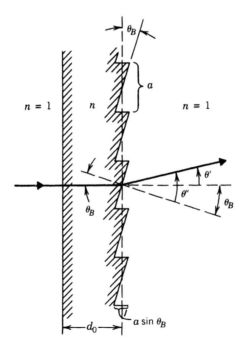

Fig. 6.32 A transmission grating that modulates the incident light by introducing a periodic phase shift.

Now introduce interference. Maxima occur for

$$\sin \theta' = m_B \frac{\lambda_B}{a} \tag{6.126}$$

where λ_B is the wavelength and m_B the order for which the grating is to be optimized. A reasonable choice for the parameters of our wedge grating would require that the angle of deviation calculated in Eq. (6.125) match up with the desired interference maximum in Eq. (6.126) thus,

$$\frac{m_B \lambda_B}{a} = \sin \theta_B(n-1) \tag{6.127}$$

This should send the center of the blaze function to the mth order for the chosen λ_B.

To see this more clearly, let us model the blaze function through the Fourier transform of a complex transmission function. From Eq. (6.10) we know how to do this. We need $d(\tilde{x})$, the spatial variation of the thickness of the grating. Here we have $d(\tilde{x}) = d_0 + \tilde{x} \sin \theta_B$ for $\tilde{x}$ on the interval $-a/2$ to $+a/2$. Then

$$\tilde{\tau}_1(\tilde{x}) = \tilde{\tau}_0(\tilde{x}) \exp\left[-i \frac{2\pi}{\lambda}(n-1)\tilde{x} \sin \theta_B \right] \tag{6.128}$$

where

$$\tilde{\tau}_0(\tilde{x}) = \exp\left[-i \frac{2\pi}{\lambda}(n-1) d_0 \right]$$

is defined for $|x| \leqslant a/2$ only. This is equivalent to the product of a box function (of constant complex amplitude and width a) and an extended phase-shift function.

Let

$$u_B \equiv \frac{-(n-1)}{\lambda} \sin \theta_B \tag{6.129}$$

then the phase-shift function is $e^{+i2\pi u_B \tilde{x}}$. The convolution theorem tells us that $T_1(u)$ will be the convolution of the transform of the box function, $a \operatorname{sinc}(\pi u a)$, and the transform of the phase-shift function, $\delta(u - u_B)$ or

$$T_1(u) = \int_{-\infty}^{\infty} a \operatorname{sinc}(\pi u' a) \, \delta(u - u' - u_B) \, du'$$

$$= a \operatorname{sinc}[\pi(u - u_B)a] \tag{6.130}$$

The blaze function then becomes

$$B(u) = \frac{\operatorname{sinc}^2[\pi(u - u_B)a]}{\operatorname{sinc}^2(-\pi u_B a)} \tag{6.131}$$

This has the same form as Eq. (6.124) for the unblazed grating except that the pattern is shifted to place the maximum in the blaze function at

$$u = u_B \tag{6.132}$$

From Eq. (6.126) recall that interference maxima occur for $\sin \theta' = m\lambda/a$. If we want the blaze function to be largest on a given maxima for $\lambda = \lambda_B$, then u_B must be

$$-u_B = \frac{-m_B}{a}$$

By substitution from the definition of u_B, this becomes

$$\frac{(n-1)}{\lambda_B} \sin \theta_B = \frac{m_B}{a}$$

which is the same as Eq. (6.127), which was developed using concepts from geometrical optics. The optimum wavelength will be

$$\lambda_B = \frac{(n-1)a \sin \theta_B}{m_B} \tag{6.133}$$

Using Eq. (6.131) in Eq. (6.123) instead of Eq. (6.124) produces an asymmetric pattern with much more useful energy available in the dispersion region. To see this more clearly, identify

$$\delta_B \equiv -2u_B a = \frac{2\pi(n-1)a \sin \theta_B}{\lambda} \tag{6.134}$$

then

$$B(\delta) = \frac{\operatorname{sinc}^2\left[\dfrac{\delta - \delta_B}{2}\right]}{\operatorname{sinc}^2\left[\dfrac{\delta_B}{2}\right]}$$

where

$$\sin \theta_z' = \frac{\lambda}{2\pi a} \delta \tag{6.135}$$

Eq. (6.123) becomes

$$S'(\delta) = \left[\frac{S'(0)}{\operatorname{sinc}^2\left(\dfrac{\delta_B}{2}\right)}\right]\left[\frac{D'}{R_0'}\right]^2 \operatorname{sinc}^2\left[\frac{\delta - \delta_B}{2}\right]\left[\frac{\sin\left(\dfrac{N\delta}{2}\right)}{N \sin\left(\dfrac{\delta}{2}\right)}\right]^2 \tag{6.136}$$

The first factor is constant for fixed δ_B. The second factor is slowly varying with δ. The third and fourth are sketched as functions of $\sin \theta' = \lambda\delta/2\pi a$ in Fig. 6.33. When $\lambda = \lambda_B$, $\lambda_B/2$, $\lambda_B/3$, and so on, the peak of the diffraction pattern coincides with an interference maximum. All other interference maxima are eliminated by zeros in the diffraction pattern. This is illustrated in Fig. 6.33 for various values λ at or near λ_B and $\lambda_B/2$.

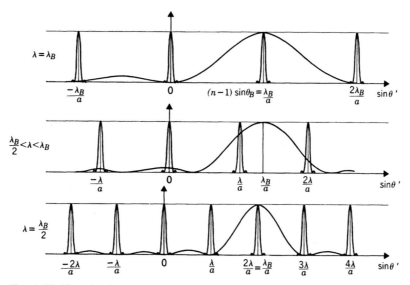

Fig. 6.33 Flux-density pattern for the interference grating showing the effect of blazing, produced by a slit profile that shifts the phase of the incident beam.

At the peak of the mth-order interference maximum where $\delta = 2\pi m$, the flux density is determined by

$$\text{sinc}^2\left[\frac{\delta - \delta_B}{2}\right]$$

But we can rewrite δ_B using Eq. (6.133)

$$\delta_B = 2\pi m_B \frac{\lambda_B}{\lambda}$$

Then we obtain

$$B(\delta) \propto \text{sinc}^2\left(\pi m - \pi m_B \frac{\lambda_B}{\lambda}\right) \tag{6.137}$$

We may use Eq. (6.137) to predict the relative energy in various orders m as a function of λ. Usually we work in the m_Bth order over the range

$$\frac{1}{\lambda} = \frac{1}{\lambda_B}\left(1 - \frac{1}{2m_B}\right) \quad \text{to} \quad \frac{1}{\lambda} = \frac{1}{\lambda_B}\left(1 + \frac{1}{2m_B}\right)$$

At the extreme ends of this range for $m_B = 1$ the diffraction factor is reduced from its maximum value of unity to

$$\text{sinc}^2\left(\frac{\pi}{2}\right) = 0.41$$

D. Diffraction-Limited Performance

1. Fraunhofer Diffraction with Lenses. The finite size of the Airy disk or similar diffraction patterns will place an ultimate limit on the performance of optical instruments. Before we discuss some specific examples, we will show how *Fraunhofer diffraction can occur in the image plane of an optical instrument with a point source.* In the cases examined so far, the distance R'_0 was so large that all light rays from the diffracting aperture to the observation point P' were essentially parallel, as shown in Fig. 6.34a. A linear variation of the phase proportional to $\alpha'\tilde{x} + \beta'\tilde{y}$ (that is, linear in the direction cosines of these rays) was found.

These parallel rays propagating in direction $(\alpha', \beta', \gamma')$ can also be brought together in a finite distance with the help of a lens L_2, as shown in Fig. 6.34b. The point P' having coordinates (x', y', f_2) with respect to the second principal point of the lens, where f_2 is the focal length, will be a geometrical focus for all rays with direction cosines $\alpha' = x'/R'_0$, $\beta = y'/R'_0$, with $R'_0 = \sqrt{f_2^2 + x'^2 + y'^2}$. If the aperture is close to the lens, the resulting pattern is mathematically the same as that

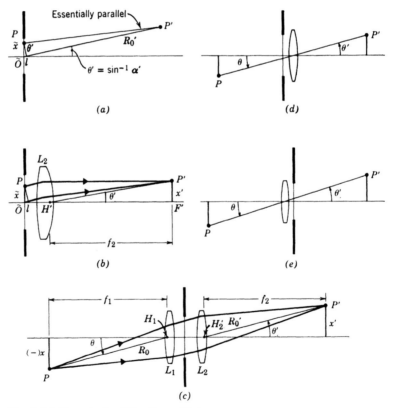

Fig. 6.34 Fraunhofer diffraction with lenses.

produced with no lens, but it now involves this very much smaller value of R_0' that need not satisfy the lensless Fraunhofer condition, Eqs. (6.18).

A more detailed discussion of this sort of problem will be given in Chapter 7, where the preceding statements will be proved more rigorously. It turns out that the remarks about the effect of the lens apply equally well if the aperture is placed just to the right of the lens as if it is placed just to the left. Diffraction occurs because the wavefront emerging from the system is essentially part of a spherical surface.

Another lens L_1 may be similarly used to bring the point source P in from what is essentially infinity while preserving the preceding mathematical results. The emerging wavefront is the same partial sphere. The appropriate variables here are the direction cosines $\alpha = -x/R_0$, $\beta = -y/R_0$, and the distance $R_0 = \sqrt{f_1^2 + y^2 + x^2}$. The source at P is placed in the focal plane of L_1, as shown in Fig. 6.34c. Here L_1 and L_2 are both very close to the diffracting aperture. After passing through L_1, the light from P is collimated parallel to the direction of the vector $\overline{\mathbf{PH}}_1 = (-x, -y, f_1) = R_0(\alpha, \beta, \gamma)$.

The combination effect of two lenses is to produce an "image" of P at P'. (Note that $\overline{\mathbf{PH}}$ and $\overline{\mathbf{H'P'}}$ are parallel if the index of refraction is the same on both sides of the system.) But instead of a true point image, what is observed near P' is the Fraunhofer or far-field diffraction pattern of the aperture, for example, the Airy disk with rings if the aperture is a circle. The effect is the same if the two lenses are combined into a single lens on either side of the aperture, as shown in Fig. 6.34d and e.

2. Longitudinal Displacement of the Aperture. The aperture we are discussing is the aperture stop of the system. In addition to its geometrical role in limiting the angular spread of rays from the source point that may pass through the system, it is now seen to play a dominant role in the diffraction properties of the system, a role that is maintained in complicated multilens systems.

We now quote part of a general result that describes the effects of moving the aperture from the lens toward the image P', as shown in Fig. 6.35. If the aperture size is changed so that it has the same angular shape as seen from P' (that is, so that

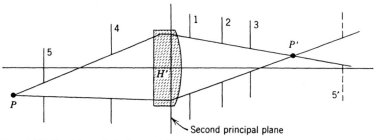

Fig. 6.35 Positions 1, 2, 3, 4, 5 label possible locations for the diffracting aperture. 5' is the image of 5.

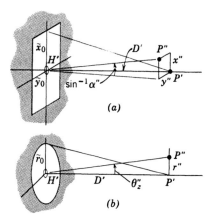

Fig. 6.36 Two shapes that might apply to the exit pupil.

it has a constant projection from P' on any convenient plane, say the second principal plane at H' of the lens), then the intensity distribution in the diffraction pattern of the aperture will be unchanged. (The phases in various parts of the pattern will change as the aperture is moved, however.) This result is also true if the aperture is to the left of the lens in position 4 or 5. Then its image in image space, that is, the exit pupil, must have the proper projection on the H' plane. In all cases, then, we may say that: *If the projection of the exit pupil through the point P' onto the second principal plane remains the same, the intensity distribution near P' will be unchanged.*

Thus it is the size and shape of this projected exit pupil that determine the intensity distribution in the "image" of P at P'. Projected pupils are shown explicitly in Fig. 6.36 for two aperture shapes. The geometrical point image is a distance D' away from the second principal plane. Here we have assumed $\theta_z = 0$. Surrounding this point is a diffraction pattern characteristic of the shape of the pupil. The flux density at P'', near P' due to pupil (a) will be

$$S''(P'') \propto \text{sinc}^2(2\pi u \tilde{x}_0)\, \text{sinc}^2(2\pi v \tilde{y}_0)$$

with

$$u\tilde{x}_0 = -\frac{x''}{D'}\frac{\tilde{x}_0}{\lambda} = -\frac{\tilde{x}_0}{D'}\frac{x''}{\lambda}$$

and

$$v\tilde{y}_0 = -\frac{y''}{D'}\frac{\tilde{y}_0}{\lambda} = -\frac{\tilde{y}_0}{D'}\frac{y''}{\lambda} \tag{6.138}$$

Here (x''/D') and (y''/D') represent the angular coordinates of the observation point P'' as seen from the principal point H'. On the other hand, we note that for small $\tilde{x}_0$ and $\tilde{y}_0$, $(\tilde{x}_0/D')$ and $(\tilde{y}_0/D')$ represent the angular half-width of the aperture as seen from P'. For the circular pupil (b) we have

$$S''(P'') \propto \left[\frac{2J_1(w)}{w}\right]^2$$

with

$$w = \frac{2\pi \tilde{r}_0 \sin \theta''_z}{\lambda} \cong 2\pi \frac{\tilde{r}_0}{D'} \frac{r''}{\lambda} = 2\pi \theta''_z \frac{r_0}{\lambda} \qquad (6.139)$$

Here θ''_z is the angular coordinate of P' as seen from the principal point, and from the other point of view for small $\tilde{r}_0$ we may interpret $(\tilde{r}_0/D')$ as the angular radius of the aperture as seen from P'.

With this general background we now turn to a discussion of specific systems.

3. Telescopic Resolving Power. The finite size of the Airy disk will place an ultimate limit on the angular resolving power of a telescope. Suppose, for example, that we are looking at a double star through a telescope having an objective with focal length f and aperture diameter d. The two components of the double star will be assumed to be equally intense and to have an angular separation $\Delta\phi$ as shown in Fig. 6.37a. The diffraction pattern appearing in the focal plane of the objective will consist of the superposition of two Airy disks having an angular separation $\Delta\phi$ together with their surrounding rings.

For normal visual observation these diffraction patterns will be magnified without appreciable additional diffraction by the eyepiece, because the aperture stop of the overall system is the telescope objective.

They will be resolved as two patterns if $\Delta\phi$ is much larger than the angular width of the disk and not resolved if $\Delta\phi$ is much smaller. The critical value of $\Delta\phi$ for which resolution is just possible is usually taken to be that given by Lord Rayleigh. It occurs when $\Delta\phi$ equals the angular radius of the Airy disk, which from Eq. (6.56) gives

$$\Delta\phi_{\min} = \frac{0.61\lambda}{\tilde{r}_0} = \frac{1.22\lambda}{d} \qquad (6.140)$$

A photograph of this double diffraction pattern at the Rayleigh limit is shown in

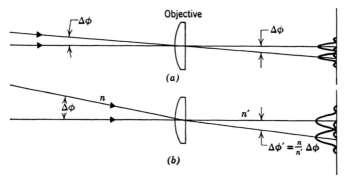

Fig. 6.37 Diffraction pattern produced in the image plane as a result of two distant stars: (a) equal indices of refraction on both sides of the lens; (b) unequal indices of refraction.

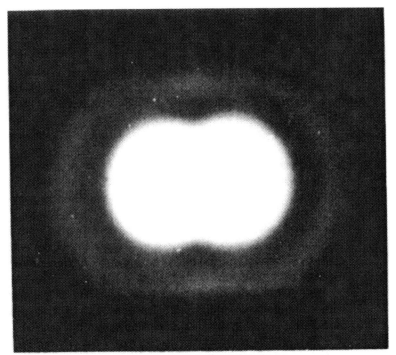

Fig. 6.38 Image of a double star at the Rayleigh limit.

Fig. 6.38. Many telescopes are not of high enough optical quality to yield this "diffraction-limited" resolution. Even if they are, atmospheric fluctuations may cause the image to move slightly in a random way, so that the practical value of $\Delta\phi_{\min}$ may be greater than that of Eq. (6.140)

If the index of refraction n differs from unity, λ in Eq. (6.140) is assumed to be the reduced wavelength $\lambda = \lambda_0/n$. Thus, for the situation in Fig. 6.37b, before the lens, the angle would be

$$\Delta\phi = \frac{1.22\lambda_0}{n}$$

whereas after the lens it would be

$$\Delta\phi' = \frac{1.22\lambda_0}{n'}$$

4. Telescopic Light-Gathering Power. For visual observation using both the telescope and the eye, the exit pupil of the telescope is the same size as the pupil of the observer's eye (Fig. 6.39). For the combined system of telescope plus eye, the aperture stop may be considered to be either the objective lens or the eye pupil—it makes no difference—and the diffraction pattern on the retina of the eye will be the same Airy disk that would be obtained without the telescope.

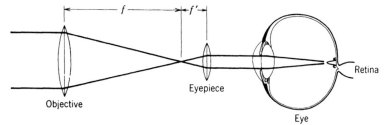

Fig. 6.39 Telescope plus eye as a combined optical system.

Thus the telescope cannot change the size of the retinal diffraction pattern from a distant point object. What it can do is increase the radiant flux density or irradiance (in W/mm^2) or the luminous flux density or illuminance in lm/mm^2 falling on the retina in the Airy disk. With the unaided eye, the flux passing through the pupil is focused on the disk, whereas with the telescope the flux passing through its objective is focused onto the disk, and this is greater by the ratio of the area of the objective to the area of the pupil, that is, by the ratio of the squares of the diameters,

$$\left(\frac{d_{obj}}{d_{pupil}}\right)^2 = m_\alpha^2$$

where $m_\alpha = -f/f'$ is the magnification of the telescope. The telescope has m_α^2 times more *light-gathering power* than the unaided eye.

We can say that the use of a telescope for visual astronomical observations has two important advantages: (1) it increases the separation of diffraction patterns of individual stars on the retina by the ratio m_α; and (2) it increases the flux in each pattern by the ratio m_α^2. If the stars are seen on a continuous background such as the daytime sky, then the irradiance in the retinal image of this background will be independent of the magnification m_α (m_α^2 times more flux is incident on m_α^2 times more retinal area). Thus the use of a telescope will increase the contrast of the irradiance in the star's Airy disk over that of the background by the ratio m_α^2. Bright stars then become visible in daytime.

5. Spectrograph. Diffraction ultimately limits the resolving power of a prism spectrograph, monochromator, or similar instrument of the type shown in Fig. 6.40. Because of its high cost, the prism itself usually forms the aperture stop of such instruments. Light from the entrance slit is collimated by the collimating lens L_1, dispersed by the prism, and then focused on the photographic plate or multichannel photon detector by a camera lens L_2 or, in a monochromator, light is focused on an exit slit by a lens as in Fig. 6.40.

If the entrance slit were replaced by a pinhole, and if the light were truly monochromatic, the light would be exactly collimated and deviated through a well-defined angle θ_D that depends on the wavelength λ. In the focal plane of lens L_2 we would observe the Fraunhofer diffraction pattern of the rectangular prism as

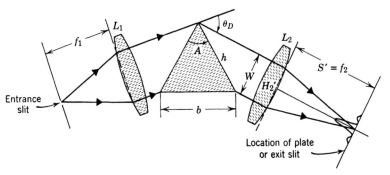

Fig. 6.40 Essential elements of a prism spectrograph.

projected onto a plane through H'_2 normal to the final direction of propagation. This pattern would be something like that in Fig. 6.18.

With a *narrow slit* instead of a *pinhole* at the entrance, we obtain a superposition of infinitely many such patterns displaced along the direction of the slit that is out of the paper in Fig. 6.40. The resulting flux density pattern will vary in the direction normal to the slit exactly as does the "single slit" pattern of Fig. 6.14. The effective "slit width" of this pattern is the projected height of the prism denoted by W in Fig. 6.40. Thus the angular distance $\Delta\theta_z = \Delta x'/2S'$ from the center of the diffraction pattern to the first zero, which is also the minimum resolvable angle by the Rayleigh criterion, is given by Eq. (6.39) with $R'_0 = S'$, $2\tilde{x}_0 = W$ to be

$$\Delta\theta_2 = \frac{\lambda}{W} \tag{6.141}$$

Suppose that the prism is operated near minimum deviation so that the deviation angle θ_D is close to the minimum value $\theta_{D,\min}$, which is given in terms of the prism angle A and the refractive index n by Eq. (2.98)

$$n = \frac{\sin[(A + \theta_{D,\min})/2]}{\sin(A/2)} \tag{6.142}$$

The angular spread $\Delta\theta_D$ implies an uncertainty in the wavelength determination of a given spectral line of an amount $\Delta\lambda = |d\lambda/d\theta_D|\Delta\theta_D$, where $d\lambda/d\theta_D$ is the reciprocal of the angular dispersion of the instrument. The spectroscopic resolving power $\lambda/\Delta\lambda$ is then given by

$$\frac{\lambda}{\Delta\lambda} = W\left|\frac{d\theta_D}{d\lambda}\right|$$

The wavelength dependence of the deviation angle θ_D is given implicitly for θ_D near $\theta_{D,\min}$ by the $n(\lambda)$ dependence through Eq. (6.142). In fact, we can write $d\theta_D/d\lambda \simeq d\theta_{D,\min}/d\lambda$ and obtain

$$\left[\sin\frac{A}{2}\right]\frac{dn}{d\lambda} \simeq \frac{1}{2}\left[\cos\left(\frac{A + \theta_D}{2}\right)\right]\frac{d\theta_{D,\min}}{d\lambda}$$

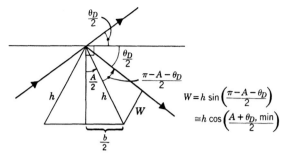

$$W = h \sin\left(\frac{\pi - A - \theta_D}{2}\right)$$

$$\cong h \cos\left(\frac{A + \theta_{D,\, min}}{2}\right)$$

Fig. 6.41

Therefore,

$$\frac{\lambda}{\Delta\lambda} = W\left|\frac{d\theta_{D,min}}{d\lambda}\right| = \frac{2W\sin\dfrac{A}{2}}{\cos\left(\dfrac{A + \theta_{D,min}}{2}\right)}\left|\frac{dn}{d\lambda}\right|$$

From Fig. 6.41 we see that the slant height h of the prism is given by

$$h = \frac{W}{\cos\left(\dfrac{A + \theta_{D,min}}{2}\right)}$$

Thus we have $\lambda/\Delta\lambda = 2h(\sin A/2)\,(dn/d\lambda)$. But $h(\sin A/2)$ can be seen from Fig. 6.41 to equal $b/2$, where b is the length of the prism base. Thus the resolving power near minimum deviation is simply

$$\frac{\lambda}{\Delta\lambda} = b\frac{dn}{d\lambda} \tag{6.143}$$

6. Microscope with Incoherent Illumination. So far we have assumed that the light from the two point sources to be resolved is incoherent. This will be the case for self-luminious sources. In a microscope, the object is illuminated by a condensing system and is usually not self-luminous. The transverse coherence length of the light illuminating the object is usually larger than the separation between the two points we wish to resolve. We must then treat the light from such an object as coherent. The discussion of resolving power changes somewhat in this case and will be postponed until we discuss the Abbe theory of image formation.

But for a microscope with incoherent illumination we can find the minimum resolvable separation as follows. Consider Fig. 6.42, which shows the objective lens. The objects P_I and P_{II} are separated a distance x and are imbedded in a medium with index of refraction n (in an oil-immersion microscope n will be larger than 1).

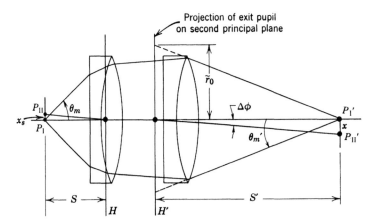

Fig. 6.42 Geometry pertinent to the calculation of the resolving power of a microscope.

The images P_I' and P_{II}' are found in air. The limiting angular aperture is θ_m', as seen from the image. The Abbe sine condition (section 4.3A4) gives

$$nx \sin \theta_m = x' \sin \theta_m' \tag{6.144}$$

Now the limiting angular resolution is given by Eq. (6.140) as

$$\Delta\phi_{min} = \frac{x'_{min}}{S'} = \frac{0.61\lambda}{\tilde{r}_0} \tag{6.145}$$

Because of the large magnification produced by the objective, we have $S' \gg S$; hence, θ_m' will be small even if θ_m is not. Thus we can write.

$$\theta_m' \approx \frac{\tilde{r}_0}{S'} \tag{6.146}$$

Thus, from Eqs. (6.145) and (6.146) we obtain

$$x'_{min} = \frac{0.61\lambda}{\theta_m'} \tag{6.147}$$

Then Eq. (6.144) becomes

$$nx_{min} \sin \theta_m = x'_{min} \sin \theta_m' \approx x'_{min} \theta_m' = 0.61\lambda$$

or

$$x_{min} = \frac{0.61\lambda}{n \sin \theta_m} \tag{6.148}$$

The quantity $n \sin \theta_m$ appearing here [as in Eq. (2.100) in the discussion of dielectric waveguides] is called the *numerical aperture*. It can be made as large as 1.6 for an immersion objective. With $\lambda = 560$ nm, this will give $x_{min} = 210$ nm.

REFERENCES

Arfken, G. *Mathematical Methods for Physicists*. Academic Press, New York, 1970.

Baker, B. B., and E. J. Copson. *The Mathematical Theory of Huygens' Principle*. Oxford University Press, London. 1969.

Born, Max, and Emil Wolf. *Principles of Optics*. Pergamon Press, Oxford, 1980.

Cagnet, Michel, Maurice Françon, and Jean Claude Thrierr. *Atlas of Optical Phenomena*. Springer-Verlag, Berlin, 1962.

Duffieux, P. M. *The Fourier Transform and Its Application to Optics*. Wiley, New York, 1983.

Fowles, Grant R. *Introduction to Modern Optics*. Holt, Rinehart and Winston, New York, 1968.

Françon, Maurice. *Diffraction: Coherence in Optics*. Pergamon Press, Oxford, 1966.

Goodman, Joseph W. *Introduction to Fourier Optics*. McGraw-Hill, New York, 1968.

Jackson, John D. *Classical Electrodynamics*. Wiley, New York, 1975.

James, J. F., and R. S. Sternberg. *The Design of Optical Spectrometers*. Chapman and Hall, Ltd. London, 1969.

Kline, Morris, and Irwin W. Kay. *Electromagnetic Theory and Geometrical Optics*. Interscience, New York, 1965.

Lighthill, M. J. *Introduction to Fourier Analysis and Generalized Functions*. Cambridge University Press, Cambridge, 1958.

Mertz, Lawrence. *Transformations in Optics*. Wiley, New York, 1965.

Stone, J. M. *Radiation and Optics*. McGraw-Hill, New York, 1963.

Watson, G. N. *Theory of Bessel Functions*. Cambridge University Press, Cambridge, 1956.

PROBLEMS

Section 6.1 General Concepts of Diffraction

1. In far-field diffraction the inclination factor is set equal to unity. If this is to introduce errors no larger than 10%, find the corresponding limitation on the diffraction geometry. In practice, we may neglect this factor even outside of this range in many cases because the angular variation of the inclination factor is weak.

2. What is the proper transmission function to be used for an aperture that is covered by a filter that reduces the flux density according to the following form

$$\tilde{S}(\bar{r}) = \tilde{S}(0)\exp(-\pi b^2 r^2)$$

3. The output from an idealized source at a wavelength of 488 nm forms a uniform collimated beam 6 cm in diameter. It is focused by a well-corrected lens to a diffraction limited "point." A 4-cm diameter circular stop is just behind the lens. If the factor C in Eq. (6.2) is i/λ (we will prove this in Chapter 7), find the flux density at the paraxial focal point. [Hint: Find an appropriate $\tilde{\sigma}_2$ in Fig. 6.9].

Section 6.2 Far-Field Diffraction

4. An experimenter is studying Fraunhofer diffraction by a slit. The slit height is $2\tilde{x}_0 = 1$ mm. The wavelength is 500 nm. How large does the distance to the observation plane have to be if the magnitude of the neglected terms in the linearized phase factor in the diffraction integral is less than 0.01 rad?

5. Equation (5.44) described the pattern of flux density due to interference by N parallel very narrow slits in the aperture plane. Starting from this expression for a region of total width $b = (N - 1)d = $ constant, let N go to infinity while requiring d to approach zero. In this way the region of width b eventually becomes entirely open. Show that this leads to an expression for the flux density in the diffraction pattern due to a wide slit of width b. This approach is equivalent to considering cylindrical Huygens' wavelets in the description of diffraction for an opening that varies in one dimension.

6. Show that the maxima in the flux density of the Fraunhofer diffraction pattern of a slit occur at values of u that are approximately the same as the maxima in $\sin^2(2\pi u\tilde{x}_0)$ when u is large.

7. Show that the flux density of the ℓth minor maximum in the diffraction pattern for a narrow slit in the far-field limit can be approximately evaluated by

$$S_\ell = S_0[(\ell + \tfrac{1}{2})\pi]^{-2}$$

8. Consider diffraction by a long, narrow slit of width $2\tilde{x}_0$. Construct phasor diagrams depicting the far-field diffraction integral for the following values of $u\tilde{x}_0$: $0, \tfrac{1}{6}, \tfrac{1}{3}, \tfrac{1}{2}, \tfrac{2}{3}$. Make your construction accurate in units of $\tilde{x}_0$. From the diagram determine the electric field intensity ratio $E(u)/E(0)$.

9. Consider the far-field diffraction pattern by a square aperture of sides $2\tilde{x}_0$. What is the expression for the flux density along the line $x' = y'$?

10. The flux density at the center of the diffraction pattern produced by a long slit 0.1 mm wide on a screen 5 m away by light of wavelength 500 nm is 15 μW/cm^2. Find the flux density at the following positions away from the center of the pattern: 6.25 mm, 12.5 mm, 25 mm.

11. A rectangular aperture is illuminated at normal incidence by parallel light with wavelength of 500 nm. The dimensions of the aperture are 1 × 3 mm. What are the dimensions of the main maximum of the diffraction pattern formed on a screen 50 m away oriented parallel to the aperture?

12. A point source that emits white light from 400 nm to 700 nm is 2 m from an aperture that is 2-mm square. Describe the major features of the diffraction pattern, paying particular attention to the wavelength dependence.

13. Another way to do the integral for far-field diffraction from a circular aperture is to stay in rectangular coordinates. Show that by dividing the integration up into parallel strips the field at the observation point is proportional to

$$\int_0^1 \sqrt{1 - t^2} \cos(tw)\, dt$$

where $w = 2\pi\rho\tilde{r}_0/(\lambda R_0')$. Compare your result with Eq. (6.51) to determine another representation of the Bessel function of first order.

14. Discuss the local variation in the diffraction pattern in the far-field limit at high diffraction angles. As a model system, consider an infinitely long slit parallel to the y axis that is illuminated by collimated light at normal incidence. Consider the region on the observation plane that is in the vicinity of $x' = R_0'$, where R_0' is also the distance between the slit plane and the observation plane. This is at an angle of 45°. Show that the problem can be analyzed by considering diffraction through an effective slit that has a width equal to the projection of the actual slit on the direction of observation.

15. A plane wave of wavelength 600 nm illuminates a long, narrow slit of width 10^{-4} m producing a far-field diffraction pattern that is described by

$$S(u) = S(0)\, \text{sinc}^2(2\pi u\tilde{x}_0)$$

where $2\tilde{x}_0$ is the slit width and $u = (\alpha - \alpha')/\lambda$.

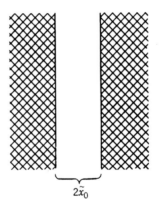

$2\tilde{x}_0$

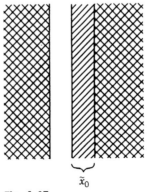

$\tilde{x}_0$

Fig. 6.43

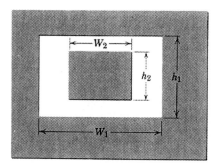

Fig. 6.44

(a) Find $\Delta\theta'_z$ (in radians), the angle at the slit that subtends the first three maxima (identified by the second minimum on either side of the central maximum).

(b) How does this angle change if the slit width is increased 10 times?

(c) What is the effect of covering half of the opening with a neutral density filter that cuts the power through that half by a factor of 4. The filter has width $\tilde{x}_0$ and splits the opening into two segments with two different transmission characteristics (Fig. 6.43).

The following problems can be answered with the help of Babinet's principle:

16. Describe analytically the flux density in the Fraunhofer diffraction pattern of an open rectangle with a centered rectangular obstruction. Sketch qualitatively the results. See Fig. 6.44.

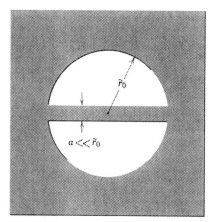

Fig. 6.46

17. Describe analytically the flux density in the Fraunhofer diffraction pattern of the annular aperture shown in Fig. 6.45. Discuss the qualitative effect on the height and width of the main maximum when $\tilde{r}_1 \ll \tilde{r}_2$.

18. Describe quantitatively the flux density in the far-field diffraction pattern of the aperture in Fig. 6.46 in regions away from the center of the pattern where without the bar the flux density would be essentially zero. Sketch the appearance of the two-dimensional pattern.

19. Same as Problem 18, but with the aperture in Fig. 6.47.

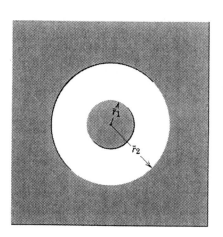

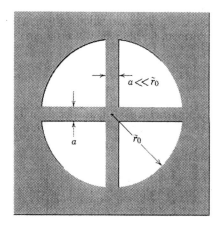

Section 6.3   Fourier Analysis

20. Design a "triangular" function sequence that forms a representation of the delta function.

21. Prove by direct calculation the shifting theorem for the inverse transform, Eq. (6.87).

22. Evaluate the Fourier transform of a derivative

$$\frac{df(x)}{dx}$$

in terms of the transform of the original function

$$F(u) = \mathscr{F}[f(x)]$$

23. Determine the Fourier cosine transform of the step function

$$f(x) = \begin{Bmatrix} 1 & x > 0 \\ 0 & x < 0 \end{Bmatrix}$$

24. Calculate the Fourier transform of

$$\exp(-\pi b^2 r^2)$$

You will need the definite integral

$$\int_{-\infty}^{\infty} e^{-y^2}\,dy = \sqrt{\pi}$$

What is the relationship between the full width at half maximum for the function and for its transform?

25. Calculate the Fourier transform of

$$f(x) = \begin{Bmatrix} e^{-xb} & x > 0 \\ 0 & x < 0 \end{Bmatrix}$$

26. Find the Fourier series representation for the infinitely periodic sawtooth function for which the local function is

$$f_1(x) = f_0\left(\frac{x}{a} + \frac{1}{2}\right), \qquad -\frac{a}{2} < x < \frac{a}{2}$$

27. Find the Fourier series representation for the infinitely periodic square wave function for which the local function is

$$f_1(x) = \begin{cases} f_0, & \dfrac{-a}{4} < x < \dfrac{a}{4} \\[2mm] 0, & \dfrac{-a}{2} < x < \dfrac{-a}{4};\ \dfrac{a}{4} < x < \dfrac{a}{2} \end{cases}$$

Show that the terms associated with odd values of n vanish. Show how this can lead to a form for $f(x)$ that is written as a series of cosine functions.

28. Derive an explicit form for Parseval's theorem in the case of an infinitely periodic function $f(x)$.

29. Use the convolution theorem to evaluate the Fourier transform of $\sin(2\pi u_0 x) \cdot \cos(2\pi u_0 x)$.

30. Determine the Fourier transform of the function

$$f(x) = U(x)\cos^2\left(\frac{4\pi x}{x_0}\right)$$

where $U(x)$ is the unit amplitude "box" function extending from $x = -x_0$ to x_0

Section 6.4   Examples of Fourier Analysis in Diffraction

31. Prove that far-field diffraction patterns that are produced by transmission functions that do not introduce phase changes will always display inversion symmetry in the observation plane. This is true independent of the shape or density of the aperture.

32. The flux density at the center of the diffraction pattern in the far-field limit due to a circular aperture of radius $\tilde{r}_0$ is found to be S_0. If two such openings are introduced in the aperture with separation $a > 2\tilde{r}_0$, find the new flux density at the center of the pattern.

33. Use the Fourier analysis approach to far-field diffraction to find the relative flux density in a two-slit pattern produced by slits of width $2\tilde{x}_0$ separated by center-to-center spacing $8\tilde{x}_0$ when one slit is covered by a thin glass plate of thickness d and index of refraction n. Express your result in coordinates appropriate to the interference plane and referenced to the perpendicular bisector of the slit aperture. Assume any parameters not provided.

34. Consider a grating made up of a series of slits of width $d/4$ separated a distance d.

(a) Discuss the flux-density pattern observed on the focal plane of a lens just beyond the grating with light of wavelength λ for a total of 2 slits, 4 slits, and 100 slits. Include a discussion of the positions and widths of main maxima and subsidiary maxima. Are any orders missing?

(b) What is the effect on the observed pattern of varying the angle of incidence of the light?

(c) Discuss semiquantitatively the effect on the patterns in **(a)** if the incident light, instead of being accurately collimated, was collimated into a beam of total angular width $\Delta\theta = \lambda/10d$.

35. The Fraunhofer diffraction pattern formed by three long parallel slits consists of principal maxima and subsidiary maxima. If the distance between slit centers is equal to twice the slit width (all slits being of equal width), find the flux density at the subsidiary maximum closest to the central maximum in terms of the flux density at the center of the pattern.

36. Find the flux density distribution in the far-field diffraction pattern formed by a grating with $4N$ slits in which every fourth slit is covered (including the last slit). Assume that the slits are infinitesimally narrow and are separated by distance d.

37. Find the relative flux density in the third and the fourth orders for light having a wavelength of 500 nm diffracted by a grating whose slit width and separation are 2 μm and 3 μm, respectively.

38. Use the convolution theorem to find the analytical form of the diffraction pattern of the aperture pictured (Fig. 6.48). Assume the far-field geometry and any parameters that are required.

39. Describe quantitatively the far-field diffraction pattern observed with an aperture consisting of two narrow slits of equal width oriented at right angles as in Fig. 6.49.

Fig. 6.49

40. Describe quantitatively the field in the Fraunhofer diffraction pattern of the aperture shown in Fig. 6.50. The slits are identical. Make a rough sketch of the two-dimensional distribution of flux density. Assume that the aperture is just in front of a lens of focal length $10^6\lambda$.

41. If you focus your eye at a point source of light emitting monochromatic radiation (say at a wavelength of 500 nm, 100 m distant) and hold up a handkerchief stretched in a plane in front of your eye perpendicular to the light rays with the threads running horizontal and vertical, you will see a rectangular array of bright spots instead of a single point source. If the apparent separation of the bright spots is 0.3 m horizontal and 0.25 m vertical, determine:

(a) The horizontal and vertical separations of the threads of the handkerchief (your eyes will be focused essentially at infinity).

(b) The number of spots that can be seen.

42. Consider a grating with spacing d for which the ruled area is circular with a radius much larger than the groove spacing. Discuss the quantitative behavior of

Fig. 6.48

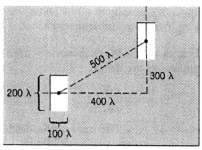

Fig. 6.50

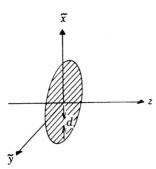

Fig. 6.51

the interference pattern for such a grating for light incident along the z axis as a function of the angle θ'_z (see Fig. 6.51).

43. Calculate the blaze angle of a 600 line/mm grating that is to be used in first order at 750 nm. What would be the theoretical resolving power of a 52-mm-wide grating of this type? Assume $n = 1.6$.

44. Find the Fraunhofer diffraction pattern at a distance $R' \gg a, b$ at wavelength λ in a plane parallel to an aperture plane containing a triangular opening defined by the lines $\tilde{y} = a$, $\tilde{x} = b\tilde{y}/a$, $\tilde{x} = 0$.

45. A "cosine grating" is one with a constant transmission as a function of $\tilde{y}$ and a transmission depending on $\tilde{x}$ as

$$\tilde{t}(\tilde{x}) = \frac{A}{2}(1 + B \cos \gamma \tilde{x}), \quad |A| < 1, |B| < 1$$

Let the grating extend a distance $2\tilde{x}_0$ in the $\tilde{x}$ direction and let $N = \dfrac{\tilde{x}_0 \gamma}{\pi}$ be the total integral number of repetitions of the cosine pattern. For light of wavelength λ, calculate the relative flux density as a function of the diffraction angle θ'_z. How many orders are seen?

46. A narrow slit is covered with an attenuating filter with a power transmission profile of

$$\tilde{T}(\tilde{x}) = \sin^2\left(\frac{\pi \tilde{x}}{\tilde{x}_0}\right)$$

over the width from $-x_0$ to x_0. Determine the far-field diffraction pattern to which such a slit would lead.

47. The flux density in the paraxial focal plane at the center of the diffraction pattern produced by a positive thin lens of radius a and focal length f that is illuminated by a plane wave of wavelength λ is found to be S_0. Now a circular obstruction is placed over the center of the lens so as to block the central portion inside radius b. Derive an expression for the new flux density at the center of the diffraction pattern.

48. A neutral density filter with a Gaussian profile attenuates the beam over a circular aperture according to Problem 24. Calculate the far-field diffraction pattern due to this configuration. Note that it is the flux density, not the electric field, that is Gaussian.

49. If the diameter of the entrance pupil of the eye is 5 mm, determine the distance at which the two headlights of an approaching car can be resolved by the observer. Assume a mean wavelength of 550 nm and that the headlights are separated by a distance of 4 ft.

50. The objective of a telescope is 12 cm in diameter. At what distance would two green (wavelength of 540 nm) objects be barely resolved using the Rayleigh criterion if they are separated by 30 cm?

51. What is the theoretical angular resolution of the 200-in. diameter Palomar telescope? Take an effective value of the wavelength as 550 nm. (Atmospheric fluctuations prevent large telescopes from reaching their theoretical resolution.)

52. Estimate the resolution in lines/mm required of a photographic plate to be used at the focus of a diffraction-limited telescope objective 6 in. in diameter if it is to reveal any structure in the diffraction pattern at the "image" of a star.

53. In the text we argued that for visual observation with a telescope at normal magnification the contrast of the flux density in the Airy disk of a star image over that of a continuous background is proportional to $m^2 \propto d^2_{obj}$. Show that this proportionality to d^2_{obj} also holds for a photographic telescope.

54. In a spectrograph such as that shown in Fig. 6.40 the effective width W of the prism is 40 mm, and the focal lengths of L_1 and L_2 are 300 and 150 mm, respectively. Estimate an upper limit on the width of the entrance slit that will not appreciably reduce the ultimate resolution of the instrument.

55. The dispersion of transparent media can often be approximated with the "Cauchy" formula as outlined in Problem 2.54. For barium flint glass, $n = 1.59825$ at 486.1 nm and $n = 1.60870$ at 298.8 nm. From these data you can determine the constants in the dispersion formula. Calculate the resolving power at 400 nm of a spectrograph that uses a prism of this glass with a base 50-mm long.

56. An oil immersion objective is designed for use with light having a wavelength of 400 nm so as to enable the resolution of objects that are separated by 200 nm. Is this possible? If so, find the numerical aperture of the objective. If not, explain why.

7 Diffraction II

The conditions for far-field diffraction without the aid of lenses are severe. We require R_0 and R_0' to be very much larger than $\tilde{x}^2/\lambda$ or $\tilde{y}^2/\lambda$. We rationalized in Chapter 6 that this formalism could be made practical by using lenses to bring the source and observation planes in closer to the aperture while still maintaining the far-field conditions at the aperture (Fig. 6.34). Through the following developments we will justify this idea by explicitly including the lens in the diffraction process. In doing so, we acquire the capability to discuss a much wider variety of diffraction applications.

Before proceeding with the effect of lenses, we want to improve on our linearized-phase diffraction integral because we will need a more accurate approximation of the true phase. This will allow us to treat *near field diffraction* (without lenses) in which the source and observation planes need not be extremely far from the aperture.

7.1 Fresnel Transformations

The basic method in this discussion will be repeated application of a form of the Fresnel-Kirchhoff diffraction integral. This tells us how the electric field distribution in a given transverse plane is related to that in another plane. We assume the light to have perfect transverse and longitudinal coherence. This can be achieved by using light from a laser that is operating in a single transverse mode.

Rather than the linear approximation of Chapter 6, we will use here the *parabolic approximation* in which spherical waves are replaced by paraboloids. When applied to lens action, this approximation yields the wave equivalent of the paraxial theory in geometrical optics. It must be extended to include the diffraction theory of aberrations, but it can be generalized rather easily to large aperture systems that satisfy the Abbe sine condition Eq. (4.103).

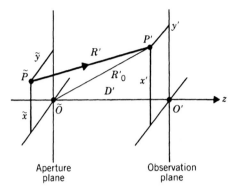

Fig. 7.1 Geometry that applies to propagation from the aperture plane to the observation plane.

We continue to treat the problem by ignoring the polarization properties of light. However, we should point out that these effects cannot be ignored if more accurate calculations are to be done with large aperture systems.

A. General Transformation

We consider first the geometry illustrated in Fig. 7.1. As established in Chapter 6, $\tilde{P}(\tilde{x}, \tilde{y})$ is a typical point in the aperture plane and the observation point is $P'(x', y')$. Although we have again chosen the aperture to be a plane, this choice is one of convenience. There are certain special cases where an equally convenient choice would be a surface that was part of the incident wave surface that, for a point source, would be a section of a sphere. We will stick with the plane because it lends itself to a more general treatment.

The electric field at $\tilde{P}$ on the incident side of the aperture is $\tilde{E}(\tilde{P})$. After the aperture at $\tilde{P}$ this field is modified by the transmission function of the aperture to yield, $\tilde{E}(\tilde{P})\tilde{\tau}(\tilde{P})$. The transmission function describes the extent of the open part of the aperture and also includes the influence of semitransparent portions and variable thickness dielectric contributions (lenses). Figure 7.2 illustrates several types of apertures that can be described with this kind of transmission function.

We consider the modified electric field at the aperture plane to be a source of secondary Huygens wavelets. Then Eq. (6.7), the Fresnel-Kirchhoff diffraction integral (without the inclination factor) will give us the approximate field at P'

$$E'(P') = C \iint \tilde{E}(\tilde{P})\tilde{\tau}(\tilde{P}) \frac{e^{-ikR'}}{R'} \, d\tilde{x} \, d\tilde{y} \tag{7.1}$$

This can be written as an integral transform by defining the "kernel" $h(P_1 \rightarrow P_2)$

$$h(P_1 \rightarrow P_2) \equiv C \frac{e^{-ikR_{12}}}{R_{12}}$$

$$R_{12} = \overline{P_1 P_2} \tag{7.2}$$

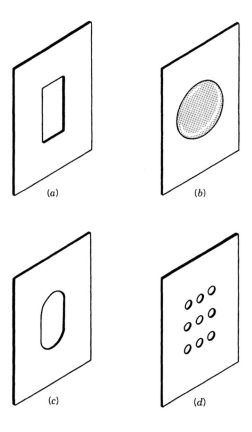

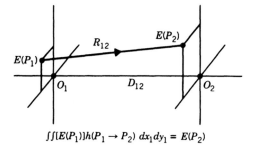

Fig. 7.2 Typical aperture configurations with different transmission functions. (*a*) Rectangular opening. (*b*) Circular lens. (*c*) Oval opening. (*d*) Array of circular openings.

Then Eq. 7.1 takes on the form

$$E'(P') = \iint \tilde{E}(\tilde{P})\tilde{\tau}(\tilde{P})h(\tilde{P} \to P') \, d\tilde{x} \, d\tilde{y} \tag{7.3}$$

Figure 7.3 shows the operation of the integral transform in the general case where the propagation is between two arbitrary planes. The transformation acts on the

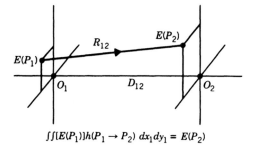

$$\iint[E(P_1)]h(P_1 \to P_2) \, dx_1 dy_1 = E(P_2)$$

Fig. 7.3 Illustration of the kernel that describes propagation from the first plane to the second plane. The integration is performed over the surface of the first plane. This implies that the field at all points on the first plane contributes to the field at a given point on the second plane.

field at the first plane, changing it into the field at a point on the second plane. In the general case we have

$$E(P_2) = \iint E(P_1)h(P_1 \to P_2)\, dx_1\, dy_1 \tag{7.4}$$

Equation (7.1) is also valid if P_1 and P_2 are exchanged. This describes the inverse propagation situation. That is,

$$h(P_2 \to P_1) = -C\frac{e^{+ikR_{12}}}{R_{12}} \tag{7.5}$$

is the appropriate kernel for use in a transformation that would tell us what field was required at the first plane in order that a particular field distribution be found at the second plane.

$$E(P_1) = \iint E(P_2)h(P_2 \to P_1)\, dx_2\, dy_2 \tag{7.6}$$

Returning now to the special case of interest in diffraction, we may apply a transformation on the field in the source plane to generate the field on the incident side of the aperture plane. The situation is shown in Fig. 7.4. Our source distribution at $P(x, y)$ can be described by a source field $E(P)$ and a source transmission function $\tau(P)$. Then, according to Eq. (7.4)

$$\tilde{E}(\tilde{P}) = \iint E(P)\tau(P)h(P \to \tilde{P})\, dx\, dy \tag{7.7}$$

We can substitute Eq. (7.7) into Eq. (7.3) to produce an expression for the field at P' in terms of the field at P.

$$E'(P') = \iint \left[\iint E(P)\tau(P)h(P \to \tilde{P})\, dx\, dy \right] \tilde{\tau}(\tilde{P})h(\tilde{P} \to P')\, d\tilde{x}\, d\tilde{y}$$

$$= \iint E(P)\tau(P) \left[\iint h(P \to \tilde{P})\tilde{\tau}(\tilde{P})h(\tilde{P} \to P')\, d\tilde{x}\, d\tilde{y} \right] dx\, dy \tag{7.8}$$

Equation (7.8) has the same form as the general integral transformation

$$E'(P') = \iint E(P)\tau(P)h_\tau(P \to P')\, dx\, dy \tag{7.9}$$

This will be true if we define the overall propagation kernel

$$h_\tau(P \to P') \equiv \iint h(P \to \tilde{P})\tilde{\tau}(\tilde{P})h(\tilde{P} \to P')\, d\tilde{x}\, d\tilde{y}$$

$$= C^2 \iint \tilde{\tau}(\tilde{P})\frac{e^{-ik(R+R')}}{RR'}\, d\tilde{x}\, d\tilde{y} \tag{7.10}$$

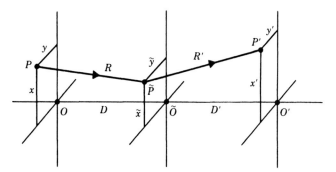

Fig. 7.4 Diffraction geometry showing the source plane, the aperture plane, and the observation plane.

where

$$R = \overline{P\tilde{P}} \quad \text{and} \quad R' = \overline{\tilde{P}P'}$$

Equation (7.9) includes the effect of the transmission function at the aperture $\tilde{\tau}(\tilde{P})$.

This formalism can be used to describe the transformation of a *general* field at the x-y plane by the Fresnel–Kirchhoff theory to a point in the observation plane. If the source is a monochromatic *point* of strength A at $P_0(x_0, y_0)$, then the field at the aperture plane must be

$$\tilde{E}(\tilde{P}) = \frac{A}{R} e^{i(\omega t - kR)}, \ R = \overline{P_0 \tilde{P}} \tag{7.11}$$

Because Eq. (7.7) should also give the same result, we need to identify a function for $E(P)\tau(P)$ that behaves like a point source. If we define

$$E(P)\tau(P) = \frac{A}{C} e^{i\omega t} \delta(x - x_0)\delta(y - y_0) \tag{7.12}$$

and substitute this into Eq. (7.7), then we arrive at the correct expression at $\tilde{P}$. Therefore we will use Eq. (7.12) when we need to represent a problem involving a point source.

When we combine Eq. (7.12) with the overall propagation kernel of Eq. (7.10) in the integral transformation of Eq. (7.9), we find

$$E'(P') = \iint \frac{A}{C} e^{i\omega t} \delta(x - x_0)\delta(y - y_0) C^2 \iint \tilde{\tau}(\tilde{P}) \frac{e^{-ik(R + R')}}{RR'} \, d\tilde{x} \, d\tilde{y} \, dx \, dy$$

$$= CA \, e^{i\omega t} \iint \tilde{\tau}(\tilde{P}) \frac{e^{-ik(R + R')}}{RR'} \, d\tilde{x} \, d\tilde{y} \tag{7.13}$$

which is the same result with which we started in Chapter 6, Eq. (6.15), the Fresnel–Kirchhoff formula for diffraction of the light from a point source by an aperture.

B. Phase Approximations

The propagation and diffraction information is carried by the kernel $h_\tau(P \to P')$ defined in Eq. (7.10). The factor

$$\frac{e^{-ik(R+R')}}{RR'} \tag{7.14}$$

in the integrand in Eq. (7.10) is the result of the application of Huygens' principle. This phase and amplitude behavior results from the product of two spherical waves, one from the source plane and the other from the aperture plane. To make analytical headway with this we need to approximate this factor by a simpler form.

1. Linear Approximation. In Chapter 6 we required $\tilde{P}$ to be very close to $\tilde{O}$ in comparison with the distances D and D'. We did not require, however, that the source point P and the observation point P' be close to the respective origins O and O'. This was an advantage because we are frequently far from the "straight through" geometry when using a grating. Referring to Fig. 6.12a, we identified R_0 and R_0', obtaining, for example

$$R \simeq R_0 - \frac{x\tilde{x} + y\tilde{y}}{R_0} + \frac{\tilde{x}^2 + \tilde{y}^2}{2R_0} - \frac{(x\tilde{x} + y\tilde{y})^2}{2R_0^3} \tag{7.15}$$

Because $\tilde{x}$ and $\tilde{y}$ remain small compared with R_0, we dropped the last two terms. This left us with a phase that was linear in the aperture plane coordinates. We showed how this linear phase was a characteristic of plane waves. Thus, the linearization amounts to treating the spherical waves in expression (7.14) like plane waves.

2. Parabolic Approximation. To improve the approximation, we should keep higher order terms in Eq. (7.15). This leads to an unwieldy result because of the $\tilde{x}\tilde{y}$ cross term obtained by multiplying out

$$\frac{(x\tilde{x} + y\tilde{y})^2}{2R_0^3}$$

The cross term can be eliminated only by assuming that x and y are small. When we do this, we should also expand R_0 about D. This is equivalent to treating $x, y, \tilde{x}$, and $\tilde{y}$ on an equal footing and expanding R about D. Thus,

$$R = \sqrt{D^2 + (\tilde{x} - x)^2 + (\tilde{y} - y)^2}$$
$$\simeq D + \frac{(\tilde{x} - x)^2}{2D} + \frac{(\tilde{y} - y)^2}{2D} - \frac{1}{8}\frac{[(\tilde{x} - x)^2 + (\tilde{y} - y)^2]^2}{D^3} \tag{7.16}$$

We intend to drop the last term in Eq. (7.16). It must therefore be much less than λ, and almost always is in the cases of interest. We then have

$$R \simeq D + \frac{(\tilde{x} - x)^2}{2D} + \frac{(\tilde{y} - y)^2}{2D} \tag{7.17}$$

In the amplitude factor we replace R by D because the denominator is slowly varying.

Inserting these same "parabolic" approximations into Eq. (7.2) for the general case we get

$$h(P_1 \rightarrow P_2) = \frac{C}{D_{12}} \exp(-ikD_{12}) \exp\left\{ \frac{-ik}{2D_{12}} [(x_2 - x_1)^2 + (y_2 - y_1)^2] \right\} \tag{7.18}$$

When used in the integral of Eq. (7.4) this leads to what we call the *Fresnel transformation*.

$$E(P_2) = \frac{C}{D_{12}} e^{-ikD_{12}} \iint E(P_1) \exp\left\{ \frac{-ik}{2D_{12}} [(x_2 - x_1)^2 + (y_2 - y_1)^2] \right\} dx_1 \, dx_2 \tag{7.19}$$

It is often more convenient to multiply out and combine the factors in the exponential of Eq. (7.18),

$$-ik\left[D_{12} + \frac{x_1^2 + y_1^2}{2D_{12}} + \frac{x_2^2 + y_2^2}{2D_{12}} - \frac{(x_1 x_2 + y_1 y_2)}{D_{12}} \right]$$

Referring to Fig. 7.5, we can see that

$$R_{12,0} = (D_{12} + x_2^2 + y_2^2)^{1/2}$$

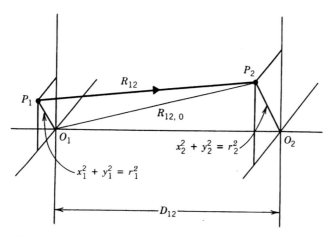

Fig. 7.5 General geometry for the Fresnel transformation between plane 1 and plane 2.

To the same level of approximation that we are already using, this is

$$R_{12,0} \simeq D_{12} + \frac{x_2^2 + y_2^2}{2D_{12}}$$

Therefore the argument of the exponentials becomes

$$-ik\left[R_{12,0} + \frac{r_1^2}{2D_{12}} - \frac{x_1 x_2 + y_1 y_2}{D_{12}}\right]$$

where $r_1^2 = x_1^2 + y_1^2$.

The *alternative form* for Eq. (7.18) is

$$h(P_1 \to P_2) = \frac{C}{D_{12}} \exp(-ikR_{12,0}) \exp\left(\frac{-i\pi}{\lambda D_{12}} r_1^2\right)$$

$$\times \exp\left[-i2\pi\left(\frac{-x_1 x_2 - y_1 y_2}{\lambda D_{12}}\right)\right] \qquad (7.20a)$$

or

$$h(P_1 \to P_2) = \frac{C}{D_{12}} \exp(-ikD_{12}) \exp\left(\frac{-ikr_1^2}{2D_{12}}\right) \exp\left(\frac{-ikr_2^2}{2D_{12}}\right)$$

$$\times \exp\left[-i2\pi\left(\frac{-x_1 x_2 - y_1 y_2}{\lambda D_{12}}\right)\right] \qquad (7.20b)$$

It is easy to see from this what the parabolic approximation has added to the phase factor. Equation (7.20) is essentially the same as the propagation kernel that would yield the linear form in the far-field situation, provided $r_1^2 \ll \lambda D_{12}$. In the present situation, where this condition is not met, we must include the correction factor of

$$\exp\left(\frac{-i\pi}{\lambda D_{12}} r_1^2\right)$$

For the particular case of Fig. (7.4) we must make the same approximation for both factors in expression (7.14). Thus we need

$$R' \simeq D' + \frac{(x' - \tilde{x})^2}{2D'} + \frac{(y' - \tilde{y})^2}{2D'}$$

in the phase factor and $R' \simeq D'$ in the amplitude factor.

In the parabolic approximation, then, we can rewrite Eq. (7.10) as

$$h_r(P \to P') \simeq \frac{C^2}{DD'} e^{-ik(D+D')} \iint \tilde{\tau}(\tilde{P})$$

$$\times \exp\left\{-ik\left[\frac{[(\tilde{x} - x)^2 + (\tilde{y} - y)^2]}{2D} + \frac{[(x' - \tilde{x})^2 + (y' - \tilde{y})^2]}{2D'}\right]\right\} d\tilde{x}\, d\tilde{y}$$

$$(7.21)$$

for the appropriate kernel in the overall Fresnel transformation.

3. Standard Form. The integral for the overall Fresnel transformation kernel in Eq. (7.21) is difficult to evaluate in its present form. We want to manipulate this equation, according to common practice, into one of two alternative forms, depending on the nature of the transmission function $\tilde{\tau}(\tilde{P})$. If the transmission function includes the action of a lens, then it will be most useful to follow a procedure similar to that which resulted in Eq. (7.20). We will do that later in this chapter. If the transmission function describes a simple opening by establishing the limits for the integration only, then a different form will be more convenient.

With that end in mind, we collect length factors in the arguments of the exponentials in Eq. (7.21). These are an approximation for $R + R'$.

$$R + R' \simeq D + D' + \frac{1}{2}\left[\frac{(\tilde{x} - x^2)}{D} + \frac{(\tilde{x} - x')^2}{D'}\right] + \frac{1}{2}\left[\frac{(\tilde{y} - y)^2}{D} + \frac{(\tilde{y} - y')^2}{D'}\right] \quad (7.22)$$

It is convenient to use the line connection P with P' as the reference (see Fig. 7.6). This line meets the aperture plane at $\tilde{P}_u$ (u for undeviated). We may write the equation of this line in terms of its "x" and z coordinates in two different ways, the first uses P and $\tilde{P}_u$

$$\tilde{x}_u = \left(\frac{D}{D' + D}\right)(x' - x) + x \quad (7.23a)$$

the second uses $\tilde{P}_u$ and P'

$$\tilde{x}_u = \left(\frac{D'}{D + D'}\right)(x - x') + x' \quad (7.23b)$$

We proceed by forming $(\tilde{x} - \tilde{x}_u)$ from Eq. (7.23a) and squaring it

$$(\tilde{x} - \tilde{x}_u)^2 = (\tilde{x} - x)^2 + \frac{(x - x')^2}{(D' + D)^2} D^2 + 2\frac{(\tilde{x} - x)(x - x')D}{(D' + D)} \quad (7.24)$$

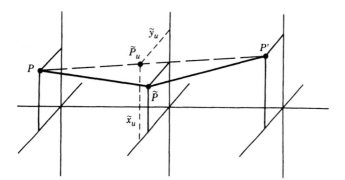

Fig. 7.6 Points in the aperture plane are most conveniently referenced with respect to the intercept of the line connecting the source point with the observation point in the case where the aperture contains an opening (as opposed to a lens, for instance).

We also may form $(\tilde{x} - \tilde{x}_u)$ from Eq. (7.23b) and calculate the square of that:

$$(\tilde{x} - \tilde{x}_u)^2 = (\tilde{x} - x')^2 + \frac{(x - x')^2}{(D' + D)^2} D'^2 - 2\frac{(\tilde{x} - x')(x - x')D'}{(D' + D)} \quad (7.25)$$

Now divide Eq. (7.24) by D and Eq. (7.25) by D' and add the two equations

$$\frac{(\tilde{x} - \tilde{x}_u)^2}{D} + \frac{(\tilde{x} - \tilde{x}_u)^2}{D'} = \frac{(\tilde{x} - x)^2}{D} + \frac{(\tilde{x} - x')^2}{D'} + \frac{(x - x')^2}{D' + D} - 2\frac{(x - x')^2}{D + D'}$$

This can be rewritten as

$$\frac{(\tilde{x} - x)^2}{D} + \frac{(\tilde{x} - x')^2}{D'} = \left[\frac{1}{D} + \frac{1}{D'}\right](\tilde{x} - \tilde{x}_u)^2 + \frac{(x - x')^2}{D + D'} \quad (7.26)$$

The left side of Eq. (7.26) is the same as the first term in brackets in Eq. (7.22).

Now we repeat this procedure for the equations of the line from P to P' in terms of its "y" and z coordinates.

With the definition

$$\frac{1}{d} \equiv \left[\frac{1}{D} + \frac{1}{D'}\right] \quad (7.27)$$

we can write $R + R'$ as

$$R + R' \cong D + D' + \frac{1}{2}\left[\frac{(x - x')^2}{D + D'} + \frac{(y - y')^2}{D + D'}\right] + \frac{1}{2d}[(\tilde{x} - \tilde{x}_u)^2 + (\tilde{y} - \tilde{y}_u)^2] \quad (7.28)$$

The last term in Eq. (7.28) is expressed in terms of coordinates in the aperture plane referred to the point P_u. The remaining terms in Eq. (7.28) can be seen to be, at the same level of approximation, equal to the length of line $\overline{PP'}$:

$$(\overline{PP'}) = [(D + D')^2 + (x' - x)^2 + (y' - y)^2]^{1/2}$$

$$= (D + D')\left[1 + \frac{(x - x')^2}{(D + D')^2} + \frac{(y - y')^2}{(D + D')^2}\right]^{1/2}$$

$$\cong D + D' + \frac{1}{2}\left[\frac{(x - x')^2}{(D + D')^2} + \frac{(y - y')^2}{(D + D')^2}\right] \quad (7.29)$$

Thus

$$R + R \simeq (\overline{PP'}) + \frac{1}{2d}[(\tilde{x} - \tilde{x}_u)^2 + (\tilde{y} - \tilde{y}_u)^2] \quad (7.30)$$

In the denominator of Eq. (7.21) we make one more approximation

$$\frac{1}{DD'} = \frac{(D + D')}{DD'} \frac{1}{(D + D')} \simeq \frac{1}{d} \frac{1}{(\overline{PP'})} \quad (7.31)$$

With these changes Eq. (7.21) assumes a simpler form

$$h_\tau(P \rightarrow P') \cong \frac{e^{-ik(\overline{PP'})}}{(\overline{PP'})} \frac{C^2}{d} \iint \tilde{\tau}(\tilde{x}, \tilde{y})$$

$$\times \exp\left\{ \frac{-ik}{2d} \left[(\tilde{x} - \tilde{x}_u)^2 + (\tilde{y} - \tilde{y}_u)^2 \right] \right\} d\tilde{x} \, d\tilde{y} \qquad (7.32)$$

This can be used in Eq. (7.9) instead of Eq. (7.10), which it approximates. This gives us for a point source at $P_0(x_0, y_0)$ by use of Eq. (7.12)

$$E'(P') = E'_{na}(P') \frac{C}{d} \iint \tilde{\tau}(\tilde{x}, \tilde{y}) \, e^{-(ik/2d)[(\tilde{x} - \tilde{x}_u)^2 + (\tilde{y} - \tilde{y}_u)^2]} \, d\tilde{x} \, d\tilde{y} \qquad (7.33)$$

Where

$$E'_{na}(P') = \frac{A \, e^{i(\omega t - k(\overline{P_0 P'}))}}{(\overline{P_0 P'})} \qquad (7.34)$$

is the field at P' due to the *point source* in the absence of the aperture. We will call the integral in Eq. (7.33) the *Fresnel integral*.

It is convenient and conventional to define dimensionless coordinates η_x and η_y where

$$\eta_x \equiv \sqrt{\frac{2}{\lambda d}} (\tilde{x} - \tilde{x}_u) \qquad (7.35a)$$

$$\eta_y \equiv \sqrt{\frac{2}{\lambda d}} (\tilde{y} - \tilde{y}_u) \qquad (7.35b)$$

Then Eq. (7.33) becomes

$$E'(P') = \frac{C\lambda}{2} E'_{na}(P') \iint \tilde{\tau}(\eta_x, \eta_y) \, e^{-(i\pi/2)(\eta_x^2 + \eta_y^2)} \, d\eta_x \, d\eta_y \qquad (7.36)$$

The aperture function is now expressed in terms of η_x and η_y. It serves to limit the range of the integration as determined by Eqs (7.35).

7.2   Fresnel Diffraction

We will now discuss several special cases in which the Fresnel integral can be easily evaluated. The general situation is much more complex. However, the examples we study form an important class of problems. The examples differ only in the form that the transmission function takes.

In each case the source is a point at $P = P_0$ in the x, y plane. The aperture, in the $\tilde{x}, \tilde{y}$ plane is at least partially open and, for now, does not contain lenses or transmission filters. We require the electric field E' and the flux density S' at point

P' in the x', y' or observation plane. Our starting point will usually be Eq. (7.36), the standard form of the Fresnel integral.

In the limit that the aperture becomes very large or is removed entirely, $\tilde{\tau} = 1$. We can then immediately evaluate the integral in Eq. (7.36)

$$\iint\limits_{-\infty}^{-\infty} e^{-(i\pi/2)(\eta_x^2 + \eta_y^2)} \, d\eta_x \, d\eta_y = \int_{-\infty}^{-\infty} e^{-(i\pi/2)\eta_x^2} \, d\eta_x \int_{-\infty}^{\infty} e^{-(i\pi/2)\eta_y^2} \, d\eta_y$$

$$= \sqrt{\pi} \sqrt{\frac{2}{i\pi}} \sqrt{\pi} \sqrt{\frac{2}{i\pi}}$$

$$= \frac{2}{i} \tag{7.37}$$

This tells us what we must choose for the constant C that we have carried through from the beginning of Chapter 6.

$$C = \frac{i}{\lambda} \tag{7.38}$$

From now on we will use this substitution instead of C. The Fresnel integral can then be expressed in terms of the "no aperture" field.

$$E'(P') = \frac{i}{2} E'_{na}(P') \iint \tilde{\tau}(\eta_x, \eta_y) \, e^{-(i\pi/2)(\eta_x^2 + \eta_y^2)} \, d\eta_x \, d\eta_y \tag{7.39}$$

A. Rectangular Aperture

Because the transmission function in this case can be factored

$$\tilde{\tau}(\eta_x, \eta_y) = \tilde{\tau}_x(\eta_x)\tilde{\tau}_y(\eta_y)$$

we can discuss the "x" and "y" dependence separately. The Fresnel integral becomes a product of one-dimensional integrals.

$$E'(P') = \frac{i}{2} E'_{na}(P') I_x I_y \tag{7.40}$$

where

$$I_x \equiv \int_{-\infty}^{\infty} \tilde{\tau}_x(\eta_x) \, e^{-(i\pi/2)\eta_x^2} \, d\eta_x \tag{7.41a}$$

and

$$I_y \equiv \int_{-\infty}^{\infty} \tilde{\tau}_y(\eta_y) \, e^{-(i\pi/2)\eta_y^2} \, d\eta_y \tag{7.41b}$$

If $S'_{na}(P')$ represents the flux density with no aperture, then the flux density at P' in the presence of the aperture will be

$$S'(P') = S'_{na}(P')\frac{|I_x|^2|I_y|^2}{4} \qquad (7.42)$$

1. Normalized Limits. Figure 7.7 illustrates two typical rectangular apertures that conform to equations

$$\tilde{\tau}_x = \begin{cases} 1, & \tilde{x}_1 \leq \tilde{x} \leq \tilde{x}_2 \\ 0, & \text{otherwise} \end{cases}$$

$$\tilde{\tau}_y = \begin{cases} 1, & \tilde{y}_1 \leq \tilde{y} \leq \tilde{y}_2 \\ 0, & \text{otherwise} \end{cases}$$

In Figure 7.7b the edges identified at $\tilde{x}_2$, $\tilde{y}_1$, and $\tilde{y}_2$ have been taken to be, ∞, $-\infty$, and ∞, respectively. This represents the diffraction situation for an aperture that is a partial plane, open for $\tilde{x} \geq \tilde{x}_1$.

Often the most convenient method of describing Fresnel diffraction refers

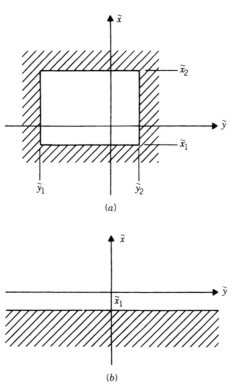

(a)

(b)

Fig. 7.7 (a) Rectangular aperture coordinates. (b) Half-plane obstruction.

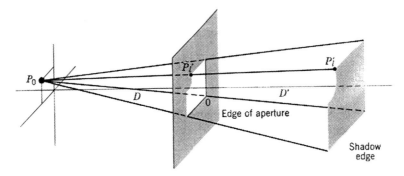

Fig. 7.8 Projected coordinates have a one-to-one relationship with points in the aperture plane.

everything to the plane of observation, rather than to the plane of the aperture. So, instead of $\tilde{P}$, we move to $P'(x', y')$ where by analogy with Eq. (7.23),

$$x' = \tilde{x}\,\frac{D' + D}{D} - x_0\,\frac{D'}{D} \tag{7.43a}$$

and

$$y' = \tilde{y}\,\frac{D' + D}{D} - y_0\,\frac{D'}{D} \tag{7.43b}$$

If P', which is specified by Eqs. (7.43), is the observation point, then $(\tilde{x}, \tilde{y})$ in those equations is the same as $(\tilde{x}_u, \tilde{y}_u)$. Instead of the limiting point $\tilde{P}_l$ in the plane of the aperture, we now consider its projection P'_l having coordinates (x'_l, y'_l) in the observation plane. This is shown in Fig. 7.8. The coordinates of P'_l are found by substituting the coordinates of $\tilde{P}_l$ into Eqs. (7.43). Here P'_l is at the edge of the geometrical shadow of the aperture produced by the light from the point source at P_0. We may then write

$$(x'_l - x') = \frac{D' + D}{D}\,(\tilde{x}_l - \tilde{x}_u)$$

and

$$(y'_l - y') = \frac{D' + D}{D}\,(\tilde{x}_l - \tilde{x}_u)$$

where x', y' refer to the observation point P'.

In terms of the dimensionless coordinates η_x and η_y, these become

$$\eta_{x_l} = \frac{1}{F}\,(x'_l - x') \tag{7.44a}$$

$$\eta_{y_l} = \frac{1}{F}\,(y'_l - y') \tag{7.44b}$$

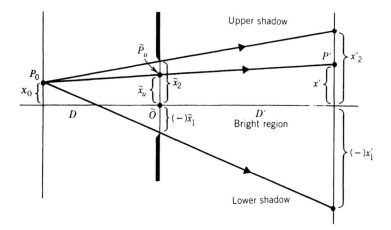

Fig. 7.9

where

$$F \equiv \left[\frac{\lambda D'(D + D')}{2D} \right]^{1/2} \tag{7.44c}$$

is a scale factor having units of length.

The bright regions of the shadow in the observation plane (Fig. 7.9) will be defined by $x'_1 \leq x' \leq x'_2$, $y'_1 \leq y' \leq y'_2$, where, for instance,

$$x'_1 = \tilde{x}_1 \frac{D' + D}{D} - x \frac{D'}{D}$$

with similar expressions for x'_2, y'_1 and y'_2. Therefore, through Eqs. (7.42) and (7.44), we establish limits on the integrals in Eqs. (7.41) in terms of the observation plane coordinates.

$$\eta_{x_1} = \frac{x'_1 - x'}{F} \tag{7.45a}$$

$$\eta_{x_2} = \frac{x'_2 - x'}{F} \tag{7.45b}$$

$$\eta_{y_1} = \frac{y'_1 - y'}{F} \tag{7.45c}$$

$$\eta_{y_2} = \frac{y'_2 - y'}{F} \tag{7.45d}$$

Here $P'(x', y')$ is the observation point under consideration.

2. Fresnel Integrals. The integrals I_x and I_y in Eqs. (7.41) may be more easily evaluated between the limits imposed by Eqs. (7.45) if we write them as follows:

$$I_x = I(\eta_{x_2}) - I(\eta_{x_1}) \tag{7.46a}$$

$$I_y = I(\eta_{y_2}) - I(\eta_{y_1}) \tag{7.46b}$$

with

$$I(\eta) \equiv \int_0^\eta e^{-(i\pi/2)u^2} \, du \tag{7.46c}$$

The integral in Eq. (7.46c) is called the *complex Fresnel integral*. It can be split into real and imaginary parts as follows:

$$I(\eta) = \mathscr{C}(\eta) + i\mathscr{S}(\eta) \tag{7.47a}$$

where

$$\mathscr{C}(\eta) = \int_0^\eta \cos\!\left(\frac{\pi}{2} u^2\right) du \tag{7.47b}$$

and

$$\mathscr{S}(\eta) = -\int_0^\eta \sin\!\left(\frac{\pi}{2} u^2\right) du \tag{7.47c}$$

These are evaluated in Table 7.1 for a selected range of η. With this information the electric field strength and the flux density may be calculated for Fresnel diffraction by a simple aperture or obstruction.

3. Cornu Spiral. The integral in Eqs. (7.46) has a graphical interpretation in terms of a phasor curve in the complex plane. This is the same concept as was presented in section 6.1B2. We approximate Eq. (7.46c) by

$$I(\eta) \simeq \sum_{m=0}^{N-1} \Delta u \, e^{i\delta(m)}$$

where

$$\delta(m) = \frac{-\pi(m\,\Delta u)^2}{2}$$

As shown in Fig. 7.10, this will be the phasor sum of N incremental phasors, each of

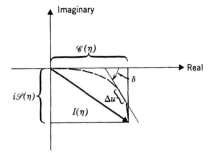

Fig. 7.10 Incremental phasors that represent components of the complex Fresnel integral.

Table 7.1. Values of the Complex Fresnel Integral

η	$\mathscr{C}(\eta)$	$\mathscr{S}(\eta)$	η	$\mathscr{C}(\eta)$	$\mathscr{S}(\eta)$
0.00	0.0000	−0.0000	4.50	0.5261	−0.4342
0.10	0.1000	−0.0005	4.60	0.5673	−0.5162
0.20	0.1999	−0.0042	4.70	0.4914	−0.5672
0.30	0.2994	−0.0141	4.80	0.4338	−0.4968
0.40	0.3975	−0.0334	4.90	0.5002	−0.4350
0.50	0.4923	−0.0647	5.00	0.5637	−0.4992
0.60	0.5811	−0.1105	5.05	0.5450	−0.5442
0.70	0.6597	−0.1721	5.10	0.4998	−0.5624
0.80	0.7230	−0.2493	5.15	0.4553	−0.5427
0.90	0.7648	−0.3398	5.20	0.4389	−0.4969
1.00	0.7799	−0.4383	5.25	0.4610	−0.4536
1.10	0.7638	−0.5365	5.30	0.5078	−0.4405
1.20	0.7154	−0.6234	5.35	0.5490	−0.4662
1.30	0.6386	−0.6863	5.40	0.5573	−0.5140
1.40	0.5431	−0.7135	5.45	0.5269	−0.5519
1.50	0.4453	−0.6975	5.50	0.4784	−0.5537
1.60	0.3655	−0.6389	5.55	0.4456	−0.5181
1.70	0.3238	−0.5492	5.60	0.4517	−0.4700
1.80	0.3336	−0.4508	5.65	0.4926	−0.4441
1.90	0.3944	−0.3734	5.70	0.5385	−0.4595
2.00	0.4882	−0.3434	5.75	0.5551	−0.5049
2.10	0.5815	−0.3743	5.80	0.5298	−0.5461
2.20	0.6363	−0.4557	5.85	0.4819	−0.5513
2.30	0.6266	−0.5531	5.90	0.4486	−0.5163
2.40	0.5550	−0.6197	5.95	0.4566	−0.4688
2.50	0.4574	−0.6192	6.00	0.4995	−0.4470
2.60	0.3890	−0.5500	6.05	0.5424	−0.4689
2.70	0.3925	−0.4529	6.10	0.5495	−0.5165
2.80	0.4675	−0.3915	6.15	0.5146	−0.5496
2.90	0.5624	−0.4101	6.20	0.4676	−0.5398
3.00	0.6058	−0.4963	6.25	0.4493	−0.4954
3.10	0.5616	−0.5818	6.30	0.4760	−0.4555
3.20	0.4664	−0.5933	6.35	0.5240	−0.4560
3.30	0.4058	−0.5192	6.40	0.5496	−0.4965
3.40	0.4385	−0.4296	6.45	0.5292	−0.5398
3.50	0.5326	−0.4152	6.50	0.4816	−0.5454
3.60	0.5880	−0.4923	6.55	0.4520	−0.5078
3.70	0.5420	−0.5750	6.60	0.4690	−0.4631
3.80	0.4481	−0.5656	6.65	0.5161	−0.4549
3.90	0.4223	−0.4752	6.70	0.5467	−0.4915
4.00	0.4984	−0.4204	6.75	0.5302	−0.5362
4.10	0.5738	−0.4758	6.80	0.4831	−0.5436
4.20	0.5418	−0.5633	6.85	0.4539	−0.5060
4.30	0.4494	−0.5540	6.90	0.4732	−0.4624
4.40	0.4383	−0.4622	6.95	0.5207	−0.4591

amplitude Δu and phase angle $\delta(m)$. Here we arrange the sum so that $\Delta u = \eta/N$. Then the integral represents the superposition of fields at the observation point due to contributions from the portion of the aperture between the origin and the appropriate limiting η. In the limit that N goes to infinity, the phasor sum becomes a continuous curve. The total length of the curve is given by

$$\eta = \int_0^\eta du$$

The quadratic dependence of δ on the arc length variable u, which is characteristic of Fresnel diffraction, comes originally from the quadratic dependence of the optical path length difference $(R + R) - (\overline{PP'})$ on

$$(\tilde{x} - \tilde{x}_u)^2 + (\tilde{y} - \tilde{y}_u)^2$$

This is apparent from Eq. (7.30). In the far-field case, the phase angle depends linearly on the arc length and the phasor curve was a circle. Here the curve has greater curvature than a circle, leading to a "coiled up" behavior. The result is the *Cornu spiral* of Fig. 7.11.

For a given value of η, $I(\eta)$ is obtained from the spiral as shown in Fig. 7.12. The total arc length of the spiral starting from the origin is labeled by the numbers

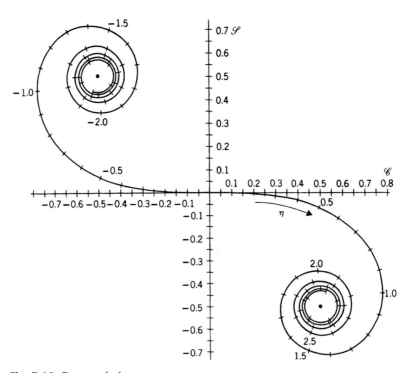

Fig. 7.11 Cornu spiral.

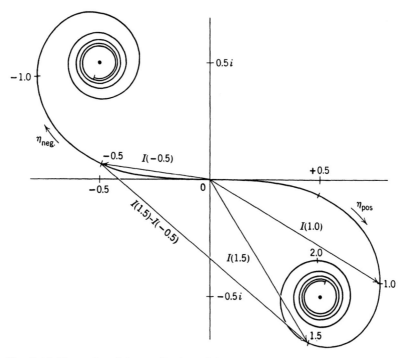

Fig. 7.12 Examples of the application of the Cornu spiral to determine the solution to the complex Fresnel integral.

on the curve. The sign convention is such that η is positive in the lower right quadrant and negative in the upper left quadrant. Then $I(\eta)$ is the complex number represented by a vector with its tail at the origin and its head at the place on the spiral marked "η," whose total arc length from the origin amounts to η units.

Then $I(\eta_2) - I(\eta_1)$ is a complex vector whose head lies at η_2 with its tail at η_1 on the spiral. If $\eta_2 > 0$ and $\eta_1 < 0$, the head of the vector is in the fourth quadrant and the tail is in the second quadrant. If $\eta_2 > 0$ and $\eta_1 > 0$, both head and tail lie in the fourth quadrant. If $\eta_2 < 0$ and $\eta_1 < 0$, both lie in the second quadrant.

The limiting values of the integrals as $\eta \to \pm\infty$ can be shown to be

$$\mathscr{C}(+\infty) = -\mathscr{C}(-\infty) = \frac{1}{2}$$

and

$$\mathscr{S}(+\infty) = -\mathscr{S}(-\infty) = \frac{1}{2} \tag{7.48}$$

Thus the "eye" of the spiral in the fourth quadrant is at

$$I(-\infty) = \mathscr{C}(+\infty) + i\mathscr{S}(+\infty) = \frac{1}{2}(1 - i) \tag{7.49a}$$

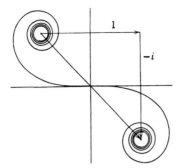

Fig. 7.13 Representation of the unobstructed wave.

and the eye in the second quadrant is at

$$I(-\infty) = \mathscr{C}(-\infty) + i\mathscr{S}(-\infty) = -\frac{1}{2}(1-i) \qquad (7.49\text{b})$$

We now turn to some limiting cases of the rectangular aperture.

4. No Aperture. The no aperture can be handled by setting $\tilde{\tau}_x$ and $\tilde{\tau}_y$ both equal to unity as was done before. Then the limits on both the x and y integral extend from $-\infty$ to ∞. From Eqs. (7.46) and (7.49) we obtain

$$I_x = I_y = I(+\infty) - I(-\infty) = (1-i)$$

The phasor representing I_x or I_y goes from eye to eye of the Cornu spiral as shown in Fig. 7.13. Equation (7.40) then gives the field at P'

$$E'(P') = \frac{i}{2} E'_{na}(P')(1-i)^2 = E'_{na}(P')$$

A result that is hardly unexpected!

The values of Eqs. (7.49) imply the following important result that was already obtained as Eq. (7.37):

$$\frac{i}{\lambda d} \int\!\!\!\int_{-\infty}^{\infty} \exp\left\{\frac{-i\pi}{\lambda d}\left[(\tilde{x} - \tilde{x}_u)^2 + (\tilde{y} - \tilde{y}_u)^2\right]\right\} d\tilde{x}\, d\tilde{y} = 1 \qquad (7.50)$$

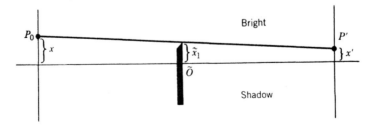

Fig. 7.14 Straight-edge barrier.

5. Straight Edge. The straight-edge barrier runs from $\tilde{x} = -\infty$ to $\tilde{x} = \tilde{x}_1$. From $\tilde{x} = \tilde{x}_1$ to $\tilde{x} = \infty$ there is no barrier. This is shown in Fig. 7.14. The geometrical bright region in the observation plane extends from $x' = x'_1$ to $x' = \infty$. Because we will have $-y'_1 = y'_2 = +\infty$, the integral I_y is still $(1 - i)$. For I_x the upper limit remains $+\infty$, but the lower limit is

$$\eta_{x_1} = \frac{x'_1 - x'}{F}$$

The head of the phasor representing I_x remains in the lower eye of the Cornu spiral. The position of the tail of the phasor depends on the lower limit η_{x_1}. If x is more positive than x'_1 by an amount that is many multiples of the scale factor

$$F = \left[\frac{\lambda D'(D + D')}{2D} \right]^{1/2}$$

then η_{x_1} is sufficiently negative to be practically $-\infty$, and $I_x \simeq (1 - i)$, giving

$$E'(P') \simeq E'_{na}(P')$$

This is represented by point A in Fig. 7.15.

As x' decreases toward x'_1, the limit of the geometrical shadow, η_{x_1} remains

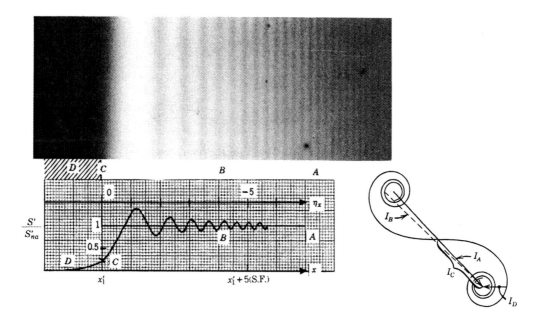

Fig. 7.15 Details of the diffraction pattern produced by a straight-edge barrier in the near-field limit.

negative but also decreases. The tail of the phasor for I_x comes out of the eye in the upper part of the Cornu spiral and traverses a larger and larger arc. In so doing, the magnitude of I_x increases and decreases above and below the value $|(1 - i)| = \sqrt{2}$. This causes fringes in the region that, by geometrical optics, should be uniformly bright. This becomes prominent at values of η_{x_1} near -5. A typical point corresponds to the behavior near B in Fig. 7.15.

When $\eta_{x_1} = 0$, then $I_x = (1/2)(1 - i)$, corresponding to the point C. This value is just half that obtained with no aperture leading to the flux density at C of

$$S'(x_1') = \frac{1}{4} S_{na}'$$

Point C corresponds exactly to the edge of the geometrical shadow.

For $x' < x_1'$, η_{x_1} is positive and the tail of the phasor for I_x is on the lower half of the spiral. As x' continues to decrease, η_{x_1} grows, and the tail spirals in toward the eye. This is region D. The magnitude of I_x decreases monotonically and for $\eta_{x_1} > +5$, say, I_x is quite small. This is well into the region of the geometrical shadow.

6. Wide Slits. Let the edges of the slits be at $\tilde{x} = \pm \tilde{x}_0$. Then the shadow edges in the observation plane are at $x' = x_1'$ and $x' = x_2'$, where

$$x_1' = -\frac{D + D'}{D} \tilde{x}_0 - \frac{D'}{D} x$$

$$x_2' = \frac{D + D'}{D} \tilde{x}_0 - \frac{D'}{D} x$$

and the shadow region has width

$$\Delta x' = x_2' - x_1' = \frac{D + D'}{D} 2\tilde{x}_0 \tag{7.51}$$

The integral for I_y has infinite limits and gives $I_y = (1 - i)$.

A "wide" slit is one that has a total arc length on the Cornu spiral

$$\Delta \eta = \eta_{x_2} - \eta_{x_1}$$

that is very large compared with unity. This condition can be rewritten using Eq. (7.45) or (7.35).

$$1 \ll (\eta_{x_2} - \eta_{x_1}) = \frac{x_2' - x_1'}{F} = 2\tilde{x}_0 \sqrt{\frac{2}{\lambda d}} \tag{7.52}$$

When x' is in the center area of the geometrical bright region, we have

$$x_2' - x' \gg F \quad \text{or} \quad \eta_{x_2} \gg 1, \quad \text{that is,} \quad \eta_{x_2} \cong +\infty$$

and

$$x' - x_1' \gg F \quad \text{or} \quad \eta_{x_1} \ll -1, \quad \text{that is,} \quad \eta_{x_1} \cong -\infty$$

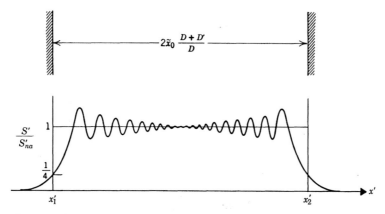

Fig. 7.16 Flux density in the Fresnel diffraction pattern for a slit. The region in the top of the figure shows the extent of the geometrical shadow.

Then I_x is essentially equal to its value for no aperture $(1 - i)$ with its phasor running practically from eye to eye on the Cornu spiral.

As x moves toward the left edge of the shadow at x_1, still in the bright region, η_{x_2} becomes even larger, and the head of the phasor stays in the lower right eye. Then n_{x_1} becomes less negative and the tail of the phasor begins to unwind from the upper left eye, just as though there were no right edge to the slit. Similarly, as x' approaches the right edge of the shadow at x'_2, the tail of the phasor for I_x remains very close to the upper left eye, and its head begins to spiral out of the lower right branch, just as we would expect if there were no left edge of the slit. The resulting diffraction pattern is essentially a superposition of two straight-edge patterns, as indicated in Fig. 7.16.

7. Narrow Slit. If the slit width $2\tilde{x}_0$ is of order $\sqrt{\lambda d/2}$ so that $\Delta \eta$ is of order unity or perhaps a bit larger, then one end of the phasor for I_x will come out of one eye of the Cornu spiral before the other end has a chance to wind up into the other eye. The form of the diffraction pattern is quite sensitive to the size of $\Delta \eta$, as shown in Fig. 7.17.

In the limiting case where $\Delta \eta \ll 1$, we go over to Fraunhofer diffraction. The arc length is so small that it must be wound up into an eye of the Cornu spiral before the chord representing I_x goes through a minimum. The extent of the diffraction pattern is much greater than that of the shadow. We can show that in the limit as $\Delta \eta \to 0$ the Fresnel integrals do give the Fraunhofer or far-field diffraction expression.

8. The "Limit" of Geometrical Optics. Consider a slit that is very wide. The requirement is [Eq. (7.52)]

$$1 \ll \Delta \eta = \eta_{x_2} - \eta_{x_1} = 2\tilde{x}_0 \left(\frac{2}{\lambda d} \right)^{1/2}$$

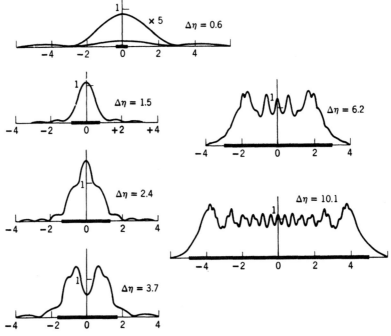

Fig. 7.17 Flux density in the Fresnel diffraction pattern for a slit as a function of the width of the slit. The bar at the bottom indicates the geometrical shadow.

No matter how large $\Delta\eta$ is, there are always departures from the predictions of geometrical optics near the edge of the geometrical shadow. This occurs for $|\eta_{x_1}|$ or $|\eta_{x_2}|$ of the order of 10, for then the oscillations in region B and the nonzero intensity in region D of Fig. 7.15 occur. These diffraction effects take place when x' is within a distance $\Delta x'$ of the shadow edge given by

$$\Delta x' \cong 10F = 10\left[\frac{\lambda D'(D + D')}{2D}\right]^{1/2} \tag{7.53}$$

Hence, for fixed D and D', $\Delta x'$ can be made arbitrarily small by decreasing the wavelength sufficiently, although the size of the oscillations in E' will not diminish. In this sense geometrical optics can be obtained from physical, or wave optics, in the limit as $\lambda \to 0$.

9. Rectangular Patterns. An expression for the flux density from Eq. (7.42)

$$S'(P') = S'_{na}(P')\frac{|I_x|^2|I_y|^2}{4}$$

When the limits on the integration are not infinite, the integral I_y will behave just like I_x discussed previously, and the two-dimensional diffraction can be obtained from the product $|I_xI_y|^2$.

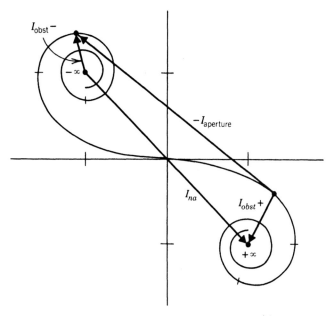

Fig. 7.18 Demonstration of Babinet's principle in Fresnel diffraction.

10. Obstructions. The case of a rectangular obstacle and its limiting configuration of a needle can be treated by an obvious generalization of the preceding methods. It is sometimes convenient to use Babinet's principle, Eq. (6.11), which can be written in the form

$$E'_{na}(P') = E'_{\text{aperture}}(P') + E'_{\text{obstruction}}(P') \tag{7.54}$$

With rectangular symmetry and the resulting factorization of the integral, we obtain a corresponding equation for the integrals I_x and I_y, for example,

$$I_{x,\text{aperture}} + I_{x,\text{obst}} = I_{x,na} = (1 - i) \tag{7.55}$$

Here we would want to use the principle by setting $I_{x,\text{obst}} = (1 - i) - I_{x,\text{aperture}}$, where $I_{x,\text{aperture}}$ is the value of the Fresnel integral for an aperture with the same dimensions as the obstruction. Fig. 7.18 demonstrates this principle. This would represent a position within, but not at the center of, the geometrical shadow of an obstruction. Note that the phasor sum of *two* components, representing I_{obst}^{-} and I_{obst}^{+}, is required to represent the open areas on the aperture plane that extends to $\tilde{x} = \pm\infty$ on either side of the obstruction.

B. Circular Aperture

Fresnel diffraction from circular apertures and obstacles has some surprising properties. An essential ingredient is the rotational symmetry about an axis through the center of the circle involved. We will therefore assume that the source

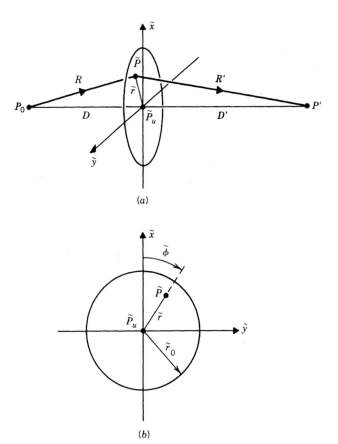

Fig. 7.19 Geometry for Fresnel diffraction by a circular aperture. (a) Perspective view from the side. (b) View of the aperture plane looking in the direction of the light propagation.

point P_0 lies on this axis, the z-axis. For simplicity we consider, initially, an observation point that is also on the optical axis. The geometry is shown in Fig. 7.19. We will need to reintroduce the inclination factor that was defined in Eq. (6.8). Due to the high degree of symmetry in this configuration, the inclination effects are more noticeable than they are for the rectangular aperture.

1. Fresnel Integral in Polar Coordinates. We start with Eq. (7.39) with the inclination factor Q inserted.

$$E'(P') = \frac{i}{2} E'_{na}(P') \iint \tilde{\tau}(\eta) Q \exp\left[\frac{-i\pi}{2}\eta^2\right] d\eta_x \, d\eta_y \qquad (7.56)$$

where $\eta^2 \equiv \eta_x^2 + \eta_y^2$.

The exponential will be written $e^{-i\delta}$, where by Eqs. (7.35) and (7.30) we may put

$$
\delta = \frac{\pi}{2}\eta^2 = \frac{\pi}{\lambda d}\left[(\tilde{x} - \tilde{x}_u)^2 + (\tilde{y} - y_u)^2\right]
$$

$$
= \frac{2\pi}{\lambda}\left[(R + R') - \left(\overline{P_0P'}\right)\right] \tag{7.57}
$$

Note that $\tilde{P}_u$ is at the center of the aperture and may be used as a convenient origin for polar coordinates $\tilde{r}$, and $\tilde{\phi}$, where

$$
\tilde{r}^2 = (\tilde{x} - x_u)^2 + (\tilde{y} - \tilde{y}_u)^2 = (\overline{\tilde{P}_u\tilde{P}})^2
$$

as shown in Fig. 7.19b. Then Eq. (7.57) becomes

$$
\delta = \frac{\pi\tilde{r}^2}{\lambda d} \tag{7.58}
$$

where d obeys Eq. (7.27). We may then scale $\tilde{r}$ by the factor $\sqrt{2/(\lambda d)}$ to obtain polar coordinates η, $\tilde{\phi}$ in terms of dimensionless variables. If we set $d\eta_x\, d\eta_y = \eta\, d\eta\, d\tilde{\phi}$, Eq. (7.56) becomes

$$
E'(P') = \frac{i}{2}E'_{na}(P')\iint\tilde{\tau}(\eta)e^{-i\delta}Q\eta\, d\eta\, d\tilde{\phi} \tag{7.59}
$$

The transmission function, which merely establishes the integration limits, will have the form

$$
\tilde{\tau}(\eta) = \begin{cases} 1, & \eta \le \tilde{r}_0\sqrt{2/(\lambda d)} \\ 0, & \eta > \tilde{r}_0\sqrt{2/(\lambda d)} \end{cases} \tag{7.60}
$$

if the aperture is a hole of radius $\tilde{r}_0$.

There is no restriction on the variable $\tilde{\phi}$, which may range from 0 to 2π independently from $\tilde{r}$. Then, because

$$
\eta\, d\eta\int_0^{2\pi} d\tilde{\phi} = 2\pi\eta\, d\eta = \pi d(\eta^2) = 2\, d\delta
$$

we obtain

$$
E'(P') = iE'_{na}(P')\int_0^{\delta_0} e^{-i\delta}Q(\delta)\, d\delta \tag{7.61}
$$

where $\delta_0 \equiv \pi\tilde{r}_0/(\lambda d)$ is the value of the phase difference at P' between a wave that comes from the edge of the aperture at $\tilde{r}_0$ and the wave that follows the z-axis.

2. Phasor Curve. Equation 7.61 is a complex integral that can be represented as a phasor curve in the complex plane. Let us examine

$$
I(\delta_0) = \int_0^{\delta_0} Q(\delta)\, e^{-i\delta}\, d\delta \tag{7.62}
$$

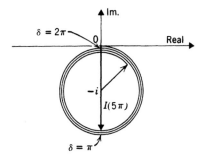

Fig. 7.20 Phasor curve for the complex Fresnel integral in the circular aperture situation. The radius of the curve decreases as the spiral coils up due to the inclination factor.

The curve is shown in Fig. 7.20. The inclination factor $Q(\delta)$ is a smooth, monotonically decreasing function of δ as shown in Fig. 7.21. If it were to keep its initial value of unity, we could immediately perform the integration in Eq. (7.61). The phasor curve would be a circle, as it is for the far-field diffraction case. Furthermore, by an argument similar to that surrounding Eq. (6.36), we can show that the circle should have a radius of one. Because the phasor curve starts out tangent to the real axis at the origin of the complex plane, with $\delta = 0$, and then curves into the fourth quadrant as the phase becomes negative, we can conclude that the center of the circle would be at the point $(0, -i)$. The slow decrease in Q will cause the curve to spiral gradually toward the center. The departure from the original circle is slow because $Q(\delta)$ is very slowly varying, especially for the first few multiples of 2π. Then $I(\delta_0)$ will initially oscillate between the extremes of zero and $-2i$.

In the limit as δ_0 gets very large, Q will tend to zero. (In this limit the angles θ_η and $\theta_{\eta'}$ approach $\pi/2$. Hence, $Q = (1/2)(\cos \theta_\eta + \cos \theta_{\eta'}) \to 0$.) The phasor curve gradually spirals in to the point $-i$, and $E'(P')$ approaches the value of $E'_{na}(P')$.

3. Fresnel Zones. It is useful to think of the open area of the diffracting aperture as divided into zones, the so-called *Fresnel zones*, which are centered at $\tilde{P}_u$. These are defined so that the phase difference δ changes by π across one zone and equals $n\pi$ at the edge of the nth zone. This is equivalent to changes in the path difference $[(R + R') - (\overline{P_0P'})]$ by $\lambda/2$ and $n\lambda/2$, respectively. An illustration of the method of constructing these zones is found in Fig. 7.22 for the case where the source point P is at infinity. The successive spheres drawn from the observation point P' have radii differing by $\lambda/2$.

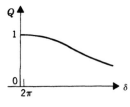

Fig. 7.21 Inclination factor versus δ showing the gradual change.

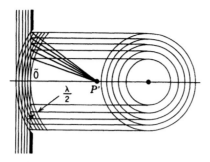

Fig. 7.22 Construction for the Fresnel zones. The source point is at infinity. The spheres about $\tilde{P}'$ have radii differing by $\lambda/2$. Their intersections with the plane of the aperture form the circular zones.

Equation (7.59) gives for the radius of the nth zone

$$\tilde{r}_n = \sqrt{n\lambda d} \tag{7.63}$$

Note that the area of a zone $\Delta\sigma$ is independent of n (at least within the quadratic approximation for the δ that we are using here):

$$\Delta\sigma = \pi(\tilde{r}_{n+1}^2 - \tilde{r}_n^2) = \pi\lambda d = \pi\tilde{r}_1^2 \tag{7.64}$$

As we integrate over a total of N zones in the diffracting aperture, the phasor curve in Fig. 7.20 makes $N/2$ revolutions. If N is not too large and there are an even number of zones inside the aperture, the integral $I(\delta_0)$ is very small; if there are an odd number of zones, $I(\delta_0)$ is close to $-2i$.

We may take the inclination factor Q into account in the following approximate way. The contribution to the integral from just the nth zone is essentially.

$$\Delta I_n \approx -2iQ_n, \quad n \text{ odd}$$

$$\Delta I_n \approx +2iQ_n, \quad n \text{ even}$$

where Q_n is the average value of the inclination factor over the nth zone. The integral $I(\delta_0)$ is then given approximately by

$$I = \sum_{n=1}^{N} \Delta I_n = -2i[Q_1 - Q_2 + Q_3 - Q_4 + \cdots \pm Q_N]$$

$$= -2i\left[\frac{Q_1}{2} + \left(\frac{Q_1}{2} - Q_2 + \frac{Q_3}{2}\right) + \left(\frac{Q_3}{2} - Q_4 + \frac{Q_5}{2}\right) \cdots \pm Q_N\right] \tag{7.65}$$

The sum has been organized in this fashion because the terms in parentheses are essentially zero. To see this consider Fig. 7.23. From Eq. (7.63) and a study of the diagram we can write

$$\cos\theta_n = D(D^2 + n\lambda d)^{-1/2}$$

and

$$\cos\theta_n' = D'(D'^2 + n\lambda d)^{-1/2}$$

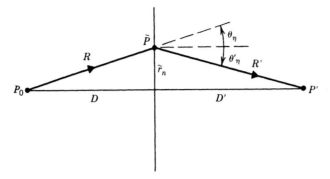

Fig. 7.23 Evaluation of the inclination factor.

Then the inclination factor at the edge of the nth Fresnel zone is

$$Q \simeq \frac{1}{2}\left[\left(1 + \frac{n\lambda d}{D^2}\right)^{-1/2} + \left(1 + \frac{n\lambda d}{D^2}\right)^{-1/2}\right]$$

This may be approximated (to the extent that λ is much smaller than D or D') by

$$Q \simeq \frac{1}{2}\left[1 - n\frac{\lambda d}{2D^2} + 1 - n\frac{\lambda d}{2D'^2}\right]$$

or

$$Q \simeq 1 - nK \tag{7.66}$$

where $K = \lambda d/(2D^2 + 2D'^2)$.

Then the terms in parentheses in Eq. (7.65) can be seen to be of the form

$$\frac{Q_m}{2} - Q_{m+1} + \frac{Q_{m+2}}{2} \simeq K\left[\frac{-m}{2} + (m+1) - \frac{(m+2)}{2}\right]$$

$$\simeq 0$$

Now we can write Eq. (7.65) as

$$I \simeq -i(Q_1 + Q_N) \qquad N \text{ odd}$$

$$I \simeq -i(Q_1 - Q_N) \qquad N \text{ even}$$

where $Q_1 \simeq 1$.

For large N, Q_N approaches zero, and I approaches $-i$, *just half the contribution from the first Fresnel zone.*

4. Off-Axis Behavior. If we move the observation point off the optical axis, then the "straight-through" optical path will no longer coincide with the z-axis. The points $\tilde{O}$ and $\tilde{P}_u$ in the plane of the aperture are then separated by distance $\tilde{r}_u$. As shown in Fig. 7.24, we can still use $\tilde{P}_u$ as the center of a polar coordinate system, only now the integration in Eq. (7.59) will have asymmetric limits.

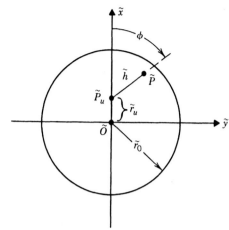

Fig. 7.24 Aperture coordinates showing the intercept of the line connecting the source and the observation point as $\tilde{P}_u$ for the case that the observation point is not on the optical axis.

The Fresnel zones will help us to understand qualitatively or semiquantitatively what happens in this case. We can show that

$$\int d\tilde{\phi} = 2\pi \qquad \text{for } \tilde{h} < (\tilde{r}_0 - \tilde{r}_u)$$

and

$$\int d\tilde{\phi} = 2 \arccos\left(\frac{\tilde{r}_u^2 + \tilde{h}^2 - \tilde{r}_0^2}{2\tilde{h}\tilde{r}_u}\right) \tag{7.67}$$

for $(\tilde{r}_0 - \tilde{r}_u) < \tilde{h} < (\tilde{r}_0 + \tilde{r}_u)$.

We thus get the full contribution from all zones of radius less than $(\tilde{r}_0 - \tilde{r}_u)$ and decreasing contributions from zones of radius between $(\tilde{r}_0 - \tilde{r}_u)$ and $(\tilde{r}_0 + \tilde{r}_u)$, as can be seen from Fig. 7.24 and Fig. 7.25.

The contribution to the arc length of the phasor curve from each of these partially obstructed zones is proportional to the unobstructed area and hence is less than the value of $Q_n\pi$ that comes from the full nth zone. The phase difference

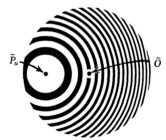

Fig. 7.25 The nature of the Fresnel zones inside a circular diffracting aperture when the center of the zones at $\tilde{P}_u$, which is along the line from the point source to the observation point, is not at the center of the aperture $\tilde{O}$. Every other zone is shown dark. The contributions from adjacent zones are of opposite sign and almost cancel.

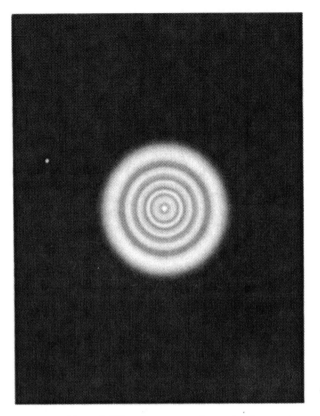

Fig. 7.26 Fresnel diffraction pattern for a circular aperture.

still changes by π across the zone. The result is that the phasor curve coils up more rapidly. The kind of diffraction pattern that is observed is indicated in Fig. 7.26.

The number of rings in the pattern will increase as the plane of observation moves toward the diffracting aperture; the pattern will expand outward. As the plane of observation moves away from the aperture, the pattern will have fewer rings and contract. Finally, when only a small fraction of a zone covers the aperture for P' on axis, we pass over to the far-field limit.

For a large aperture diameter r_0 and/or a short distance D', the total number of zones involved will be quite large. *The behavior of the diffraction pattern for $\tilde{P}_u$ near $\tilde{O}$ depends on the aperture being very accurately a circle.* If it has ragged edges with a roughness scale comparable to the width of the last Nth zone, which from Eq. (7.63) is given by

$$\tilde{r}_N - \tilde{r}_{N-1} = \Delta\tilde{r}_N = \frac{\tilde{r}_1^2}{\tilde{r}_N + \tilde{r}_{N-1}} \approx \frac{\tilde{r}_1^2}{2\tilde{r}_N} = \frac{\tilde{r}_1^2}{2\sqrt{\lambda d}}, \quad \tilde{r}_1 = \sqrt{\lambda d} \qquad (7.68)$$

then the last turn or so of the phasor curve in Fig. 7.20 will be modified, producing important changes in the diffraction pattern and tending to smooth out the rings.

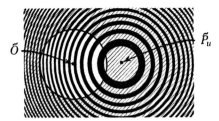

Fig. 7.27

Because any real aperture will be rough on a small enough scale, at a sufficiently small wavelength this effect will take over. In this way the limit of geometrical optics is reached as $\lambda \to 0$.

When the observation point P' moves into the geometrical shadow, the point $\tilde{P}_u$ moves into the opaque part of the aperture, as shown in Fig. 7.27. The contributions from the many partial zones that are exposed tend to cancel one another, giving a very small resultant electric field. There is one last broad bright diffraction ring formed just before the point P' passes into the shadow, because of incomplete cancellation of the contribution from the central zone. This ring appears in Fig. 7.26. Its width is about equal to that of the central zone projected onto the plane of observation at P'. This is of the order the scale factor $F = \sqrt{\lambda D'(D + D)/(2D)}$ used in our discussion of straight edges and rectangular apertures, and the first bright fringe for a straight edge was about F wide. Near the shadow edge of a large diameter circular aperture the diffraction pattern should be quite similar to that of a straight edge.

5. Circular Obstacle. Consider an accurately round circular obstacle of radius $\tilde{r}_0$. Then for P' on the axis Eq. (7.59) becomes

$$E'(P') = iE'_{na}(P')I, \quad I = \int_{\delta_0}^{\infty} e^{-i\delta}Q(\delta)\, d\delta \tag{7.69}$$

where the initial phase angle δ_0 is given by

$$\delta_0 = \frac{\pi}{\lambda d}\tilde{r}_0^2$$

In this case the section of the phasor curve in Fig. 7.20 from $\delta = 0$ to $\delta = \delta_0$ is missing. The tail of the phasor for I will be on the spiral where the phase equals δ_0. Its head will be at the center at $-i$. Thus, if δ_0 is not too large, so that Q is still approximately unity, we find that I should have an absolute value of one, and the flux density should be given by

$$S' = S'_{na}$$

Thus the flux density at the center of the diffraction pattern of a circular obstacle equals the flux density when there is no obstacle! This surprising prediction is strictly a wave phenomenon.

For this effect to be observed, the departure of the outline of the obstacle from a true circle must be much less than the width $\Delta \tilde{r}_N$ of the last or Nth Fresnel zone that is covered by the obstacle. This is given by Eq. 7.68 as

$$\Delta \tilde{r}_N = \frac{1}{2} \left(\frac{\lambda d}{N} \right)^{1/2}$$

Furthermore, the width $\Delta \tilde{r}$ of the bright spot in the plane of observation will be approximately half the width $\Delta \tilde{r}_N$ of this last zone projected from the source point onto the observation plane. That this is roughly true can be seen by sketching the phasor curve with about one-fourth of the Nth zone and three-fourths of the $(N + 1)$th zone open. From Eq. (7.68) we obtain

$$\Delta \tilde{r} \cong \frac{1}{2} \Delta \tilde{r}_N \frac{D + D'}{D} = \frac{\sqrt{2F}}{4\sqrt{N}}, \quad F = \sqrt{\frac{\lambda D'(D + D')}{2D}} \tag{7.70}$$

The resulting pattern is shown in Fig. 7.28. The spot is known as *Poisson's bright spot*, after the French mathematician who predicted it.

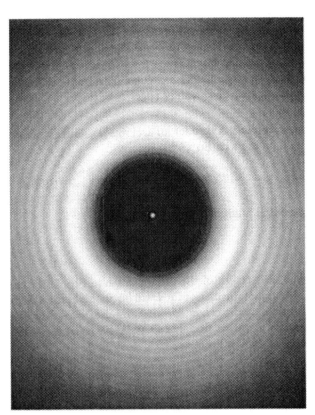

Fig. 7.28 Fresnel diffraction pattern for a circular obstacle.

There is an interesting story behind Poisson's prediction of the bright spot. In an effort to discredit the Fresnel theory of diffraction, Poisson used the theory to predict the existence of the bright spot. This, he said, violated both common sense and the results of the crude experiments that had been done up to that time. When more refined experiments were tried, the spot was indeed seen. Instead of discrediting Fresnel's theory, Poisson's prediction led to a vindication of it.

6. Zone Plates. Suppose that we have an aperture with alternating opaque and transparent circular regions that just coincide with the Fresnel zones appropriate to the values of D, D', and λ being used. This is one type of "zone plate." Suppose that such a plate is centered and used as the diffracting aperture as in Fig. 7.19. The radius of its first zone will be.

$$\tilde{r}_1 = \sqrt{\lambda d} = \left(\frac{\lambda D D'}{D + D'} \right)^{1/2} \tag{7.71}$$

as given by Eq. (7.63). If the plate has $N/2$ open zones and is then opaque after that, and if we neglect the inclination factor Q, we find that the phasor curve consists of a column of $N/2$ semicircles of unit radius as shown in Fig. 7.29. The integral I is the sum of the chords of these circles and is $-Ni$ for Fig. 7.29a and $+Ni$ for Fig. 7.29b. The field is $\pm NE'_{na}$ in the two cases, and the flux density

$$S' = N^2 S'_{na} \quad (N = \text{twice the number of open zones})$$

Now suppose that for given D, D', and d the zone plate is so large that it blocks or transmits the zones three at a time. If the center is open, it will let through the light from the first three zones, then block the light from the next three, and so on. The phasor curve of Fig. 7.29 would then become that of Fig. 7.30. The resultant phasor representing the electric field would be reduced by a factor of 3 and the flux density would be smaller by a factor of 9. Similar results would be obtained with a

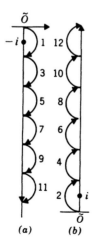

(a)   (b)

Fig. 7.29 The phasor diagram for the diffraction pattern of a zone plate having 12 zones (6 open zones). (a) The odd zones are transparent. (b) The even zones are transparent.

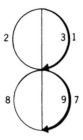

Fig. 7.30 Phasor diagram for a zone that blocks three zones at a time.

large enough zone plate that blocked or passed 5, 7, and so on zones at a time. With an even number of zones blocked and passed, however, the field is approximately zero.

The condition that the first circle in the zone plate covers n_1 zones is that its radius be given by

$$\tilde{r}_1 = \sqrt{n_1 \lambda d} \tag{7.72}$$

We may rewrite Eq. (7.72) in the form

$$\frac{n_1 \lambda}{\tilde{r}_1^2} = \frac{1}{d} = \frac{1}{D} + \frac{1}{D'} \tag{7.73}$$

where n_1 is an *odd* integer. This has the appearance of the paraxial thin-lens equation provided that the focal length is considered to be

$$f_{n_1} \equiv \frac{\tilde{r}_1^2}{n_1 \lambda} \tag{7.74}$$

The properties of the zone plate are similar to those of a simple lens except for the multiple values of f_{n_1}. For instance, if plane waves of wavelength λ are incident normally onto the zone plate, there will be a bright spot a distance f_1 to the right of the plate and other spots of decreasing brightness at distances of $f_1/3, f_1/5$, and so on.

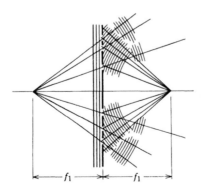

Fig. 7.31 Illustration of the simultaneous positive and negative lens action of a zone plate.

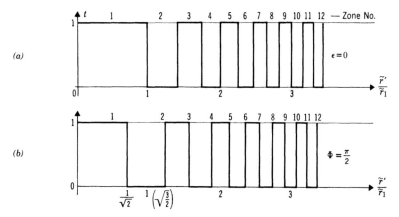

Fig. 7.32

The zone plate acts simultaneously as a positive and negative lens with focal lengths $\pm f_1$, $\pm f_1/3$, and so on. In addition to the series of spots just mentioned, formed to the right of the plate with parallel incident light, a similar series of "virtual spots" will be found at distances $f_1, f_1/3, f_1/5, \ldots$ to the left of the lens. We will try to illustrate this effect in Fig. 7.31.

Figure 7.32 shows the transmission function of the zone plate having odd-numbered zones open and having the phasor curve of Fig. 7.29a. We can also start with a partial zone open; say a fraction ε of the area of the first zone is open. The radius of its edge will be $\sqrt{\varepsilon}\tilde{r}_1$, and the phase angle across it will vary from 0 to $\varepsilon\pi$. Then we alternate opaque and open rings, each having an area πr_1^2. The first dark ring will extend from $\tilde{r}' = \sqrt{\varepsilon}\tilde{r}_1$ to $\tilde{r}' = \sqrt{\varepsilon + 1}\tilde{r}_1$; the first full bright ring from $\sqrt{\varepsilon + 1}\tilde{r}_1$ to $\sqrt{\varepsilon + 2}\tilde{r}_1$; and so on. The transmission function for the case $\varepsilon = 1/2$ is shown in Fig. 7.32b, and the phasor curve is drawn in Fig. 7.33 for a total of 12 zones. This should be compared with the phasor curves of Fig. 7.29a, b, which correspond to $\varepsilon = 0$ and $\varepsilon = 1$, respectively.

It is easily seen that the effect of the partial zone is to change the resulting electric field from NE'_{na} to $Ne^{i\pi\varepsilon}E'_{na}$ (if we set $Q = 1$), where N is the total number of zones.

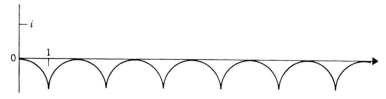

Fig. 7.33

7.3   Image Formation: Coherent Objects

The theory of image formation in physical optics centers around the phase-altering characteristics of a lens. We will consider here only coherent objects. These are defined to be objects smaller than the transverse coherence length of the light coming from them, as often occurs in microscopes. This is also the case for macroscopic objects illuminated by a laser.

The foundations for this approach were laid independently by Ernst Abbe (1840-1905) and Herman L. F. von Helmholtz (1821-1894) in the early 1870s. The basis for the modern treatment was given by Lord Rayleigh (1842-1919) in 1896. Today this is a firmly established formalism from which most optical concepts can be derived, including all of geometrical optics.

We return to the general method outlined in Section 7.1 within the parabolic approximation. We wish to consider the geometry of Fig. 7.4 with a lens in the aperture plane. This can be treated within the diffraction formalism, provided we can identify the proper transmission function that accounts, not only for the extent of the lens but also for its focusing properties. The situation is illustrated in Fig. 7.34, which is a schematic view of a simple image-forming system.

The incident field will in many cases be a plane wave. Shown in Fig. 7.35 is the propagation vector for a plane wave that is inclined with respect to the x- and y-axes by angles $\theta_{x,0}$ and $\theta_{y,0}$, respectively. The direction cosines that identify the direction of propagation of this particular plane wave are then $\alpha_0 = \cos\theta_{x,0}$ and $\beta_0 = \cos\theta_{y,0}$. In this case the incident field at the object plane will be

$$E(P) = E(x, y) = Ae^{i\omega t}e^{-ik(\alpha_0 x + \beta_0 y)} \tag{7.75}$$

Our object will be any essentially two-dimensional, partially transmitting film or screen that modulates the incident light falling on it. It will be described by a transmission function $\tau(x, y)$. The object could be a simple rectangular or circular opening or something much more complicated, like a diffraction grating or a photograph transparency. Immediately beyond the object plane the field distribu-

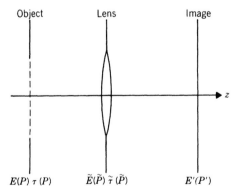

Fig. 7.34 Schematic representation of a simple image-forming system. The expressions for the optical fields on the transmitting side of each plane are shown at the bottom of the diagram.

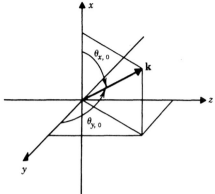

Fig. 7.35 Propagation vector that describes the direction for a plane wave.

tion will be the product of the incident field and the transmission function of the object.

There will also be cases in which we will need to study the effect of the optical system on a point source. As in Section 7.1 for a source at (x_0, y_0), this can be handled with a source distribution

$$E(P)\tau(P) = E(x, y)\tau(x, y) = \frac{A\lambda}{i} e^{i\omega t}\delta(x - x_0)\delta(y - y_0) \qquad (7.76)$$

[Note that A, the strength of the source field, must have different units in Eqs. (7.75) and (7.76).]

A. Lens Action

1. Lens Transmission Function. We neglect light losses that result from reflections from the lens surfaces. We also assume that the refractive index is the same on both sides of the lens.

The optical effect of the lens that is the most important is provided by the position-dependent phase shift caused by the variable thickness of the lens. Transmission functions of this type were originally brought up in Chapter 6. The lens transmission function takes on the form of Eq. (6.10)

$$\tilde{\tau}_L(\vec{r}) = |\tilde{\tau}_L(\vec{r})|\exp\left[\frac{-i2\pi}{\lambda}(n - 1)d(\vec{r})\right] \qquad (7.77)$$

The phase is measured relative to the case for the open aperture without the lens. The amplitude factor $|\tilde{\tau}_L(\vec{r})|$ describes the size of the lens and is equal to unity where the lens is transparent and zero elsewhere.

The phase can be determined by reconsidering Fig. 4.31 and Eq. (4.53). There we discussed the quadratic approximation for the variation in the thickness of a

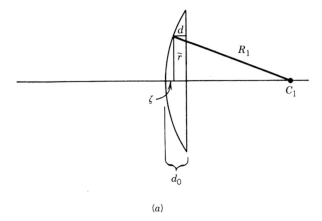

(a)

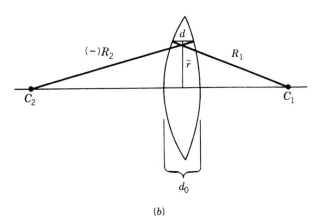

(b)

Fig. 7.36 Cross-sectional diagrams of circular lenses that modulate the incident beam by introducing a position-dependent phase shift: (a) planoconvex case; (b) biconvex case. By suitable changes of the radii, any lens shape with spherical surfaces can be modeled.

spherical refracting surface with the distance from the optical axis. The essential geometry is reproduced for a planoconvex lens in Fig. 7.36a. The distance ζ in the figure is related to $\tilde{r}$ by Eq. (4.53)

$$\zeta \simeq \frac{\tilde{r}^2}{2R_1} \tag{7.78}$$

This level of approximation is consistent with the parabolic approximation used in the Fresnel transformation. The thickness of the lens is then

$$d(\tilde{r}) = d_0 - \frac{\tilde{r}^2}{2R_1} \tag{7.79}$$

Figure 7.36*b* shows the case for a biconvex lens. It is easy to demonstrate that the thickness in this example should be

$$d(\tilde{r}) = d_0 - \frac{\tilde{r}^2}{2}\left(\frac{1}{R_1} - \frac{1}{R_2}\right) \tag{7.80}$$

These results then lead to

$$\tilde{\tau}_L(\tilde{r}) = |\tilde{\tau}_L(\tilde{r})|e^{i\phi(d_0)}\exp\left[\frac{i\pi}{\lambda f}\tilde{r}^2\right]$$

where

$$\tilde{r}^2 = \tilde{x}^2 + \tilde{y}^2 \tag{7.81}$$

for the transmission function of the general lens with focal length f. Here, as in Chapter 3, we must obey the algebraic sign convention and, as established in Eq. (3.59), we have used

$$\frac{1}{f} = (n-1)\left[\frac{1}{R_1} - \frac{1}{R_2}\right] \tag{7.82}$$

The factor $e^{i\phi(d_0)}$ is an irrelevant phase contribution that is constant and will henceforth be dropped.

2. Transformation Between Conjugate Planes. We will now derive the transformation that starts with the electric field distribution in one plane on the incident side of the lens and gives the field in the plane that, by paraxial optics, would be the conjugate plane of the object. This is handled in physical optics as a diffraction process in which the aperture plane contains the lens. The situation is illustrated in Fig. 7.37. In the parabolic approximation, the kernel that describes propagation

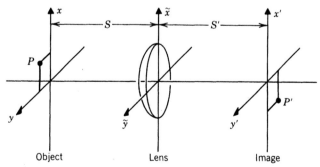

Fig. 7.37 Perspective drawing showing the coordinates in the object plane, the lens plane, and the image plane.

from the object plane through an intermediate aperture to the image plane has the form of Eq. (7.21)

$$h_t(P \to P') = \left(\frac{i^2}{\lambda^2}\right)\left(\frac{1}{SS'}\right)e^{-ik(S+S')}\iint \tilde{\tau}_L(\tilde{x}, \tilde{y})$$

$$\times \exp\left\{-ik\left[\frac{[(\tilde{x}-x)^2+(\tilde{y}-y)^2]}{2S}+\frac{[(x'-\tilde{x})^2+(y'-\tilde{y})^2]}{2S'}\right]\right\}d\tilde{x}\,d\tilde{y}$$

(7.83)

Here the aperture contains the lens, so the transmission function will be described by Eq. (7.81).

We prefer to express this in a form that is compatible with Eq. (7.20a). We collect the arguments of the exponentials and reexpress them according to the following:

$$-ik(S+S') - ik\left[\frac{(\tilde{x}-x)^2+(\tilde{y}-y)^2}{2S}+\frac{(x'-\tilde{x})+(y'-\tilde{y})^2}{2S'}\right]$$

$$= -ik(R_0+R_0') - \frac{i\pi\tilde{r}^2}{\lambda S} - \frac{i\pi\tilde{r}^2}{\lambda S'} - i2\pi\left[-\left(\frac{x}{S}+\frac{x'}{S'}\right)\frac{\tilde{x}}{\lambda}-\left(\frac{y}{S}+\frac{y'}{S'}\right)\frac{\tilde{y}}{\lambda}\right]$$

(7.84)

Figure 7.38 shows how the distances are defined. This is consistent with Fig. 7.5, only here we concentrate on isolating the terms that are proportional to $\tilde{r}^2$ from the path length due to propagation on both the incident and the transmitting sides of the lens.

This result can be simplified if we define *spatial frequencies* in analogy with Chapter 6.

$$u \equiv \frac{x}{S\lambda} \cong \frac{-\alpha}{\lambda}$$

(7.85a)

$$u' \equiv \frac{-x'}{S'\lambda} \cong \frac{-\alpha'}{\lambda}$$

(7.85b)

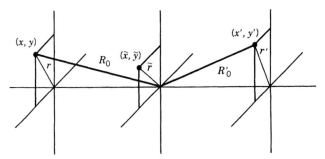

Fig. 7.38

and

$$v \equiv \frac{y}{S\lambda} \cong \frac{-\beta}{\lambda} \tag{7.85c}$$

$$v' \equiv \frac{-y'}{S'\lambda} \cong \frac{-\beta'}{\lambda} \tag{7.85d}$$

where α, β, α', and β' are the direction cosines that identify the locations of P and P' *with respect to* $\tilde{O}$, the center of the aperture plane. [See Fig. 6.12b and Eqs. (6.14)]. Then

$$\Delta u = u' - u = -\left(\frac{x}{S} + \frac{x'}{S'}\right)\frac{1}{\lambda}$$

is equivalent to u, which we used in Eq. (6.21a) and

$$\Delta v = v' - v = -\left(\frac{y}{S} + \frac{y'}{S'}\right)\frac{1}{\lambda}$$

is equivalent to v, which was used in Eq. (6.21b). Here it is useful to have explicit notation for reciprocal coordinates in the object and the image planes rather than to combine these in single parameters u and v.

With Eqs. (7.85) we can consolidate Eq. (7.84),

$$-ik(S + S') - ik\left[\frac{(\tilde{x} - x)^2 + (\tilde{y} - y)^2}{2S} + \frac{(x' - \tilde{x}) + (y' - \tilde{y})^2}{2S'}\right]$$

$$= -ik(R_0 + R'_0) - \frac{i\pi}{\lambda}\tilde{r}^2\left(\frac{1}{S} + \frac{1}{S'}\right) - i2\pi(\Delta u\tilde{x} + \Delta v\tilde{y}) \tag{7.86}$$

The first and last terms in Eq. (7.86) are very similar to those that we used in the linearized form of the phase in Chapter 6. The new feature here is the quadratic contribution of the second term. In the general case, this contribution makes the integral in Eq. (7.83) difficult to calculate in closed form. However, when the transmission function of the lens is also included, a remarkable simplification is reached. From Eq. (7.81) we find that the phase shift due to the lens appears as an exponential factor whose argument is

$$+\frac{i\pi}{\lambda}\tilde{r}^2\frac{1}{f}$$

The arguments of all the exponentials in Eq. (7.83) are then

$$-ik(R_0 + R'_0) - \frac{i\pi}{\lambda}\tilde{r}^2\left(\frac{1}{S} + \frac{1}{S'} - \frac{1}{f}\right) - i2\pi(\Delta u\tilde{x} + \Delta v\tilde{y}) \tag{7.87}$$

If the source plane and the observation plane are conjugate to each other, then the thin-lens formula of Chapter 3 allows the second factor to *vanish*. In other words, if the observation plane is the image plane, then the phase shift caused by the lens cancels the quadratic contribution to the optical path length difference

$(R + R') - (R_0 + R'_0)$. This formalism is compatible with the approximation that led to the paraxial theory in geometrical optics.

Therefore, if P and P' are images of each other, the propagation kernel of Eq. (7.83) becomes

$$h_r(P \to P') = \left(\frac{i}{\lambda}\right)^2 \left(\frac{1}{SS'}\right) e^{-ik(R_0 + R'_0)} \iint |\tilde{\tau}_L(\tilde{r})| e^{-i2\pi(\Delta u \tilde{x} + \Delta v \tilde{y})} \, d\tilde{x} \, d\tilde{y} \quad (7.88)$$

We recognize the integral as the Fourier transform of the amplitude of the lens transmission function. This is the part that describes the extent of the lens opening. Therefore,

$$h_r(P \to P') = \left(\frac{i}{\lambda}\right)^2 \left(\frac{1}{SS'}\right) e^{-ik(R_0 + R'_0)} T_L(\Delta u, \Delta v) \quad (7.89)$$

where

$$T_L = \mathscr{F}[|\tilde{\tau}_L|]$$

To find the resultant field in the image plane, we apply the propagation kernel in Eq. (7.89) to the overall field transformation integral of Eq. (7.9) to get

$$E'(P') = \left(\frac{i^2}{\lambda^2 SS'}\right) e^{-ikR'_0} \iint E(x, y)\tau(x, y) e^{-ikR_0} T_L(\Delta u, \Delta v) \, dx \, dy \quad (7.90)$$

Here we must remember that, in the general case R_0, Δu and Δv are functions of the source plane coordinates (x, y) and thus must be considered as implicit functions of the integration variables in Eq. (7.90).

This is the general result for the influence of a finite-size lens on the transformation of the electric field distribution of the object to that of the image. Further insight into this result is gained by changing the integration variables using Eqs. (7.85a and b). Therefore,

$$dx \, dy = S^2 \lambda^2 \, du \, dv$$

and

$$E(x, y)\tau(x, y) e^{-ikR_0} \equiv \mathscr{E}(x, y) = \mathscr{E}(S\lambda u, S\lambda v) \quad (7.91)$$

Here $\mathscr{E}$ is a phase-shifted source function that combines the source-dependent parts of the integrand. Now

$$E'(x', y') = i^2 \left(\frac{S}{S'}\right) e^{-ikR'_0} \iint \mathscr{E}(S\lambda u, S\lambda v) T_L(u' - u, v' - v) \, du \, dv$$

$$= \frac{e^{-ikR'_0}}{m} \{\mathscr{E}(S\lambda u, S\lambda v) \circledast T_L(u, v)\} \quad (7.92)$$

where $m = -S'/S$ is the transverse magnification.

This shows that the observed field distribution in the image plane is the convolution of the phase-shifted source function and the Fourier transform of the transmission function that describes the extent of the lens. This is a result of diffraction that occurs at the lens aperture. The focusing action of the lens causes what would be a far-field pattern to be developed closer to the lens, in the image plane. The magnification factor is required to conserve energy from the object plane to the image plane.

3. Point Source. Suppose that we have a point source $P = P_0(x_0, y_0)$. This may be represented by the source distribution of Eq. (7.76). Because, in this case, the source contribution to the convolution integral in Eq. (7.92) involves delta functions, we may simplify Eq. (7.91)

$$\mathscr{E}(S\lambda u, S\lambda v) = \left(\frac{A\lambda}{i}\right)e^{i\omega t}\, e^{-ikR_0}\delta(S\lambda u - x_0)\delta(S\lambda v - y_0) \qquad (7.93\text{a})$$

where, because the source is a point we have

$$R_0 = \overline{P_0 O} \qquad (7.93\text{b})$$

Before we can use Eq. (7.93), we must express the argument of the delta functions in terms of the integration variables in Eq. (7.92). We therefore employ the scaling relationship of Eq. (6.62) to arrive at

$$\mathscr{E}(S\lambda u, S\lambda v) = \left(\frac{A\lambda}{i}\right)e^{i\omega t}\, e^{-ikR_0}\left(\frac{1}{S^2\lambda^2}\right)\delta\left(u - \frac{x_0}{S\lambda}\right)\delta\left(v - \frac{y_0}{S\lambda}\right) \qquad (7.94)$$

This is used in Eq. (7.92) to obtain the field in the image plane.

$$
\begin{aligned}
E'(x', y') &= \left(\frac{A\lambda}{i}\right)e^{i(\omega t - k(R_0 + R_0'))}\left(\frac{1}{mS^2\lambda^2}\right) \\
&\quad \times \iint \delta\left(u - \frac{x_0}{S\lambda}\right)\delta\left(v - \frac{y_0}{S\lambda}\right)T_L(u' - u, v' - v)\, du\, dv \\
&= \left(\frac{A\lambda}{i}\right)e^{i(\omega t - k(R_0 + R_0'))}\left(\frac{1}{mS^2\lambda^2}\right)T_L\left[\left(u' - \frac{x_0}{S\lambda}\right), \left(v' - \frac{y_0}{S\lambda}\right)\right] \quad (7.95)
\end{aligned}
$$

The phase factor is the same as that which was identified in Eq. (6.23), $e^{i\phi_0}$. Also, the arguments of T_L can be reexpressed in one of several forms:

$$\left(u' - \frac{x_0}{\lambda S}\right) = -\frac{1}{\lambda}\left(\frac{x'}{S'} + \frac{x_0}{S}\right) = \frac{-1}{\lambda S}\left(x_0 - \frac{x'}{m}\right) = \frac{-1}{\lambda S'}(x' - mx_0) \quad (7.96\text{a})$$

$$\left(v' - \frac{y_0}{\lambda S}\right) = -\frac{1}{\lambda}\left(\frac{y'}{S'} + \frac{y_0}{S}\right) = \frac{-1}{\lambda S}\left(y_0 - \frac{y'}{m}\right) = \frac{-1}{\lambda S'}(y' - my_0) \quad (7.96\text{b})$$

In addition, it may be convenient to take advantage of relation:

$$mS^2 = \frac{S'^2}{m} = -SS' \qquad (7.97)$$

Following these guidelines we find an alternative form for Eq. (7.95)

$$E'(x', y') = \left(\frac{A\lambda}{i}\right)e^{i\phi_0}\left(\frac{m}{S'^2\lambda^2}\right)T_L\left[\left(\frac{-1}{\lambda S'}(x' - mx_0)\right), \left(\frac{-1}{\lambda S'}(y' - my_0)\right)\right] \quad (7.98)$$

Instead of a point image, the diffraction of light by the lens aperture causes a smeared-out image. The center of the image pattern is at $x'_0 = mx_0$ and $y'_0 = my_0$ as expected. Here $T_L = T_L(0, 0) = \sigma_L$, the area of the lens opening. The optical field will be a maximum at this position and will be given by

$$E'(x'_0, y'_0) = \left(\frac{Am}{i\lambda}\right)e^{i\phi_0}\Omega' \quad (7.99)$$

where $\Omega' \cong \sigma_L/S'^2$ is the solid angle subtended by the lens at the image (see Fig. 7.39).

Away from the paraxial image point, the field distribution is more complicated. However, because this is related to the far-field diffraction pattern of the lens opening, we can extract results from Chapter 6.

In the common case of a circular lens centered on the optical axis we have

$$|\tilde{\tau}_L(\tilde{r})| = \begin{cases} 1, & \tilde{r} \le \tilde{r}_0, \\ 0, & \text{otherwise} \end{cases} \quad (7.100)$$

where $\tilde{r} = [\tilde{x}^2 + \tilde{y}^2]^{1/2}$. This has the same form as Eq. (6.42). We can immediately write down its Fourier transform

$$T_L\left(\frac{\Delta r'}{\lambda S'}\right) = \pi \tilde{r}_0^2 \left(\frac{2J_1\left(\frac{2\pi\tilde{r}_0\,\Delta r'}{\lambda S'}\right)}{\left(\frac{2\pi\tilde{r}_0\,\Delta r'}{\lambda S'}\right)}\right) \quad (7.101)$$

In this case

$$\Delta r' = [(x' - x'_0)^2 + (y' - y'_0)^2]^{1/2} \quad (7.102)$$

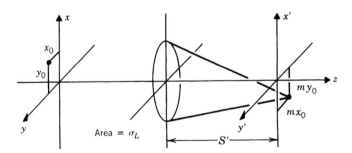

Fig. 7.39 The singularity in the geometrical theory at the focus of a converging spherical wave is removed in the wave theory because of diffraction effects.

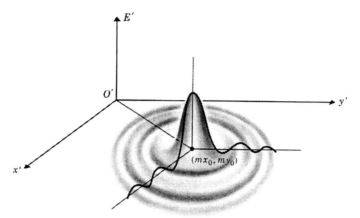

Fig. 7.40 Optical electric field in the vicinity of the image of a point object. The center of the pattern appears at $(x', y') = (x_0', y_0') = (mx_0, my_0)$.

The distance $\Delta r'$ measures the observation point relative to the paraxial image position. The field distribution is then

$$E'(x', y') = E'(x_0', y_0') \left(\frac{2J_1\left(\dfrac{2\pi \tilde{r}_0 \, \Delta r'}{\lambda S'}\right)}{\left(\dfrac{2\pi \tilde{r}_0 \, \Delta r'}{\lambda S'}\right)} \right) \tag{7.103}$$

where we have used Eq. (7.99) to express the result in terms of the field at the paraxial image point. This behavior is sketched in Fig. 7.40.

We have already used this information in the discussion of image resolution in Section 6.4D1. Each point in the object plane will be converted to a diffraction pattern in the image plane that is associated with a flux density of

$$S'(x', y') = S'(x_0', y_0') \left(\frac{2J_1\left(\dfrac{2\pi \tilde{r}_0 \, \Delta r'}{\lambda S'}\right)}{\left(\dfrac{2\pi \tilde{r}_0 \, \Delta r'}{\lambda S'}\right)} \right)^2 \tag{7.104}$$

4. Infinite Aperture. Although an infinite aperture cannot be realized in practice, it is still instructive to investigate the image produced by the unlimited lens. This will be the ideal situation in which all of the light coming in the forward direction from the object is collected and focused by the lens. We use the notation $E'_\infty(P')$ for the field at point P' in the observation plane under these conditions.

The magnitude of the lens transmission function will be

$$|\tilde{\tau}_L(\tilde{x}, \tilde{y})| = 1 \tag{7.105}$$

From the results of Chapter 6, we know that the Fourier transform of a constant is a delta function. In the present example this leads to

$$T_L(u, v) = \delta(u)\delta(v) \tag{7.106}$$

Use of this relation leads to direct simplification of Eq. (7.92) for the image-field distribution in the infinite lens limit.

$$E'_\infty(x', y') = \frac{e^{-ikR_0'}}{m} \iint \mathscr{E}(S\lambda u, S\lambda v)\delta(u - u')\delta(v - v') \, du \, dv$$

$$= \frac{e^{-ikR_0'}}{m} \mathscr{E}(S\lambda u', S\lambda v') \tag{7.107}$$

The arguments of the source function are

$$S\lambda u' = -\frac{Sx'}{S'} = \frac{x'}{m} \tag{7.108a}$$

and

$$S\lambda v' = -\frac{Sy'}{S'} = \frac{y'}{m} \tag{7.108b}$$

Together with Eq. (7.91) this leads to the infinite lens result.

$$E'_\infty(x', y') = \frac{e^{-ik(R_0 + R_0')}}{m} E\left(\frac{x'}{m}, \frac{y'}{m}\right)\tau\left(\frac{x'}{m}, \frac{y'}{m}\right) \tag{7.109a}$$

where

$$R_0' = \overline{OP'} \tag{7.109b}$$

and

$$R_0 = \overline{PO}, \quad P = \left(\frac{x'}{m}, \frac{y'}{m}\right) \tag{7.109c}$$

This result shows that, with an infinite lens, the field in the image plane at $P' = (x', y')$ is proportional to the field at the conjugate point in the object plane at $P = (x, y)$. There is also a phase shift of $-(2\pi/\lambda)(R_0 + R_0')$, to account for the distance traveled by the light, and a factor of $1/m$, to guarantee that the total flux at the image plane equals that at the object plane.

In the infinite lens case image formation is ideal, so there is a one-to-one relationship between points in the object plane and the image plane.

$$x' = mx \tag{7.110a}$$

and

$$y' = my \tag{7.110b}$$

With a finite lens we need to be more careful. Coordinates (x, y) and (x', y') all appear (through u, v and u', v') in the integrand of Eq. (7.92). However, we cannot use Eqs. (7.110) to eliminate one of the pairs of coordinates in the integrand. The field at a given point in the image plane will, in general, depend on a *region* around what (in the paraxial theory) would be the conjugate point on the object plane.

The phase factor in Eq. (7.109a) can be rewritten in another form that we will find useful later. This exploits the approximate equality

$$R_0 + R_0' \cong S + S' + \frac{r'^2}{2X'} \tag{7.111a}$$

or

$$R_0 + R_0' \cong S + S' + \frac{r^2}{2X} \tag{7.111b}$$

where $X' = S' - f$ and $X = S - f$ are the Newtonian image and object distances, respectively. With these substitutions the infinite lens result becomes

$$E_\infty'(x', y') = \frac{\exp[-ik(S + S')]}{m} \exp\left(-ik\frac{r'^2}{2X'}\right) E\left(\frac{x'}{m}, \frac{y'}{m}\right) \tau\left(\frac{x'}{m}, \frac{y'}{m}\right) \tag{7.112a}$$

or

$$E_\infty'(x', y') = \frac{\exp[-ik(S + S')]}{m} \exp\left(-ik\frac{r^2}{2X}\right) E\left(\frac{x'}{m}, \frac{y'}{m}\right) \tau\left(\frac{x'}{m}, \frac{y'}{m}\right) \tag{7.112b}$$

To better understand Eqs. (7.111) consider Fig. 7.41, which illustrates the action of some important rays in the infinite thin-lens imaging situation. Because P and P' are conjugate points, we must have equality of the optical paths $P\tilde{P}P'$ and $P\tilde{O}P'$. In addition, because parallel rays will come to a focus at F', we must also have equality of the optical paths $P\tilde{P}F'$ and $O\tilde{O}F'$. If nd is the optical path through the center of the lens (of thickness d), then

$$\text{OPL}(P\tilde{O}P') = R_0 + R_0' + nd$$

and

$$\text{OPL}(O\tilde{O}O') = S + S' + nd$$

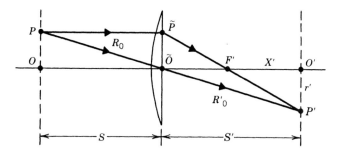

Fig. 7.41

Writing these two equations in terms of nd and equating, leads to

$$R_0 + R'_0 - \text{OPL}(P\tilde{O}P') = S + S - \text{OPL}(O\tilde{O}O') \tag{7.113}$$

This can be further reduced by writing

$$\text{OPL}(P\tilde{O}P') = \text{OPL}(P\tilde{P}F') + \overline{F'P'}$$

and

$$\text{OPL}(O\tilde{O}O') = \text{OPL}(O\tilde{O}F') + X'$$

When substituted into Eq. (7.113) these produce

$$R_0 + R'_0 - \text{OPL}(P\tilde{P}F') - \overline{F'P'} = S + S' - \text{OPL}(O\tilde{O}F') - X'$$

But because $\text{OPL}(P\tilde{P}F') = \text{OPL}(O\tilde{O}F')$, we have

$$R_0 + R'_0 = S + S' + \overline{F'P'} - X' \tag{7.114}$$

In the parabolic approximation, $\overline{F'P'}$ can be approximated as

$$\overline{F'P'} \approx X' + \frac{r'^2}{2X'}$$

When this is used in Eq. (7.114) we arrive at the final result of Eq. (7.111a). A similar argument justifies Eq. (7.111b).

5. Aberrations. We have been discussing the effects of diffraction in the parabolic limit, which is equivalent to the paraxial approximation of geometrical optics. This justified the approximation of Eq. (7.78) for the change in thickness of a spherical refracting surface with radial coordinate of $\tilde{r}$ measured from the optical axis. Chapter 4 dealt with deviations from the paraxial approximation in the geometrical theory. We compared the true wave front converging on a point image with a spherical reference surface centered on the paraxial image point (Fig. 7.42). We introduced the aberration function W as the optical path length difference between these two surfaces at a given point $\tilde{P} = (\tilde{x}, \tilde{y})$ in the exit pupil. In general, W depends on the location of the image $P' = (x', y')$ as well.

Our discussions of the diffraction theory of image formation have been based on the assumption that $W = 0$; that is, that the true wavefront is spherical. If W is small compared with $\tilde{r}$, we can imagine the perfect sphere converted into the true wave surface by shifting the phase of the spherical waves at $\tilde{P}$ by the amount $(2\pi/\lambda)/W$. This can be accomplished in the case where the lens is the aperture stop by using the factor $\exp[-i(2\pi/\lambda)W]$ in the lens transmission function. We then obtain the corrected form of Eq. (7.81)

$$\tilde{t}_L(\tilde{P}, P') = |\tilde{t}_L(r)| \exp\left(\frac{i\pi}{\lambda f} \tilde{r}^2\right) \exp\left(\frac{-i2\pi}{\lambda} W(\tilde{P}, P')\right)$$

The exponential factor that is quadratic in r will cancel with the quadratic exponential in the parabolic form of the Fresnel propagation kernel. The correc-

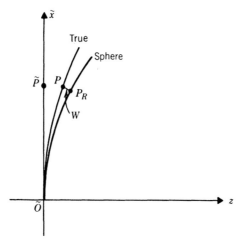

Fig. 7.42 The aberration function represents the optical path length difference between a spherical reference surface and a true wave surface. This can be included in the lens transmission function as a correlation to the, essentially paraxial, expression for the phase change introduced by the lens.

tion resulting from aberrations will remain. Therefore, when forming the Fourier transform of the lens transmission function [as required in Eq. (7.89)] we need to use

$$T_L = \mathscr{F}\left[|\tilde{\tau}_L(\tilde{P})| \exp\left(\frac{-i2\pi}{\lambda} W(\tilde{P}, P') \right) \right] \qquad (7.115)$$

instead of

$$T_L = \mathscr{F}[|\tilde{\tau}_L(\tilde{P})|]$$

This corrected form, when substituted into the subsequent equations following Eq. (7.89) will allow for the influence of the primary aberrations on the image.

This approach to aberration analysis is convenient because the primary monochromatic aberrations are classified in terms of the components of W. Alternatively, where high-speed computers are employed, the true optical path $R + R'$ in the phase of the propagation kernel [Eq. (7.10)], and the true lens thickness function [(Eq. 7.81)] for the phase of the lens effect in that kernel, can be used. Numerical integration of Eq. (7.10) then leads to a more correct but conceptually more obscure result.

B. Fourier Optics

Now that we have laid the groundwork for the theory of image formation through an analysis of diffraction by the lens, we are in a position to study these phenomena when a general object function is present. To simplify the analysis we limit the discussion to the infinite lens case. This is not a serious compromise, because we already know what the effect of a finite lens size will be. Instead of the one-to-one relationship between object points and image points, as expressed in Eq. (7.109), the finite lens contributes a fundamental uncertainty to the image point, as expressed by the convolution in Eq. (7.92).

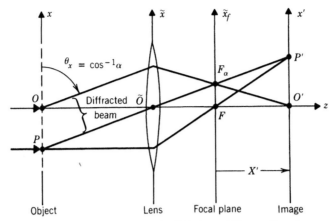

Fig. 7.43 The object diffracts the light into different directions. Each direction is associated with a point on the focal plane. The larger angles are associated with features in the object that represent fine detail. Light that cannot pass through the lens because of its finite size cannot reach the image. This will filter the image and results in a loss of detail.

Consider the situation illustrated in Fig. 7.43 where we return to the arbitrary partially transmitting object as was introduced in Fig. 7.32. Let the source be at infinity and centered on the optical axis so as to create a plane wave at the object plane. From Eq. (7.75) this identifies the form for the incident field.

$$E(x, y) = A\, e^{i\omega t} \tag{7.116}$$

With no object present, the incident parallel light beam will be focused at F. With a partially transmitting object in place, the resulting diffracted beam may be described as a superposition of many parallel beams (plane waves) propagating in various directions. Each beam will be focused onto a spot in the focal plane. The location of the spot depends on the angle of the particular plane-wave component at the object. (If the lens is finite, then these spots are broadened into Airy disks.)

Each spot can be considered to be the source for Huygens' wavelets that diverge from the spots and interfere with each other, ultimately forming an image of the original object in the image plane.

1. Fourier Decomposition of the Object. The transmission function of the object $\tau(x, y)$ can be considered to be the aperture function for a far-field diffraction process at the object plane. Accordingly, we form the Fourier transform

$$T(u, v) = \iint \tau(x, y)e^{-i2\pi(ux + vy)}\, dx\, dy \tag{7.117}$$

This is proportional to the field distribution of the diffracted light in the direction specified by

$$\alpha = -\lambda u, \quad \beta = -\lambda v$$

The object function can then be written in terms of a superposition of these diffracted plane-wave components by the inverse transform

$$\tau(x, y) = \iint T(u, v)e^{i2\pi(ux + vy)} \, du \, dv \tag{7.118}$$

Here each component has a magnitude of

$$T(u, v) \, du \, dv$$

and a phase at the object plane of

$$e^{i2\pi(ux + vy)} = e^{-ik(\alpha x + \beta y)}$$

a. Spatial Frequencies. To gain a better understanding of the significance of Eq. (7.118), consider an infinitely periodic cosine function as the object (Fig. 7.44).

$$\tau(x, y) = \tau_0 + \tau_1 \cos(2\pi u_0 x) \tag{7.119}$$

where a is the spatial period and $u_0 = 1/a$ is the *spatial frequency* of the cosine. The Fourier transform of the hypothetical object is

$$T(u, v) = \left\{ \tau_0 \delta(u) + \frac{\tau_1}{2} \left[\delta(u - u_0) + \delta(v - v_0) \right] \right\} \delta(v) \tag{7.120}$$

This shows that the object function could have been written following Eq. (7.118) as

$$\tau(x, y) = \tau_0 + \frac{\tau_1}{2} \left(e^{i2\pi u_0 x} + e^{-i2\pi u_0 x} \right) \tag{7.121}$$

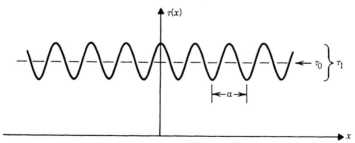

Fig. 7.44 The "cosine" object modulates the field with a well-defined spatial frequency.

Table 7.2. Fourier Decomposition of
a Cosine Object

Strength	Phase	Direction
τ_0	1	$\alpha = 0$ $\beta = 0$
$\dfrac{\tau_1}{2}$	$e^{-i2\pi u_0 x}$	$\alpha = \lambda u_0 = \dfrac{\lambda}{a}$ $\beta = 0$
$\dfrac{\tau_1}{2}$	$e^{-i2\pi u_0 x}$	$\alpha = -\lambda u_0 = \dfrac{-\lambda}{a}$ $\beta = 0$

In this example there are only three plane-wave components in the field diffracted by the object. These are itemized in Table 7.2. Note that there is a direct relationship between the direction cosine of the diffracted wave and the spatial frequency of the periodic part of the object function. This is clarified in Fig. 7.45.

b. Image Decomposition. Equation (7.112a) tells us what the image-field distribution will be in the infinite lens limit. If we substitute the form for the Fourier decomposed object into this equation we arrive at the Fourier decomposed image.

$$E'_\infty(x', y') = \left(\frac{A}{m}\right) \exp\{i[\omega t - k(S + S')]\} \exp\left(-ik\,\frac{r'^2}{2X'}\right)$$

$$\times \iint T(u, v)e^{i2\pi[u(x'/m) + v(y'/m)]}\,du\,dv \tag{7.122}$$

Because of the phase factor that depends on r'^2 this decomposition no longer represents a superposition of plane waves. The surfaces of constant phase, which were planes at the object, bear curvature in the vicinity of the image plane.

2. Field Distribution in the Focal Plane. A plane wave at the object will be brought to a focus in the focal plane of the lens. The position of the point in the focal plane depends on the direction of propagation of the initial plane wave. In Fig. 7.46 a single plane wave is shown that yields a spot at F_α. Ray $P\bar{O}F$ passes through the center of the lens undeviated, thus identifying similar triangles $PO\bar{O}$ and $F_\alpha F\bar{O}$. This shows that

$$\frac{x_f}{f} = \frac{(-)x}{S}$$

From Eq. (7.85a) we see that the direct relationship among x_f, the direction cosine

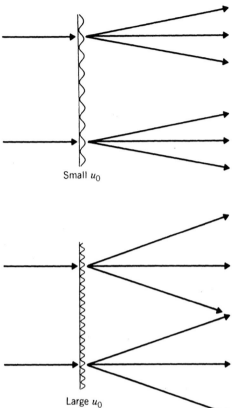

Small u_0

Large u_0

Fig. 7.45 The spatial frequency determines the angle at which the light from the object is diffracted.

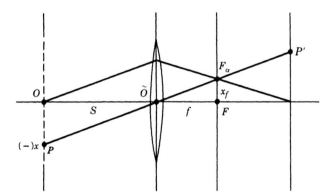

Fig. 7.46 Light diffracted by a single frequency leaves the object at a single angle that leads to a focus in the focal plane at a single spot.

of the plane wave α, and the spatial frequency u is

$$x_f = -x\frac{f}{S} = \alpha f = -u\lambda f \tag{7.123a}$$

A similar relationship may be derived for y_f.

$$y_f = y\frac{f}{S} = \beta f = -v\lambda f \tag{7.123b}$$

The object leads to a collection of diffracted plane waves, as we have seen in the previous section. Now we recognize that each of these component plane waves will be focused to a unique spot in the focal plane of the lens. Thus the field distribution in the focal plane must be related to the distribution of field intensity among the various plane waves. Because the decomposition of the object into plane waves is equivalent to the Fourier decomposition of the object function, the field distribution in the focal plane must be related to the Fourier transform of the object. Larger spatial frequencies will be represented by intensity in the focal plane that is farther from the optical axis. This is an extremely powerful concept. Let us further demonstrate it mathematically.

To generate an expression for the field distribution in the focal plane, start with the known distribution in the image plane. Now apply the backward Fresnel transformation as suggested by Eqs. (7.6) and (7.20b) to reach the focal plane.

$$E_f(x_f, y_f) = \iint E'(x', y')h(P' \to P_f)\, dx'\, dy' \tag{7.124}$$

$$h(P' \to P_f) = \left(\frac{-i}{\lambda X'}\right)\exp(ikX')\exp\left(\frac{ik}{2X'}[(x_f - x')^2 + (y_f - y')^2]\right)$$

$$= \left(\frac{-i}{\lambda X'}\right)\exp(ikX')\exp\left(\frac{ikr_f^2}{2X'}\right)\exp\left(\frac{ikr'^2}{2X'}\right)$$

$$\times \exp\left[-i2\pi\left(\frac{x'x_f + y'y_f}{\lambda X'}\right)\right] \tag{7.125}$$

and

$$r_f^2 = x_f^2 + y_f^2$$

Before proceeding with this operation it is useful to modify some of the factors in Eq. (7.125) by changing X'. To do this, recall the thin-lens equation

$$\frac{1}{S} + \frac{1}{S'} = \frac{1}{f}$$

which can be rewritten in another form

$$\frac{1}{S} = \frac{S' - f}{fS'}$$

This shows that

$$X' = \frac{S'}{S} f$$

or, using the transverse magnification,

$$X' = -mf \qquad (7.126)$$

We introduce this into the last phase factor in Eq. (7.125) to get

$$\exp\left[-i2\pi\left(\frac{x'x_f + y'y_f}{\lambda X'}\right)\right] = \exp\left[-i2\pi\left[\frac{x'}{m}\left(\frac{-x_f}{\lambda f}\right) + \frac{y'}{m}\left(\frac{-y_f}{\lambda f}\right)\right]\right] \qquad (7.127)$$

From Eq. (7.123) we see that the terms in the argument of the exponential involve the spatial frequencies. Therefore, Eq. (7.127) becomes

$$\exp\left[-i2\pi\left(\frac{x'x_f + y'y_f}{\lambda X'}\right)\right] = \exp\left[-i2\pi\left(u\frac{x'}{m} + v\frac{y'}{m}\right)\right] \qquad (7.128)$$

We use this form in the backward propagator kernel of Eq. (7.125) and also modify the amplitude factor with Eq. (7.126) to get

$$h(P' \rightarrow P_f) = \left(\frac{i}{\lambda fm}\right) \exp(ikX') \exp\left(\frac{ikr_f^2}{2X'}\right) \exp\left(\frac{ikr'^2}{2X'}\right) \exp\left[-i2\pi\left(u\frac{x'}{m} + v\frac{y'}{m}\right)\right] \qquad (7.129)$$

For the known image field we will use the infinite lens result from Eqs. (7.112a) and (7.116). This introduces the scaled image variables x'/m and y'/m. To perform the required integration in Eq. (7.124), make the variable changes $x'' = x'/m$ and $y'' = y'/m$ for the dummy integration variables. In the infinite lens situation this is equivalent to performing the integration in the object plane.

The kernel in Eq. (7.124) is what we have determined for the backward propagator from the image plane to the focal plane, Eq. (7.129). When all of these are combined we find that the distance terms in the exponentials can be simplified, $S + S' - X' = S - f$. We also see that the exponential that is quadratic in r' is canceled out.

These changes lead to the result for the field in the focal plane in the infinite lens limit:

$$E_{f\infty}(x_f, y_f) = \left(\frac{iA}{\lambda f}\right) \exp[i(\omega t - k(S + f))] \exp\left(\frac{ikr_f^2}{2X'}\right)$$

$$\times \iint \tau(x'', y'')e^{-i2\pi(ux'' + vy'')} dx'' dy''$$

$$= \left(\frac{iA}{\lambda f}\right) \exp[i(\omega t - k(S + f))] \exp\left(\frac{ikr_f^2}{2X'}\right) T(u, v) \qquad (7.130)$$

where

$$u = -x_f/\lambda f \quad \text{and} \quad v = -y_f/\lambda f \qquad (7.131)$$

This proves that the field distribution in the focal plane is proportional to the Fourier transform of the object transmission function.

The distribution of flux density in the focal plane will be proportional to the absolute square of the field. This no longer contains any of the exponential factors. It is only in these factors that information about the location of the object is contained. Therefore, the pattern of flux density in the focal plane is independent of object position. This can be readily verified in the laboratory.

Note that there is a further simplification when the object is at the first focal plane of the lens. Then $S = f$, S' and $X' = \infty$, and the phase factors reduce to

$$e^{-ik2f}$$

We then have a simple expression for the focal-plane field distribution.

$$E_{f\infty}(x_f, y_f) = \left(\frac{iA}{\lambda f}\right) e^{i(\omega t - 2kf)} T(u, v) \qquad (7.132)$$

3. Square Transmission Grating Object. At this point we could specify the field distribution in the focal plane for an arbitrary object. It is particularly enlightening, however, to study a grating with a "square" profile, as shown in Fig. 7.47. If the grating is infinite in the extent in the x-direction (physically unrealistic but useful as an example), then the transmission function can be written, according to Eq. (6.122), as a Fourier series.

$$\tau(x, y) = \sum_{n=-\infty}^{\infty} T_n \exp\left(i2\pi n \frac{x}{a}\right) \qquad (7.133)$$

The Fourier coefficients T_n are related to the Fourier transform of the local function describing one cycle of the grating

$$T_n \equiv \frac{1}{|a|} T_1\left(\frac{n}{a}\right) = \frac{1}{|a|} \int_{-a/2}^{a/2} \tau_1(x) \exp\left(-i2\pi n \frac{x}{a}\right) dx \qquad (7.134)$$

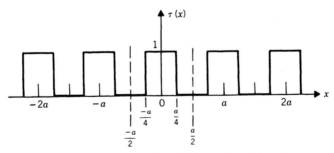

Fig. 7.47 An object that consists of alternating strips of transparent and opaque regions is represented by a transmission function that is a square wave.

The local transmission function restricts the integral in Eq. (7.134), leading to the result

$$T_1(u) = \frac{a}{2} \text{sinc}\left(\pi u \frac{a}{2} \right)$$

or

$$T_n = \frac{1}{2} \text{sinc}\left(n \frac{\pi}{2} \right) \tag{7.135}$$

When this is evaluated for all n we find

$$n = 0, \qquad T_0 = \tfrac{1}{2}$$

$$n = \text{even}, \quad T_n = 0$$

$$n = \text{odd}, \quad T_n = \frac{(-1)^{(|n| - 1)/2}}{\pi |n|}$$

or

$$T_n = +\frac{1}{\pi}, -\frac{1}{3\pi}, +\frac{1}{5\pi} \ldots, \quad \text{for } n = \pm 1, \pm 3, \pm 5, \ldots \tag{7.136}$$

We can use this information and Eq. (6.120) to write down the Fourier transform of the square grating object in terms of the spatial frequencies u and v.

$$T(u, v) = T_1(u)\delta(v) \sum_{n=-\infty}^{\infty} \delta(ua - n)$$

$$= \sum_{n=-\infty}^{\infty} T_n \delta\left(u - \frac{n}{a} \right)\delta(v) \tag{7.137}$$

Because the field distribution in the focal plane will be proportional to the Fourier transform, Eq. (7.137) shows that the focal plane will contain a series of spots along the x_f axis. These will be located at $x_f = -uf\lambda$ for the allowed values of u or

$$x_f = 0, \pm \lambda \frac{f}{a}, \pm 3\lambda \frac{f}{a}, \pm 5\lambda \frac{f}{a} \cdots \tag{7.138}$$

Each of these represents a Fourier component of the object.

The object transmission function can be rewritten using Eqs. (7.136) in Eq. (7.133)

$$\tau(x, y) = \frac{1}{2} + \frac{1}{\pi} \left[e^{i2\pi(x/a)} + e^{-i2\pi(x/a)} \right]$$

$$- \frac{1}{3\pi} \left[e^{i6\pi(x/a)} + e^{-i6\pi(x/a)} \right] + \cdots$$

$$= \frac{1}{2} + \frac{2}{\pi} \cos\left(2\pi \frac{x}{a} \right) - \frac{2}{3\pi} \cos\left(6\pi \frac{x}{a} \right) + \cdots \tag{7.139}$$

4. Spatial Filtering. The image-field distribution is specified, in Eq. (7.122), in terms of the inverse Fourier transform of the Fourier transform of the object function. Now we see that the Fourier decomposed object is physically distinct in the focal plane of the lens. The field distribution in the focal plane is directly proportional to the same Fourier transform that enters into Eq. (7.122). If we insert a "filter" in the form of a thin, partially transmitting barrier at the focal plane, we can modify the Fourier transform of the object. The altered transform must then be entered into Eq. (7.122). In this way the final image may be modified. In mathematical terms,

$$T(u, v) \rightarrow \tau_f(u, v)T(u, v)$$

so

$$E'_\infty(x', y') = \left(\frac{A}{m}\right) \exp[i(\omega t - k(S + S'))] \exp\left(-ik\frac{r'^2}{2X'}\right)$$

$$\times \iint \tau_f(u, v)T(u, v)e^{i2\pi[u(x'/m) + v(y'/m)]} \, du \, dv \qquad (7.140)$$

The filter may simply block part of the focal plane, thus eliminating some of the Fourier components in the reconstructed image.

To see how this works, consider the square transmission grating of the previous section. If we design a filter that is opaque beyond $x_f = \pm\lambda f/a$, then only the zero frequency and the lowest nonzero frequency components of the Fourier spectrum are passed (see Fig. 7.48). From Eq. 7.139 we see that this operation leaves us with an average term and a cosine term that have the same periodicity as the original function. Instead of the square transmission grating in the image, we are left with a cosine.

$$E'_\infty(x', y') = \frac{A}{m} \exp\{i[\omega t - k(S + S')]\} \exp\left(\frac{-ik}{2X'}r'^2\right)\left[\frac{1}{2} + \frac{2}{\pi}\cos\left(2\pi\frac{x'}{ma}\right)\right] \quad (7.141)$$

As shown in Fig. 7.49, the flux density will roughly approximate the square transmission function. The reason the image is so poor is that the higher Fourier components, which are required to reproduce the detail accurately, have been eliminated.

In general, as the region of the focal plane allowed to transmit light through to the image is increased, the image will more accurately reproduce the original

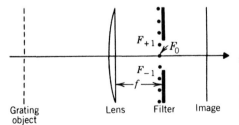

Fig. 7.48 A spatial filter in the focal plane blocks some of the higher Fourier components. These are then eliminated from the integral in Eq. (7.140) for the field in the image plane.

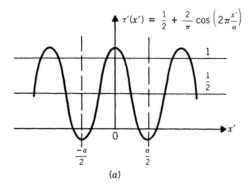

$$\tau'(x') = \frac{1}{2} + \frac{2}{\pi} \cos\left(2\pi\frac{x'}{a}\right)$$

(a)

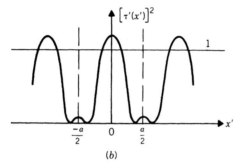

$$\left[\tau'(x')\right]^2$$

(b)

Fig. 7.49 Representation of the image of the square grating object when only the central three Fourier components are allowed to reach the image plane. The flux density approximates the object. It has the proper frequency and general shape; however, the details (particularly the abrupt corners) are missing.

object. In a real optical system there must be a limit to how far we may go off axis. Hence, beyond a certain order the higher diffraction spots will always be blocked.

Another interesting illustration of spatial filtering is shown in Fig. 7.50 where the central spot and the next two spots on either side are removed. This is accomplished with an obstruction that is opaque up to and including $x_f = \pm\lambda f/a$. In the image plane the field will be missing the lowest order Fourier components. Those missing are the same as the components that were present in the previous example. We end up with a square pattern minus the offset cosine. The results for the field intensity and for the flux density are shown in Fig. 7.51. Note that the pattern is strongest at the values of x_f associated with abrupt changes in the original object. This is due to the higher Fourier components that are allowed to

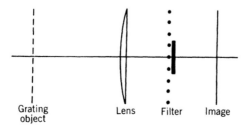

Grating object   Lens   Filter   Image

Fig. 7.50 A spatial filter that blocks the central three Fourier components in the focal plane.

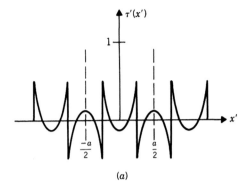

(a)

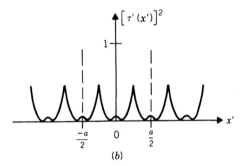

(b)

Fig. 7.51 The filtered image of the square grating object when the lowest Fourier components are missing. The flux density is now largest in the parts of the image where abrupt changes occur (on the corners of the square grating profile). The periodicity is twice that of the original object because the grating changes twice per cycle. The background level is missing.

pass through the filter. The low-order components contribute the background and general shape of the square grating. These are missing in the filtered result of Fig. 7.51.

C. Applications of Image Formation

The diffraction theory of image formation is the foundation of many modern applications of optics. In this section we examine some of the more important examples. In all of these we still assume that the source delivers light that is perfectly coherent. The characteristics of laser-beam optics are sufficiently unique that we deal with this topic first.

1. Gaussian Beam Optics. We have already discussed some of the characteristics of lasers in Chapter 5, where the longitudinal modes were described. There we used the ray picture, neglecting the transverse distribution of the field. Here we examine the character of the modes in more detail.

a. Cavity Modes. Our object is to specify the conditions under which the optical electric field exists in a standing wave within the laser cavity. We can imagine a certain distribution of the electric field just in front of one of the mirrors, and then by the Fresnel transformation calculate the field just in front of the other

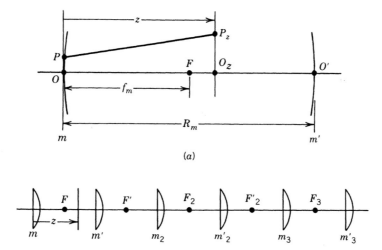

Fig. 7.52 Confocal resonant cavity. (*a*) Actual configuration of the mirrors at the ends of the cavity. (*b*) Lens analog that is useful in the discussion of multiple trips between mirrors. The function of the lenses is to alter the phase relationship across the transverse extent of the beam. This is the same thing that the mirrors do, with the added effect of changing the beam direction.

mirror. This process of calculation can then be repeated until a definite field pattern, or mode, emerges that reproduces itself at a given mirror, except possibly for small diffraction and reflection losses. If we can excite the optical field into such a mode, it will persist in the region between the mirrors for a relatively long time, while oscillating at its characteristic frequency. These optical modes are analogous to the modes of a microwave cavity, and the oscillations they undergo are analogous to the oscillations of an LC resonant circuit.

One of the simplest laser cavities to analyze is provided by the confocal configuration of Fig. 7.52. Two spherical mirrors, each of radius of curvature R_m are located a distance R_m apart. Our general condition for the occurrence of a mode of the cavity is that the field distribution at M reproduce, within a factor of ± 1, after reflection off M'. The situation is equivalent to the series of lenses shown in Fig. 7.52*b*. Here, rather than reversing direction on reaching M', a hypothetical wavefront is modified by the lens and continues on to M_2, which is the lens analog of the starting mirror in Fig. 7.52*a*.

We apply the Fresnel transformation to identify the form of the field at P_z, given the starting field at P.

$$E_z(P_z) = \iint E(P)h(P \to P_z)\, dx\, dy \qquad (7.142)$$

where from Eq. (7.20b)

$$h(P \to P_z) = \frac{i}{\lambda z} e^{-ikz} \exp\left(\frac{-ik}{2z} r_z^2\right) \exp\left(\frac{-ik}{2z} r^2\right) \exp\left[-i2\pi\left(\frac{-xx_z - yy_z}{\lambda z}\right)\right]$$

(7.143)

and where $r_z^2 = x_z^2 + y_z^2$, $P_z = (x_z, y_z)$. Thus

$$E_z(P_z) = \frac{i}{\lambda z} e^{-ikz} \exp\left(\frac{-ik}{2z} r_z^2\right) \iint E(x, y) \exp\left(\frac{-ik}{2z} r^2\right) e^{-i2\pi(u_z x + v_z y)} \, dx \, dy \quad (7.144)$$

where

$$u_z \equiv \frac{-x_z}{\lambda z} \quad \text{and} \quad v_z \equiv \frac{-y_z}{\lambda z} \tag{7.145}$$

are Fourier variables.

The integral in Eq. (7.144) is

$$\mathscr{F}\left[E(x, y) \exp\left(\frac{-ik}{2z} r^2\right)\right]$$

the Fourier transform of the phase-shifted starting field.

We need to examine the field at the end of the cavity where $z = R_m$. In addition, we must apply the field-modification factor that occurs at M' when the wavefront is reflected from the spherical mirror. In our lens analog we can model this effect by using Eq. (7.81), the lens transmission function. We ignore the finite extent of the lens. Therefore the influence of the mirror is described by the factor.

$$\tau'(P') = \exp\left(\frac{i\pi}{\lambda f_m} r'^2\right) = \exp\left(\frac{ik}{R_m} r'^2\right) \tag{7.146}$$

where $f_m = R_m/2$ [Eq. (3.35) for a mirror]. Naturally, this leads to a wavefront curvature modification that is in the "wrong" direction. However, in the expanded picture (Fig. 7.52) of the operation of the laser cavity, propagation of the wave between M' and M_2 is interpreted as a reflected wave.

We proceed by substituting R_m for z, and $P'(x', y')$ for $P_z(x_z, y_z)$ in Eq. (7.144) and multiplying by Eq. (7.146). When this is done, one of the phase factors in front of the integral in Eq. (7.144) combines with the mirror transmission function. This leads to the relatively simple result

$$E'(x', y') = \frac{ie^{-ikR_m}}{\lambda R_m} \exp\left(\frac{ikr'^2}{2R_m}\right) \mathscr{F}\left[E(x, y) \exp\left(\frac{-ik}{2R_m} r^2\right)\right] \tag{7.147}$$

The condition for a standing wave is that

$$E'(x', y') = \pm E(x, y)$$

The simplest solution to this requirement begins with the elimination of the factor of $\exp[(-ik/2R_m)r^2]$ in the argument of the Fourier transform operation. This is accomplished provided

$$E(x, y) = A(x, y) \exp\left(\frac{+ik}{2R_m} r^2\right)$$

where $A(x, y)$ is an amplitude that depends on the transverse coordinates. This form is already present in Eq. (7.147) for the transformed field. The other factors in Eq. (7.147) establish the following condition,

$$A(x', y') = \{\pm ie^{-ikR_m}\}\left\{\left(\frac{1}{\lambda R_m}\right)\mathscr{F}[A(x, y)]\right\} \tag{7.148}$$

The factor in the first curly brackets in Eq. (7.148) must be unity. This requires that

$$ie^{ikR_m} = \pm 1$$

or

$$kR_m - \frac{\pi}{2} = v\frac{2\pi R_m}{c} - \frac{\pi}{2} = n\pi \tag{7.149}$$

which imposes a limitation on the allowed frequencies that can resonate within the cavity. The separation between adjacent allowed "longitudinal modes" is [consistent with Eq. (5.147)]

$$\Delta v = \frac{c}{2R_m} \tag{7.150}$$

The transverse character of the allowed modes comes from the second curly bracket in Eq. (7.148), which requires that

$$\frac{1}{\lambda R_m} \mathscr{F}[A(x, y)] = A(x', y') \tag{7.151}$$

Eq. (7.151) invokes an amplitude function for the optical field that will be the Fourier transform of itself. One of the possibilities has already been encountered in Chapter 6, the Gaussian function. There we found

$$\mathscr{F}[\exp(-\pi b^2 r^2)] = b^{-2} \exp(-\pi \rho^2 b^{-2}) \tag{7.152}$$

Here the parameter b would need to be $(\lambda R_m)^{-1/2}$. Then if

$$A(x, y) = A_0 e^{i\omega t} \exp\left(\frac{-\pi r^2}{\lambda R_m}\right) \tag{7.153}$$

we would have

$$\mathscr{F}[A(x, y)] = A_0 e^{i\omega t} \exp\left(\frac{-\pi r'^2}{\lambda R_m}\right)\lambda R_m \tag{7.154}$$

as required, where

$$\rho^2 = u^2 + v^2 = \frac{r'^2}{\lambda^2 R_m^2}$$

Therefore, provided the conditions for the longitudinal mode in Eq. (7.149) have been met, a solution to the problem entails

$$E(x, y) = A_0 \, e^{i\omega t} \exp\left(\frac{-\pi r^2}{\lambda R_m}\right) \exp\left(\frac{ikr^2}{2R_m}\right)$$

$$= A_0 \, e^{i\omega t} \exp\left[-\pi r^2\left(\frac{i}{\lambda R_m}\right)(-i-1)\right] \tag{7.155}$$

and

$$E'(x', y') = A_0 \, e^{i\omega t} \exp\left(\frac{-\pi r'^2}{\lambda R_m}\right) \exp\left(\frac{ikr'^2}{2R_m}\right) \tag{7.156}$$

as allowed forms for the fields at the two mirrors. This shows that the transverse variation of the field will depend on r like a Gaussian function. This also shows that the allowed wavefronts at the mirrors will be curved. The constant phase surfaces are spheres. The factor $\exp(ikr^2/2R_m)$ is the parabolic approximation for the transverse variation of the phase of a spherical wave of radius R_m. The center of curvature of the constant phase surface at M will be at the center of M' and vice versa. The wavefronts match the mirror surfaces at reflection (Fig. 7.53).

Now that we understand the character of the field at the ends of the cavity we can use this information as input to determine the field at an arbitrary position between the mirrors through Eq. (7.144). We use the second form of Eq. (7.155) as the starting field. Then at a distance from M,

$$E_z(P_z) = A_0 \, e^{i\omega t}\left(\frac{i}{\lambda z}\right)e^{-ikz} \exp\left(\frac{-ikr_z^2}{2z}\right)$$

$$\times \mathscr{F}\left\{\exp\left[-\pi r^2\left(\frac{i}{\lambda z}\right)\left(1 - \frac{z}{R_m} - i\frac{z}{R_m}\right)\right]\right\}$$

$$= A_0 \, e^{i\omega t} e^{-ikz} \exp\left(\frac{-ikr_z^2}{2z}\right)$$

$$\times \left(1 - \frac{z}{R_m} - i\frac{z}{R_m}\right)^{-1} \exp\left[\left(\frac{ik}{2z}\right)r_z^2\left(1 - \frac{z}{R_m} - i\frac{z}{R_m}\right)^{-1}\right] \tag{7.157}$$

An important special case is $z = R_m/2 = f_m$, the center of the cavity at F. Here $[1 - (z/R_m) - i(z/R_m)]^{-1} = (1 + i)$. Then Eq. (7.157) becomes

$$E_{f_m}(P_{f_m}) = A_0 \, e^{i\omega t} e^{-ik(R_m/2)}(1 + i) \exp\left(\frac{-kr_{f_m}^2}{R_m}\right) \tag{7.158}$$

where $P_{f_m} = (x_{f_m}, y_{f_m})$ and $r_{f_m}^2 = x_{f_m}^2 + y_{f_m}^2$ in the center of the cavity at the focal plane of the two mirrors. The spherical phase factor has vanished. Here the surfaces of constant phase are *planes*, as they must be by symmetry.

For a symmetrical cavity, the beam will be smallest at the center of the cavity for this mode. Here the contour of the beam is called a "waist."

Let us *redefine our reference plane, where $z = 0$, to occur at the beam waist.*

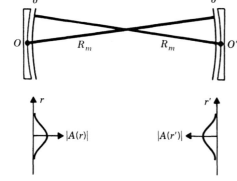

Fig. 7.53 The allowed form for the field distribution in a symmetrical cavity has a Gaussian amplitude behavior. Other modes are also possible. Here we illustrate the simplest.

There in general the field distribution goes like

$$E(x, y) = A_0 \, e^{i\omega t} \exp\left(\frac{-kr^2}{2z_0}\right)$$

$$= A_0 \, e^{i\omega t} \exp\left(\frac{-r^2}{w_0^2}\right) \tag{7.159}$$

where $2z_0$ is the length of the beam waist and

$$w_0 = \sqrt{\frac{\lambda z_0}{\pi}} \tag{7.160}$$

is the radius of the beam waist. At $r = w_0$ the field has decreased by a factor of $1/e$ from its maximum value at $r = 0$. A longitudinal distance z away from the waist the field is given by a Fresnel transformation.

$$E(x', y'; z) = A_0 \, e^{i\omega t}\left(\frac{i}{\lambda z}\right) e^{-ikz} \exp\left(\frac{-ikr'^2}{2z}\right)$$

$$\times \mathscr{F}\left\{\exp\left[-\pi r^2\left(\frac{i}{\lambda z}\right)\left(1 - i\frac{z}{z_0}\right)\right]\right\} \tag{7.161}$$

The integral is the Fourier transform of a Gaussian function with

$$u_z = \frac{-x'}{\lambda z}$$

$$v_z = \frac{-y'}{\lambda z} \tag{7.162}$$

as the Fourier variables. This leads to

$$E(x', y'; z) = A_0 \, e^{i\omega t} \, e^{-ikz}\left(\frac{z_0}{(z_0 - iz)}\right) \exp\left(\frac{-kr'^2}{2}\frac{1}{(z_0 - iz)}\right) \tag{7.163}$$

To help interpret this equation, define the following quantities:

$$w(z) \equiv w_0 \left[1 + \left(\frac{z}{z_0} \right)^2 \right]^{1/2} \tag{7.164}$$

$$R(z) \equiv z + \frac{z_0^2}{z} \tag{7.165}$$

$$\phi(z) \equiv \tan^{-1}\left(\frac{z}{z_0} \right) \tag{7.166}$$

Then Eq. (7.163) can be written in the form

$$E(x', y'; z) = A_0 \, e^{i\omega t} \frac{w_0}{w} \exp\left(-\frac{r'^2}{w^2} - \frac{ikr'^2}{2R} - ikz + i\phi \right) \tag{7.167}$$

We first note that the absolute value of the field is given by

$$|E(x', y'; z)| = A_0 \frac{w_0}{w} \exp\left(\frac{-r'^2}{w^2} \right) \tag{7.168}$$

Hence the magnitude of the field is still a Gaussian function. The term $i\phi(z)$ slowly changes from zero at the waist to $\pi/2$ at infinity. More rapid phase changes are provided by the $-i[kr'^2/2R(z)]$ term. This allows the interpretation of $R(z)$ as the radius of curvature of the constant phase surface at z. One can show that the wavefronts intersect at right angles with the surfaces where the ratio $|E(x', y'; z)/E(0, 0; z)|$ is constant. The surface where that constant is $1/e$ is usually taken as the definition of the "size" of the Gaussian beam, as shown schematically in Fig. 7.54. This surface and other surfaces like it are hyperboloids of revolution.

In the limit of large z, we have

$$R \to z \quad \text{and} \quad w \to \theta z$$

where

$$\theta = \frac{w_0}{z_0} = \frac{\lambda}{(\pi w_0)} \tag{7.169}$$

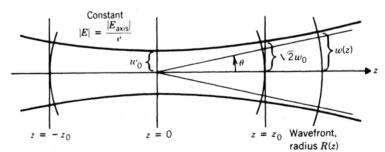

Fig. 7.54 Detail of the waist at the diffraction-limited convergence of a Gaussian beam.

is the angle between the asymptote to the hyperboloid where the ratio

$$|E(x', y'; z)/E(0, 0; z)|$$

is $1/e$. This angle represents the half-angle of divergence of the laser beam in the far-field limit. This limit corresponds to the limit of Fraunhofer diffraction, where the angular width of the beam is proportional to the wavelength divided by the beam width.

The preceding description has been applied to the lowest-order Gaussian beam, where the field depends only on the distance from the axis. Higher-order Gaussian beams have nodal surfaces as functions of x and y, the transverse coordinates. The higher-order modes occupy more lateral space and have sign variations in the phase from one portion of the transverse pattern to the next. For these reasons, it is common to establish conditions where only the lowest-order mode is allowed to resonate. This can be done by limiting the cross section of the beam with an aperture. As an example, consider a confocal cavity with the following parameters:

$$2z_0 = R_m = 10 \text{ cm}, \quad \lambda = 632.8 \text{ nm}$$

Eq. (7.160) identifies the waist size to be 0.1 mm, a very small radius for the lowest-order mode. A laser cavity with such a design having bore diameter of its discharge tube of 1 mm would lase in a higher mode than the Gaussian mode.

In the more general case of a symmetric resonator, where identical spherical mirrors of radii R_m are placed a distance L_m apart, stability can be achieved under different conditions. The wavefront curvature at the mirrors must match the mirror curvature. Therefore, from Eq. (7.165) with $R(z) = R_m$ at $z = L_m/2$, the beam-waist length parameter can be found

$$z_0 = \sqrt{(R_m - \tfrac{1}{2}L_m)\tfrac{1}{2}L_m} \tag{7.170a}$$

Through Eq. (7.160) the beam-waist radius is

$$w_0 = \left(\frac{\lambda}{\pi}\right)^{1/2} [(R_m - \tfrac{1}{2}L_m)(\tfrac{1}{2}L_m)]^{1/4} \tag{7.170b}$$

At the end mirrors the beam radius is

$$w\left(\tfrac{1}{2}L_m\right) = w_0\sqrt{\frac{R_m}{\left(R_m - \tfrac{1}{2}L_m\right)}} \tag{7.170c}$$

If one of the mirrors is partially transmitting, then some of the cavity radiation will leak out forming the output of the laser. This is shown in Fig. 7.55. Also shown in the diagram is an example of the "half-symmetric" cavity, for which the parameters in Eqs. (7.170) also apply.

For a numerical example, consider an argon–ion laser with a half-symmetric cavity design. If $R_m = 5$ m, $L_m = 2.3$ m, and $\lambda = 514.5$ nm, we find

$$z_0 = 2.1042 \text{ m}, \; w_0 = 0.5867 \text{ mm}, \; w(\tfrac{1}{2}L_m) = 0.6686 \text{ mm}, \; \theta = 0.2788 \text{ mrad}$$

w_0 is the beam radius at the flat mirror, $W(\tfrac{1}{2}L_m)$ is the radius at the spherical mirror.

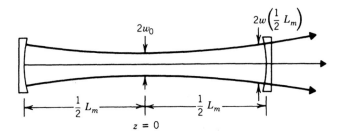

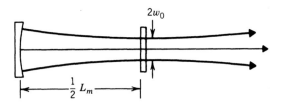

Fig. 7.55 The symmetrical cavity with confocal end mirrors and the half-symmetric design in which the flat mirror is placed at the beam waist. In both examples, the mirror is partially transmitting so that some of the coherent radiation is able to leak out.

The beam divergence, 2θ, amounts to 0.56 mrad. If the spherical mirror is the output end, the beam diameter there is 1.34 mm.

b. Focusing of Gaussian Beams. Consider the lowest-order Gaussian beam that originated, for instance, in a laser and is now propagating outside it. This beam will be characterized by its wavelength, waist radius, and waist length. The location of the beam waist can be determined from the radii of curvature of the mirrors in the laser cavity. A symmetrical cavity will have a waist in the center of the cavity. A half-symmetric cavity has a waist at the plane mirror.

Now let this beam be incident on a lens of focal length f located at a distance S from the beam waist, as shown in Fig. 7.56. We now show that in the parabolic approximation the beam transmitted by the lens is still a Gaussian beam.

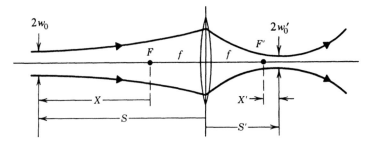

Fig. 7.56 Focusing a Gaussian beam.

The electric field at the lens is given by Eq. (7.163) with $z = S$. After the lens, the lens transmission function (neglecting the finite extent of the lens) contributes a phase factor of

$$\tau_L = \exp\left(\frac{ikr^2}{2f}\right)$$

Therefore, the field at the transmitting side of the lens is

$$E_L(x, y) = A_0\, e^{i\omega t}\, e^{-ikS}\, \frac{z_0}{(z_0 - iS)}\, \exp\left\{\frac{-kr^2}{2}\left[\frac{1}{z_0 - iS} - \frac{i}{f}\right]\right\} \qquad (7.171)$$

This will have the required form for a Gaussian beam with a length parameter z_0' and a beam waist located a distance S' to the right of the lens, provided that the term in square brackets in Eq. (7.71) equals

$$(z_0' + iS')^{-1}$$

Equating these two expressions gives the following complex "image formation" equation for Gaussian beams

$$\frac{1}{S + iz_0} + \frac{1}{S' - iz_0'} = \frac{1}{f} \qquad (7.172)$$

When we multiply Eq. (7.172) by the product of the three denominators and take the real and imaginary parts of the result we obtain two equations that may be solved for z_0' and S' (or $X' = S' - f$) in terms of z_0 and S (or $X = S - f$). The solutions are:

$$X' = \frac{f^2 X}{X^2 + z_0^2}; \quad \text{or } S' = f + \frac{f^2(S - f)}{(S - f)^2 + z_0^2} \qquad (7.173)$$

$$z_0' = z_0\frac{X'}{X} = z_0\frac{f^2}{X^2 + z_0^2} = \frac{z_0 f^2}{(S - f)^2 + z_0^2} \qquad (7.174)$$

Equation (7.174) together with Eq. (7.160) imply that the new radius of the beam waist is

$$w_0' = \sqrt{\frac{\lambda z_0'}{\pi}} = w_0\sqrt{\frac{f^2}{X^2 + z_0^2}} \qquad (7.175)$$

An interesting special case occurs when the beam waist is at the left focal point F of the lens. Then $X = 0$, and the length parameter and waist radius obey the simple relations

$$z_0' = \frac{f^2}{z_0}; \quad w_0' = \frac{f}{z_0} = \theta f \qquad (7.176)$$

where $\theta = w_0/z_0$ is the half-angle of divergence of the original Gaussian beam.

Equations (7.173 through 7.175) also apply for negative X, which implies that the original beam waist is to the right of F. In that case X' is also negative, implying that the "image" beam waist is to the left of F'. The equations also apply for a negative, or diverging, lens with its negative value of f. In such a case the focal points are reversed. F' is to the left of the lens and F to the right.

We illustrate the use of Eqs. (7.173 through 7.175) with a numerical example, as shown schematically in Fig. (7.57). The laser is the same one used in the previous example with $z_0 = 2.1042$ m and $w_0 = 0.5867$ mm. If a lens of 100-mm focal length is placed 150 mm to the right of the spherical output mirror, then $S = 1.3$ m, $X = 1.2$ m, and we find

$$X' = 2.045 \text{ mm}, \quad z_0' = 3.59 \text{ mm}, \quad w_0' = 24.22 \text{ } \mu\text{m}$$

It is instructive to compare these results with other predictions. Geometrical optics would predict an aberration-free image spot of infinitesimal radius in the focal plane at $S' = f$. If a uniform, collimated beam of wavelength 514.5 nm were to pass through a circular aperture of 0.6 mm radius just in front of the lens, a far-field diffraction pattern of the aperture would be observed in the right focal plane. This would consist of the Airy disk, surrounded by weaker rings. The radius of the Airy disk, conventionally taken to be that of the first dark ring, is given by Eq. (6.65) as

$$r_{\text{disk}} = 0.61 \frac{f}{r_{\text{aperture}}} = 52.3 \text{ } \mu\text{m}$$

in this case.

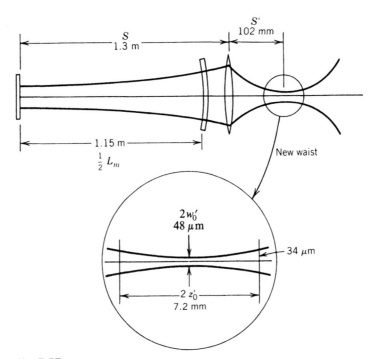

Fig. 7.57

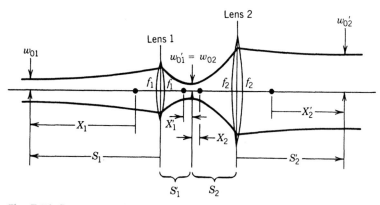

Fig. 7.58 Beam expander.

Two lenses may be used together to produce a laser beam expander, as shown schematically in Fig. 7.58. The first lens has focal length f_1, and the intermediate Gaussian beam has properties described by Eqs. (7.173 through 7.175) with subscripts 1 inserted everywhere. The second lens has a larger focal length, f_2. The parameters of the final Gaussian beam obeys Eqs. (7.173 through 7.175) with subscripts 2 inserted everywhere.

For a given value of focal length and placement of the first lens, an expanded beam with the largest value of waist radius w'_{02} will have the largest value of the length parameter z'_{02}. For a given value of f_2 of the focal length of the second lens, z'_{02} will be a maximum if $X_2 = 0$. This implies that the intermediate waist should be positioned in the left focal plane of the second lens. Then $X'_2 = 0$ and the final waist is located at the right focal plane of that lens. The radius of the final beam waist is then obtained by applying Eq. (7.175) twice and (7.174) once:

$$w'_{02} = w'_{01} \frac{f_2}{z_{02}} = w_0 \frac{f_2}{f_1} \sqrt{\frac{X_1^2 + z_{01}^2}{z_{01}^2}} \tag{7.177}$$

If it can be arranged that $X_1 = 0$, then $X'_1 = 0$, and

$$w'_{02} = w_{01} \frac{f_2}{f_1} \tag{7.178}$$

In this case the separation of the lenses would be if $f_1 + f_2$, and the beam expander behaves like an astronomical telescope. In practice, X'_1 is small compared with $f_1 + f_2$ and only small adjustments are necessary to optimize the beam expander. Care must be exercised when using two positive lenses as a beam expander for high-power lasers. In air the field strength at the beam waist can be large enough to induce breakdown. Under these circumstances it is best to use the Galilean beam expander. This is composed of a positive lens and a negative lens as in Fig. 3.41 with the negative lens first and the algebraic sum of the focal lengths still equal to the lens separation.

2. Image Processing. The concept of spatial filtering was introduced in section 7.B4. There it was demonstrated that manipulation of the field in the focal plane of a lens is equivalent to modification of the Fourier transform of an object that is in front of the lens. The image is related to the inverse transform of the field distribution in the focal plane. We illustrated the concept with specific mathematical examples that corresponded to processing of the image of a square grating. Here we wish to expand on the topic of spatial filtering with more concrete examples.

a. Low-Pass Filter. Figure 7.48 shows how a spatial filter can be used to pass only the central part of the Fourier spectrum of the object. The resulting image in Figure 7.49 lacks detail. This concept can be used to advantage in optical setups where high spatial frequencies are undesirable.

A low-pass filter consists of a very small aperture that blocks all of the high frequency components of the object so that the image is "smoothed out." This is commonly employed to "clean up" the output of a laser (Fig. 7.59), which, when operated in the lowest-order transverse mode, has a Gaussian transverse profile. In actual practice the profile is "noisy" because of inhomogeneities in the glass used for windows and mirrors within the laser and because of dust. The actual field distribution has a "rough" part owing to these effects superimposed on the smooth ideal distribution and may be written

$$E_{\text{actual}} = E_{\text{ideal}} + \Delta E$$

As shown in Fig. 7.60*a*, ΔE varies rapidly and randomly over distances small compared with w_0. Let $u_0 \gg 1/w_0$ be a characteristic spatial frequency for fluctuation in ΔE. Whereas the Fourier transform $\mathscr{F}[E_{\text{ideal}}]$ contains spatial frequencies from zero to about $1/w_0$, the transform $\mathscr{F}[\Delta E]$ will contain much higher frequencies centered about u_0. The resultant distribution in flux density is sketched in Fig. 7.60*b*. We readily see that we can remove the Fourier components

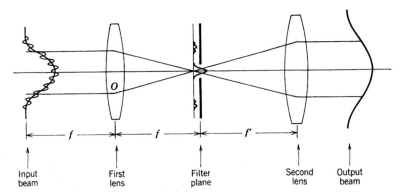

Fig. 7.59 Spatial filter used to remove the high spatial frequencies associated with a nonideal Gaussian beam.

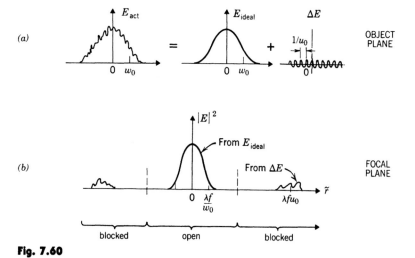

Fig. 7.60

of the noise $\mathscr{F}[\Delta E]$ if we insert a pinhole in the second focal plane of the first lens such that the hole has a radius that lies between $\lambda f/w_0$ and $\lambda f u_0$. The field distribution in the second focal plane of the second lens will then be a smooth Gaussian corresponding to E_{ideal} alone. Photographs and photometric scans of the flux density of laser beams with and without spatial filtering are presented in Fig. 7.61.

A low-pass spatial filter is frequently employed in conjunction with a beam expander. In the schematic example shown in Fig. 7.58 the spatial filter would be placed at the waist in between the two lenses. The resulting beam is broader in cross section than the input beam and has most of the high frequency noise removed.

b. Image Enhancement. Unwanted structure or spurious structure, such as scratches that have regularities, in a transparency can often be removed by spatial filtering. The transparency is used as an object in the setup of Fig. 7.43. The unwanted structure must have a Fourier transform narrowly limited as a function of the spatial frequencies u and v. Then the region in the filtering plane (the focal plane) corresponding to these frequencies can be blocked and the spurious structure removed from the final image. For instance, consider a series of parallel lines, not necessarily equally spaced, oriented at an angle θ from the horizontal, as shown in the transparency of Fig. 7.62a. Each line will yield a far-field diffraction pattern quite like that from a long, narrow slit. This will be centered on the axis of the system and will be elongated in the direction θ, from the vertical, as shown in Fig. 7.62b. To block these Fourier components we need an obstruction in the shape of two narrow wedges that do not quite meet at F, as shown in Fig. 7.62c.

Actual results of spatial filtering of this type are shown in the photographs of Fig. 7.63. The top photograph is a picture of the surface of the moon as telemetered to earth by an unmanned lunar orbiter. The "grating" of equally spaced lines is an

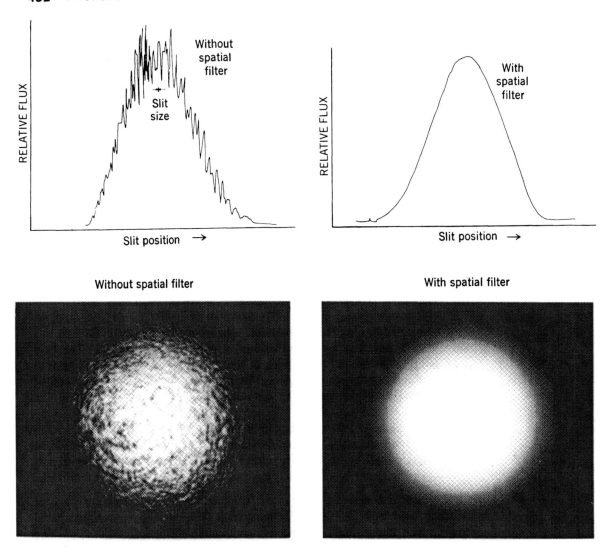

Fig. 7.61 The effect of spatial filtering on a uniphase laser beam. The plots show the results of scanning across the beam with a narrow slit in front of a photometer. (Illlustrations courtesy of Spectra-Physics, Inc.)

artifact of the telemetering process. Fig. 7.64 shows the diffraction pattern in the filtering plane when a negative transparency of the moon photograph is illuminated as in Fig. 7.43. The equally spaced large dots represent diffraction maxima resulting from the "grating" in the moon photograph and can be blocked in the filtering plane by the obstruction of Fig. 7.62c with $\theta = \pi/2$. A positive photograph can be made in the image plane that is the original minus the unwanted grating (Fig. 7.63b).

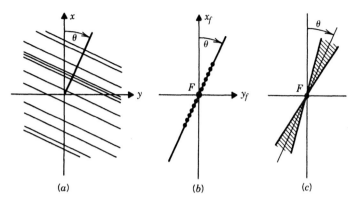

Fig. 7.62 Object transparency with oriented lines: (*a*) object; (*b*) diffraction pattern in the focal plane; (*c*) filter that will block the light associated with the regular characteristic of the object, thus eliminating that feature of the image.

Spatial filtering can also be used to remove the halftone lattice that is a characteristic of some printing procedures. Careful examination of the photographic reproductions of both pictures in Fig. 7.61 will reveal the regular lattice on which spots of different size are placed. As the density of the figure goes from light to dark, the spots merge. Our eye integrates this pattern of spots to yield the impression of continuous range of gray. In fact, the image is produced entirely with black or white.

In some cases it may be possible to cover the gray scale if the final recording medium is capable of rendering a continuous range of tones. This would be desirable in the photographic or analog electronic detection of a halftone object. It would be necessary to use the object as a transparency as we have done before. The diffraction pattern would consist of a disordered distribution resulting from the characteristics of the ideal object as well as a regular pattern that would be caused by the halftone lattice. In the case of Fig. 7.61 the pattern would consist of an array of spots on a square grid oriented at an angle of 45° with respect to the page. To remove this from the light going to the image we would have to block all of the spots on the grid. To do this would require a specially constructed filter that could be made in a separate process by exposing a sheet of film in the filter plane that was illuminated by the diffraction pattern of a halftone object of uniform density. The developed negative would be dark at the same place where the halftone diffraction pattern illuminated the film. When inserted in the focal plane during the processing of the original object, the halftone spatial frequencies would be eliminated from the image. This would produce a "softened" image with a continuous range of gray. However, because the filter also removes some of the spatial frequencies that might be required to reproduce all of the features of the object, the final image may take on a "blurred" quality. The result may still be more desirable than the original halftone object.

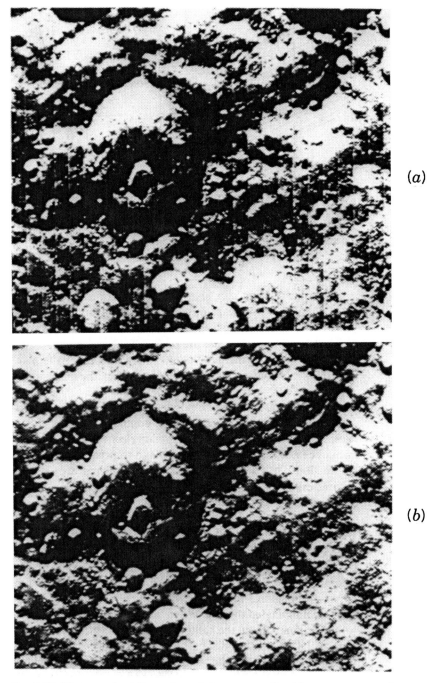

(a)

(b)

Fig. 7.63 Photograph of the surface of the moon radioed from an unmanned lunar orbiter: (a) before spatial filtering; (b) after spatial filtering. (Photographs courtesy of Conductron Corp.)

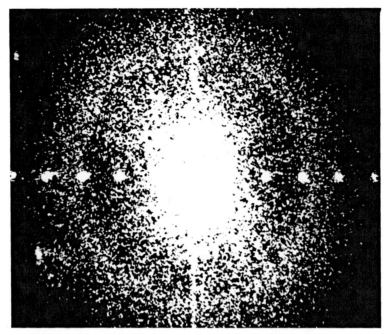

Fig. 7.64 Diffraction pattern in the focal plane due to an object transparency that is the unfiltered moon photograph of Fig. 7.63(a). (Courtesy of Conductron Corp.)

c. Phase-Contrast Microscopy. Many biological and other microscope specimens are almost completely transparent, so that they do not modulate the amplitude of a light beam passing through them; only the phase is modulated. The transmission function of a microscope slide containing such specimens has the form

$$\tau(x, y) = e^{-i\delta(x, y)} \tag{7.179a}$$

where

$$\delta = \frac{2\pi}{\lambda} \left[n(x, y) - 1 \right] d(x, y) \tag{7.179b}$$

Here n and d represent the local values of the refractive index and thickness of the specimen. Because the human eye responds to amplitude, not phase, these samples will appear invisible with an ordinary microscope. The problem is to convert somehow the *phase contrast* represented by Eq. (7.179) into amplitude contrast.

Let the microscope be illuminated coherently. Figure 7.43 will then apply to the objective lens of the microscope. The image it forms at the image plane then acts as an object for the eyepiece, not shown. If the object has a transmission function given by Eq. (7.179), the flux density in the image plane will be constant, and the object will be invisible. The object can be rendered visible if a spatial filter in the form of a *phase plate* is inserted in the focal plane of the objective. If the illumination is provided by a point source (as it is in a microscope), the filter plane is the *image plane* for the *point source*. As the source distance moves to infinity the filter plane becomes the focal plane.

The analysis is simplified if we assume that $\delta(x, y)$ is much less than unity, as is quite often the case. Then we may write, to first order in δ,

$$\tau(x, y) \simeq 1 - i\delta(x, y) \tag{7.180}$$

The first term (1) in Eq. (7.180) yields the undiffracted spot at F; the second term $(-i\delta)$ yields all the diffracted spots such as F_α. Each of these spots is actually an Airy disk with rings, but this is not relevant to our argument.

Without spatial filtering the electric field in the image plane is given by Eq. (7.112) with

$$\tau\left(\frac{x'}{m}, \frac{y'}{m}\right) \cong 1 - i\delta\left(\frac{x'}{m}, \frac{y'}{m}\right) \tag{7.181}$$

and

$$|\tau|^2 \cong 1$$

The phase plate is made so that it changes the phase of the spot at F by an angle $-\delta_f$ with respect to that of all the diffracted spots at F_α. One way of doing this is shown in Fig. (7.65a). The first term in Eq. (7.181) is then changed from unity to $e^{-i\delta_f}$.

Suppose that we have a quarter-wave phase plate (see Chapter 9) so that $\delta_f = \pi/2$. Then

$$e^{-i\delta_f} = e^{-i\pi/2} = -i$$

and (7.181) becomes

$$\tau\left(\frac{x'}{m}, \frac{y'}{m}\right) = i\left[-1 - \delta\left(\frac{x'}{m}, \frac{y'}{m}\right)\right] \tag{7.182}$$

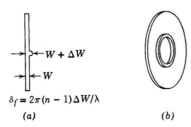

$\leftarrow W + \Delta W$

$\leftarrow W$

$\delta_f = 2\pi(n-1)\Delta W/\lambda$

(a)   (b)

Fig. 7.65 The filter in a phase-contrast microscope shifts the phase of the zero-order Fourier component by $\lambda/2$. If the incident beam at the object is a plane wave, then the filtering is done in the focal plane of the objective. If the incident beam comes from a finite distance source, then the filter plane is the image plane of the source. If the source is a point, then the phase filter has a single raised area as in (a). If the source is a ring (common in these types of instruments), then the phase filter must also be a ring as in (b).

Now to first order in δ the flux density will be proportional to

$$|\tau|^2 = 1 + 2\delta\left(\frac{x'}{m}, \frac{y'}{m}\right) \qquad (7.183)$$

We see that the phase contrast has indeed been converted to amplitude contrast. The object is now visible.

The visibility of the object may be improved if some attenuation $\tau < 1$ is introduced along with the phase change $-\delta_f$ at F. Then for a general value of δ_f we have

$$\tau\left(\frac{x'}{m}, \frac{y'}{m}\right) \cong \tau e^{i\delta_f} - i\delta\left(\frac{x'}{m}, \frac{y'}{m}\right)$$

$$= e^{-i\delta_f}\left[\tau - e^{i((\pi/2) + \delta_f)}\delta\left(\frac{x'}{m}, \frac{y'}{m}\right)\right] \qquad (7.184)$$

and

$$|\tau|^2 \approx \tau^2 + \delta^2 - 2\tau\delta\cos\left(\frac{\pi}{2} + \delta_f\right) \qquad (7.185)$$

This will yield maximum contrast when $\delta_f = \pm\pi/2$ and $\tau \approx |\delta|$.

Practical phase-contrast microscopes often use a condenser in such a way that the effective source has the shape of a thin ring rather than a point. The raised part of the phase plate must match the image of this source in the filter plane and will then be ring-shaped, as shown in Fig. 7.65b. Photomicrographs are shown in Fig. 7.66 that clearly reveal the improvement in visual contrast when the phase-contrast method is used.

d. Resolution with Coherent Illumination. In Chapter 6 where the discussion of diffraction limited resolution was first brought up we assumed that the two-point objects in question emitted light that was not mutually coherent. This is a reasonable assumption for light from two distant stars. Each star acts like a point source that can lead to diffraction at the lens aperture. However, emission from the sources is not correlated.

In a microscope the situation may be quite different. When a very small source is defocused by the condenser to produce a parallel beam at the object, we obtain transverse coherence over the entire field of view if the source is sufficiently small. In this case the resolution of two points in the object plane must be approached by considering the two points as a single object with structure.

The resolving power of a coherently illuminated optical system is quite difficult to define. An early criterion was developed by Abbe as an outgrowth of his diffraction theory of the image formation of a periodic object. He argued that the zero- and first-order diffraction spots are necessary to preserve a semblance of the original periodic signal. Figure 7.49 shows that this is indeed the case.

Suppose that a microscope objective is set up to operate as shown in Fig. 7.43 and that the aperture stop of the system is in the focal plane of the objective. If it

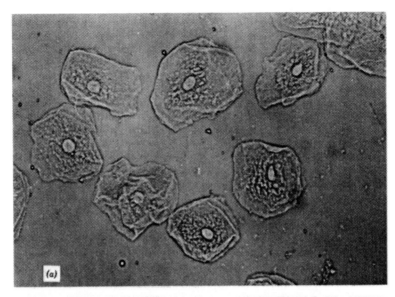

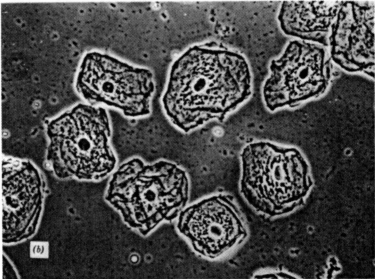

Fig. 7.66 Epithelial cells of the mouth as revealed by two kinds of microscopy: (a) ordinary bright field; (b) phase-contrast. (Photographs courtesy of Carl Zeiss, Inc.)

has a radius r_f, we have for the transmission function that must be inserted into Eq. (7.140) the following:

$$\tau_f(u, v) = \tau_f\left(\frac{-x_f}{\lambda_f}, \frac{-y_f}{\lambda_f}\right) = \begin{cases} 1, & u^2 + v^2 \leq w^2 \\ 0, & u^2 + v^2 > w^2 \end{cases} \qquad (7.186)$$

This has the effect of blocking all spatial frequencies u, v that satisfy

$$u^2 + v^2 > w^2$$

where

$$w = \frac{r_f}{\lambda f}$$

The integral in Eq. (7.140) is then limited to the domain

$$\sqrt{u^2 + v^2} < w$$

$$a > \frac{1}{w} = \frac{\lambda f}{r_f} \simeq \frac{\lambda}{\theta_{z,m}} \qquad (7.187)$$

Equation (7.187) represents one expression for the limiting resolution of a coherently illuminated system. We can show that for large angle optical systems obeying the sine condition (section 4.3A4) we should replace $\theta_{z,m}$ by $\sin \theta_{z,m}$. If the result is expanded to cover the case where the index of refraction at the object is n, the minimum resolvable value of the grating spacing is

$$a_{min} = \frac{\lambda}{n \sin \theta_{z,m}} \qquad (7.188)$$

This expression is often taken as the definition of the resolution of a microscope under coherent illumination at normal incidence, but it has a clear significance only for a periodic object. Note that a_{min} in Eq. (7.188) is 1.6 times greater than the value

$$x_{min} = \frac{0.61\lambda}{n \sin \theta_{z,m}}$$

given for incoherent illumination in Eq. (6.148).

If we are willing to suffer a loss of contrast and other sometimes undesirable effects, resolution can be improved by oblique illumination with coherent light. Suppose that the zero-order diffracted spot at F in Fig. 7.46 is no longer on axis, but is moved down until it is just inside the lower rim of the aperture to where the F_{-1} spot used to be. Then if the second-order spot is now just inside the upper rim, we have doubled the limiting spatial frequency and therefore halved the minimum resolvable object size to

$$a_{min} = \frac{0.50\lambda}{n \sin \theta_{z,m}} \qquad (7.189)$$

which is 22% better than the value for incoherent illumination.

With this oblique arrangement all diffraction spots of negative order are blocked. This will tend to reduce contrast, unless the zero-order is artificially attenuated. The asymmetry in the missing orders may produce other effects, but we will not pursue them here.

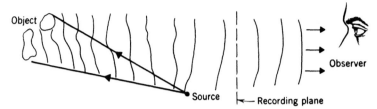

Object
Source
Recording plane
Observer

Fig. 7.67

3. Holography. In ordinary photography the image of the object is arranged to fall on the film where a two-dimensional map of the flux density distribution in the image plane is recorded. To view the photograph we observe the film directly or by transmission and thus simulate in our eyes the appearance of the original object. This method cannot reproduce the true optical field distribution that was originally present in the image plane because the information about phase of the field is lost in the photographic process. The film responds only to the time-averaged flux density.

In holography one reconstructs the wavefronts themselves. They contain all the optical information contained in the original light waves coming from the object. This is indicated schematically in Fig. 7.67, which shows an object being illuminated by light from a source. The scattered light waves then fall onto the eyes of an observer. If we have some way of recording these waves at, say, the plane shown dashed in the figure and then reproducing them at a later time so that the same waves proceed toward the observer, then what he or she will see is an image of the original object. However, in contrast to the case with normal photography, the image will appear in three dimensions with the proper relative position of its various parts. In the ideal situation the observer would not be able to tell if the light were coming from a reconstructed wavefront or from the original object.

a. Recording the Phase. The required phase record is produced with photographic film in a special way. The result is a spatial record in a particular plane of the relative phase relationships in the light coming from the object. Phase measurements are most effectively handled by interferometry. The optical beam containing the desired information is caused to interfere with a reference beam that is characterized by a featureless wavefront. The resultant flux density is sensitive to the phase difference between the information beam and the reference beam. The concept is illustrated in Fig. 7.68. The interference pattern in the recording plane is a record of the phase and amplitude distribution of the light coming from the object. A film plate exposed in this plane will bear a permanent record of this distribution. Equation (6.24) provides the quantitative expression for the spatial distribution of the time-averaged flux density in the recording plane,

$$\tilde{S} = \tilde{S}_1 + \tilde{S}_r + 2\sqrt{\tilde{S}_1\tilde{S}_r}\cos(\tilde{\phi}_r - \tilde{\phi}_1) \qquad (7.190)$$

where $\tilde{S}_1$ is the flux density in the beam coming from the object, $\tilde{S}_r$ is the flux density in the reference beam, and $\tilde{\phi}_r - \tilde{\phi}_1$ is the phase difference between the two

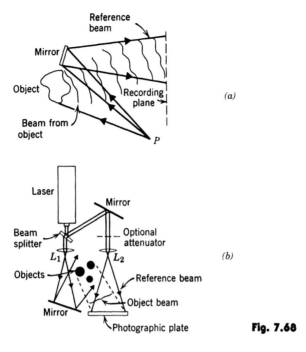

Reference beam

Mirror

Object

Recording plane

(a)

Beam from object

P

Laser

Mirror

Beam splitter

Optional attenuator

L_1 L_2

(b)

Objects

Reference beam

Object beam

Mirror

Photographic plate

Fig. 7.68

beams at the recording plane. The flux densities and phases are functions of $\tilde{x}$, $\tilde{y}$, the coordinates in the recording plane.

At this point it is convenient to adjust the relative strengths so that the reference beam is significantly brighter than the information beam ($\tilde{S}_r \gg \tilde{S}_1$). A three to fivefold inequality is sufficient in practice. This allows us to write an approximate form for Eq. (7.190) that is easier to handle.

$$\tilde{S} \cong \tilde{S}_r\left[1 + 2\sqrt{\frac{\tilde{S}_1}{\tilde{S}_r}}\cos(\tilde{\phi}_r - \tilde{\phi}_1)\right]$$

$$\cong \tilde{S}_r\left[1 + 2\left(\frac{\tilde{A}_1}{\tilde{A}_r}\right)\cos(\tilde{\phi}_r - \tilde{\phi}_1)\right] \tag{7.191}$$

This is the distribution of density to which the photographic film is exposed, where

$$\tilde{S}_1 = |\tilde{E}_1|^2, \quad \tilde{S}_r = |\tilde{E}_r|^2 \quad \text{with}$$

$$\tilde{E}_r(\tilde{x}, \tilde{y}) = \tilde{A}_r\, e^{i(\omega t + \tilde{\phi}_r(\tilde{x}, \tilde{y}))} \tag{7.192}$$

and

$$\tilde{E}_1(x, y) = \tilde{A}_1(x, y)e^{i(\omega t + \tilde{\phi}_1(\tilde{x}, \tilde{y}))} \tag{7.193}$$

For photographic film the relationship between the incident flux density and the film transmission after development is given by

$$\tau = K\tilde{S}^{-\gamma} \tag{7.194}$$

The variables K and γ depend on the length of the exposure and the characteristics of the film. Because Eq. (7.191) has the form

$$\tilde{S} = \tilde{S}_r(1 + \Delta)$$

where $\Delta \ll 1$, we can write

$$\tau = K[\tilde{S}_r(1 + \Delta)]^{-\gamma}$$

$$\cong KS_r^{-\gamma}(1 - \gamma\Delta)$$

$$\cong \tau_0\left[1 - 2\gamma\left(\frac{\tilde{A}_1}{\tilde{A}_r}\right)\cos(\tilde{\phi}_r - \tilde{\phi}_1)\right] \qquad (7.195)$$

where

$$\tau_0 = KS_r^{-\gamma}$$

This plate so developed is called a *hologram* and was first given this name by Dennis Gabor, the originator of the technique, in 1948. A photomicrograph of a hologram is shown in Fig. 7.69b. The small scale fringes carry the information.

b. Reconstructing the Wavefront. To obtain the exact reconstruction of the original beam from the object, we put the developed plate back in its original place and illuminate it with the original reference beam. In other words, we keep the original geometry of Fig. 7.68 but eliminate the original object. The light incident on the plate will have a field distribution given by Eq. (7.192).

Using Eq. (7.195) we obtain the electric field transmitted by the plate

$$E_t = \tau E_r = \tau_0\left[1 - 2\gamma\left(\frac{\tilde{A}_1}{\tilde{A}_r}\right)\cos(\tilde{\phi}_r - \tilde{\phi}_1)\right]\tilde{A}_r\, e^{i(\omega t + \tilde{\phi}_r)}$$

$$= \tau_0\left[1 - \gamma\left(\frac{\tilde{A}_1}{\tilde{A}_r}\right)(e^{i(\tilde{\phi}_r - \tilde{\phi}_1)} + e^{-i(\tilde{\phi}_r - \tilde{\phi}_1)})\right]\tilde{A}_r\, e^{i(\omega t + \tilde{\phi}_r)}$$

or

$$\tilde{E}_t = \tau_0\tilde{E}_r - \gamma\tau_0\tilde{E}_1\, e^{i2(\tilde{\phi}_r - \tilde{\phi}_1)} - \gamma\tau_0\tilde{E}_1 \qquad (7.196)$$

The first term in Eq. (7.196) is what the plate would transmit if it had been exposed only to the reference beam. The third term is proportional to the original electric field and represents the reconstructed light waves. If only it were present, an observer would see a virtual image of the original object in its original position. The second term often gives a real image of the object, as we will see later. It and the first term can be separated from the desired third term because of the different directions traveled by the beams they represent.

c. Point-Source Geometry. To describe the effect of the second term in Eq. (7.196) and to handle cases where the hologram is viewed in a different geometry from that of the reference beam to which it was exposed, it is convenient to change our point of view. We now concentrate on light from a point source as the reference beam and a particular point on the object. This leads to a spherical reference wavefront interfering with the wavefront from a particular part of the object. To describe the influence of the entire object we must superimpose the fields from all points on the object with their proper phases and amplitudes. Instead of Eq. (7.191), in analogy with Eq. (6.34) we have

$$\tilde{S} \cong \tilde{S}_r \left[1 + \left(\frac{2}{A_r} \right) \sum_n A_n \cos(\tilde{\phi}_r - \tilde{\phi}_n) \right] \qquad (7.197)$$

where the sum is over all points on the object. This leads to the more general form for Eq. (7.196). If the field from the object is written as

$$E_1 \rightarrow \sum_n E_n = e^{i\omega t} \sum_n A_n e^{i\tilde{\phi}_n} \qquad (7.198)$$

we have

$$\tilde{E}_t \cong \tau_0 \tilde{E}_r - \gamma \tau_0 e^{i2\tilde{\phi}_r} \sum_n E_n e^{-i2\tilde{\phi}_n} - \gamma \tau_0 \sum_n E_n \qquad (7.199)$$

for the transmitted field in the case where the reconstruction beam is identical to the reference beam. In the limit where the points are very close together the sum becomes an integral. We will not go that far with our theory but will develop the ideas necessary to understand the reconstruction of a point. The actual object will require a field that is a sum of the results due to a collection of points.

Consider Fig. 7.70, where the recording plane is identified. The origin of our coordinate system is at $\tilde{O}$ with $\tilde{x}$, and $\tilde{y}$ in the plane and z perpendicular to the plane. In exposing the hologram we are interested in the phase of a reference wave from a point at $P_r = (x_r, y_r)$ in a plane perpendicular to and a distance D_r along z as shown. In the parabolic approximation the field at $\tilde{P}$ in the recording plane due to the reference point source will be

$$E_r = \left(\frac{A_r}{R_r} \right) e^{i\omega t} e^{-ikR_r} \cong \left(\frac{A_r}{D_r} \right) e^{i\omega t} e^{i\tilde{\phi}_r}$$

where

$$\tilde{\phi}_r \cong -kR_{0,r} - \left(\frac{k\tilde{r}^2}{2D_r} \right) + k \left(\frac{x_r\tilde{x} + y_r\tilde{y}}{D_r} \right)$$

$$\cong -kR_{0,r} - \left(\frac{k\tilde{r}^2}{2D_r} \right) - k(\alpha_r\tilde{x} + \beta_r\tilde{y}) \qquad (7.200)$$

We have made use of Eqs. (7.85a and c) to define the direction cosines for a ray from P_r to $\tilde{P}$.

$$\alpha_r = -\frac{x_r}{D_r}, \quad \beta_r = -\frac{y_r}{D_r}$$

In addition to this wavefront, at $\tilde{P}$ we need the field from the point on the object,

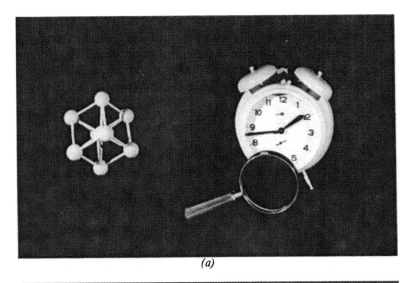

(a)

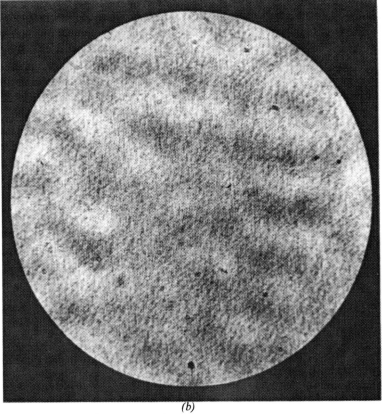

(b)

Fig. 7.69 *(continued on next page)*

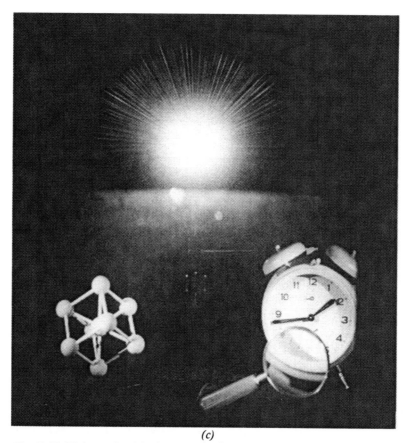

(c)

Fig. 7.69 Holography: (*a*) photograph of a three-dimensional object; (*b*) photomicrograph of hologram; (*c*) holographic reconstruction. (Photographs courtesy of Conductron Corp.)

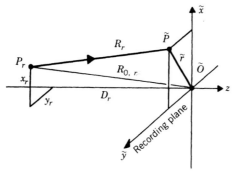

Fig. 7.70 Geometry for specification of the phase from a reference point source as measured in the recording plane.

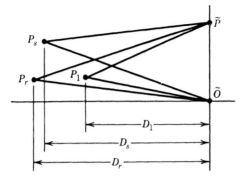

Fig. 7.71 Schematic representation of the reference point, the object point, and the point source used in the reconstruction. These are P_r, P_1, and P_s, respectively.

P_1. This is schematically shown in Fig. 7.71. With the appropriate subscripts we find the field at $\tilde{P}$ due to the object point at P_1 to have the same form at Eq. (7.200).

$$\tilde{E}_1 \cong \left(\frac{A_1}{D_1}\right) e^{i\omega t}\, e^{i\tilde{\phi}_1}$$

where

$$\tilde{\phi}_1 \cong -kR_{0,1} - \left(\frac{k\tilde{r}^2}{2D_1}\right) - k(\alpha_1\tilde{x} + \beta_1\tilde{y}) \tag{7.201}$$

The transmission function of the developed hologram will be given by Eq. (7.195) where $(\tilde{\phi}_r - \tilde{\phi}_1)$ is

$$\tilde{\phi}_r - \tilde{\phi}_1 = -k(R_{0,r} - R_{0,1}) - \left(\frac{k\tilde{r}^2}{2}\right)\left(\frac{1}{D_r} - \frac{1}{D_1}\right)$$
$$- k[(\alpha_r - \alpha_1)\tilde{x} + (\beta_r - \beta_1)\tilde{y}] \tag{7.202}$$

We now consider reconstruction where the point source P_s for the reconstruction beam is not oriented exactly the same way as was the source P_r for the reference beam. We further let the wavelength $\lambda' = 2\pi/k' = 2\pi c/\omega'$ of the light used for reconstruction be different from that used to expose the hologram ($\lambda = 2\pi/k = 2\pi c/\omega$). With the source distance and direction cosine denoted by D_s and α_s, the incident electric field at the hologram will then be

$$\tilde{E}_s \cong \frac{A_s}{D_s}\, e^{i\omega' t}\, e^{i\tilde{\phi}_s}$$

with

$$\tilde{\phi}_s \cong -k'R_{0,s} - \left(\frac{k'\tilde{r}^2}{2D_s}\right) - k'(\alpha_s\tilde{x} + \beta_s\tilde{y}) \tag{7.203}$$

and the transmitted light will be given in analogy with Eq. (7.196) by

$$\tilde{E}_t \cong \tau_0 \tilde{E}_s - \gamma\tau_0 \, e^{i\omega' t} \left(\frac{A_1}{D_1} \frac{A_s}{D_s} \frac{D_r}{A_r}\right) e^{i(\tilde{\phi}_r - \tilde{\phi}_1 + \tilde{\phi}_s)}$$

$$- \gamma\tau_0 \, e^{i\omega' t} \left(\frac{A_1}{D_1} \frac{A_s}{D_s} \frac{D_r}{A_r}\right) e^{i(\tilde{\phi}_1 - \tilde{\phi}_r + \tilde{\phi}_s)} \tag{7.204}$$

The spatial-dependent phase of the third term in Eq. (7.204) may be written approximately

$$(\tilde{\phi}_1 - \tilde{\phi}_r + \tilde{\phi}_s) \equiv -k'R' \cong -k'R'_0 - \left(\frac{k'\tilde{r}^2}{2} \frac{1}{D'}\right) - k'(\alpha'\tilde{x} + \beta'\tilde{y}) \tag{7.205}$$

This is equivalent to the phase of light from a point source P' having direction cosines α', β' and distance D' where

$$\frac{1}{D'} = \frac{k}{k'}\left(\frac{1}{D_1} - \frac{1}{D_r}\right) + \frac{1}{D_s} = \frac{\lambda'}{\lambda}\left(\frac{1}{D_1} - \frac{1}{D_r}\right) + \frac{1}{D_s} \tag{7.206}$$

and

$$\alpha' = \frac{\lambda'}{\lambda}(\alpha_1 - \alpha_r) + \alpha_s; \quad \beta' = \frac{\lambda'}{\lambda}(\beta_1 - \beta_r) + \beta_s \tag{7.207}$$

If P_r coincides with P_s and λ with λ', then P' will coincide with P_1.

If the object point is moved laterally a distance $\Delta x_1 \cong \Delta\alpha_1 D_1$, then P' moves laterally a distance $\Delta x' \cong \Delta\alpha' D'$. We may then define a transverse magnification that to first order in α_1 and $\tilde{x}/D_1$ is given by

$$m \equiv \frac{\Delta x'}{\Delta x_1} = \frac{\lambda' D'}{\lambda D} \tag{7.208}$$

as indicated in Fig. 7.72.

A small change in D_1 will result in a corresponding change in D' obtained by differentiating Eq. (7.206). For the longitudinal magnification we then obtain

$$\frac{\Delta D'}{\Delta D_1} = \frac{\lambda' D'^2}{\lambda D_1^2} \tag{7.209}$$

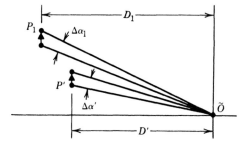

Fig. 7.72 Demonstration of the transverse magnification between a point object and its reconstructed image.

A similar treatment can be given the second term in Eq. (7.204). The spatial part of its phase is given by

$$(\tilde{\phi}_r - \tilde{\phi}_1 + \tilde{\phi}_s) \equiv -k'R'' \cong -k'R_0'' - \left(\frac{k'\tilde{r}^2}{2}\frac{1}{D''}\right) - k'(\alpha''\tilde{x} + \beta''\tilde{y}) \quad (7.210)$$

This corresponds to light from a point source at P'' with direction cosines

$$\alpha'' = \frac{\lambda'}{\lambda}(\alpha_r - \alpha_1) + \alpha_s, \quad \beta'' = \frac{\lambda'}{\lambda}(\beta_r - \beta_1) + \beta_s \quad (7.211)$$

and distance D'' given by

$$\frac{1}{D''} = \frac{\lambda'}{\lambda}\left(\frac{1}{D_r} - \frac{1}{D_1}\right) + \frac{1}{D_s} \quad (7.212)$$

In this case the angular orientation of the light tends to be opposite to that from P'. For, if both α_r and α_s are zero, we have

$$\alpha'' = -\alpha_1\lambda'/\lambda = -\alpha'$$

The transverse magnification for this image is

$$m = \frac{\Delta x''}{\Delta x_1} = \frac{D''}{D_1}\frac{\Delta\alpha''}{\Delta\alpha_1} = -\frac{\lambda'D''}{\lambda D_1} \quad (7.213)$$

and the longitudinal magnification is

$$\frac{\Delta D''}{\Delta D_1} = -\frac{\lambda'}{\lambda}\frac{D''^2}{D_1^2} \quad (7.214)$$

d. Fourier Transform Holography. In Fourier transform holography the object and reference sources are equidistant from the recording plane, so that $D_1 = D_r$. The spatial-dependent part of the phase difference in Eq. (7.202) has only the linear terms,

$$-k[(\alpha_r - \alpha_1)\tilde{x} + (\beta_r - \beta_1)\tilde{y}] \quad (7.215)$$

The hologram consists of equidistant sinusoidal fringes with a separation

$$\Delta\tilde{x} = \frac{\lambda}{\alpha_1 - \alpha_r} \quad (7.216)$$

Equations (7.206) and (7.212) then give us

$$D' = D'' = D_s$$

The situation is particularly simple when $\lambda = \lambda'$ and $\alpha_r = \alpha_s = 0$. Then $\alpha' = \alpha_1$ and $\alpha'' = -\alpha_1$. Figure 7.73 shows schematically the two images of a three-point object P_1, P_2, P_3. Note that the three-dimensional reconstruction of the double-primed objects has the wrong perspective. That is, P_2'' would appear to be in front of P_3'', even if it is larger. (For point P_3 the condition $D_3 = D_r$ is not satisfied, and

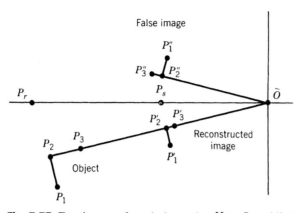

Fig. 7.73 Fourier transform holography. Here $D_s = 1/2 D_r$.

$(\tilde{\phi}_1 - \tilde{\phi}_r)$ will contain a term quadratic in $\tilde{r}$.) This is characteristic of the "false image" coming from the second term in Eq. (7.204). The reconstructed image P'_1, P'_2, P'_3 is distorted. The transverse magnification is one-half (from Eq. (7.208)), but longitudinal magnification is one-fourth (from Eq. (7.209)).

The name "Fourier transform" is applied to this type of holography because the transmission function $\tau(\tilde{x}, \tilde{y})$ of the hologram is related to the Fourier transform of the electric field at the object.

e. Fresnel Transform Holography in Parallel Light. When the condition $D_1 = D_r$ is not satisfied, so that there are terms in $\tilde{\phi}_1 - \tilde{\phi}_r$ that vary with $\tilde{r}^2$, we have what is known as Fresnel transform holography. A simple example occurs when both reference and reconstruction beams are parallel, that is $D_r = D_s = \infty$. For simplicity, we again assume that $\lambda' = \lambda$, and $\alpha_r = \alpha_s = 0$. Then $\alpha' = \alpha_1$ and $\alpha'' = -\alpha_1$. We also have $D' = D_1$ and $D'' = -D_1$. This means that the "false image", P'', is on the opposite side of the recording plane. This implies that the light actually comes to a focus at P''. The situation is illustrated in Fig. 7.74. Again notice the wrong perspective in the "false image," P''_1, P''_2, P''_3.

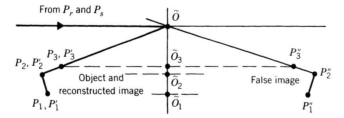

Fig. 7.74 Fresnel transform holography. Here D_s and D_r are both located a very large distance to the left of the drawing.

REFERENCES

Abramson, Nils H. *The Making and Evaluation of Holograms.* Academic Press, New York, 1981.

Arfken, G. *Mathematical Methods for Physicists.* Academic Press, New York, 1970.

Born, Max, and Emil Wolf. *Principles of Optics.* Pergamon Press, Oxford, 1980.

Cagnet, Michel, Maurice Françon, and Jean-Claude Thrierr. *Atlas of Optical Phenomena.* Springer-Verlag, Berlin, 1962.

Cathey, W. T. *Optical Information Processing and Holography.* Wiley Interscience, New York, 1974.

Collier, R. J., C. B. Burckhardt, and L. H. Lin. *Optical Holography.* Academic Press, New York, 1971.

De Velis, John B., and George O. Reynolds. *Theory and Application of Holography.* Addison-Wesley, Reading, Mass., 1967.

Duffieux, P. M. *The Fourier Transform and Its Application to Optics.* Wiley, New York, 1983.

Fowles, Grant R. *Introduction to Modern Optics.* Holt, Rinehart and Winston, New York, 1968.

Françon, Maurice. *Diffraction; Coherence in Optics.* Pergamon Press, Oxford, 1966.

Françon, Maurice. *Optical Image Formation and Processing.* Academic Press, New York, 1979.

Goodman, Joseph W. *Introduction to Fourier Optics.* McGraw-Hill, New York, 1968.

Kline, Morris, and Irwin W. Kay. *Electromagnetic Theory and Geometrical Optics.* Interscience, New York, 1965.

Lengyel, Bela A. *Introduction to Laser Physics.* Wiley, New York, 1966.

Linfoot, E. H. *Fourier Methods in Optical Image Evaluation.* Focal Press, London, 1964.

Martin, L. C. *The Theory of the Microscope.* Elsevier, New York, 1965.

Mertz, Lawrence. *Transformations in Optics.* Wiley, New York, 1965.

Pearcey, T. *Table of the Fresnel Integral to Six Decimal Places,* Cambridge University Press, Cambridge, 1956.

Rossi, Bruno. *Optics.* Addison-Wesley, Reading, Mass., 1957.

Smith, Howard M. *Principles of Holography.* Wiley, New York, 1975.

Smith, William V., and Peter P. Sorokin. *The Laser.* McGraw-Hill, New York, 1966.

Soroko, Lev Markovich. *Holography and Coherent Optics.* Plenum Press, New York, 1980.

Stark, Henry. *Applications of Optical Fourier Transforms.* Academic Press, New York, 1982.

Steward, E. G. *Fourier Optics.* Wiley, New York, 1983.

Stone, J. M. *Radiation and Optics.* McGraw-Hill, New York, 1963.

Stroke, George W. *An Introduction to Coherent Optics and Holography.* Academic Press, New York, 1969.

Verdeyen, Joseph T. *Laser Electronics.* Prentice-Hall, Englewood Cliffs, N.J., 1981.

Yu, Francis T. S. *Optical Information Processing.* Wiley, New York, 1982.

PROBLEMS

Section 7.1 Fresnel Transformations

1. Estimate the requirements on the displacement $(x - \tilde{x}) \simeq (y - \tilde{y}) \simeq (\tilde{x} - x') \simeq (\tilde{y} - y')$ for $D = D'$, so that the last terms in Eq. (7.16) may be safely neglected. Do this numerically for the case $D = D' = 2$ m, $\lambda = 500$ nm.

2. How does the form of the Fresnel transformation change if the surface of integration is changed to be defined by a sphere centered on P'?

3. Is it possible to design a propagator for the far-field limit of the Fresnel transformation? The far-field situation holds when P_1 and P_2 are separated by a distance that is large compared with the range of the integration over the $x_1 y_1$ plane. Discuss your answer.

4. Fill in the steps between equations so as to demonstrate explicitly that the propagator $h(P_1 \rightarrow P_2)$ in the parabolic approximation takes the equivalent forms represented by Eqs. (7.18), (7.20a), and (7.20b).

5. Describe a propagator that is an approximation of Eq. (7.2) and that carries the next order beyond the parabolic approximation in the specification of R_{12}.

6. In an imaginary situation assume that you need to find the transverse distribution of flux density in a coherent beam at a location that is inaccessible. You are able to measure the flux density at a position that is at a distance D before the desired location. You know that in this position the wavefront is a plane wave and possesses the flux distribution given by

$$\left\{ \begin{array}{l} A^2 \cos^2\left(\dfrac{2\pi x_1}{a}\right) \cos^2\left(\dfrac{2\pi y_1}{a}\right), \quad \text{for } x_1, y_1 \text{ between } \pm \dfrac{a}{4} \\[2mm] \qquad\qquad \text{zero,} \quad \text{otherwise} \end{array} \right\}$$

Use the Fresnel transformation to determine the flux density on the optical axis at the desired location D beyond the measurement plane.

7. Suppose that we have two coherent optical systems using light of the same wavelength. System 2 has an electric field in the plane $z = 0$ that is the same as that for system 1, but with a scale change, that is

$$E_2(x, y, 0) = E_1(mx, my, 0)$$

Show that, except for a constant phase factor, a Fresnel transformation gives

$$E_2(x', y', L) = E_1(mx', my', m^2 L)$$

Section 7.2   Fresnel Diffraction

8. In the laboratory we can observe the Fresnel diffraction pattern of a straight edge using the setup in Fig. 7.75, with a diffuse source behind a narrow slit parallel to the straight edge. Explain how this will work by considering the open region of the slit to be composed of many incoherent point sources. Describe semiquantitatively the effect on the diffraction pattern of each of the following acting separately.

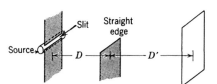

Fig. 7.75

(a) A small spread $\Delta\lambda$ in the wavelength.

(b) A finite width w in the slit.

(c) A small angular misalignment between slit and straight edge.

(d) A rough edge on the straight edge with a variation Δx.

Your answers can be made more precise if you use the results of Problem 7.13.

9. A monochromatic plane wave with wavelength 400 nm is normally incident on an opaque screen in which there is a long, narrow horizontal slit of width 0.7 mm. If the incident light at the slit has a flux density of 100 mW/cm², find the flux density at the center of the diffraction pattern on a screen 4 m from the slit using Cornu's spiral. Locate the position of the first minimum using Cornu's spiral. Compare this solution with that which would be obtained using the far-field approximation, Eq. (6.35).

10. A point source of wavelength 500 nm is 8 m from a screen. Another opaque screen is placed halfway between the source and the first screen. This second screen contains a square hole of sides 2 mm that is centered on a line perpendicular to both screens and passing through the source. Use Cornu's spiral to determine the location of the first two minima closest to the center of the pattern. Using the tables for the Fresnel integrals, calculate the flux density at the center of the pattern and at the first minimum if the point source emits 50 W of radiation.

11. A plane monochromatic wave with a wavelength of 400 nm is incident perpendicularly on an opaque plane screen bounded by a straight edge. Using Cornu's spiral, determine the positions of the first three maxima and the minima between them in the diffraction pattern observed on a plane parallel to the screen at a distance of 2 m.

12. A straight wire 1 mm in diameter lies along the path of a plane light wave of wavelength 500 nm, perpendicular to the direction of propagation. Plot the distribution of flux density on a screen 2 m from the wire. Mark the edges of the geometrical shadow.

13. The behavior of the Fresnel diffraction pattern of a straight edge in the bright region far from the geometric shadow may be obtained by studying the Fresnel integral Eq. (7.46c) with the lower integration limit replaced by minus infinity and the upper limit considered to be large. Let $\eta = \eta_0 + \Delta\eta$, where $\eta_0 \gg 1$ and $\Delta\eta \ll 1$. Show that one can then obtain an approximate expression for the integral that oscillates corresponding to a sinusoidal oscillation of $S(\eta)$ about S_{na} with amplitude $\Delta S = S(\eta) - S_{na} = S_{na}[\sqrt{2/(\pi\eta_0)}]$ and with a period of $2/\eta_0$.

14. A collimated beam of light wavelength 546 nm is incident normally on a thin opaque screen containing a circular aperture of diameter 5 mm. Find the positions of the axial points of maximum and minimum flux density.

15. Consider diffraction by the obstruction shown in Fig. 7.76. The incident radiation comes from a distant point source and has a wavelength of 500 nm. The observations are made at P' on the axis of the obstruction a distance of 2 m away. Find the flux density.

16. What is the electric field at an axial point in the observation plane if the aperture is an open section of wedge angle γ? (see Fig. 7.77.)

17. Monochromatic plane waves of wavelength λ are incident normally on a circular opening in a screen of

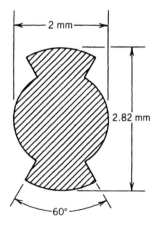

Fig. 7.76

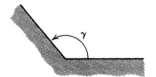

Fig. 7.77

radius $\tilde{r}_0$, and they then fall onto an observation plane a distance D' away. Assume that $\tilde{r}_0 = \sqrt{2N\lambda D'}$ where N is an integer. Let P' be the point on the observation plane that is in the center of the geometrical bright region.

(a) How many half-period zones are there in the opening as seen from P'? Call this answer M. In terms of M: What is the ratio of the electric field at P' to the electric field with no screen at all?

(b) When a zone plate is put in the opening that blocks out every other half-period zone and lets all the light from the other zones through?

(c) When a zone plate is put in the opening that shifts the phase of every other half-period zone by π radians and does not affect the other zones?

(d) When a perfect lens is put in the opening that focuses the rays at P'? You may neglect inclination factors.

18. A monochromatic point source P emits light of wavelength 600 nm. It falls onto an aperture $\tilde{A}$ 10 cm away and then onto a screen 20 cm beyond $\tilde{A}$.

(a) What is the value of $\tilde{r}_1$, the radius of the first Fresnel half-period zone at $\tilde{A}$?

(b) The aperture $\tilde{A}$ is a circle of radius 1 cm. How many Fresnel half-period zones does it contain?

(c) A zone plate with every other zone blocked out and with the radius of its first zone $= \tilde{r}_1$ is placed at $\tilde{A}$. This focuses the light from P onto the screen. The screen is moved toward $\tilde{A}$. Where will additional images be formed?

19. Show that for small values of δ the inclination factor $Q(\delta)$ in Eq. (7.61) may be written

$$Q(\delta) \approx 1 - \frac{\lambda}{4\pi} \frac{D^2 + (D')^2}{DD'(D + D')} \delta$$

20. Estimate a value for the integral $I(\delta_0)$ in Eq. (7.62) by integrating by parts. Now use the approximation of Problem 7.19 to obtain an explicit expression for $I(\delta_0)$.

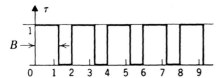

Fig. 7.78

21. Consider a zone plate that has the transmission function shown in Fig. 7.78 (plotted versus $\tilde{r}^2/\tilde{r}_1^2$). Each open region has a width of $B < 2$. Sketch the phasor curve at an axial point P' when such a zone plate is used as an aperture with $\tilde{r}_1 = \sqrt{\lambda d}$. What is the field at P' if there are N zones in the plate? (N is even). When $B \neq 1$ the zone plate can give a focus at a point P' for which $\tilde{r}_1 = \sqrt{2\lambda d}$, unlike the zone plate of Fig. 7.32b. Explain.

Section 7.3 Image Formation

22. Assume an optical system that is defect free (and for this example remains unspecified) creates a spherical wave that converges on point P'_0. Let an aperture be introduced a distance L' from P'_0. Use the Fresnel transformation formalism to determine the functional form for the field distribution in the vicinity of P'_0 on an observation plane parallel to the aperture plane. Use the parabolic approximation for the phase of the spherical wave at the aperture plane.

23. A small disk-shaped object of radius a is illuminated by a plane wave and imaged by a lens of radius b that is placed a distance $2f$ from the object. The configuration is that of Fig. 7.37 except that the object is at the origin of the object plane. Using Eq. (7.92) write down the integral that needs to be calculated in order that the field distribution in the image plane can be determined. Simplify the integral as much as possible. Now use the series expansion for the Bessel function to further simplify your result.

$$J_1(w) = \sum_{n=0}^{\infty} \frac{(-1)^n}{n!(n+1)!} \left(\frac{w}{2}\right)^{2n+1}$$

24. Using Eq. (7.92), find the analytical form for the optical electric field in the image of a square hole of sides a centered on a perpendicular to the z axis. The hole is illuminated from behind by a monochromatic plane wave. The lens has a focal length of f and an aperture defined by a square hole (whose sides are parallel to the object hole and are of size $b \gg a$) in a screen that is in contact with the lens. Assume that the object plane is $2f$ from the lens.

25. Derive Eq. (7.130), the field distribution in the focal plane, by a forward Fresnel transformation directly from the object, rather than by the backward transformation from the image, as was done in the text.

26. Prove the equivalencies expressed in Eq. (7.96a) for the arguments of the Fourier transform of the pupil transmission function in the relation for the electric field at the image plane due to a point source.

27. Derive the justification for Eq. (7.111b) in steps similar to those presented in the text to justify Eq. (7.111a). A diagram similar to Fig. 7.41 would be helpful.

28. Derive an expression for the field distribution in the focal plane of an infinite aperture lens that is illuminated by a monochromatic plane wave in the case where a transparent wedge is the object. Let the wedge be a square of sides a in the x-y plane and have an increasing thickness as y increases. Assume that the wedge is made of a material with index of refraction n. The wedge should be positioned at a distance $2f$ from the lens, where f is the focal length of the lens. Use the parabolic approximation for spherical wavefronts.

29. Consider the situation shown in Fig. 7.43 with illumination by a plane wave of wavelength λ. Let the object be a thin lens with focal length f' and radius a. In the case where we can neglect the size of the aperture of the imaging lens in Fig. 7.43 (because of its large size), compute the intensity and phase of the optical field in the image plane as a function of position in that plane.

30. Determine the modifications that must be made in the diffraction theory of image formation in the parabolic approximation if the lens is cylindrical rather than spherical. Is it possible to recover the spherical lens result with two different cylindrical lenses whose axes are perpendicular?

31. In the grating illumination experiment shown in Fig. 7.48, find the distribution of flux density in the focal plane of the lens when the object is similar to the regular grating of Fig. 7.47 except that the "boxes" centered on odd multiples of $\pm a$ are missing. This is an infinite grating with box functions centered on 0, $\pm 2a$, $\pm 4a \ldots$, each box having width $a/2$.

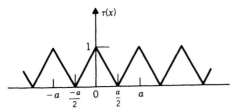

Fig. 7.79

32. The grating illumination experiment pictured in Fig. 7.48 is repeated with a transmission function at the object given by the infinite sawtooth waveform, Fig. 7.79. Find the expression for the flux density in the object. Determine the location and size of a spatial filter that will permit the zero order and the first two nonzero order spots in the pattern to pass, as shown in Fig. 7.48. Calculate the spatial distribution of flux density in the resulting image.

33. A photographic slide is used as an object in a system that is illuminated by a normally incident plane wave of wavelength λ. A lens is used to produce an image as shown in Fig. 7.43. The optical electric field is modulated by the object transmission function so as to bear the functional form

$$E(x) = E_0 \cos^2(2\pi u_0 x)$$

There is no variation in the y direction. Find the field distribution in the focal plane of the lens. Design a spatial filter that will alter the optical electric field so as to yield a flux density in the image that is given by $S'(x') = S'_0 \sin^4(2\pi u_0 x'/m)$.

34. Consider the object (a transparency) that is defined by the sum of these two functions:

$$\tau_a(x) = \begin{cases} |\tau_a|, & -b/2 \le x \le b/2 \\ 0, & \text{otherwise} \end{cases}$$

$$\tau_b(x) = \begin{cases} |\tau_b|\cos(20\pi x/b), & -b/2 \le x \le b/2 \\ 0, & \text{otherwise} \end{cases}$$

$$|\tau_b| \ll |\tau_a|$$

If this sum is the representation of the intensity of the *electric field* in the object, design a spatial filter that will produce an image that is a better rendition of the "box" part of the object. (The filter should remove the cosine function with as little distortion of the box function as

possible.) Assume that the lenses in your system have infinite apertures.

35. Describe how an "optical computer" might be used to identify all of the "a's" in a paragraph of text.

36. Quantitatively determine the form of the transmission function of a coating that, when applied to a lens, would remove the Airy rings in the image of a point source, leaving only the central spot.

37. Prove that a laser cavity can maintain a transverse mode that has the following dependence

$$E(x', y'; z) = A_0\, e^{i\omega t} 2^{3/2}\, \frac{x' w_0}{[w(z)]^2}$$

$$\times \exp\left[\frac{-r'^2}{[w(z)]^2} - i\,\frac{kr'^2}{2R(z)} - ikz + i2\phi(z) \right]$$

This is called the TEM$_{10}$ mode. It is characterized by two lobes of opposite phase on either side of the y-z plane.

38. Consider the half-symmetric laser cavity designed to operate with wavelength 524.5 nm. Its length is 2.3 m and its mirror radius is 5 m. Determine the beam radius in the middle of the cavity. If the plane mirror is the output mirror, find the beam radius a distance 5 m outside of the cavity.

39. A helium–neon laser (wavelength 632.8 nm) has a beam divergence of 1 mrad and a beam diameter of 1.0 mm at the exit mirror of its symmetric cavity. Design a beam expander that will produce a new beam with a diameter that does not fall below 1 cm over a length of 2 m.

40. A half-symmetrical cavity is designed to operate at 632.8 nm. The spherical mirror has a radius of 1 m. The plane mirror is 0.75 m from the spherical mirror and is designed to be the output mirror. Find the beam size and divergence at the output. What will the size of the spot be on the spherical mirror within the cavity?

41. In the *Schlieren* technique for examining the phase of an object, a spatial filter is introduced into an imaging system similar to Fig. 7.43 so that one half of the focal plane is blocked. That is, the filter function is

$$\tau_f(u, v) = \begin{cases} 1, & u > 0 \\ 0, & u \le 0 \end{cases}$$

With an object of the form given by Eq. (7.180) (small phase variations in a transparent object as might be expected from density variations in the air patterns within a wind tunnel), describe the characteristics of the image.

42. An oil-immersion objective is used with coherent illumination at 500 nm. If the numerical aperture is 1.2, find the minimum size of an object that can be resolved. Compare this with the resolution for incoherent illumination.

43. Show that a photographic negative of a hologram made with unit magnification will give the same observable image as the original hologram.

44. Suppose a hologram is made using light of wavelength λ with a reference source-to-hologram distance D_r. A photographic copy is now made at magnification m. The copy is then used with a reconstruction source of wavelength $\lambda' = m\lambda$, same angle, and distance $D_s = mD_r$. Show that a perfect reconstructed virtual image is formed that is magnified in all dimensions by a factor m.

45. A particular hologram is designed so that the reference and reconstruction beams make an angle of about 45° with the hologram plate. The object and reconstructed image are in a direction roughly normal to the hologram plane. What is the required resolution of the hologram plate if laser light of 633 nm wavelength is used to make the hologram? Explain why no "false image" is observed in this case.

46. Describe the nature of the reconstructed images when a hologram is exposed as in Fig. 7.73 but reconstructed as in Fig. 7.74 and vice versa. Assume $\lambda = \lambda'$.

47. Show that with a point object at P_1 and a point reference source at P_r, the fringe pattern on the hologram is obtained as the intersection in its plane of a family of hyperpoloids of revolution having P_1 and P_r as foci.

8 Coherence

In the previous chapters of this book we have dealt with monochromatic light that was perfectly coherent. That is, the phase differences among the component beams in the interference or the contributing Huygens' wavelets in diffraction remained constant during the time of measurement. In practice, perfect coherence is never possible. Any real source will only be partially coherent. This is related to a finite bandwidth about the average optical frequency of the source. In addition, for any extended source (even if it is monochromatic) whose component parts are not in perfect phase with one another, the radiated optical field will display a degree of incoherence because of the time-dependent superposition of fields from each component.

These effects are related to each other. The first is classified as *temporal incoherence*, the second is *spatial incoherence*. We deal with them separately for the most part; however, both involve the time over which the component parts of a resultant optical field maintain well-defined phase relationships.

Our notation in this chapter differs slightly from that used in other parts of the book. The fields in this chapter are, unless otherwise specified, real, whereas in the rest of the book we use complex fields to represent the actual optical disturbance and then take the real part of the complex field when required here, rather than using the exponential notation to describe the phases of optical fields, we use the cosine formalism. In this way we avoid complications in the temporal averages, which involve products of the fields.

8.1 Temporal Coherence

In Young's double-slit experiment, two light beams obtained from the original source are brought together after they have divided and have traveled different optical path lengths. A similar result is achieved in the Michelson interferometer. In either case, if light of a well-defined frequency is present, we observe regular interference fringes as a function of the optical path difference. If more than one frequency is present, we must add the flux density that would be produced by each acting separately. Then if, for example, the total flux density is divided equally between two frequencies, we see a beat pattern ranging in a regular periodic way from fringes of maximum contrast to fringes of zero contrast. We will show that, in general, the observed fringe pattern is uniquely determined by the distribution of flux density among the various component frequencies, and we will describe the method of recovering the spectrum from the interference pattern.

A. Introduction to Temporal Coherence

1. Fundamental Assumption. The basic fact that simplifies our considerations is that under ordinary circumstances, light signals of different frequency do not interfere. The issue of interference versus noninterference can be clarified by examining the simplest possible case. Consider a fictitious optical signal that is the superposition of two pure monochromatic components:

$$E = E_i + E_j$$

$$E_i = A_i \cos(2\pi v_i t + \varphi_i) = A_i \cos \phi_i$$

$$E_j = A_j \cos(2\pi v_j t + \varphi_j) = A_j \cos \phi_j$$

Here A_i, A_j, φ_i and φ_j are constants. We assume that the experimentally determined flux density at time t_0 is the average of the instantaneous flux density over the interval from $t_0 - T/2$ to $t_0 + T/2$, where T is the "measurement time." The instantaneous flux density is a constant times E^2. We take the average, as before

$$\langle E^2 \rangle = \frac{1}{T} \int_{t_0 - T/2}^{t_0 + T/2} E^2(t) \, dt$$

We thus obtain

$$\langle E^2 \rangle' = \langle E_i^2 \rangle + \langle E_j^2 \rangle + 2\langle E_i E_j \rangle$$

$$= \tfrac{1}{2}A_i^2 + \tfrac{1}{2}A_j^2 + 2A_i A_j \langle \cos \phi_i \cos \phi_j \rangle \qquad (8.1)$$

The factors of 1/2 come from taking the averages of $\cos^2 \phi_i$ and $\cos^2 \phi_j$, as discussed in Chapters 1 and 5. If there is to be any interference, it must come from the third term. Using a trigonometric identity, we rewrite that term as

$$2A_i A_j \langle \cos \phi_i \cos \phi_j \rangle = A_i A_j \langle \cos(\phi_i - \phi_j) \rangle + A_i A_j \langle \cos(\phi_i + \phi_j) \rangle$$

For optical frequencies,

$$\cos(\phi_i + \phi_j) = \cos[2\pi(v_i + v_j)t + \varphi_i + \varphi_j]$$

oscillates about 10^{15} times per second and averages to zero no matter how short the measuring time T is. The difference term

$$\cos(\phi_i - \phi_j) = \cos[2\pi(v_i - v_j)t + \varphi_i - \varphi_j]$$

ordinarily oscillates many times during a measurement also, because usually we have $(v_i - v_j)T \gg 1$. The interference term is then zero.

2. Monochromatic and Quasimonochromatic Light.
The electric field from a truly monochromatic source would be sinusoidal and would have a constant amplitude A and phase φ:

$$E(t) = A \cos(2\pi vt + \varphi) \tag{8.2}$$

Real sources are not truly monochromatic. Their output can be represented as a superposition of terms like Eq. (8.2). A continuous distribution of frequencies is needed to describe this superposition mathematically.

$$E(t) = \int_0^\infty A(v) \cos[2\pi vt + \varphi(v)] \, dv \tag{8.3a}$$

Here $A(v) \, dv$ is the amplitude of the field in a band of frequencies from v to $v + dv$. $A(v)$ is a continuous function called the spectral field amplitude.

If the frequencies v involved in this superposition are concentrated about a certain frequency v_0, the light is called *quasimonochromatic*. Then it becomes possible to write the optical field in the form

$$E(t) = A(t) \cos[2\pi v_0 t + \varphi(t)] \tag{8.3b}$$

where $A(t)$ and $\varphi(t)$ are slowly varying functions compared with $\cos 2\pi v_0 t$. That is, they change only slightly during one period $1/v_0$.

An important experimental quantity is the frequency distribution function or spectrum that can be determined by spectroscopy. We must therefore examine an idealized spectroscopy experiment.

3. Spectrometers and Spectral Functions.
Without interference between fields having different frequencies, two independent optical sources will contribute separately to the measured flux. The squares of the individual electric fields simply add

$$\langle E^2 \rangle = \langle E_i^2 \rangle + \langle E_j^2 \rangle$$

To generalize, we say that when frequencies v_i, v_j, $v_k, \ldots$ are present in the measured beam, the flux density will be proportional to

$$\langle E^2 \rangle = \langle E_i^2 \rangle + \langle E_j^2 \rangle + \langle E_k^2 \rangle + \cdots$$

This leads to

$$S_0 = S_i + S_j + S_k + \cdots \tag{8.4}$$

where S_i is the flux that would be detected if only frequency v_i were present in the beam and S_0 is the total flux density at all frequencies. Because real sources have a continuous distribution of frequency, we must generalize Eq. (8.4) further to

$$S_0 = \int_0^\infty S(v) \, dv \qquad (8.5)$$

Here $S(v) \, dv$ can be interpreted as the amount of optical flux density in the frequency range from v to $v + dv$. Here $S(v)$ is called the *spectral flux density*. It can be measured by passing the beam through an ideal spectrometer. This can be defined as an instrument that passes all light in the frequency range v to $v + \Delta v$ and none else. If Δv is small enough, this will approximate $S(v) \, dv$.

Spectrometers of the ordinary type work because they contain an element such as a prism to disperse light spatially according to its frequency. A prism accomplishes this because the refractive index and hence the deviation angle depend on frequency—normally higher frequencies are deviated more than lower frequencies. This is illustrated in Fig. 8.1. If the slit widths are small, the output will be proportional to $S(v)$ (except for complications because of diffraction-limited resolving power, which we now ignore). Gratings or other devices can also be used to achieve the desired dispersion. In real situations, the efficiency of the instrument,

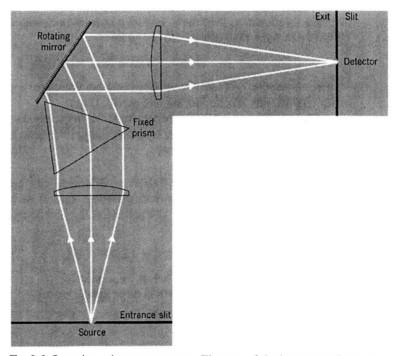

Fig. 8.1 Scanning prism spectrometer. The part of the instrument between the source and the detector is the monochromator. The spectrum is moved across the exit slit by moving the mirror, which is called a *Wadsworth mirror*.

including that of the detector, must be known if an absolute measure of $S(v)$ is to be obtained. Most of the time we are willing to settle for a relative measurement of $S(v)$. It will be described by the *normalized spectral distribution function $P(v)$* (having units of v^{-1})

$$P(v) = \frac{S(v)}{S_0} \qquad (8.6)$$

This function is normalized to unity:

$$\int_0^\infty P(v)\, dv = 1 \qquad (8.7)$$

The types of curves of $S(v)$ to be expected are indicated in Fig. 8.2.

The traditional method of measuring $S(v)$ is that just discussed. A prism or grating is employed to disperse the light as a function of v. The signal produced when the dispersed light reaches a detector is then recorded electrically. Another method, that of *interference spectroscopy*, determines $S(v)$ from the fringes produced in a Michelson interferometer or similar instrument. In practice this method

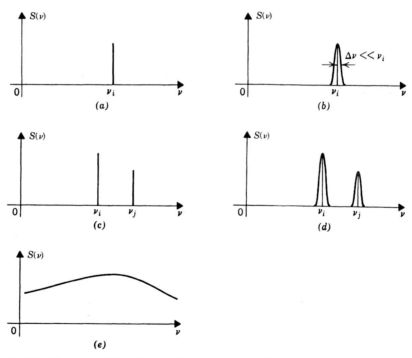

Fig. 8.2 The spectral flux density for several types of sources: (*a*) ideal monochromatic source; (*b*) quasimonochromatic source; (*c*) idealized two-frequency source; (*d*) real two-frequency source; (*e*) broadband frequency distribution as from a thermal (blackbody) source.

has proved particularly valuable in the far-infrared region of the spectrum, but there have also been specialized applications in the visible and near infrared. Interference spectroscopy also forms an ideal basis for a discussion of temporal coherence.

B. Interference Spectroscopy

One interference spectrometer is a Twyman-Green interferometer (Fig. 5.33). This is a Michelson interferometer modified to use collimated light. Equation (5.69) provides the phase shift for light of one wavelength λ_i or frequency ν_i reflected from the two mirrors:

$$\delta = \frac{4\pi \, d}{\lambda_i} = 2\pi\nu_i\left(\frac{2d}{c}\right)$$

Here d is the effective separation along the optical path of mirror I_2 from mirror I_1.

If only one frequency ν_i is present at the source and the beam splitter divides the incident beam equally, then for the time-averaged flux density at the detector from Eq. (5.25a) we can write

$$S = S_i[1 + \cos \delta]$$

$$= S_i\left[1 + \cos\left(\frac{4\pi\nu_i d}{c}\right)\right] \tag{8.8}$$

where S_i is the sum of the flux density from both beams. It is convenient to express the phase shift in the units of time. We therefore define

$$\tau \equiv \frac{2d}{c} \tag{8.9}$$

which represents the retardation time suffered by the beam in arm 2 of the interferometer relative to the beam in arm 1.

Equation (8.8) may then be written

$$S(\tau) = S_i[1 + \cos(2\pi\nu_i\tau)] \tag{8.10}$$

When considered as a function of τ, S consists of an average term S_i plus an oscillatory term

$$S_{\text{osc}} = S_i \cos(2\pi\nu\tau)$$

which averages to zero as τ is varied. The oscillatory term may be normalized by dividing by S_i to obtain an important quantity

$$\gamma(\tau) = \frac{S_{\text{osc}}}{S_i} \tag{8.11}$$

For the present example,

$$\gamma(\tau) = \cos(2\pi\nu_i\tau) \tag{8.12}$$

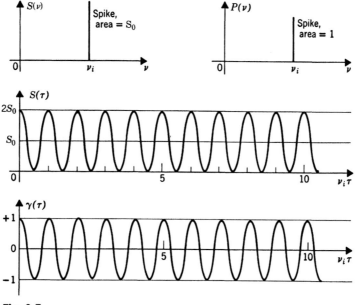

Fig. 8.3

For this single-frequency case the spectral distribution function $P(v)$ is an infinite spike having unit area at $v = v_i$; that is, $P(v)$ is a delta function.

$$P(v) = \delta(v - v_i) \tag{8.13}$$

The spectral flux density is simply

$$S(v) = S_i P(v) \tag{8.14}$$

The functions $S(v)$, $P(v)$, $S(\tau)$, and $\gamma(\tau)$ are sketched in Fig. 8.3.

1. Two-Frequency Source. Consider a source that emits light of two frequencies, v_i and v_j having total flux densities S_i and S_j. After a beam from this source is passed through the interferometer the detected flux density will be the sum of the interference effects resulting from each frequency separately.

$$S(\tau) = S_i[1 + \cos(2\pi v_i \tau)] + S_j[1 + \cos(2\pi v_j \tau)] \tag{8.15}$$

Again this consists of a τ-independent sum

$$S_0 = S_i + S_j \tag{8.16}$$

plus an oscillatory term

$$S_{\text{osc}}(\tau) = S_i \cos(2\pi v_i \tau) + S_j \cos(2\pi v_j \tau)$$

which may be written

$$S_{\text{osc}}(\tau) = S_0 \gamma(\tau) \tag{8.17a}$$

provided

$$\gamma(\tau) = \rlap{/}{\mu}_i \cos(2\pi v_i \tau) + (1 - \rlap{/}{\mu}_i) \cos(2\pi v_j \tau) \tag{8.17b}$$

with

$$\rlap{/}{\mu}_i \equiv \frac{S_i}{S_0}$$

the fraction of the total flux density at v_i, and $(1 - \rlap{/}{\mu}_i)$ the fraction of the flux density at v_j.

The flux density at the detector has the same form found for the single-frequency case, namely,

$$S(\tau) = S_0[1 + \gamma(\tau)] \tag{8.18}$$

but with $\gamma(\tau)$ now given by Eq. (8.17b).

For this case the normalized spectral distribution function is a sum of two delta functions

$$P(v) = \rlap{/}{\mu}_i \delta(v - v_i) + (1 - \rlap{/}{\mu}_i) \delta(v - v_j) \tag{8.19}$$

and the spectral flux density is still $S(v) = S_0 P(v)$.

A comparison of Eq. (8.17b) with (8.19) shows that the cosine terms and the delta functions have the same coefficients.

Figure 8.4a shows the behavior of the relevant functions for the case $\rlap{/}{\mu}_i = 1/2$; Fig. 8.4b shows what happens when $\rlap{/}{\mu}_i \neq 1/2$. $V(\tau)$ also shown in these figures is a quantitative measure of fringe contrast and will be discussed later.

If something should happen to destroy the interference—for example, a misalignment of one of the mirrors of the interferometer—there would be no fringes, γ would be zero, and the detector flux $S(\tau)$ would be simply a constant S_0. Then we learn nothing about the nature of the light in the original beam. Thus it is the function $\gamma(\tau)$ that really contains the useful information in this experiment.

The "beats" in Figs. 8.4 come about through the alternate reinforcement or cancellation of the two cosine terms at the different frequencies. Reinforcement occurs whenever $2|v_j - v_i|\tau$ is an even integer; then for a limited range of τ both cosines oscillate together between ± 1, giving a γ that also oscillates between ± 1. Cancellation occurs whenever $2|v_j - v_i|\tau$ is an odd integer and is complete if $\rlap{/}{\mu}_i = 1/2$; otherwise the fringes are still present but with reduced strength.

The interference fringes in Figs. 8.3 and 8.4 continue periodically for all values of the retardation time τ. This behavior is unreal and is a property of the discrete nature of the distribution functions that we have used thus far.

2. General Case, Multiple Frequencies. In the general case the spectral flux density will be described by a continuous function $S(v)$ of the type shown in Figs. 8.2b, d, or e. The sum in Eq. (8.15) then generalizes to an integral

$$S(\tau) = \int_0^\infty S(v)[1 + \cos(2\pi v \tau)] \, dv \tag{8.20}$$

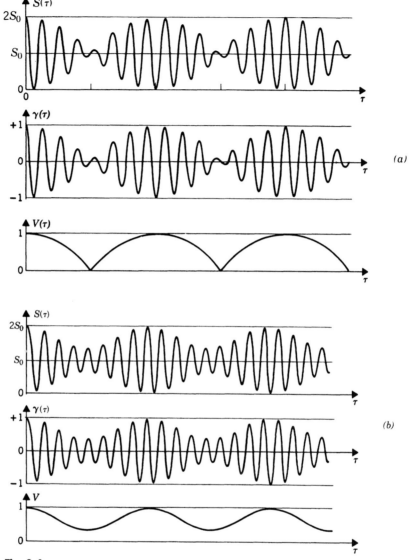

Fig. 8.4

Again this result consists of a constant term

$$S_0 = \int_0^\infty S(v)\, dv$$

plus an oscillatory term

$$S_{\text{osc}}(\tau) = \int_0^\infty S(v) \cos(2\pi v \tau)\, dv$$

[Note that $S(v)\, dv$ has the same units as $S(\tau)$.]

We can again rewrite Eq. (8.20) in the form of Eq. (8.18)

$$S(\tau) = S_0[1 + \gamma(\tau)]$$

provided

$$\gamma(\tau) = \int_0^\infty P(v) \cos(2\pi v\tau) \, dv \tag{8.21}$$

is the normalized *oscillatory term* and where $P(v)$ is defined by Eq. (8.6),

$$P(v) = \frac{S(v)}{S_0}$$

Refer back to Eq. (6.82), where the real Fourier transform was introduced. There we derived

$$f(x) = \int_0^\infty F_c(u) \cos(2\pi ux) \, du + \int_0^\infty F_s(u) \sin(2\pi ux) \, du$$

where $F_c(u)$ and $F_s(u)$ are the cosine and sine Fourier transforms, respectively, of $f(x)$. When $f(x)$ is an even function of x, then $F_s(u) = 0$. This leaves us with exactly the same form as Eq. (8.21). Therefore, with $x \to \tau$, $u \to v$, $f \to \gamma$, and $F_c \to P$, we see that $P(v)$ is the cosine Fourier transform of $\gamma(\tau)$. We can recover $P(v)$ from experimental knowledge of $\gamma(\tau)$ by taking the inverse transform of $\gamma(\tau)$.

According to Eq. (6.81a)

$$P(v) = 2\int_{-\infty}^\infty \gamma(\tau) \cos(2\pi v\tau) \, d\tau$$

$$= 4\int_0^\infty \gamma(\tau) \cos(2\pi v\tau) \, d\tau \tag{8.22}$$

The inversion is usually performed by a computer. There is a maximum value of τ given by $\tau_{max} = 2 \, d_{max}/c$, and this introduces some complications that will be discussed later in one of the problems.

C. Properties of $\gamma(\tau)$

1. Limits. From its defining equation (8.21) we see that $\gamma(\tau)$ is always unity at $\tau = 0$. Because $P(v)$ is nonnegative, if we multiply it by $\cos(2\pi v\tau)$ and integrate, we can never obtain a result greater in magnitude than the value obtained at $\tau = 0$. Thus

$$|\gamma(\tau)| \leqslant 1 \quad \text{for all } \tau$$

$$\gamma(0) = 1 \tag{8.23}$$

2. Superposition. Because $\gamma(\tau)$ is given as an integral transform of $P(v)$ and because the integral of a sum is the sum of the integrals, a normalized spectral distribution function $P(v)$ that is a superposition of individual distribution functions will have a

$\gamma(\tau)$ that is a similar superposition. Let

$$P(v) = \sum_j C_j P_j(v) \tag{8.24}$$

where each of the $P_j(v)$ is normalized to unity and where

$$\sum_j C_j = 1 \tag{8.25}$$

Then if $\gamma(\tau)$ is the normalized oscillatory term for $P(v)$ and $\gamma_j(\tau)$ is that for $P_j(v)$, we have

$$\gamma(\tau) = \sum_j C_j \gamma_j(\tau) \tag{8.26}$$

3. $\gamma(\tau)$ for Quasimonochromatic Light. For quasimonochromatic light the spectral distribution function is peaked about $v = v_0$ and tends to zero rather rapidly as v departs much from the vicinity of v_0. If Δv is a measure of the range of nonzero values of P, that is, from $v_0 - \Delta v/2$ to $v + \Delta v/2$, then we must have $\Delta v \ll v_0$. It is helpful to write P in the form

$$P(v) = D(v - v_0) \tag{8.27}$$

where $D(\mu)$ is small except for $-\Delta v/2 < \mu < \Delta v/2$. The normalization of P then gives

$$1 = \int_0^\infty P(v)\, dv = \int_{-v_0}^\infty D(\mu)\, du \approx \int_{-\infty}^\infty D(\mu)\, d\mu \tag{8.28}$$

Then for γ we obtain

$$\gamma(\tau) = \int_0^\infty P(v) \cos(2\pi v\tau)\, dv = \int_0^\infty D(v - v_0) \cos(2\pi v\tau)\, dv$$

Let $\mu = v - v_0$. Then we have approximately

$$\gamma(\tau) = \int_{-v_0}^\infty D(\mu) \cos[2\pi(v_0 + \mu)\tau]\, d\mu \approx \int_{-\infty}^\infty D(\mu) \cos[2\pi(v_0 + \mu)\tau]\, d\mu$$

$$= \cos(2\pi v_0 \tau) \int_{-\infty}^\infty D(\mu) \cos(2\pi \mu\tau)\, d\mu$$

$$- \sin(2\pi v_0 \tau) \int_{-\infty}^\infty D(\mu) \sin(2\pi \mu\tau)\, d\mu \tag{8.29}$$

We define functions $D_c(\tau)$ and $D_s(\tau)$:

$$D_c(\tau) = \int_{-\infty}^\infty D(\mu) \cos(2\pi \mu\tau)\, d\mu \tag{8.30a}$$

$$-D_s(\tau) = \int_{-\infty}^\infty D(\mu) \sin(2\pi \mu\tau)\, d\mu \tag{8.30b}$$

[Note that $D_c(0) = 1$, $D_s(0) = 0$.] Then we may write

$$\gamma(\tau) = D_c(\tau)\cos(2\pi\nu_0\tau) + D_s(\tau)\sin(2\pi\nu_0\tau) \tag{8.31}$$

This result may also be written in the form

$$\gamma(\tau) = U(\tau)\cos[2\pi\nu_0\tau + \Phi(\tau)] \tag{8.32}$$

$$U(\tau)\cos\Phi(\tau) = D_c(\tau) \tag{8.33a}$$

$$-U(\tau)\sin\Phi(\tau) = D_s(\tau) \tag{8.33b}$$

$$U(\tau) = +\sqrt{D_c(\tau)^2 + D_s(\tau)^2} \tag{8.33c}$$

Because according to Eq. (8.23) $|\gamma(\tau)| \leqslant 1$, we always have

$$0 \leqslant U(\tau) \leqslant 1 \tag{8.34}$$

The functions D_c, D_s, U, and Φ are all slowly varying compared with $\cos(2\pi\nu_0\tau)$ or $\sin(2\pi\nu_0\tau)$. It is difficult to give a general justification for this remark that is both rigorous and simple. The basic idea is that rapid oscillations in D_c or D_s would require that they be superpositions of rapidly oscillating cosine or sine components. But because $D(\mu)$ is limited to a range of $\pm\Delta\nu/2$ that is much smaller than ν_0, the resulting D_c and D_s cannot oscillate nearly as rapidly as $\cos(2\pi\nu_0\tau)$ or $\sin(2\pi\nu_0\tau)$.

From $\gamma(\tau)$ we can calculate the normalized spectral distribution function using Eq. (8.22).

$$P(\nu) = 4\int_0^\infty D_c(\tau)\cos(2\pi\nu_0\tau)\cos(2\pi\nu\tau)\,d\tau$$

$$+4\int_0^\infty D_s(\tau)\sin(2\pi\nu_0\tau)\cos(2\pi\nu\tau)\,d\tau$$

$$= 2\int_0^\infty D_c(\tau)\cos[2\pi(\nu + \nu_0)\tau]\,d\tau$$

$$+2\int_0^\infty D_c(\tau)\cos[2\pi(\nu - \nu_0)\tau]\,d\tau$$

$$+2\int_0^\infty D_s(\tau)\sin[2\pi(\nu + \nu_0)\tau]\,d\tau$$

$$-2\int_0^\infty D_s(\tau)\sin[2\pi(\nu - \nu_0)\tau]\,d\tau$$

or

$$D(\mu) = 2\int_0^\infty D_c(\tau)\cos[2\pi\mu\tau]\,d\tau$$

$$-2\int_0^\infty D_s(\tau)\sin[2\pi\mu\tau]\,d\tau \tag{8.35}$$

where the last step follows from the rapid variation in the sinusoidal functions in the first and third terms of the previous equation.

4. Symmetric Quasimonochromatic Spectrum. For quasimonochromatic light with a spectral distribution function $P(v)$ that is symmetric about $v = v_0$, we have $D(\mu) = D(-\mu)$. Now $\sin(2\pi\mu\tau)$ is an odd function of μ, and the product of an odd function and an even function is an odd function. This means that the contribution to the integral

$$\int_{-\infty}^{\infty} D(\mu) \sin(2\pi\mu\tau) \, d\mu$$

from positive values of μ will cancel that from negative values of μ, giving $D_s(\tau) = 0$. Then we have simply

$$\gamma(\tau) = D_c(\tau) \cos(2\pi v_0 \tau)$$

Examples of spectral distribution functions and D's are shown for symmetric and asymmetric cases in Fig. 8.5. Thus, $D(\mu)$ is simply $P(v)$ centered on zero frequency.

5. Contrast. The measured intensity is given by

$$S(\tau) = S_0[1 + \gamma(\tau)]$$

For quasimonochromatic light with γ given by Eq. (8.32) this becomes

$$S(\tau) = S_0[1 + U(\tau) \cos[2\pi v_0 \tau + \Phi(\tau)]] \tag{8.36}$$

which is to be compared with Eq. (8.10), the result for the idealistic but unachievable purely monochromatic source. There will be oscillations of γ between the limits $\pm U(\tau)$, and, because U is slowly varying in comparison with these oscillations, we may consider them to be well-defined "fringes." In the fringe pattern the intensity varies between a maximum of

$$S(\tau)_{\text{max}} = S_0[1 + U(\tau)] \tag{8.37a}$$

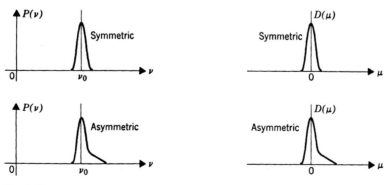

Fig. 8.5

and the adjacent minimum of

$$S(\tau)_{\min} = S_0[1 - U(\tau)] \tag{8.37b}$$

The *visibility* $V(\tau)$ of the pattern is defined by

$$V(\tau) = \frac{S(\tau)_{\max} - S(\tau)_{\min}}{S(\tau)_{\max} + S(\tau)_{\min}} \tag{8.38}$$

In the present case, equations (8.37) and (8.38) then yield simply

$$V(\tau) = U(\tau) \tag{8.39}$$

6. $\gamma(\tau)$ for a Two-Frequency Source. For two monochromatic lines at frequencies v_i and v_j, Eq. (8.17b) gives us for $\gamma(\tau)$:

$$\gamma(\tau) = p_i \cos(2\pi v_i \tau) + (1 - p_i) \cos(2\pi v_j \tau)$$

Now suppose that $v_i > v_j$, and write

$$v_i = v_0 + \frac{\Delta v}{2}$$

$$v_j = v_0 - \frac{\Delta v}{2}$$

where v_0 is the average $1/2(v_i + v_j)$ and Δv is the difference,

$$\Delta v = v_i - v_j$$

We then readily obtain

$$\gamma(\tau) = \cos(\pi \Delta v \tau) \cos(2\pi v_0 \tau) - (2p_i - 1) \sin(\pi \Delta v \tau) \sin(2\pi v_0 \tau)$$

$$= U(\tau) \cos[2\pi v_0 \tau + \Phi(\tau)] \tag{8.40}$$

with

$$U(\tau) = + \sqrt{1 - 4p_i(1 - p_i) \sin^2(\pi \Delta v \tau)} \tag{8.41}$$

and $\Phi(\tau)$ is obtainable from

$$U(\tau) \cos \Phi(\tau) = \cos(\pi \Delta v \tau) \tag{8.42a}$$

$$U(\tau) \sin \Phi(\tau) = (2p_i - 1) \sin(\pi \Delta v \tau) \tag{8.42b}$$

For $\Delta v \ll v_0$, Eq. (8.40) gives a slowly varying beat pattern, plotted in Fig. 8.4a for $p_i = 1/2$ and in Fig. 8.4b for $p_i \neq 1/2$. The visibility $V(\tau) = U(\tau)$ given by Eq. (8.41) is also plotted in Figs. 8.4. The visibility periodically changes from $V_{\max} = 1$ to $V_{\min} = |1 - 2p_i|$.

7. $\gamma(\tau)$ for a Gaussian Distribution Source. When the Michelson interferometer is used with any real optical signal, there is a finite range of retardation time $\tau = 2d/c$ over which interference fringes can be observed. Eventually the maxima in the interference fringes from one frequency fill in between the minima from another frequency, and the normalized oscillatory function goes to zero.

This effect will be illustrated by a specific example. Assume quasimonochromatic light with a spectral distribution function in the form of a Gaussian

$$P(v) = D(v - v_0) = \frac{1}{\Delta v} \exp\left[-\pi\left(\frac{v - v_0}{\Delta v}\right)^2\right] \tag{8.43}$$

or

$$D(\mu) = \frac{1}{\Delta v} \exp\left[-\pi\left(\frac{\mu}{\Delta v}\right)^2\right]$$

with

$$\mu = v - v_0$$

The parameter Δv characterizes the width of the line because $P(v)$ is down to $1/e$ of its maximum value when $v = v_0 \pm \Delta v/\sqrt{\pi}$. For the light to be quasimonochromatic it is necessary that $v/\Delta v \gg 1$. When this condition is met, integrals over positive v may be extended to negative v without risk.

Because the function $D(\mu)$ is symmetric about $\mu = 0$ in this case, Eq. (8.35) holds, and the normalized oscillatory term that describes the interference pattern is

$$\gamma(\tau) = D_c(\tau) \cos(2\pi v_0 \tau)$$

where $D_c(\tau)$ is given by Eq. (8.30a):

$$D_c(\tau) = \int_{-\infty}^{\infty} D(\mu) \cos(2\pi\mu\tau)\, d\mu = \text{Re}\left\{\int_{-\infty}^{\infty} D(\mu) e^{i2\pi\mu\tau}\, d\mu\right\}$$

$$= \text{Re}\{\mathscr{F}^{-1}[D(\mu)]\} \tag{8.44}$$

We have already encountered the Fourier transform and the inverse transform of a Gaussian. These relationships are documented in Table 6.2. With $x \to \tau$, $u \to \mu$, and $b \to \Delta v$ we find the inverse transform of $D(\mu)$

$$\mathscr{F}^{-1}[D(\mu)] = \exp[-\pi(\Delta v \tau)^2] \tag{8.45}$$

and

$$\gamma(\tau) = \exp[-\pi(\Delta v \tau)^2] \cos(2\pi v_0 \tau) \tag{8.46}$$

The behavior of $D(\mu)$ and $\gamma(\tau)$ for this case is sketched in Fig. 8.6. Note that the envelope $\exp[-\pi(\Delta v \tau)^2]$ is slowly varying in comparison with the cosine factor, because Δv is much smaller than v_0. For small values of τ (compared with $1/\Delta v$) we have approximately

$$\gamma(\tau) \approx \cos(2\pi v_0 \tau) \tag{8.47}$$

This, of course, is just the single-frequency result of Eq. (8.12). To see the effects of the finite spread Δv in the spectral distribution, we must use retardation times τ

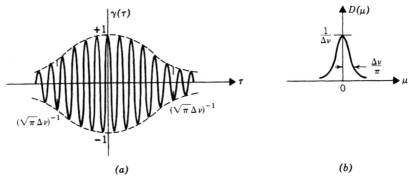

Fig. 8.6 Normalized oscillatory term (*a*) and spectral distribution function (*b*) for a quasimonochromatic line with a Gaussian shape.

of the order of $1/\Delta\nu$, for then the envelope is noticeably reduced—in this case by the factor $1/e^\pi$.

We may use this to define the *coherence time* τ_c for this example

$$\tau_c \approx \frac{1}{\Delta\nu} \tag{8.48}$$

The relation (8.48) is a general one connecting the coherence time and the frequency spread and will be discussed more fully later.

8.2 Statistical Optics

At the end of section 8.1 we considered a specific example of a continuous spectral distribution function for quasimonochromatic light and showed by explicit calculation that the contribution of the various frequencies to the pattern of interference fringes eventually caused a decrease in the oscillatory part of the pattern. This occurred when the the retardation time τ was comparable to the reciprocal of the width of the frequency distribution. This loss of fringe contrast is a general property of all true light sources, and its interpretation as a filling-in of the fringes from one frequency by those of another frequency is a perfectly valid one.

There is another way of looking at the loss of interference in a Michelson interferometer. In this second method we consider the statistical aspects of the time dependence of the light signal. For quasimonochromatic light, the light signal may be written in the form of Eq. (8.3):

$$E(t) = A(t) \cos[2\pi\nu_0 t + \varphi(t)]$$

All light signals have a random nature. Certain properties cannot be predicted exactly, but mean values can be defined. Real quasimonochromatic light fields will have randomly varying amplitudes $A(t)$ and phases $\varphi(t)$. This is also true in other regions of the electromagnetic spectrum. Of course $A(t)$ is not always fluctuating—radio frequency oscillators and some lasers give rise to fields with constant amplitudes—but the phase fluctuates because of inevitable drifts in the oscillator itself. These drifts, which are usually of thermal origin, can be reduced by lowering the temperature of certain components of the source, but ultimately for very stable systems these fluctuations will be a symptom of the fundamental quantum nature of light and matter.

Because the phase $\varphi(t)$ [and usually the amplitude $A(t)$] is not a well-defined function of time, it must be described by its statistical properties. Consider, for instance, its mean-square change (mean-square fluctuation) after a time interval Δt, which we write in the form

$$\langle [\Delta \varphi(\Delta t)]^2 \rangle = \langle [\varphi(t + \Delta t) - \varphi(t)]^2 \rangle = \langle [\varphi(t) - \varphi(t - \Delta t)]^2 \rangle$$

The average $\langle ... \rangle$ is over t for fixed Δt. To keep the present discussion simple, we assume that the average change in phase

$$\langle \Delta \varphi(\Delta t) \rangle = \langle \varphi(t + \Delta t) - \varphi(t) \rangle$$

is zero. Then positive fluctuations in φ are just as likely as negative fluctuations. Let τ_c be that value of Δt that gives a mean-square fluctuation $\langle \Delta \varphi^2 \rangle$ of one radian^2.

Now imagine this light used in a Michelson interferometer. The light in one arm is retarded by the time interval $\tau = 2d/c$ with respect to that in the other arm. If the retardation time is close to τ_c as just defined, there will be a 1-radian root-mean-square phase difference between the light in the two arms. This will substantially reduce the interference that can be observed when the two beams are brought together.

We now develop a quantitative description of the process just mentioned. The topic of mean-square phase fluctuations is developed quantitatively in the problems.

A. The Autocorrelation Function

1. Derivation. If the optical electric field intensity at the detector coming from one arm is $E(t)$, then that from the other arm is $E(t - \tau)$, namely E at the earlier time $t - \tau$. The detector measures an optical flux density proportional to the time average of the total squared and is

$$S(\tau) \propto \langle [E(t) + E(t - \tau)]^2 \rangle = \langle [E(t)]^2 \rangle + \langle [E(t - \tau)]^2 \rangle + 2\langle E(t)E(t - \tau) \rangle \tag{8.49}$$

The time averages are defined as usual

$$\langle [E(t)]^2 \rangle \equiv \frac{1}{T} \int_{t_0 + T/2}^{t_0 - T/2} [E(t)]^2 dt \tag{8.50}$$

We assume that T is so large that the integral is independent of t_0. This is equivalent to taking the limit as $T \to \infty$.

The terms $\langle [E(t)]^2 \rangle$ and $\langle [E(t - \tau)]^2 \rangle$ are proportional to the respective flux densities received in the beams from each arm and for a steady source are each proportional to $S_0/2$ where S_0 is the total flux density detected if interference does not occur. The term $2\langle E(t)E(t-\tau)\rangle$ contains all the interference effects. We normalize it by dividing by $\langle[E(t)]^2\rangle$

$$\gamma(\tau) = \frac{\langle E(t)E(t-\tau)\rangle}{\langle[E(t)]^2\rangle} \tag{8.51}$$

Then we have

$$S(\tau) = S_0[1 + \gamma(\tau)]$$

This last equation has the same form as Eq. (8.18) and is deliberately written using the same function $\gamma(\tau)$. In fact, the two definitions of $\gamma(\tau)$, Eq. (8.21) and (8.51), are equivalent. When written in the form of Eq. (8.51), $\gamma(\tau)$ is called the *normalized autocorrelation function* of the field $E(t)$. It describes the mean correlation of E with itself at an earlier time. We have already introduced the autocorrelation function in abstract form in Chapter 6. There we were dealing with complex functions.

In many treatments of partial coherence, use is made of the *analytical signal* $E^c(t)$. This is a complex function of time that is handled the same way we have used the complex notation in the past, namely

$$E(t) = \mathrm{Re}\{E^c(t)\}$$

A complex normalized autocorrelation function can then be defined

$$\gamma^c(\tau) = \frac{\langle E^c(t)E^{c*}(t-\tau)\rangle}{\langle|E^c(t)|^2\rangle} = \frac{\hat{\gamma}(\tau)}{\hat{\gamma}(0)} \tag{8.52}$$

where $\hat{\gamma}$ is the autocorrelation integral of Eq. (6.100). The real function of Eq. (8.51) can be shown to be

$$\gamma(\tau) = \mathrm{Re}\{\gamma^c(\tau)\} \tag{8.53}$$

To demonstrate this, we write the spectral flux density in terms of the amplitude of the analytical signal at frequency v

$$S(v) = K|A(v)^2| \tag{8.54}$$

where K is a constant. $A(v)$ is the Fourier component of the analytical signal.

$$E^c(t) = \int_{-\infty}^{\infty} A(v)e^{i2\pi vt}\,dv = \mathscr{F}^{-1}[A] \tag{8.55a}$$

$$A(v) = \int_{-\infty}^{\infty} E^c(t)e^{-i2\pi vt}\,dt = \mathscr{F}[E^c] \tag{8.55b}$$

Equation (6.98) defines the correlation

$$\hat{\gamma}(\tau) = \int_{-\infty}^{\infty} E(t)E^*(t-\tau)\,dt \tag{8.56}$$

This is related to the Fourier transform of E through the *Weiner–Khintchine theorem* of Eq. (6.102).

$$\mathscr{F}[\hat{\gamma}(\tau)] = |\mathscr{F}[E^c]|^2 = |A(v)|^2 = K^{-1}S(v) \qquad (8.57)$$

Therefore,

$$\hat{\gamma}(\tau) = K^{-1}\mathscr{F}^{-1}[S(v)] = K^{-1}\int_{-\infty}^{\infty} S(v)e^{i2\pi v\tau}\,dv \qquad (8.58)$$

We treat negative frequencies by defining $S(v) = S(-v)$; then

$$\gamma^c(\tau) = \frac{\hat{\gamma}(\tau)}{\hat{\gamma}(0)} = \frac{\displaystyle\int_0^{\infty} S(v)e^{i2\pi v\tau}\,dv}{\displaystyle\int_0^{\infty} S(v)\,dv}$$

$$= \int_0^{\infty} P(v)e^{i2\pi v\tau}\,dv \qquad (8.59)$$

Thus

$$\gamma(\tau) = \text{Re}\{\gamma^c(\tau)\} \qquad (8.60)$$

and we have shown that

$$\frac{\langle E(t)E(t-\tau)\rangle}{\langle [E(t)]^2\rangle} = \int_0^{\infty} P(v)\cos(2\pi v\tau)\,dv \qquad (8.61)$$

2. $\gamma(\tau)$ for a General Quasimonochromatic Source. Previously when we were concentrating on the spectral distribution function we defined quasimonochromatic light as light having a distribution function peaked at a certain frequency v_0, and then showed that the normalized oscillatory interference term could be written in the form of Eq. (8.31) or (8.32).

$$\gamma(\tau) = D_c(\tau)\cos(2\pi v_0\tau) + D_s(\tau)\sin(2\pi v_0\tau)$$
$$\gamma(\tau) = U(\tau)\cos[2\pi v_0\tau + \Phi(\tau)]$$

We now want to show that these same results follow from our new point of view of concentrating on the statistical signal.

We start with

$$E(t) = A(t)\cos[2\pi v_0 t + \varphi(t)]$$

The numerator of Eq. (8.51) for $\gamma(\tau)$ then becomes

$$\langle E(t)E(t-\tau)\rangle = \langle A(t)A(t-\tau)\cos A\cos B\rangle$$

where

$$A = 2\pi v_0 t + \varphi(t), \quad B = 2\pi v_0[t-\tau] + \varphi(t-\tau)$$

Use of the identity $\cos A \cos B = 1/2 \cos(A + B) + 1/2 \cos(A - B)$ gives $\cos A \cos B$
$= 1/2 \cos[2\pi v_0[2t - \tau] + \varphi(t) + \varphi(t - \tau)] + 1/2 \cos[2\pi v_0 \tau + \varphi(t) - \varphi(t - \tau)]$.

The first term oscillates rapidly with t and gives zero when the time average is taken. This leaves

$$\text{Numerator of } \gamma(\tau) = \tfrac{1}{2}\langle A(t)A(t - \tau) \cos[2\pi v_0 \tau + \varphi(t) - \varphi(t - \tau)]\rangle$$

The denominator of $\gamma(\tau)$ is simply the numerator at $\tau = 0$, which is $1/2\langle [A(t)]^2\rangle$. Thus,

$$\gamma(\tau) = \frac{\langle A(t)A(t - \tau) \cos[2\pi v_0 \tau + \varphi(t) - \varphi(t - \tau)]\rangle}{\langle [A(t)]^2\rangle} \tag{8.62}$$

Equation (8.62) expresses precisely what we previously tried to say imprecisely with words. It relates the interference pattern, represented by $\gamma(\tau)$, to the mean of a trigonometric function of the phase difference $[\varphi(t) - \varphi(t - \tau)]$. This is not surprising, because the phase is only defined to within an integral multiple of 2π, and only trigonometric functions of it are meaningful.

A qualitative interpretation of Eq. (8.62) may be given as follows. If the retardation time τ is short enough so that there is little fluctuation in phase between the times t and $t - \tau$, the cosine factor may be well approximated by $\cos 2\pi v_0 \tau$ and taken in front of the averaging integral $\langle\ldots\rangle$. It is the nature of light signals that a small phase fluctuation implies a small amplitude fluctuation, so that in this case $A(t - \tau) \approx A(t)$, and Eq. (8.62) is approximately

$$\gamma(\tau) = \cos 2\pi v_0 \tau$$

This represents simply the normalized oscillatory interference pattern for a monochromatic source.

When the rms phase fluctuations are of the order 1 radian or more, the cosine factor in Eq. (8.62) will fluctuate wildly and will average to something considerably smaller than $\cos(2\pi v_0 \tau)$. The interference has then been considerably reduced.

When the cosine in Eq. (8.62) is expanded, Eq. (8.31) is obtained, but with

$$D_c(\tau) = \frac{\langle A(t)A(t - \tau) \cos[\varphi(t) - \varphi(t - \tau)]\rangle}{\langle [A(t)]^2\rangle} \tag{8.63a}$$

$$-D_s(\tau) = \frac{\langle A(t)A(t - \tau) \sin[\varphi(t) - \varphi(t - \tau)]\rangle}{\langle [A(t)]^2\rangle} \tag{8.63b}$$

Equations (8.32 and 8.33) follow as before. Equations (8.30) then say that the cosine and sine Fourier transform of the function $D(\mu)$ are given by the right sides of Eqs. (8.63).

This is as far as we can go unless we know more about the functions A and φ. For a radio signal or for the light signal from an ideal single-mode laser, the amplitude A may be a constant. Then it will drop out of the averages in the preceding discussions, and the various functions γ, D_c, D_s, U, and Φ will depend only on the phase fluctuation through expressions such as

$$D_c(\tau) = \langle \cos[\varphi(t) - \varphi(t - \tau)] \rangle$$

and

$$-D_s(\tau) = \langle \sin[\varphi(t) - \varphi(t - \tau)] \rangle$$

3. $\gamma(\tau)$ for Thermal Light. At the other extreme from single-mode laser light is another kind of quasimonochromatic light known as *thermal light*. The use of the word "thermal" does not imply that the light comes from a thermal blackbody source; it refers to the statistical properties of the fluctuating light amplitude, which are the same as those of blackbody light. Thermal light results from the superposition of the light signals from a large number of independent microscopic sources such as individual atoms or molecules. The resulting signal is what is known in the theory of probability as a *Gaussian random variable*. For this reason, thermal light is often called *Gaussian light*.

It is not appropriate for us to study in detail the statistical aspects of thermal light, but we can discuss a model that will bring out many of its important properties. Assume that the electric field is the sum of the fields from a large number of independent quasimonochromatic oscillators having the same dominant frequency ν_0. They will be excited by collisions or other mechanisms, emit radiation for a finite time, be reexcited, reemit, and so on. Each of these events occurs at a random time, and there is no loss of generality if we assume each oscillator to be excited and emit light only once.

To visualize what happens, consider the complex description where the light signal E is given as the real part of a complex analytic signal E^c. The complex signal from an individual source will be of the form

$$E_j^c = A_j e^{i\varphi_j} e^{i2\pi\nu_0 t}$$

All complex signals have the common factor $e^{i2\pi\nu_0 t}$, which causes the phasor diagram in the complex plane to rotate counterclockwise with angular frequency $2\pi\nu_0$. The total signal at any instant will be the real part of $Re^{i2\pi\nu_0 t}$ where R has amplitude A and phase φ and is the sum

$$R = Ae^{i\varphi} = \sum_j A_j e^{i\varphi_j}$$

This is schematically shown in Fig. 8.7a. At a later time, different atoms are radiating with phases φ_j completely unrelated to the earlier phases. For steady sources the average number of oscillators is constant, but the magnitude A of the resultant R and its phase φ will have changed unpredictably, as in Fig. 8.7b. The time for an important fluctuation in the amplitude and phase of R, the coherence time, is essentially the duration of coherent radiation from an individual oscillator.

Because the individual oscillators that contribute to a thermal light signal are independent, and because the Michelson interferometer experiment is done under

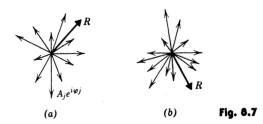

(a)   (b)   **Fig. 8.7**

conditions not capable of detecting interference between independent sources, the resulting interference pattern is essentially a superposition of patterns produced by each individual oscillator. In other words, *the light signal from each atomic or molecular oscillator interferes only with itself.* The optical field will be of the form

$$E(t) = \sum_{j=1}^{N} E_j(t) = \sum_{j=1}^{N} A_j(t) \cos[2\pi\nu_0 t + \varphi_j(t)] \tag{8.64}$$

If we introduce the autocorrelation function for the jth oscillator,

$$\gamma(j; \tau) = \frac{\int_{-\infty}^{+\infty} A_j(t) A_j(t - \tau) \cos\left[2\pi\nu_0\tau + \varphi_j(t) - \varphi_j(t - \tau)\right] dt}{\int_{-\infty}^{+\infty} \left[A_j(t)\right]^2 dt.} \tag{8.65}$$

The autocorrelation function for the entire source is

$$\gamma(\tau) = \frac{\sum_{j=1}^{N} \left\{ \int_{-\infty}^{\infty} [A_j(t)]^2 \, dt \right\} \gamma(j; \tau)}{\sum_{j=1}^{N} \int_{-\infty}^{+\infty} [A_j(t)]^2 \, dt} \tag{8.66}$$

This expression is a weighted average of the individual autocorrelation functions. The weighting factor

$$\int_{-\infty}^{\infty} [A_j(t)]^2 \, dt$$

is proportional to the total *energy* emitted by the jth oscillator.

If the individual oscillators are identical to the extent that each has the same $\gamma(j; \tau)$ we obtain simply

$$\gamma(\tau) = \gamma(j; \tau) \tag{8.67}$$

In this case the macroscopic autocorrelation function, which is the normalized oscillatory part of the interference pattern and can be measured directly, simply represents the autocorrelation function of the field from an *individual microscopic oscillator.* This result is sometimes known as *Campbell's theorem.*

B. Examples for Model "Thermal" Sources

The preceding results can be applied to simple models of the oscillators in radiating systems to calculate the oscillator autocorrelations $\gamma(j;\tau)$, the overall autocorrelation function, and the spectral distribution function. We make the *simplifying assumption* that the *phase* φ in the quasimonochromatic light *is a constant*. Then the amplitude A_j acts as a simple envelope modulating the fast cosine oscillations. Then Eq. (8.62) gives

$$\gamma(\tau) = \gamma(j;\tau) = \cos(2\pi v_0 \tau) \frac{\langle A_j(t)A_j(t-\tau)\rangle}{\langle [A_j(t)]^2\rangle} \tag{8.68}$$

We drop the index j, as it is understood that we are now referring to a signal from an invididual oscillator or to a collection of identical oscillators. Equation (8.68) contains the factor $\cos(2\pi v_0 \tau)$ characteristic of ideal monochromatic light times the autocorrelation function for the envelope or amplitude $A(t)$:

$$\gamma_A(\tau) = \frac{\langle A(t)A(t-\tau)\rangle}{\langle [A(t)]^2\rangle} \tag{8.69}$$

Thus, in the interference experiment $\gamma_A(\tau)$ acts as a slowly varying envelope for the $\cos(2\pi v_0 \tau)$ oscillations. Because the phase is constant, Eq. (8.63a) gives the results $D_c(\tau) = \gamma_A(\tau)$ and $D_s(\tau) = 0$, and by Eq. (8.30a) we find that the envelope autocorrelation function obeys

$$D_c(\tau) = \gamma_A(\tau) = \int_{-\infty}^{\infty} D(\mu) \cos(2\pi\mu\tau)\, d\mu$$

For the spectral distribution function, the inversion formula of Eq. (8.35) leads to

$$D(\mu) = 2\int_0^{\infty} D_c(\tau) \cos(2\pi\mu\tau)\, d\tau \tag{8.70}$$

1. Truncation Broadening. Suppose that the oscillator emits a sinusoidal signal for a limited time interval. This is sketched in Fig. 8.8. The amplitude $A(t)$ is constant for $|t| \leqslant t_1$, and zero otherwise. Because Eq. (8.69) is independent of the magnitude of the envelope $A(t)$, we take it to be either zero or unity, as shown in Fig. 8.8b. The function $A(t-\tau)$ is simply $A(t)$ shifted to the right a distance τ. This gives for $A(t) \cdot A(t-\tau)$ the behavior as shown in Fig. 8.8c. The numerator of Eq. (8.69) is proportional to the area of the shaded rectangle, which is $(2t_1 - |\tau|)$ for $|\tau| \leqslant 2t_1$, and zero otherwise. The denominator is the numerator at $\tau = 0$. This gives

$$D_c(\tau) = \gamma_A(\tau) = 1 - \frac{|\tau|}{2t_1}, \quad |\tau| \leqslant 2t_1$$

$$= 0, \quad\quad\quad\quad |\tau| > 2t_1 \tag{8.71}$$

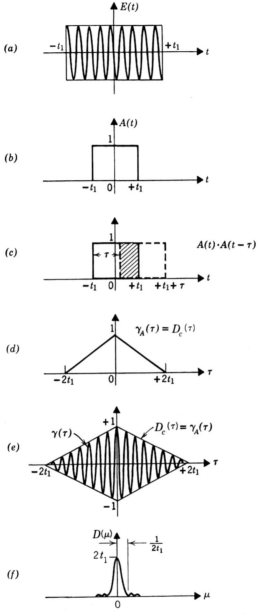

Fig. 8.8 (a) A square pulse; (b) its envelope; and (d) the autocorrelation function of the envelope. (c) Shows the autocorrelation principle for one value of τ. (e) Normalized autocorrelation function; (f) frequency distribution for a sine wave modulated by a square envelope.

The result is the triangle shown in Fig. 8.8d. The full autocorrelation function is plotted in Fig. 8.8e.

The spectral distribution function is obtained from Eq. (8.70).

$$D(\mu) = 2 \int_0^{2t_1} \left(1 - \frac{\tau}{2t_1}\right) \cos(2\pi\mu\tau)\, d\tau = \frac{2t_1 \sin^2(2\pi\mu t_1)}{(2\pi\mu t_1)^2} \tag{8.72}$$

and is plotted in Fig. 8.8f.

2. Collision Broadening. In high-pressure gas discharge lamps the quasimonochromatic spectral line width is often limited by what is known as *collision* or *pressure broadening*. The mechanism is as follows. Each radiating atom or molecule emits a cosine wave of frequency v_0 with a square envelope. The envelope should really be a slow exponential function because of the natural radiation damping of the oscillator, but we can often approximate it by a constant when the natural radiative lifetime is much longer than the collision time. The emission process is interrupted because of the collisions with other molecules; it then continues with little if any correlation of the phase angle. The time between collision t_c then equals the emission time $2t_1$ of the previous case, and the autocorrelation function for those atoms with collision time t_c is

$$\gamma(t_c; \tau) = D_c(t_c; \tau) \cos(2\pi v_0 \tau)$$

with

$$D_c(t_c; \tau) = \left(1 - \frac{|\tau|}{t_c}\right), \quad t_c \geq |\tau|$$

$$= 0, \qquad\qquad t_c < |\tau|$$

If all molecules are assumed to emit light with the same amplitude, the weight factor

$$\int_{-\infty}^{\infty} [A_j(t')]^2\, dt' = A_j^2 t_c$$

to be used in Eq. (8.66) is proportional to t_c. The sum over j must be replaced by the integral over t_c:

$$\sum_j (\cdots) \to \int_0^{\infty} (\cdots) \rho(t_c)\, dt_c$$

Here $\rho(t_c)\, dt_c$ is the collision probability in the interval t_c to $t_c + dt_c$. And $\rho(t_c)$ is given by

$$-\frac{1}{N(0)} \frac{dN(t_c)}{dt_c}$$

where $N(t_c)$ is the number of molecules that have not yet made collisions at time t_c. It obeys the differential equation $dN/dt_c = -N/\bar{t}_c$ and has the solution $N =$

$N(0)e^{-t_c/\bar{t}_c}$. Then

$$\not{p}(t_c)\, dt_c = \left(e^{-t_c/\bar{t}_c}\right)\frac{dt_c}{\bar{t}_c} \tag{8.73a}$$

where

$$\bar{t}_c = \int_0^\infty t_c \not{p}(t_c)\, dt_c \tag{8.73b}$$

is the mean time between collisions.

For the autocorrelation function we thus obtain

$$\gamma(\tau) = D_c(\tau)\cos(2\pi v_0 \tau) \tag{8.74a}$$

where

$$D_c(\tau) = \frac{\displaystyle\int_0^\infty D_c(t_c;\tau)t_c \not{p}(t_c)\, dt_c}{\displaystyle\int_0^\infty t_c \not{p}(t_c)\, dt_c}$$

For positive τ we have

$$D_c(\tau) = \int_\tau^\infty \left(1 - \frac{\tau}{t_c}\right)\left(\frac{t_c}{\bar{t}_c}\right)e^{-t_c/\bar{t}_c}\frac{dt_c}{\bar{t}_c} = e^{-\tau/\bar{t}_c}$$

Because $D_c(-\tau) = D_c(\tau)$, we have

$$D_c(\tau) = e^{-|\tau|/\bar{t}_c}$$

The final result is

$$\gamma(\tau) = e^{-|\tau|/\bar{t}_c}\cos(2\pi v_0 \tau) \tag{8.76}$$

The functions $D_c(\tau)$ and $\gamma(\tau)$ are sketched in Fig. 8.9b.

Equation (8.70) leads to the normalized spectral distribution function, which can be rewritten in another form.

$$D(\mu) = 2\int_0^\infty e^{-|\tau|/\bar{t}_c}\cos(2\pi\mu\tau)\, d\tau$$

$$= 2\,\mathrm{Re}\left\{\int_0^\infty e^{-|\tau|/\bar{t}_c}e^{-i2\pi\mu\tau}\, d\tau\right\}$$

This allows us to use Table 6.2 for the Fourier transform of an exponentially decreasing function.

$$D(\mu) = 2\,\mathrm{Re}\left\{\frac{1}{\bar{t}_c^{-1} + i2\pi\mu}\right\}$$

$$= \frac{2\bar{t}_c^{-1}}{\bar{t}_c^{-2} + 4\pi^2\mu^2} \tag{8.77}$$

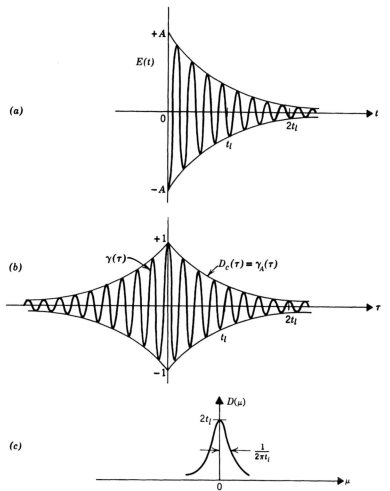

Fig. 8.9 Electric field (a), normalized autocorrelation function (b), and frequency distribution function (c) for a sine wave modulated by an exponential envelope. Curves (b) and (c) also apply to a collision broadened line for which $t_c = t_l$.

The function $D(\mu)$ in Eq. (8.77) is a Lorentzian function normalized to have unit area. It has a maximum value of $2\bar{t}_c$ at $\mu = 0$ or $\nu = \nu_0$; its half-width at half maximum is $(2\pi\bar{t}_c)^{-1}$. It is sketched in Fig. 8.9c.

We can write $\bar{t}_c$ as the mean free path ℓ divided by the mean velocity $\bar{v}$ if the molecules are in thermal equilibrium at a temperature T:

$$\bar{v} = \left(\frac{8k_B T}{\pi m}\right)^{1/2}$$

Here k_B is Boltzmann's constant, and m is the molecular mass. The mean free path is given by

$$\ell = \frac{1}{\sigma N}$$

where σ is the collision cross section and N is the number of gas molecules per unit volume. If an ideal gas law is assumed, N is given by

$$N = \frac{P}{k_B T}$$

where P is the pressure in the gas. The mean collision time is then

$$\bar{t}_c = \frac{\ell}{\bar{v}} = \frac{1}{\sigma P} \left(\frac{\pi m k_B T}{8} \right)^{1/2} \tag{8.78}$$

3. Lifetime Broadening. If allowed to radiate undisturbed, an excited oscillator will radiate a cosine wave having an exponential envelope:

$$E(t) = A e^{-t/\bar{t}_\ell} \cos(2\pi v_0 t + \varphi), \qquad t \geq 0$$
$$= 0, \qquad\qquad\qquad\qquad t < 0 \tag{8.79}$$

The parameter $\bar{t}_\ell$ then represents the natural radiative lifetime of the excited oscillator. The autocorrelation function of the envelope is readily calculated with a result that is the same as Eq. (8.76)

$$\gamma(\tau) = e^{-|\tau|/\bar{t}_\ell} \cos(2\pi v_0 \tau) \tag{8.80}$$

The spectral frequency-distribution function produced by a collection of such oscillators, each with the same $\bar{t}_\ell$, is then given by Eq. (8.77) with $\bar{t}_c$ replaced by $\bar{t}_\ell$. These results are illustrated in Fig. 8.9. Ordinarily $\bar{t}_\ell$ is much longer than $\bar{t}_c$, so that lifetime broadened lines are usually not seen in practice.

4. Doppler Broadening. Another important source of line width is the Doppler effect. When it dominates the broadening, such as in low-pressure gas discharges, the physical picture is as follows. Each atom or molecule emits a sharp, quasi-monochromatic line at frequency v_0 *in its own rest frame*. But if the molecule is moving with respect to the observer with a relative velocity v measured along the line connecting it to the observer, the observer will see a frequency

$$v = v_0 \left(1 + \frac{v}{c} \right)$$

If the molecules are in the thermal equilibrium at temperature T, the velocities v will be distributed according to the Maxwell distribution law

$$p(v)\, dv = \sqrt{\frac{m}{2\pi k_B T}} \, \exp\left(\frac{-mv^2}{2k_B T} \right) dv$$

This means that the spectral distribution function will be given by

$$D(\mu) = \frac{1}{\Delta v} \exp\left[-\pi\left(\frac{\mu}{\Delta v}\right)^2 \right]$$ (8.81)

with

$$\Delta v = \frac{v_0}{c} \sqrt{\frac{2\pi k_B T}{m}}$$

where

$$\frac{\Delta v}{v_0} \ll 1$$

This is the same form as Eq. (8.43), a Gaussian centered on v_0 with $1/e$ width $\Delta v / \sqrt{\pi}$. The autocorrelation function is given by Eq. (8.46).

$$\gamma(\tau) = \exp[-\pi(\Delta v \tau)^2] \cos(2\pi v_0 \tau)$$ (8.82)

These functions are plotted in Fig. 8.6.

C. Coherence Time and Frequency Spread

The examples of section B have some general features in common. In the case of lifetime broadening, or of collision broadening, the respective mean time of emission $\bar{t}_\ell$, or $\bar{t}_c$, furnishes a measure of the coherence time τ_c. The quasimono-chromatic spectral frequency distribution function has a width that is of order τ_c^{-1}. The Doppler broadened case behaves in a similar way, although the individual oscillators emit for a much longer time than τ_d. The frequency spread is still given by $\overline{\Delta v} \sim \tau_c^{-1}$, and $\tau_c \sim \tau_d$ is a measure of how long $\gamma(\tau)$ stays nonzero.

We are now in a position to define quantitatively the notions of coherence time τ_c and frequency spread $\overline{\Delta v}$. There is more than one way to do this. We choose a method due to Mandel:*

$$\tau_c \equiv 4 \int_0^\infty [\gamma(\tau)]^2 \, d\tau$$ (8.83)

$$\overline{\Delta v} \equiv \frac{1}{\displaystyle\int_0^\infty [P(v)]^2 \, dv}$$ (8.84)

For quasimonochromatic light where the autocorrelation function takes the form

$$\gamma(\tau) = U(\tau) \cos[2\pi v_0 \tau + \Phi(\tau)] = D_c(\tau) \cos(2\pi v_0 \tau) + D_s(\tau) \sin(2\pi v_0 \tau)$$

* L. Mandel, *Proc. Phys. Soc.* (London), 74;233, (1959).

and where $D_c(\tau)$ and $D_s(\tau)$ are slowly varying with respect to the cosine, Eq. (8.83) simplifies to

$$\tau_c = 2 \int_0^\infty [U(\tau)]^2 \, d\tau = 2 \int_0^\infty \{[D_c(\tau)]^2 + [D_s(\tau)]^2\} \, d\tau$$

This result is easily established once it is realized that the $\cos^2(2\pi v_0 \tau)$ and $\sin^2(2\pi v_0 \tau)$ oscillations will average to $1/2$ in a time interval over which U, Φ, D_c, and D_s are practically constant. Starting with Eq. (8.21)

$$\gamma(\tau) = \int_0^\infty P(v) \cos 2\pi v \tau \, dv$$

a direct calculation using the delta function representation Eq. (6.66) will then give

$$4 \int_0^\infty [\gamma(\tau)]^2 \, d\tau = \int_0^\infty [P(v)]^2 \, dv = \int_{v_0}^\infty [D(\mu)]^2 \, d\mu \qquad (8.85)$$

or by Eqs. (8.83) and (8.84)

$$\tau_c = \frac{1}{\overline{\Delta v}}$$

Expressions for τ_c and $\overline{\Delta v}$ are easily computed for the examples just discussed. For lifetime or collision broadening, we have a Lorentzian line shape,

$$D(\mu) = \frac{\dfrac{2}{\bar{t}_\ell}}{\left(\dfrac{1}{\bar{t}_\ell}\right)^2 + 4\pi^2 \mu^2}$$

with an exponential envelope on the autocorrelation function,

$$\gamma(\tau) = e^{-|\tau|/\bar{t}_\ell} \cos(2\pi v_0 \tau)$$

Then

$$\tau_c = \bar{t}_\ell, \quad \overline{\Delta v} = \frac{1}{\bar{t}_\ell} \qquad (8.86a)$$

For Doppler broadening where the line shape is Gaussian,

$$D(\mu) = (\Delta v)^{-1} \exp\left[-\pi\left(\frac{\mu}{\Delta v}\right)^2\right]$$

the envelope of the autocorrelation function is also Gaussian:

$$\gamma(\tau) = \exp[-\pi(\Delta v \tau)^2] \cos(2\pi v_0 \tau)$$

We can easily obtain

$$\tau_c = \frac{1}{\Delta v} \frac{1}{\sqrt{2}}, \quad \overline{\Delta v} = \sqrt{2} \, \Delta v \qquad (8.86b)$$

Table 8.1. Temporal Coherence Summary

	$P(v)$ or $D(\mu)$	$\gamma(\tau)$
Monochromatic	$\delta(v - v_i)$	$\cos(2\pi v_i \tau)$
Dual monochromatic	$f_i\delta(v - v_i) + (1 - f_i)\delta(v - v_j)$	$f_i\cos(2\pi v_i\tau) + (1 - f_i)\cos(2\pi v_j\tau)$
General	$4\displaystyle\int_0^\infty \gamma(\tau)\cos(2\pi v\tau)\,d\tau$	$\displaystyle\int_0^\infty P(v)\cos(2\pi v\tau)\,dv$
Quasimonochromatic $\mu = v - v_0$	$2\displaystyle\int_0^\infty D_c(\tau)\cos(2\pi\mu\tau)\,d\tau - 2\int_0^\infty D_s(\tau)\sin(2\pi\mu\tau)\,d\tau$	$D_c(\tau)\cos(2\pi v_0\tau) + D_s(\tau)\sin(2\pi v_0\tau)$ or $U(\tau)\cos[2\pi v_0\tau + \Phi(\tau)]$
Gaussian Doppler broadened	$\dfrac{1}{\Delta v}\exp\left[-\pi\left(\dfrac{\mu}{\Delta v}\right)^2\right]$	$\exp[-\pi(\Delta v\tau)^2]\cos(2\pi v_0\tau)$
Collision broadened Lifetime broadened	$\dfrac{2\bar{t}_c^{-1}}{\bar{t}_c^{-2} + 4\pi^2\mu^2}$	$e^{-|\tau|/\bar{t}_c}\cos(2\pi v_0\tau)$
Truncation broadened	$2t_1\,\mathrm{sinc}^2(2\pi\mu t_1)$	$\left(1 - \dfrac{|\tau|}{2t_1}\right)\cos(2\pi v_0\tau)$
Statistical $\varphi = \mathrm{const}$	$2\displaystyle\int_0^\infty \gamma_A(\tau)\cos(2\pi\mu\tau)\,d\tau$	$\gamma_A(\tau)\cos(2\pi v_0\tau)$

$$D_c(\tau) \equiv \int_{-\infty}^\infty D(\mu)\cos(2\pi\mu\tau)\,d\mu; \quad D_s(\tau) \equiv -\int_{-\infty}^\infty D(\mu)\sin(2\pi\mu\tau)\,d\tau$$

$$U(\tau) \equiv \{[D_c(\tau)]^2 + [D_s(\tau)]^2\}^{1/2}; \quad \tan\Phi(\tau) \equiv \frac{-D_s(\tau)}{D_c(\tau)}$$

$$\gamma_A(\tau) \equiv \langle A(t)A(t - \tau)\rangle/\langle[A(t)]^2\rangle$$

D. Temporal Coherence Summary

As a review of this section we present Table 8.1, which shows the relationships between the normalized autocorrelation function $\gamma(\tau)$ and the normalized spectral distribution function $P(v)$ or $D(\mu)$ for the examples that were discussed.

8.3 Spatial Coherence

We may interpret the coherence time τ_c as a measure of how long the phase in a quasimonochromatic light beam retains a "memory" of the value it had at an earlier time. If the beam is perfectly collimated, we may move along a distance

$$\ell_\ell = c\tau_c = \frac{c}{\Delta v}$$

and say that the phase remains correlated over this distance, which is called the *longitudinal coherence length.* For such a collimated beam the phase is the same on any surface oriented perpendicular to the direction of propagation. We say that there is perfect transverse coherence in the beam.

For an incompletely collimated beam, the transverse coherence is no longer perfect. To describe it we introduce the *transverse coherence length* ℓ_t. This represents a distance in a plane transverse to the main propagation direction over which the phases at two points remain correlated. The individual phases may fluctuate, but the two fluctuate together. Because for two points separated by a distance less than or about equal to ℓ_t there is a definite phase relationship. Interference effects should result if the light from these two points can be brought together. In principle we can do this with Young's experiment.

A spatially incoherent beam can be created by a collection of uncorrelated point sources. We will examine light from such a source as it interacts with the double-slit interferometer.

A. Young's Experiment

In the original description of Young's geometry in Chapter 5 we considered a line source and two parallel slits (Fig. 5.5). Here we limit the source initially to a point and then generalize to an extended area. The situation is pictured in Fig. 8.10 for a source point in the x–z plane.

The fringes are perpendicular to the plane of the drawing in Fig. 8.10. If the source point is displaced *out* of the plane of the drawing, the fringes will remain unchanged. Thus we may treat any two-dimensional distribution of source points as an equivalent one-dimensional distribution obtained by projecting the original distribution versus x and y onto the x-axis. We assume for now that the light is quasimonochromatic with a central frequency $\nu_0 = c/\lambda_0$ and will later generalize the results to cover a distribution of frequencies. The width of the spectral distribution must cover a sufficiently large frequency range to ensure that our distribution of point sources is mutually incoherent. That is, the mutual coherence time for the source must be shorter than the measurement time.

Let a single source located at P_1 be identified by angle θ_1 as shown in Fig. 8.10a. We number the slits L_1 and L_2 in order opposite to that used in Chapter 5 to ensure that the retardation time here will have the same sign as in the Michelson interferometer. We examine the light at P', identified by angle θ'. The contribution from slit L_1 to the total optical field at the observation point is $E_1(t)$, whereas that from slit L_2 is $E_2(t) = E_1(t - \tau)$. The retardation time τ is the interval that the phase of beam 2 is delayed with respect to that of beam 1.

$$\phi_2 - \phi_1 = -2\pi\nu_0\tau$$

Using Eq. (5.31) with the new numbering scheme for the slits this is

$$\phi_2 - \phi_1 = -\frac{2\pi a}{\lambda_0}(\sin\theta' - \sin\theta_1)$$

$$\approx -\frac{2\pi\nu_0 a}{c}(\theta' - \theta_1) \tag{8.87}$$

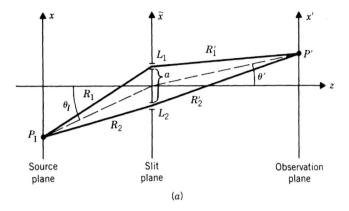

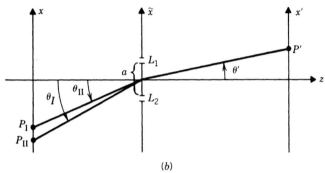

Fig. 8.10 (a) Young's geometry for a point source. (b) Young's geometry for a line source on the x axis between points P_I and P_{II}.

Therefore the retardation time associated with the source at P_I is

$$\tau_I = \frac{a(\theta' - \theta_I)}{c} \tag{8.88}$$

Equation (8.10) expresses the flux density as a function of the retardation time for quasimonochromatic light.

$$S(\tau_I) = S_I[1 + \cos(2\pi\nu_0\tau_I)] \tag{8.89}$$

1. Two-Point Sources. If point sources with the same frequency but having no mutual coherence are present at *both* P_I and P_{II} as in Fig. 8.10b, the interference patterns will add, and there may be smoothing or loss of contrast of the fringes depending on the relative strength of the sources and on their locations. We add the flux densities to obtain.

$$S = S_I[1 + \cos(2\pi\nu_0\tau_I)] + S_{II}[1 + \cos(2\pi\nu_0\tau_{II})] \tag{8.90}$$

This has the same form as Eq. (8.15). There the arguments of the cosine terms were

$2\pi\nu_i\tau$ and $2\pi\nu_j\tau$. We defined the visibility of the fringes (or the fringe contrast) V through Eq. (8.38).

$$V = \frac{S_{max} - S_{min}}{S_{max} + S_{min}}$$

where S_{max} and S_{min} are the values of the flux density for an adjacent maximum-minumum pair at an average value of the retardation τ. We saw that, for the two-frequency case, the visibility was given by Eq. (8.41)

$$V(\tau) = \sqrt{1 - 4\not p_i(1 - \not p_i)\sin^2(\pi\Delta\nu\tau)}$$

where $\pi\Delta\nu\tau = \pi(\nu_i - \nu_j)\tau$ (for $\nu_i > \nu_j$), and $\not p_i = S_i/S_0$, $S_0 = S_i + S_j$.

We make the same definition of visibility here. For a general spatial distribution of mutually incoherent quasimonochromatic point sources V will depend on both θ', the off-axis observation angle, and a, the slit separation. Let us fix θ' and vary the slit separation. Then we make the following substitutions to transform the result for temporal incoherence into the visibility for the spatially incoherent situation:

$$\pi\Delta\nu\tau \rightarrow \pi\nu_0\Delta\tau$$

where

$$\Delta\tau = \tau_I - \tau_{II} = \frac{a}{c}(\theta_{II} - \theta_I) \quad \text{(for } \theta_{II} > \theta_I) \tag{8.91}$$

$$\not p_i \rightarrow \not p_I = S_I/S_0$$

where

$$S_0 = S_I + S_{II} \tag{8.92}$$

The visibility then becomes

$$V(a) = \sqrt{1 - 4\not p_I(1 - \not p_I)\sin^2\left[\pi\frac{\nu_0}{c}a(\theta_{II} - \theta_I)\right]} \tag{8.93}$$

This shows how the spatial distribution of source brightness affects the interference pattern. As a function of a, V varies between

$$V_{max} = 1$$

when

$$\pi\frac{\nu_0}{c}a(\theta_{II} - \theta_I) = \pi \times \text{(integer)}$$

that is, when

$$a = \frac{\lambda_0}{(\theta_{II} - \theta_I)} \times \text{(integer)} \tag{8.94}$$

and

$$V_{\min} = |2\mu_1 - 1| = \frac{|S_I - S_{II}|}{S_I + S_{II}} \tag{8.95}$$

when

$$\pi \frac{\nu_0}{c} a(\theta_{II} - \theta_I) = \frac{\pi}{2} \times (\text{odd integer})$$

that is, when

$$a = \frac{\lambda_0}{(\theta_{II} - \theta_I)} \times (\text{half integer}) \tag{8.96}$$

For a fixed double point source we find that the visibility of the two-slit interference pattern reaches its first minimum when a is given by

$$a_1 = \frac{\lambda_0}{2(\theta_{II} - \theta_I)} \tag{8.97}$$

Thus, if we can measure a_1, the angular separation is given by

$$(\theta_{II} - \theta_I) = \frac{\lambda_0}{(2a_1)}$$

This variation of the visibility V with slit separation can be used to determine the nature of the source or distribution of sources. For instance, if V alternates periodically between a maximum of unity and a minimum $V_{\min}$, then we may conclude that we have two point sources, whose projected angular separation $\theta_{II} - \theta_I$ can be obtained from Eq. (8.97) and whose relative flux density can be obtained from Eq. (8.95) as the ratio

$$\frac{S_I}{S_{II}} = \frac{1 - V_{\min}}{1 + V_{\min}}$$

2. Continuous Source. We now apply these arguments to a continuous distribution of source points. A simple case is a uniform line source lying between P_I and P_{II} in Fig. 8.10b. This can result from the projection of a uniform x-y rectangle onto the plane of the paper. For this type of source we can define a function $S(\theta)\, d\theta$ to represent the total flux density at the screen coming from the part of the source located between θ and $\theta + d\theta$. In a similar way we define the normalized angular distribution function

$$i(\theta) = \frac{S(\theta)}{S_0} \tag{8.98}$$

where the total flux density with no interference is

$$S_0 = \int_{-\infty}^{\infty} S(\theta)\, d\theta \tag{8.99}$$

We show the limits on the angular integration as $\pm \infty$ although in practice the range of θ will be quite small. The normalized angular distribution function has the property

$$\int_{-\infty}^{\infty} i(\theta)\, d\theta = 1 \qquad (8.100)$$

The flux density in the two-slit interference pattern coming from the part of the source at θ of width $d\theta$ is given by a generalization of Eq. (8.90)

$$dS = S(\theta)\left\{1 + \cos\left[\frac{2\pi v_0 a(\theta' - \theta)}{c}\right]\right\} d\theta$$

or

$$dS = S_0 i(\theta)\left\{1 + \cos\left[\frac{2\pi v a(\theta' - \theta)}{c}\right]\right\} d\theta$$

The total flux density received at an angle θ' is then

$$S(\theta', a) = S_0 \int_{-\infty}^{\infty} i(\theta)\left\{1 + \cos\left[\frac{2\pi v_0 a(\theta' - \theta)}{c}\right]\right\} d\theta$$

$$= S_0[1 + \gamma_{12}] \qquad (8.101)$$

where γ_{12} is the normalized correlation function

$$\gamma_{12} = \int_{-\infty}^{\infty} i(\theta) \cos\left[\frac{2\pi v_0 a(\theta' - \theta)}{c}\right] d\theta \qquad (8.102)$$

This is a function of both θ' and a.

Let us identify

$$\tau' = \frac{a\theta'}{c} \qquad (8.103)$$

as the retardation time by which the phase of the light from slit L_2 is delayed with repect to that from slit L_1 because of propagation between the slit plane and the observation plane. The argument of the cosine in Eq. (8.102) becomes

$$\left[\frac{2\pi v_0 a(\theta' - \theta)}{c}\right] = \left[2\pi v_0 \tau' - \frac{2\pi a\theta}{\lambda_0}\right]$$

Our analysis is simplified by defining $\gamma_{12}^c(\tau', a)$, the *complex degree of coherence* for a *quasimonochromatic* source as

$$\gamma_{12}^c(\tau', a) = \int_{-\infty}^{\infty} i(\theta) e^{i(2\pi v_0 \tau' - 2\pi a\theta/\lambda_0)} d\theta$$

$$= e^{i2\pi v_0 \tau'} \int_{-\infty}^{\infty} i(\theta) e^{-i2\pi a\theta/\lambda_0}\, d\theta \qquad (8.104)$$

Then the normalized correlation function becomes

$$\gamma_{12}(\tau', a) = \text{Re}\{\gamma^c_{12}(\tau', a)\} \tag{8.105}$$

The remaining integral in Eq. (8.104) is the Fourier transform of the normalized source distribution function

$$\int_{-\infty}^{\infty} i(\theta) e^{-i2\pi a\theta/\lambda_0} \, d\theta = \mathscr{F}\left[i(\theta)\right] \equiv I\left(\frac{a}{\lambda_0}\right) \tag{8.106}$$

This leads to

$$\gamma^c_{12}(\tau', a) = e^{i2\pi v_0\tau'} I\left(\frac{a}{\lambda_0}\right) \tag{8.107a}$$

$$= \left|I\left(\frac{a}{\lambda_0}\right)\right| e^{i[2\pi v_0\tau' - \psi(a)]} \tag{8.107b}$$

where

$$I = |I| e^{-i\psi}$$

expresses the complex Fourier transform in polar form.

We wish to find the visibility of the fringes in the interference pattern of the Young's device where

$$S = S_0[1 + \gamma_{12}]$$

with

$$\gamma_{12}(\tau', a) = \text{Re}\{\gamma^c_{12}\} = \left|I\left(\frac{a}{\lambda_0}\right)\right| \cos[2\pi v_0\tau' - \psi(a)]$$

For a fixed value of θ', the observation angle, this is equivalent to

$$\gamma_{12}(a) = \left|I\left(\frac{a}{\lambda_0}\right)\right| \cos\left[\frac{2\pi a\theta'}{\gamma_0} - \psi(a)\right] \tag{8.108}$$

For quasimonochromatic light, as a the slit spacing is changed over a small range, the oscillations in the cosine due to the $2\pi a\theta'/\lambda_0$ term provide the most rapid variation in $\gamma_{12}(a)$. These are the fringes. They are modulated in strength by $I(a/\lambda_0)$ and in phase by $\psi(a)$, both of which are weak functions of a. If we consider that these functions are essentially constant over one cycle of the fringe pattern, then, for adjacent fringes, the flux density changes between the extremes of

$$S_{\text{max}}(a) = S_0\left[1 + \left|I\left(\frac{a}{\lambda_0}\right)\right|\right]$$

and

$$S_{\text{min}}(a) = S_0\left[1 - \left|I\left(\frac{a}{\lambda_0}\right)\right|\right]$$

The visibility is given by

$$V(a) = \frac{S_{max} - S_{min}}{S_{max} + S_{min}} = \left| I\left(\frac{a}{\lambda_0}\right) \right| \tag{8.109}$$

This is a quantitative measure of the degree of mutual spatial coherence for an extended quasimonochromatic source with a relatively narrow frequency distribution.

We see the relationship between spatial coherence theory and the theory of far-field diffraction. This relationship comes through the Fourier transform of the object function. The relationship is formalized in the *van Cittert-Zernike* theorem, which in this application states: As long as the source is nearly monochromatic and small compared with the distance to the interferometer, and provided the slit spacing remains small compared with that distance, then the visibility of the fringes will be equal to the absolute value of the Fourier transform of the source distribution function.

$$V = |\mathcal{F}[i]| \tag{8.110}$$

The behavior of V for typical examples will now be discussed.

a. Uniform Line Source. Consider a source of angular height $\Delta\theta$ centered about $\theta = 0$. This gives rise to a uniform source-distribution function with the following form:

$$i(\theta) = \begin{cases} \dfrac{1}{\Delta\theta}, & -\dfrac{\Delta\theta}{2} \leq \theta \leq \dfrac{\Delta\theta}{2} \\ 0, & \text{otherwise} \end{cases} \tag{8.111}$$

According to Eq. (8.106), we must find the Fourier transform of this one dimensional "box function." This was dealt with in Chapter 6 in association with diffraction by a single slit. By direct calculation or referring to Table 6.2 we arrive at

$$I\left(\frac{a}{\lambda_0}\right) = \text{sinc}\left(\frac{\pi a \Delta\theta}{\lambda_0}\right) \tag{8.112}$$

Here the visibility is given by

$$V(a) = \left| \text{sinc}\left(\frac{\pi a \Delta\theta}{\lambda_0}\right) \right| \tag{8.113}$$

This function is plotted in Fig. 8.11a (see also Fig. 8.13a). The value of a at which V is first equal to zero is given by

$$a_1 = \frac{\lambda_0}{\Delta\theta} \tag{8.114}$$

just twice the value given in Eq. (8.97) for a double-point source.

If we intend this line source to be used instead of a single-point source and have

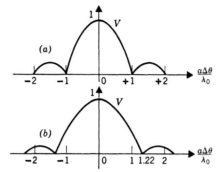

Fig. 8.11 Visibility of the fringe from (a) a uniform line source, (b) a uniform disk.

the same optical effect, then values of a and $\Delta\theta$ must be chosen so that V in Eq. (8.109) is very close to unity; that is, we would want

$$\frac{\pi a \Delta\theta}{\lambda_0} \ll 1$$

which is equivalent to the condition

$$a\Delta\theta \ll \lambda$$

We can generalize this result somewhat and say that *when the condition $a\Delta\theta \ll \lambda$ is satisfied, an extended source of angular height $\Delta\theta$ will act as if it were a point source.*

b. Uniform Disk Source. The behavior of V for a disk-shaped source (or a spherical Lambert source) of angular diameter $\Delta\theta$ is quite similar to that for a line source. The Fourier transform of a disk-shaped function is related to the first-order Bessel function, a fact that we discovered in Chapter 6. We find

$$V(a) = 2\left|\frac{J_1\left(\dfrac{\pi a \Delta\theta}{\lambda_0}\right)}{\dfrac{\pi a \Delta\theta}{\lambda_0}}\right| \tag{8.115}$$

This is plotted in Fig. 8.11b. The value of a for which V is first equal to zero turns out to be

$$a_1 = 1.22 \frac{\lambda_0}{\Delta\theta} \tag{8.116}$$

3. Michelson Stellar Interferometer. The phenomena just described were used by Michelson to measure the angular separation of double stars. He covered the objective lens of an astronomical telescope with a mask containing two slits, as shown in Fig. 8.12a. Because of the low light levels involved, he had to use white light. The effective frequency spread was that of his detector, and it was large,

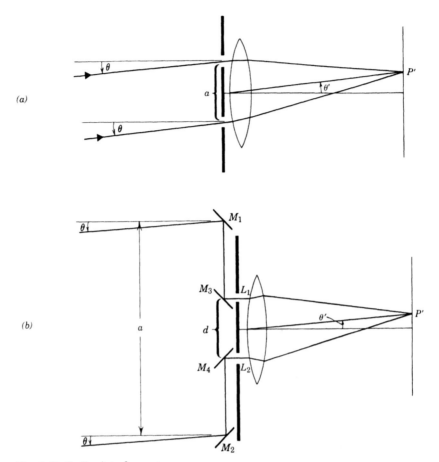

Fig. 8.12 Stellar interferometer.

perhaps one-half of the frequency itself. In spite of this, the autocorrelation function $\lambda(\tau)$ may be approximated by $\cos(2\pi\nu_0\tau)$ for one or two cycles, and a few fringes could be seen near zero order. Their visibility was studied as a function of a. For a double star the value of a at the first minimum of visibility gives a value for the angular separation $\Delta\theta$ of the stars from Eq. (8.97), $\Delta\theta = \lambda_0/2a_1$, and the flux density ratio may be obtained from the value of $V_{\min}$ in the form $(1 - V_{\min})/(1 + V_{\min})$. (The slits, of course, must be perpendicular to the line joining the two stars for this to hold.) Michelson was also able to measure the angular diameter of planetary satellites by measuring a_1 and using Eq. (8.116).

No known star has a large enough angular diameter to be measured by the apparatus shown in Fig. 8.12a with its limited range of a. Michelson extended the effective width of the slits to more than 6 m in his *stellar interferometer*, which is shown in Fig. 8.12b. Parallel light from a distant star incident at an angle θ will arrive at M_1 a time $a\theta/c$ later than at M_2. This retardation will be the same at the slits if $\overline{M_1M_3L_1} = \overline{M_2M_4L_2}$. At point P' on the screen (where a photographic plate or an eyepiece is inserted) located at an angle θ', there will be an additional retardation of opposite sign of amount $d\theta'/c$. The net retardation is thus

$$\tau = \left(\frac{d\theta'}{c} - \frac{a\theta}{c}\right) \qquad (8.117)$$

For this apparatus the fringe separation depends only on d, not a. But the fringe *contrast* depends on a, not d. We need to replace a by d in Eq. (8.103); however, the expressions for the visibility in Eqs. (8.93), (8.109), (8.113), and (8.115) are still valid as functions of a.

Using the interferometer, we can clearly distinguish a double star with its sinusoidal $V(a)$ dependence [Eq. (8.93)] from a large single star that has an a dependence shown in Fig. 8.11b. Once the type of source is determined, its angular size can be determined from Eq. (8.97) or Eq. (8.116). Only the largest stars can be measured by this technique, namely, the so-called red giants. An example is Betelgeuse, with an angular diameter of 0.047 sec of arc and an actual diameter of 4.1×10^8 km, greater that the diameter of the earth's orbit around the sun!

The mirrors in the stellar interferometer must be mounted extremely carefully, so that path difference changes caused by mechanical vibration and thermal expansion are held to a small fraction of a wavelength. This becomes impossible for values of a larger than a few meters. The intensity fluctuation interferometers recently developed by Hanbury Brown and Twiss enable much larger values of a to be used. Some information is sacrificed, however, because these new techniques measure only V. This will be discussed in section 8.4.

B. Finite Frequency Spread

Thus far in our treatment of spatial coherence the quasimonochromatic property of the light sources was used only to assure that the experiment was performed so that the various points in the extended source had complete mutual incoherence. In other words, we had to be sure that the measurement time was longer than the coherence time. Let us now see what other effects result from a finite spread in frequency.

1. Derivation. Following the method of section 8.1, we assume that the normalized frequency distribution function $P(\nu)$ describes the spectral composition of the source. In the various equations given in this section, in which ν_0 appeared, we now have to multiply the appropriate terms by $P(\nu)$ and integrate over all frequencies.

We start with Eq. (8.104) for the complex degree of coherence, which we rewrite as

$$\gamma^c_{12} = \int_{-\infty}^{\infty} i(\theta)e^{i2\pi\nu\tau}\,d\theta \qquad (8.118)$$

where

$$\tau = \frac{a}{c}(\theta' - \theta) = \tau' - \frac{a\theta}{c}$$

When a spread of frequencies is present this becomes

$$\gamma_{12}^c = \int_0^\infty P(v) \int_{-\infty}^\infty i(\theta) e^{i2\pi v\tau} \, d\theta \, dv$$

$$= \int_{-\infty}^\infty i(\theta) \left[\int_0^\infty P(v) e^{i2\pi v\tau} \, dv \right] d\theta$$

We recognize from section 8.2, Eq. (8.59) that the expression in brackets

$$\gamma^c(\tau) = \int_0^\infty P(v) e^{i2\pi v\tau} \, dv$$

is the complex normalized autocorrelation function. Using it, we may then write the complex degree of coherence as

$$\gamma_{12}^c = \int_{-\infty}^\infty i(\theta) \gamma^c(\tau) \, d\theta$$

$$= \int_{-\infty}^\infty i(\theta) \gamma^c\left(\tau' - \frac{a\theta}{c} \right) d\theta \qquad (8.119)$$

2. Quasimonochromatic Frequency Spread. We now employ the autocorrelation function for quasimonochromatic light [Eq. (8.32)], which is related to the *complex* autocorrelation function for quasimonochromatic light

$$\gamma^c(\tau) = U(\tau) e^{i[2\pi v_0\tau + \Phi(\tau)]} \qquad (8.120)$$

To proceed we must make the additional assumption that U and Φ vary slowly with τ compared with the changes contributed by $e^{i2\pi v_0\tau}$.

Let the source be of width $\Delta\theta$, centered about $\theta = 0$. Then our requirement is that U and Φ remain essentially constant when τ changes by $a\Delta\theta/c$. For this we need a coherence time τ_c much greater than $a\Delta\theta/c$. Now a will be of the order of the transverse coherence length ℓ_t. Thus

$$a \approx \ell_t \lesssim \frac{\lambda_0}{\Delta\theta}$$

and our requirement becomes

$$\tau_c \gg \frac{a\Delta\theta}{c} \approx \left(\frac{\lambda_0}{\Delta\theta} \right)\left(\frac{\Delta\theta}{c} \right) = \frac{\lambda_0}{c} = \frac{1}{v_0}$$

This last inequality is the usual one

$$\Delta v = \frac{1}{\tau_c} \ll v_0$$

used to define quasimonochromatic light. In practice it need not be a strict requirement, because the Michelson stellar interferometer works with "white" light, when $\Delta v \approx v_0/2$. Under these conditions we may replace τ by τ' in U and Φ. Now these functions are no longer dependent on θ and can be removed from the integral of Eq. (8.119).

$$\gamma_{12}^c = U(\tau')e^{i\Phi(\tau')} \int_{-\infty}^{\infty} i(\theta)e^{i2\pi v_0\tau} \, d\theta$$

We proceed in a manner similar to the technique following Eq. (8.109). We write $\tau = \tau' - a\theta/c$ in the exponent. Then

$$2\pi v_0\tau = 2\pi v_0\tau' - \frac{2\pi a\theta}{\lambda_0}$$

This leads to

$$\gamma_{12}^c(\tau', a) = U(\tau')e^{i[2\pi v_0\tau' + \Phi(\tau')]} \int_{-\infty}^{\infty} i(\theta)e^{-i2\pi a\theta/\lambda_0} \, d\theta$$

$$= U(\tau')e^{i[2\pi v_0\tau' + \Phi(\tau')]}I\left(\frac{a}{\lambda_0}\right)$$

$$= U(\tau')\left|I\left(\frac{a}{\lambda_0}\right)\right|e^{i[2\pi v_0\tau' + \Phi(\tau') - \psi(a)]}$$

Then the normalized correlation function becomes

$$\gamma_{12}(\tau', a) = U(\tau')\left|I\left(\frac{a}{\lambda_0}\right)\right|\cos\left[2\pi v_0\tau' + \Phi(\tau') - \psi(a)\right] \qquad (8.121)$$

The visibility for fixed θ' in this more general situation, in analogy with the nearly monochromatic case (Eq. 8.109), is

$$V(a) = U\left(\frac{a\theta'}{c}\right)\left|I\left(\frac{a}{\lambda_0}\right)\right| \qquad (8.122)$$

A measurement of the visibility $V(a)$ will give us the absolute value of $I(a/\lambda_0)$. The phase angle $\psi(a/\lambda_0)$ can also be obtained if accurate measurements of the flux density in the interference pattern are available for retardation times τ near zero. With $I(w)$ determined, we can reconstruct $i(\theta)$ using the inverse transform

$$i(\theta) = \int_{-\infty}^{\infty} I(w)e^{i2\pi w\theta} \, dw \qquad (8.123)$$

According to Eq. (8.121), the zero-order fringe of maximum flux density will be shifted if $\psi \neq 0$ from the position $\tau' = 0$ to one where

$$\tau' = \frac{a\theta'}{c} = \frac{1}{2\pi v_0}\psi\left(\frac{a}{\lambda_0}\right) \qquad (8.124)$$

This allows us, in principle to determine $\psi(a/\lambda_0)$.

A typical visibility curve is shown in Fig. 8.13a. This is Fig. 8.11a replotted and

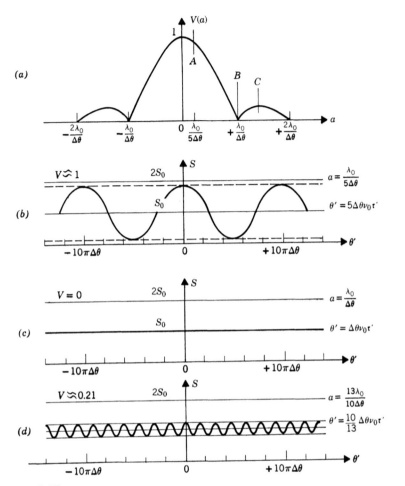

Fig. 8.13

applies to a uniform line source of angular width $\Delta\theta$. The corresponding interference pattern is shown in parts b through d of Fig. 8.13 for three different values of a. The flux density is plotted as a function of $\theta' = c\tau/a$.

From Eq. (8.120) we can develop an important property of the complex degree of coherence. Because $I(0) = 1$, we have

$$\gamma^c_{12}(\tau', 0) = \gamma^c(\tau') = U(\tau')e^{i[2\pi\nu_0\tau' + \Phi(\tau')]} \tag{8.125}$$

It is also easy to see that because $U(0) = 1$ and $\Phi(0) = 0$ we have

$$\gamma^c_{12}(0, a) = I\left(\frac{a}{\lambda_0}\right)$$

We can then write

$$\gamma^c_{12}(\tau', a) = \gamma^c(\tau')\gamma^c_{12}(0, a) \tag{8.126}$$

This is called the "reduction" of the complex degree of coherence. From this we see that the visibility is

$$V(a) = \left| \gamma^c_{12} \left(0, a \right) \right|$$ (8.127)

C. Transverse Coherence

The previous discussion of spatial coherence has been based on an analysis of the two-beam interference experiment. We can also approach these concepts from the statistical optics perspective. We developed a similar point of view during our description of temporal coherence.

1. Transverse Coherence Length. We now consider the two-slit experiment from a different but equally correct point of view, namely, that of correlations between the fluctuations in the fields at L_1 and L_2. Fringes with full visibility ($V = 1$) will occur only if these fluctuations have a definite correlation, that is, only if the light at L_1 and L_2 in Fig. 8.10 is *mutually coherent.*

For extended sources we saw that $V(a)$ decreases toward zero, perhaps with a few oscillations. Small values of V correspond to a lack of mutual coherence. It is useful to define the transverse coherence length ℓ_t such that for $a \gg \ell_t$, $V \approx 0$ and there is no coherence; and for $a \ll \ell_t$, $V \approx 1$ and there is coherence; and for $a \sim \ell_t$, V is somewhere between 0 and 1 and there is partial coherence. There is no unique definition of ℓ_t, but one possibility is discussed in the problems. This length ℓ_t is quite distinct from the longitudinal coherence length $\ell_\ell = c\tau_c$, where τ_c is the coherence time of the light.

The transverse coherence length is finite because the light is emitted from a finite-sized distribution of sources. Even if the light beams from all the individual sources were to arrive in phase at L_1, they could not all arrive in phase at L_2 unless a were very small because of the different propagation distances. When a is small, the phase at L_1 is completely correlated with that at L_2 (that is, the phase difference is a well-defined, constant quantity). When a equals the transverse correlation length, this phase difference has average fluctuations in time of about ± 1 rad. (These fluctuations occur in time intervals of order τ_c sec. A detector whose response time T is much less than τ_c could still record an interference pattern on the screen for a greater than the transverse coherence length, but this pattern would fluctuate with time and shift, on the average, by one-half fringe in a time equal to τ_c. Hence there will be no long-term interference pattern.)

For a line source or a disk we can take the transverse coherence length ℓ_t to be about equal to a_1 in Eq. (8.114) or Eq. (8.116). In general, we obtain the important result

$$\ell_t \approx \frac{\lambda}{\Delta\theta}$$ (8.128)

for this length, where $\Delta\theta$ is the angular width of the source. Exact definitions of ℓ_t and $\Delta\theta$ usually yield exact equality: $\ell_t = \lambda/\Delta\theta$.

2. Normalized Correlation Function. The electric field at the observation point P' in Fig. 8.10 at time t' will, in general, be proportional to the electric field $E_1(t' - R_1'/c)$ at L_1 at a time R_1'/c earlier plus the field $E_2(t' - R_2'/c)$ at L_2 at a time R_2'/c earlier. We have

$$E(P', t') \propto E_1\left(t' - \frac{R_1'}{c}\right) + E_2\left(t' - \frac{R_2'}{c}\right)$$

The difference in retardation times will be denoted by τ'

$$\tau' = \frac{R_2' - R_1'}{c} \approx \frac{a\theta'}{c}$$

For convenience we introduce the time variable $t = t' - R_1'/c$, which represents the arrival time at L_1. In terms of it we obtain

$$E\left(P', t + \frac{R_1'}{c}\right) = E_1(t) + E_2(t - \tau')$$

The flux density at P' is proportional to the time average of E^2. Thus, as we have done before, we form

$$\langle E^2 \rangle \approx \langle E_1(t) \rangle^2 + \langle E_2(t - \tau') \rangle^2 + 2\langle E_1(t)E_2(t - \tau') \rangle$$

and the flux density follows:

$$S = S_1 + S_2 + 2\sqrt{S_1 S_2} \gamma_{12}$$

where

$$\gamma_{12} = \frac{\langle E_1(t)E_2(t - \tau') \rangle}{\sqrt{\langle E_1^2 \rangle \langle E_2^2 \rangle}} \tag{8.129}$$

is the normalized correlation function of fields E_1 and E_2. Its independent variables are both the retardation time τ' and the slit separation a. Because $E_2(t - \tau')$ becomes a constant times $E_1(t - \tau')$ as $a \to 0$, we must have

$$\lim_{a \to 0} \gamma_{12}(\tau', a) = \gamma(\tau') \tag{8.130}$$

where $\gamma(\tau')$ is the normalized autocorrelation function. If $S_1 = S_2$ we have

$$S = S_0(1 + \gamma_{12}) \tag{8.131}$$

the same form as Eqs. (8.18) and (8.101).

If the fields are represented by complex analytical signals $E_1^c(t)$ and $E_2^c(t - \tau')$, we can define the complex normalized correlation function

$$\gamma_{12}^c = \frac{\langle E_1^c(t)E_2^c{}^*(t - \tau') \rangle}{\sqrt{\langle |E_1^c|^2 \rangle \langle |E_2^c|^2 \rangle}} \tag{8.132}$$

Then

$$\gamma_{12} = \text{Re}\{\gamma_{12}^c\} \tag{8.133}$$

The complex function defined in Eq. (8.132) is the same as that defined in Eq. (8.119). The visibility V is then the absolute value $|\gamma_{12}^c|$.

3. γ_{12}^c for an Extended Source. Let us consider an extended source composed of a collection of distant point sources making angle θ_j with the normal to the plane of our two slits L_1 and L_2. The complex analytical signal at L_1 and L_2 will be given by sums of the form

$$E_1^c(t) = \sum_{j=1}^{N} E_{1,j}^c(t)$$

$$E_2^c(t - \tau) = \sum_{j=1}^{N} E_{2,j}^c(t - \tau')$$

where, for instance, E_{2j}^c is the instantaneous complex analytical signal at L_2 resulting from the jth source. The complex normalized correlation function γ_{12}^c is to be calculated from Eq. (8.132). Here the interferometer splits the components equally $(S_1 = S_2)$. The averaging time is assumed long compared with the coherence time of the light. This means that averages of the form

$$\langle E_{1,j}^c(t)E_{2,k}^{c*}(t - \tau')\rangle$$

will give zero unless $j = k$, because the phase difference will change by many radians during the measurement. The light from the jth source can then interfere only with itself. Equation (8.132) gives

$$\gamma_{12}^c = \frac{\sum_j \langle E_{1,j}^c(t)E_{2,j}^{c*}(t - \tau')\rangle}{\sum_j \langle |E_{1,j}^c(t)|^2\rangle}$$

From Fig. 8.14 we see that for light from a distant point source at angle θ there is an extra propagation time $a\theta/c$ for the light to reach L_1, thus

$$E_{1,j}^c(t) = E_{2,j}^c\left(t - \frac{a\theta}{c}\right) \tag{8.134}$$

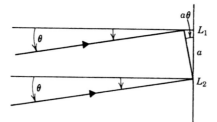

Fig. 8.14

This gives

$$\gamma_{12}^c = \frac{\sum_j \left\langle E_{2,j}^c\left(t - \frac{a\theta_j}{c}\right)E_{2,j}^{c*}(t - \tau')\right\rangle}{\sum_j \left\langle \left|E_{2,j}^c\left(t - \frac{a\theta_j}{c}\right)\right|^2\right\rangle}$$

For long averaging times T, a steady state is achieved in which each average is independent of t. This means that we can replace t by $t + constant$ in each term without changing the result. Thus we may replace t by $t + a\theta_j/c$ and obtain

$$\gamma_{12}^c = \frac{\sum_j \left\langle E_{2,j}^c(t)E_{2,j}^{c*}\left(t - \tau' + \frac{a\theta_j}{c}\right)\right\rangle}{\sum_j \left\langle |E_{2,j}^c(t)|^2\right\rangle} \tag{8.135}$$

The complex normalized autocorrelation function $\gamma_j^c(\bar{\tau}')$ for the jth source is defined by

$$\gamma_j^c(\bar{\tau}') = \frac{\langle E_{2,j}^c(t)E_{2,j}^{c*}(t - \bar{\tau}')\rangle}{\langle |E_{2,j}^c(t)|^2\rangle}$$

If we assume that the spectral distribution functions for all the sources are the same, they will have a common $\gamma_j^c(\bar{\tau}') = \gamma^c(\bar{\tau}')$.

Equation (8.135) then becomes

$$\gamma_{12} = \frac{\sum_j \langle |E_{2,j}^c(t)|^2\rangle\gamma^c\left(\tau' - \frac{a\theta_j}{c}\right)}{\sum_j \langle |E_{2,j}^c(t)|^2\rangle} \tag{8.136}$$

This is a weighed average of the various $\gamma^c(\tau' - a\theta_j/c)$. The weight factor

$$\frac{\langle |E_{2,j}^c(t)|^2\rangle}{\sum_j \langle |E_{2,j}^c(t)|^2\rangle}$$

gives the relative flux density at L_2 and hence at the observing screen if one slit is closed.

For a continuous distribution the sum goes over to the integral

$$\gamma_{12}^c = \int_{-\infty}^{\infty} i(\theta)\gamma^c\left(\tau' - \frac{a\theta}{c}\right)d\theta \tag{8.137}$$

which is identical with Eq. (8.119) for the complex degree of coherence because $\tau' = a\theta'/c$.

8.4   Fluctuations

Thermal light results from the superposition of the signals from a great many independent optical sources. For quasimonochromatic thermal light the instantaneous amplitude and phase of the resultant signal will fluctuate because of the randomly varying phase (and perhaps amplitude) relationships among the individual component signals. A large fluctuation in the phase of the resultant signal is likely to be accompanied by a large fluctuation in the amplitude. Thus we are led to expect that the behavior of the amplitude fluctuations will tell something about the phase fluctuations or, more precisely, about the correlation functions for the optical signals.

This expectation is correct for thermal light, as we will show explicitly following. If the light has mutual coherence at two points P_1 and P_2 in space, there is a positive correlation in flux-density fluctuations. If there is a positive fluctuation about the mean at P_1, then there is more likelihood of a positive fluctuation about the mean at P_2 than of a negative fluctuation. Similarly, a negative fluctuation at P_2 will accompany a negative fluctuation at P_1 more often than will a positive fluctuation at P_2.

We can use this effect to measure spatial (and temporal) coherence. In astronomy the spatial coherence of the light from a stellar source depends on its size and shape. The technique of correlation of amplitude fluctuations can be used to great advantage to measure the angular diameter of stars.

A. Correlation Interferometry

1. Experimental Arrangement. A photomultiplier or other energy-sensitive detector of light will give an output signal proportional to the optical flux incident onto it averaged over a time interval T. We will abandon the precise terminology and refer to the flux as "intensity." We assume that we have full spatial coherence across the light-sensitive area of each detector. It is then possible to speak of a well-defined "instantaneous" intensity $S(t)$ at a detector given by an extremely fast average of E^2. We assume at first that the averaging time T is much less than the coherence time τ_c of the light. We block the average component of the detector output, which will be proportional to the long-term average intensity $\langle S \rangle$, and detect only the fluctuating component, which is proportional to the intensity fluctuation (Fig. 8.15)

$$\Delta S(t) = S(t) - \langle S \rangle \tag{8.138}$$

Formerly when we wanted to study the mutual coherence between two points P_1 and P_2 we set up pinholes or slits and tried a Young interference experiment (Fig. 8.10). Now we put detectors PM1 and PM2 at P_1 and P_2, as shown in Fig. 8.16a. The output from PM2 is delayed τ sec by passing through a variable delay line. The resulting signals, which will be proportional to $\Delta S_1(t)$ and $\Delta S_2(t - \tau)$ are then fed into a correlator, which is a device that takes the long-term average of their

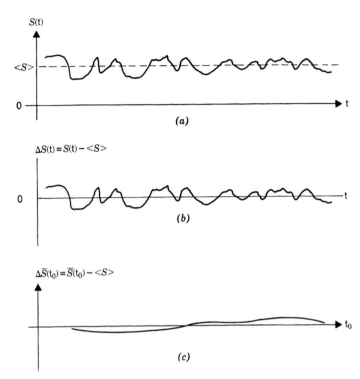

Fig. 8.15 Schematic representation of (*a*) the instantaneous flux density $S(t)$; (*b*) the instantaneous flux density fluctuations $\Delta S(t)$; and (*c*) the smoothed flux density fluctuations.

product. Hence its output is proportional to

$$\langle \Delta S_1(t)\, \Delta S_2(t - \tau)\rangle$$

and this is studied as a function of the independent variables τ and a.

If it is desired to go to the limit $a = 0$, the arrangement of Fig. 8.16*b* may be used. At $a = 0$ the output is then proportional to the autocorrelation function $\langle \Delta S(t)\, \Delta S(t - \tau)\rangle$ of the intensity fluctuations, because then $S_1(t) = S_2(t) = S(t)$.

2. Quasimonochromatic Thermal Light. The calculation of the correlation function is quite involved and will not be given here. One expresses the light signal as a superposition of signals from many independent sources and proceeds to calculate ΔS and the desired correlation function. Under the assumption that all sources have the same spectral distribution function, the result is

$$\langle \Delta S_1(t)\, \Delta S_2(t - \tau)\rangle = \langle S\rangle^2 |\gamma^c_{12}(\tau, a)|^2 \tag{8.139}$$

Here γ^c_{12} is the "complex degree of coherence" of Eq. (8.119). By the reduction property Eq. (8.126)

$$\gamma^c_{12}(\tau, a) = \gamma^c(\tau)\gamma^c_{12}(0, a)$$

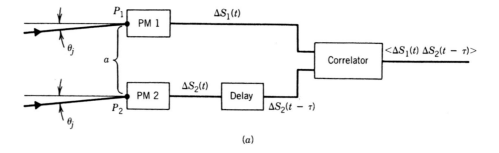

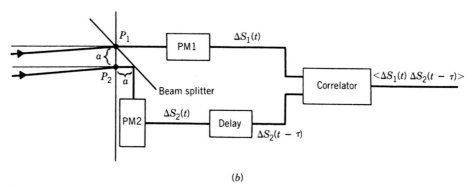

Fig. 8.16 Devices to correlate flux density fluctuation at P_1 and P_2. (a) P_1 and P_2 separated; $a > 0$. (b) P_1 and P_2 identical as $a \to 0$.

From Eqs. (8.121) and (8.127) we have the result

$$|\gamma^c_{12}(\tau, a)| = U(\tau)V(a)$$

Thus, Eq. (8.139) may also be written

$$\langle \Delta S_1(t) \, \Delta S_2(t - \tau) \rangle = \langle S \rangle^2 U^2 V^2 \qquad (8.140)$$

This is the basic result relating intensity-fluctuation correlations to coherence properties.

3. Long Averaging Time. There is a practical difficulty neglected in the derivation of Eq. (8.140). It stems from our inability to perform the required "instantaneous" average for the intensity assumed in the preceding derivation. The electronic circuits simply cannot respond rapidly enough. Instead, we are more likely to measure a smoothed intensity given by an average similar to

$$\bar{S}(t_0) \equiv \frac{1}{T} \int_{-T/2}^{T/2} S(t + t_0) \, dt$$

where $S(t + t_0)$ is the instantaneous average used previously. The deviation $\Delta \bar{S}$ of $\bar{S}$ from its long-term average $\langle S \rangle$ is much smoother, as shown in Fig. 8.15. It represents the signal correlated by the apparatus of Fig. 8.16.

$$\langle \Delta \bar{S}_1(t_0) \, \Delta \bar{S}_2(t_0 - \tau) \rangle \equiv \langle [\bar{S}_1(t_0) - \langle S \rangle][\bar{S}_2(t_0 - \tau) - \langle S \rangle] \rangle$$

where $\langle \cdots \rangle$ now represents a very long-term average over T.

When these changes are made, it can be shown that Eq. (8.140) must be replaced by the expression

$$\frac{\langle \Delta \bar{S}_1(t_0) \, \Delta \bar{S}_2(t_0 - \tau) \rangle}{\langle S \rangle^2} = [V(a)]^2 \frac{\xi(T, \tau)}{T} \tag{8.141}$$

where

$$\xi(T, \tau) = \frac{1}{T} \int_{-T}^{+T} (T - |t|)[U(\tau + t)]^2 \, dt \tag{8.142}$$

In most real situations the measurement time T is much longer than the coherence time τ_c. Then Eq. (8.142) gives an approximate result

$$\langle \Delta \bar{S}_1(t_0) \, \Delta \bar{S}_2(t_0 - \tau) \rangle = \begin{cases} 0, & \tau > T \\ \langle S \rangle^2 [V(a)]^2 (\tau_c/T), & \tau < T \end{cases} \tag{8.143}$$

These last relations hold except when $|(|\tau| - T)| \sim \tau_c$.

The reduction in Eq. (8.143) by the factor $\tau_c/T \ll 1$ makes the smoothed fluctuations much more difficult to observe than the instantaneous fluctuations given by Eq. (8.140). This conclusion is also suggested by a comparison of Fig. 8.15c to 8.15b. But note that Eq. (8.143) says that we can still measure the visibility function

$$V(a) = \left| I\left(\frac{a}{\lambda_0}\right) \right|$$

4. Examples. The correlation of intensity fluctuations in a light beam was first demonstrated experimentally by R. Hanbury Brown and R. Q. Twiss in 1956. Their apparatus was similar to that shown in Fig. 8.16b. The source was a mercury lamp–filter combination with a coherence time τ of about 1 nsec, and the electronics had an effective integration time T of about 40 nsec. They verified the behavior predicted by Eq. (8.143).

Once they had established the existence of correlated intensity fluctuations using laboratory light sources, Hanbury Brown and Twiss applied the technique to stellar interferometry. The arrangement is essentially that of Fig. 8.16a, with the phototubes placed at the focal points of two large mirrors as shown in Fig. 8.17. (Mirror diameters up to 7 m have subsequently been used.) The mirrors form images of the desired star at the two photocathodes. These images need not be of high optical quality. The telescope separation a may be varied by moving them on rails. Very long baselines are practical.

In the Michelson stellar interferometer we must hold dimensional tolerances to a fraction of a wavelength; otherwise, the interference pattern "washes out." In an intensity-fluctuation interferometer the requirements are enormously relaxed. Mechanical dimensions should be maintained to within a small fraction of ℓ_t or of

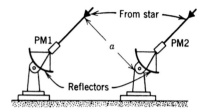

Fig. 8.17 Hanbury Brown and Twiss correlation stellar interferometer.

cT, whichever is less. Timing errors from transit time fluctuations in the electronic circuitry should be much less than T sec in duration.

B. Quantum Considerations

A second effect neglected in the treatment of section 8.4A involves the inherent quantum nature of light and of the process by which light intensities are measured. Electromagnetic radiation arrives at a detector in quanta of energy of amount hv, where h is Planck's constant. This fact causes extra fluctuations in intensity beyond those already discussed and introduces an important source of "noise."

1. Quantum Counter. For definiteness, assume that we use a photoelectric detector, such as a photomultiplier, as a quantum counter where it counts individual photoelectrons. We assume that the only source of "noise" or fluctuations in the photoelectric current is the quantum nature of the photoelectronic process or, what is the same thing, the fact that the current results from the flow of individual charged particles.

We let $S(t)$ be the short-time average flux reaching the detector. Then in a small time interval dt energy $S(t)\,dt$ has arrived at the detector. Because of the quantum nature of light, this energy arrives in quanta or "photons" of size hv, where v is the frequency of the light. Each quantum may emit a photoelectron, which we will assume will be detected with unit probability by the associated electronic circuitry. In a photomultiplier, each photoelectron is multiplied to give a charge pulse of about 10^6 electrons that can be collected in a few nanoseconds. The probability that 1 photon will yield 1 photoelectron is called the "quantum yield" α.

When flux $S(t)\,dt$ arrives at the detector, the number $dn(t)$ of photoelectrons emitted will fluctuate because of the inherent statistical nature of the photoelectric process. The "expectation value" or mean value over many experiments will be denoted by $[dn(t)]$ and is given by

$$[dn(t)] = \frac{\alpha S(t)\,dt}{hv} \tag{8.144}$$

The mean number of photoelectrons detected over a time interval equal to the averaging time T of the detector is

$$[\bar{n}(t_0)] = \int_{t_0 - T/2}^{t_0 + T/2} \frac{\alpha S(t')}{(hv)}\,dt \tag{8.145}$$

For thermal light the flux $S \sim A^2$ is a Gaussian random variable, as discussed in connection with Fig. 8.7. Detailed investigations of photon statistics have shown that in this case it is not necessary to distinguish the instantaneous value $\bar{n}(t_0)$ from the mean value $[\bar{n}(t_0)]$. The fluctuations in $\bar{n}(t_0)$ may then be treated as due entirely to fluctuations in $\bar{S}(t_0)$. We therefore proceed by using

$$\bar{n}(t_0) = \frac{\alpha T \bar{S}(t_0)}{h\nu} \tag{8.146}$$

The long-term average of Eq. (8.146) gives

$$\langle \bar{n} \rangle = \langle n \rangle = \frac{\alpha}{h\nu} T \langle \bar{S} \rangle$$

We now consider the number of counts n_1 and n_2 measured by the counters PM1 and PM2 in Fig. 8.16. The fluctuations will be given by

$$\Delta \bar{n}_1(t_0) = \bar{n}_1(t_0) - \langle n \rangle = \frac{\alpha}{h\nu} T[S_1(t_0) - \langle S \rangle]$$

$$= \frac{\alpha T}{h\nu} \Delta \bar{S}_1(t_0)$$

and

$$\Delta \bar{n}_2(t_0 - \tau) = \frac{\alpha T}{h\nu} \Delta \bar{S}_2(t_0 - \tau)$$

We may then calculate the correlation function for deviation of electron counts

$$\langle \Delta \bar{n}_1(t_0) \, \Delta \bar{n}_2(t_0 - \tau) \rangle = \left(\frac{\alpha T}{h\nu}\right)^2 \langle \Delta \bar{S}_1(t_0) \, \Delta \bar{S}_2(t_0 - \tau) \rangle$$

If we use Eq. (8.141), this becomes

$$\langle \Delta n_1(t_0) \, \Delta n_2(t_0 - \tau) \rangle = \left(\frac{\alpha T \langle S \rangle}{h\nu}\right)^2 V^2(a) \frac{\xi(T, \tau)}{T}$$

$$= \langle n \rangle^2 V^2(a) \frac{\xi(T, \tau)}{T} \tag{8.147}$$

Equation (8.147) says that for values of a and τ giving $V^2 \xi > 0$, there is a positive correlation between the excess counts at PM1 and those at PM2. Thus a count at PM1 is more likely to be accompanied by a count at PM2 than would be expected on the basis of pure chance. Such correlated counts are called *coincidences*.

2. Coincidence. Suppose that T is short enough so that the average number of counts $\langle n \rangle$ in time T is much less than unity. The function $\bar{n}_1(t_0)$ is then shown schematically in the top plot of Fig. 8.18. If a photoelectron is emitted at time t_1, then $\bar{n}_1(t_0)$ is unity over a time interval of length T centered about $t_0 = t_1$.

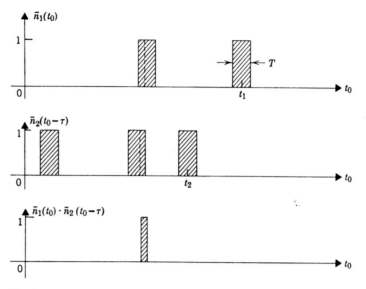

Fig. 8.18

Under these conditions the product $\bar{n}_1(t_0)\bar{n}_2(t_0 - \tau)$ then gives unity when both PM1 and PM2 have a count, and zero otherwise. Such a coincidence will occur whenever $|t_1 - t_2| \le T$ (Fig. 8.18), and the width of the overlap pulse on a plot of $\bar{n}_1(t_0)\bar{n}_2(t_0 - \tau)$ versus t_0 will be $T - |t_1 - t_2|$, which ranges from 0 to T and has an average of $T/2$. The average number of coincidences per second will therefore be

$$\rho_c = \frac{2\langle \bar{n}_1(t_0)\bar{n}_2(t_0 - \tau)\rangle}{T}$$

$$= \frac{2\langle n\rangle^2}{T}\left[1 + V^2(a)\frac{\xi(T, \tau)}{T}\right] \tag{8.148}$$

If $\bar{n}_1$ and $\bar{n}_2$ were purely random rather than correlated, there would be an accidental coincidence rate given by

$$\rho_a = \frac{2\langle n\rangle^2}{T} = 2T\rho_1^2 \tag{8.149}$$

where

$$\rho_1 = \frac{\langle n\rangle}{T} \tag{8.150}$$

is the average rate of detection of single photoelectrons. Then we can write

$$\rho_c = \rho_a[1 + \eta] \quad \text{where} \quad \eta = V^2(a)\frac{\xi(T, \tau)}{T} \tag{8.151}$$

Here $[1 + \eta]$ is the enhancement factor resulting from correlations.

If we count coincidences for T_0 sec, there would be an average of

$$\bar{N}_a = T_0 \rho_a = \frac{2T_0 \langle n \rangle^2}{T} = 2T_0 T \rho_1^2 \tag{8.152}$$

accidental coincidences and

$$\bar{N}_c = T_0 \rho_c = \bar{N}_a[1 + \eta] \tag{8.153}$$

coincidences overall.

How accurate will a given determination of N_a or N_c be? We expect the coincidences themselves to be uncorrelated random events. Then the standard deviation for a single determination of N_a or N_c is conventionally estimated to be $\sqrt{N_a}$ or $\sqrt{N_c}$, respectively. When N is not too small, approximately 67% of the individual determinations of N will be within the range $N \pm \sqrt{N}$ and more than 95% within the range $N \pm 2\sqrt{N}$.

3. Noise. To measure η within any certainty, we would like to have the difference $\bar{N}_c - \bar{N}_a$ be an order of magnitude larger than $\sqrt{N_c}$. [If N_a has been previously determined to high accuracy, then the error in $(N_c - N_a)$ will be that of N_c, namely, $\pm\sqrt{N_c}$]. This result can be expressed in the form

$$N_c - N_a = \eta \bar{N}_a \pm \sqrt{1 + \eta}\sqrt{\bar{N}_a} \tag{8.154}$$

The *signal-to-noise* ratio SNR is defined as the mean value of the signal divided by its standard deviation:

$$\text{SNR} = \frac{\eta}{\sqrt{1 + \eta}}\sqrt{\bar{N}_a} \tag{8.155}$$

In the slow-detector limit where $\tau_c \ll T$, $\eta = V^2(a)\tau_c/T \ll 1$, and we have

$$\text{SNR} = V^2(a)\left(\frac{2T_0}{T}\right)^{1/2}(\rho_1\tau_c) \tag{8.156}$$

For nonlaser thermal light, the product $\rho_1\tau_c$ will be 10^{-2} or less.

For a numerical estimate, take $\rho_1\tau_c = 10^{-3}$, $V(a) = 0.1$, $T = 10$ nsec. Then for SNR $= 10$, $T_0 = 5000$ sec ≈ 1.5 h. This is a typical value for the counting time T_0. But note that for SNR $= 1$, $T_0 = 50$ sec.

In the fast-detector limit when $T \ll \tau_c$, we have $\eta = V^2(a)U^2(\tau)$, and (8.155) can be written

$$\text{SNR} = \frac{[V(a)U(\tau)]^2}{\sqrt{1 + V^2 U^2}}\left(\frac{2T_0 T}{\tau_c^2}\right)^{1/2}(\rho_1\tau_c) \tag{8.157}$$

This uncertainty in the result of a single measurement of $(N_c - N_a)$ is due to the quantum particle nature of light and is often called *photon shot noise*. Because the same intrinsic quantum processes are involved in measurements of the correlaton of intensity fluctuations, where individual photoelectrons are not explicitly counted, this type of noise must nevertheless be present. Because a measurement of $(N_c - N_a)$ is essentially equivalent to a determination of $\langle \Delta \bar{S}_1(t') \Delta \bar{S}_2(t' - t) \rangle$, the latter will also have a signal-to-noise ratio given by Eq. (8.155), (8.156), or (8.157), where T_0 is the time interval for the determination of the average $\langle \cdots \rangle$ and T is that for determination of $\bar{S}$.

4. Degeneracy Parameter. The average flux will be given by

$$\langle \Phi \rangle = \bar{S}(\text{inc})\sigma$$

where $E_e(\text{inc})$ is the incident radiant flux density (radiant incidence) and σ is the sensitive area of the detector projected onto a plane perpendicular to the light beam. If a light-gathering mirror or lens is used, as in Fig. 8.17, then σ must be its projected area. By analogy, with Eq. (8.150), we may write

$$\rho_1 = \frac{\alpha}{h\nu} \langle \Phi \rangle = \frac{\alpha}{h\nu} \bar{S}(\text{inc})\sigma$$

If the source subtends the solid angle $\Delta\Omega$ as seen from the detector and has a total radiance L_e for light radiated in the direction of the detector, then $\bar{S}$ is simply given by

$$\bar{S} = L_e \, \Delta\Omega$$

We may then relate $\Delta\Omega$ to the so-called area of coherence σ_c at the detector by the equation

$$\sigma_c = \frac{\lambda^2}{\Delta\Omega} \tag{8.158}$$

This is a generalization to two dimensions of the notion of coherence length $\ell_t = \lambda/\Delta\theta$ for a source of angular width $\Delta\theta$. There will be a definite correlation between the electric fields at two points within the area of coherence.

Thus we may write

$$\rho_1 \tau_c = \alpha \left(\frac{L_e}{h\nu} \tau_c \right) \lambda^2 \frac{\sigma}{\sigma_c}$$

$$= \alpha g \frac{\sigma}{\sigma_c} \tag{8.159}$$

This last equality introduces the *degeneracy parameter* g. Because L_e is the rate of radiant energy emitted by the source per unit time, per unit area, and per unit solid angle, we may interpret

$$g = \left(\frac{L_e}{h\nu} \right) \tau_c \lambda^2 \tag{8.160}$$

as the average number of light quanta or photons (of a definite polarization) emitted into unit solid angle in the direction of the detector from an area of the source equal to λ^2 in a time interval equal to the coherence time τ_c. Because

$\lambda^2 = \Delta\Omega\sigma_c$, we can also interpret δ as the average number of photons that arrive on a surface of area σ_c at the detector from the entire source in the time interval τ_c, or as the number of photons at any given time that fill the "volume of coherence"

$$V_c \equiv \sigma_c \tau_c c \tag{8.161}$$

For quasimonochromatic light that is in thermal equilibrium at temperature T (blackbody radiation), the degeneracy parameter is given by a formula due to Planck,

$$g = \frac{1}{\exp\left(\dfrac{h\nu}{k_B T}\right) - 1} \tag{8.162}$$

where k_B = Boltzmann's constant. This can also be written as

$$g = \frac{1}{\exp\left(\dfrac{c_2}{\lambda T}\right) - 1} \tag{8.163}$$

$c_2 = hc/k_B = 1.438$ cm-deg. Here g is ordinarily very much less than unity.

For light from a laser, g can be as high as 10^{12}. Laser light, however, is ordinarily not thermal light in the technical sense required for the preceding discussion of fluctuations to apply. There is a method of converting light from a laser into "quasithermal light" that has both a large value of g and the right statistical properties by passing it through a moving random medium such as a moving ground-glass plate.

8.5   Image Formation: Incoherent Objects

In section 7.3 we presented the fundamentals of image formation by refraction through a spherical contour lens. There we limited the discussion to coherent objects, those smaller than the transverse coherence length of the incident light. Here we expand the formalism to include partially coherent sources.

A perfectly incoherent source does not exist. Even light from a blackbody has a coherence length at its surface equal to a mean wavelength $\bar{\lambda}$. An indirectly illuminated diffuse surface will have a coherence length not much larger than $\bar{\lambda}$ if the illuminating light covers a wide angle. If an image is formed, the coherence length increases, because the light from each point in the object forms a focal spot that is finite, because of diffraction and aberrations. The resulting coherence length has to be at least as large as the size of this focal spot, which is ordinarily much larger than $\bar{\lambda}$ times the magnification. In this case we can neglect the very small coherence length of the source itself.

This is the assumption we will make for self-luminous (or thermal) objects or for diffuse objects that are indirectly illuminated with wide-angle beams. We neglect all mutual coherence between light from two separate points in the object.

The flux density is then obtained by summing the individual contributions from all points in the object.

A. Transfer Functions

For a general linear system such as a linear electronic amplifier, a sinusoidal input will yield a sinusoidal output, with in general a frequency-dependent amplitude and phase shift. If a few reasonable assumptions are made, an optical system with incoherent illumination may be regarded as linear, and the preceding statement will then apply. A linear electronic amplifier processes signals that are functions of time; the frequencies of the individual sinusoidal components are temporal frequencies. An optical system processes signals that are functions of two spatial variables; the corresponding frequencies are spatial frequencies.

The reader should be cautioned against carrying the useful analogy between optical systems and linear electronic amplifiers too far. Optical systems connected together in series will not generally yield a system having a response function that is the product of the individual response function, as is true in the electronic case.

1. Point-Spread Function. Consider light that originates at a point $P(x_0, y_0)$ of strength A a distance S from a lens, as shown in Fig. 8.19. The field at the lens will be A/S, and the flux density will be KA^2/S^2, where K is a constant. The total flux passing through the lens will be

$$\Phi = KA^2\sigma_L/S^2 \qquad (8.164)$$

where σ_L is the area of the lens $\sigma_L = \pi r_0^2$. This will be unchanged by diffraction and by aberrations.

In the presence of aberrations and diffraction, the field in the paraxial image plane is given by Eq. (7.98), and the flux density, instead of being a delta function at x_0', y_0' will be given by

$$S(P') = \frac{\Phi}{\sigma_L} \frac{1}{\lambda^2 S'^2} \left| T_L\left(\frac{-1}{\lambda S'}(x' - x_0'), \frac{-1}{\lambda S'}(y' - y_0') \right) \right|^2 \qquad (8.165)$$

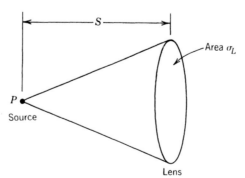

Fig. 8.19

In this result we have used Eq. (8.164) for A^2. The factor T_L is the Fourier transform of the lens opening including *deviations* from the spherical phase-shift factor (aberrations). This shows that the paraxial image "point" at x'_0, y'_0 "spreads" out. The influence of aberrations is expected to be less important than the fundamental shape of the lens opening, which leads to a diffraction pattern in the image plane. There is generally a much larger area than that of the "focal spot" over which $W(P, P')$, the aberration function, may be considered constant. It is called the *isoplanatism patch*, and its actual size depends on the strictness of our criteria. Under these circumstances x', y' enter into T_L essentially as parameters. Then the *point spread function*, defined by

$$O(x' - x'_0, y' - y'_0; P') \equiv \frac{1}{\sigma_L \lambda^2 S'^2} \left| T_L\left(\frac{-1}{\lambda S'} (x' - x'_0), \frac{-1}{\lambda S'} (y' - y'_0) \right) \right|^2 \qquad (8.166)$$

may be regarded primarily as a function of the *difference* variables $x' - x'_0$, $y' - y'_0$ with weak additional dependence on the absolute variables x', y'. The normalization of O is unity:

$$\int\!\!\!\int_{-\infty}^{\infty} O(x' - x'_0, y - y'_0; P') \, dx' \, dy' = 1 \qquad (8.167)$$

The spread function for the ideal, diffraction-limited case where there are no aberrations will be denoted by $O_0(x' - x'_0, y' - y'_0)$. Using Eq. (7.101) for a circular lens opening we obtain

$$O_0(x' - x'_0, y' - y'_0; P') = \frac{\pi \tilde{r}_0^2}{\lambda^2 S'^2} \left[\frac{2J_1(w)}{w} \right]^2 \qquad (8.168)$$

where w is defined in Eq. (7.102)

$$w = \frac{2\pi}{\lambda} \frac{\tilde{r}_0}{S'} [(x' - x'_0)^2 + (y' - y'_0)^2]^{1/2} \qquad (8.169)$$

A characteristic of this result, as well as that obtained with any simple aperture, is that

$$O_0(0, 0) = \frac{\sigma_L}{(\lambda S')^2} \qquad (8.170)$$

The distribution of flux density in the image plane described by Eq. (8.166) is shown schematically in Fig. 8.20.

With no aberrations the spread function is the Airy disk with rings centered at P'. Calculations have shown that for a small amount of spherical aberration, where the maximum value of the aberration function W is a small fraction of a wavelength, the main effect is to transfer energy from the disk to the rings. With off-axis aberrations, such as coma, the function O no longer has circular symmetry; this makes the rings nonuniform when W is less than λ. Photographs of images from a point object are shown in Fig. 8.21 when primary coma is present. It is not

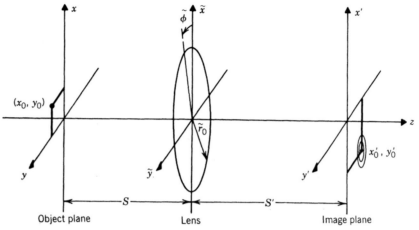

Fig. 8.20 Geometry for an analysis of image formation within the theory of transfer functions.

necessary to show images for distortion and curvature of field; they are Airy patterns with their centers displaced laterally and longitudinally, respectively. With other aberrations also, the size of the spread function can usually be reduced if the observation plane is shifted somewhat along the z-axis. For a small amount of primary spherical aberration, the spread is minimized halfway between paraxial and marginal foci. This is apparent from Fig. 8.25 in the somewhat different context of the optical-transfer function.

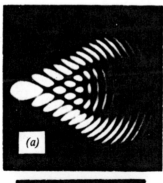

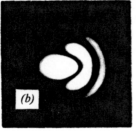

Fig. 8.21 (*a*) Monochromatic image of a point source with about five wavelengths of primary coma. (*b*) The same, but with about one half of one wavelength of primary coma. (Photographs courtesy of R. S. Longhurst.)

First a word about so-called focus errors, which are also related to curvature of field. To obtain the best focus one should shift the observation plane along the axis a distance Δz if there is a term in W proportional to $(n'\tilde{r}^2 \, \Delta z/2L'^2)$ where n' is the index after the lens. This can be written

$$W = \ell\lambda \left(\frac{\tilde{r}}{\tilde{r}_0}\right)^2$$

where λ is the wavelength in image space and where $\ell = (\Delta z/2\lambda)(\tilde{r}_0/L')^2$ gives the number of wavelengths of "focus error" at the edge ($\tilde{r} = \tilde{r}_0$) of the lens.

For general calculations with aberrations it is convenient to express W in multiples of λ as a function of the reduced radius $\tilde{r}/\tilde{r}_0$ and azimuthal angle $\tilde{\phi}$ at the lens. Thus, $W = (1/2)\lambda(\tilde{r}/\tilde{r}_0)^4$ would correspond to $1/2$ wavelength of primary spherical aberration, $W = \ell\lambda(\tilde{r}/\tilde{r}_0)^3 \cos\tilde{\phi}$ to ℓ wavelengths of primary coma. In the latter case the dimensionless coefficient ℓ is proportional to r', the displacement of P' from the axis, as shown in chapter 4. With a combination of astigmatism and curvature of field we have

$$W = \ell\lambda \left(\frac{\tilde{r}}{\tilde{r}_0}\right)^2 [\cos^2\tilde{\phi} - \mu]$$

where ℓ is proportional to r'^2. If the observation plane is the tangential image plane, then $\mu = 1$; in the sagittal image plane we have $\mu = 0$; halfway between, we have $\mu = 1/2$.

When the aberration function $W(\tilde{P}, P')$ is small compared with λ, one can take the first three terms in a series expansion of e^{-ikW} in Eq. (7.115) and derive the following approximate expression for the value of the spread function at the paraxial image point (where $x' = mx_0$, $y' = my_0$)

$$O(0, 0; P') = \frac{\sigma_P}{\lambda^2 L^2} [1 - k^2(\langle W^2\rangle - \langle W\rangle^2)] \tag{8.171}$$

where the "moments" $\langle W^n\rangle$ of W are defined by

$$\langle W^n\rangle = \frac{1}{\sigma_P} \iint [W(\tilde{P}, P')]^n |\tilde{\tau}_L(\tilde{P})| \, d\tilde{x} \, d\tilde{y} \tag{8.172}$$

Equation (8.171) describes how aberrations reduce the intensity of what would otherwise be the center of the Airy disk. Some applications of these results will be made in the problems.

2. Extended Source. The flux from a point source spreads into a focal spot described by the point spread function $O(x' - x'_0, y' - y'_0; P')$. Now consider a continuous image described by a function

$$S'_p(x'_0, y'_0)$$

such that a small region of the source conjugate to area $dx' \, dy'$ about x'_0, y'_0 radiates flux

$$d\Phi = S'_p(x'_0, y'_0) \, dx', dy'$$

through the lens. The paraxial flux density S'_p represents the prediction of paraxial geometrical optics for the flux density at the observation plane. This will spread by diffraction and aberrations and yield the flux density

$$O(x' - x'_0, y' - y'_0; P') \, d\Phi$$

in the observation plane instead of the geometrical prediction. The total flux density at the point $P'(x', y')$ is then obtained by integration:

$$S'(x', y') = \iint\limits_{-\infty}^{\infty} O(x' - x'_0, y' - y'_0; P')S'_p(x'_0, y'_0) \, dx'_0 \, dy'_0 \qquad (8.173)$$

Equation (8.173) represents $S'(P')$ in terms of a linear integral operator that acts on $S'_p(P'_0)$. It is convenient to represent $S'_p(P'_0)$ as a superposition of simple function like sines and cosines. Then $S'(P')$ will be given as the same superposition of the transformed sines and cosines.

It happens that these functions are a particularly happy choice, because transformed sine and cosine functions are still sines and cosines if they are restricted to a single isoplanatism patch, as we are about to prove.

Comparison of Eq. (8.173) with Eq. (6.91) shows that the resultant image flux density takes the form of a convolution integral. Thus,

$$S' = O \circledast S'_p \qquad (8.174)$$

3. Sinusoidal Incoherent Object. We consider a single isoplanatism patch in the observation plane. An arbitrary positive function $S'_p(P'_0)$ describing the paraxial flux density there can be expressed as a two-dimensional Fourier integral over spatial frequencies u', v'. We consider just one of the Fourier components and add to it a constant term to keep $S'_p(P'_0)$ nonnegative:

$$S'_p(x'_0, y'_0) = S_0\{1 + b \cos[2\pi(u'x'_0 + v'y'_0)]\}$$

$$= \mathrm{Re}\{S_0(1 + b \exp[-i2\pi(u'x'_0 + v'y'_0)]\} \qquad (8.175)$$

Here b is the *modulation* and is less than or equal to unity. This function is plotted schematically in two dimensions in Fig. 8.22. Polar coordinates for the spatial frequencies are

$$\rho' = \sqrt{u'^2 + v'^2} \quad \text{and} \quad \phi' = \tan^{-1}(v'/u')$$

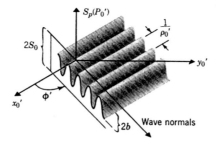

Fig. 8.22

Here ϕ' gives the direction of the "wave normals," and $1/\rho'$ gives the crest-to-crest wavelength in this direction.

The distribution of flux density at the object is the paraxial conjugate to Eq. (8.175) and is given by

$$S(x_0, y_0) = \text{const} \cdot S_p(mx_0, my_0) \qquad (8.176)$$

To obtain the actual flux density in the image plane we apply Eq. (8.173) to Eq. (8.175) and obtain

$$S'(x', y') = \text{Re}\left\{ S_0 + S_0 b \int\!\!\!\int_{-\infty}^{\infty} O(x' - x_0', y' - y_0'; P')e^{-i2\pi(ux_0' + vy_0')} \, dx_0' \, dy_0' \right\}$$

$$= \text{Re}\left\{ S_0 + S_0 b\, e^{-i2\pi(ux' + vy')} \right.$$

$$\left. \times \int\!\!\!\int_{-\infty}^{\infty} O(x' - x_0', y' - y_0'; P')e^{-i2\pi[u(x' - x_0') + v(y' - y_0')]} \, dx_0' \, dy_0' \right\}$$

$$= \text{Re}\{ S_0 + S_0 b\, e^{-i2\pi(ux' + vy')} \hat{O}(u, v; P') \} \qquad (8.177)$$

Because we stay on one isoplanatism patch, we have treated P' in Eq. (8.177) as a parameter, not as a variable to be integrated over. The function $\hat{O}$ appearing in Eq. (8.176) as a Fourier transform is called the *optical transfer function* (OTF). If we write it in the form

$$\hat{O} = \mathcal{F}[O] = |\hat{O}|e^{i\psi} \qquad (8.178)$$

we obtain

$$S'(x', y') = S_0\{1 + b|\hat{O}(u, v; P')|\cos[2\pi(ux' + vy') - \psi]\} \qquad (8.179)$$

We have just shown that the image of an incoherent sinusoidal object is sinusoidal with the same frequencies (u, v) as that of the geometrical paraxial image. The amount of modulation b will be changed by the factor $|\hat{O}|$, and the phase shifted by ψ. Because of its role in Eq. (8.178), $|\hat{O}|$ is called the *modulation transfer function* (MTF). The symbol ψ is sometimes called the *phase transfer function* (PTF).

4. Correlation Form of the Optical Transform Function (OTF). The lens transmission function that describes the extent of the lens and aberrations is called the *pupil function*. It differs from the full lens transmission function in that the phase shift leading to a parabolic wavefront distortion is missing.

$$\tilde{\tau}_p(\tilde{P}; P') = |\tilde{\tau}_L(\tilde{P})|\exp\left[\frac{-i2\pi}{\lambda} W(\tilde{P}, P')\right] \qquad (8.180)$$

where for a simple pupil $|\tilde{\tau}_L(\tilde{P})|$ is defined to be unity in the opening, and zero otherwise. We use the geometrical theory of aberrations to provide W.

We then want the OTF $\hat{O}$ in terms of the pupil function. The result of a short calculation, reserved for a problem, gives

$$\hat{O}(u, v; P') = \int\!\!\!\int_{-\infty}^{\infty} \tilde{\tau}_p(\tilde{x}, \tilde{y}; P')\tilde{\tau}_p^*(\tilde{x} - \lambda S'u, \tilde{y} - \lambda S'v; P')\frac{d\tilde{x}\,d\tilde{y}}{\sigma_L} \qquad (8.181)$$

This formula expresses $\hat{O}$ as the two-dimensional normalized autocorrelation function of the complex pupil function. For a simple pupil we may write

$$\hat{O}(u, v; P') = \int\!\!\!\int_{\Sigma_{pc}} \exp\left\{-ik[W(\tilde{x}, \tilde{y}; P') - W(\tilde{x} - \lambda S'u, \tilde{y} - \lambda S'v; P')\right\}\frac{d\tilde{x}\,d\tilde{y}}{\sigma_L}$$

$$(8.182)$$

where Σ_{pc} is the intersection of the pupil opening with itself when displaced by $(\lambda S'u, \lambda S'v)$.

Note that for the diffraction limited case, that is when there are no aberrations and $W = 0$, Eq. (8.182) gives $\hat{O}$ as simply the ratio of the area of this intersection to the area of the pupil, σ_L.

It can be shown that the MTF of a system is never greater than the OTF of an ideal aberration-free system with the same aperture.

B. Examples of Optical Transfer Functions

1. OTF for a Circular Aperture: Diffraction Limited. Besides acting as an upper bound on the MTF for real systems with aberrations, the diffraction-limited OTF is of interest in its own right, because in some cases, such as "diffraction-limited" telescope objectives (on or near axis), it is closely approached. From the definition of $|\tilde{\tau}_L(\tilde{x}, \tilde{y})|$ we recognize that the product $|\tilde{\tau}_L(\tilde{x}, \tilde{y})| \cdot |\tilde{\tau}_L(\tilde{x} - \lambda S'u, \tilde{y} - \lambda S'v)|$ gives unity if and only if the points $(\tilde{x}, \tilde{y})$ and $(\tilde{x} - \lambda S'u, \tilde{y} - \lambda S'v)$ are both in the lens opening. The integral for the OTF then represents the area of the intersection of the lens opening with itself displaced from the origin to $(\tilde{x} = \lambda S'u, \tilde{y} = \lambda S'v)$. This is shaded in Fig. 8.23a. Thus $\hat{O}_0(u_0, v_0)$ represents this shaded area divided by the area of the lens $\sigma_L = \pi\tilde{r}_0^2$. It is clear that the result can depend only on the distance $a = \lambda S'\rho = \lambda S'\sqrt{u^2 + v^2}$ that separates the centers of the two circles. The desired area is four times that shaded in Fig. 8.23b, and the OTF is given by

$$\hat{O}_0(u, v) = \frac{4}{\pi\tilde{r}_0^2}\int_{a/2}^{\tilde{r}_0} \tilde{y}\,d\tilde{x} = \frac{4}{\pi\tilde{r}_0^2}\int_{a/2}^{\tilde{r}_0}\sqrt{\tilde{r}_0^2 - \tilde{x}^2}\,d\tilde{x}$$

$$= \begin{cases} \dfrac{2}{\pi}\left\{\cos^{-1}\left(\dfrac{a}{2\tilde{r}_0}\right) - \dfrac{a}{2\tilde{r}_0}\left[1 - \left(\dfrac{a}{2\tilde{r}_0}\right)^2\right]^{1/2}\right\}, & 0 < \dfrac{a}{2\tilde{r}_0} \leq 1 \\[4mm] 0, & \dfrac{a}{2\tilde{r}_0} \geq 1 \end{cases} \qquad (8.183)$$

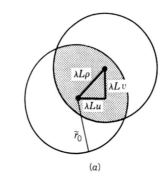

(a)

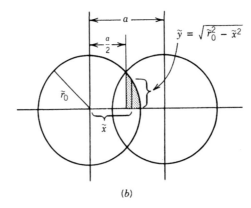

(b)

Fig. 8.23

The maximum spatial frequency that is transmitted by the system corresponds to $a = 2\tilde{r}_0$ and is

$$\rho_m = \frac{2\tilde{r}_0}{\lambda S'} \tag{8.184}$$

Thus

$$\hat{O}_0(u, v) = \frac{2}{\pi}\left\{\cos^{-1}\left(\frac{\rho}{\rho_m}\right) - \frac{\rho}{\rho_m}\left[1 - \left(\frac{\rho}{\rho_m}\right)^2\right]^{1/2}\right\}, \rho = \sqrt{u^2 + v^2} \tag{8.185}$$

This function is plotted as the dashed line in Figs. 8.25.

We should emphasize that $\hat{O}_0$ describes the loss of contrast in an ideal diffraction-limited system. For instance, when $\hat{O}_0 = 1/4$, we would have the situation shown in Fig. 8.24, if the object were a fully modulated sinusoidal test pattern. If the spatial frequency in image space is ρ, it is $m\,\rho$ in object space, where m is the transverse magnification.

2. OTF for Real Systems. Some remarks about OTF's for real systems are appropriate. There has been an extensive literature on the measurement of the OTF's for real lenses and the calculation of OTF's for lenses with aberrations. We

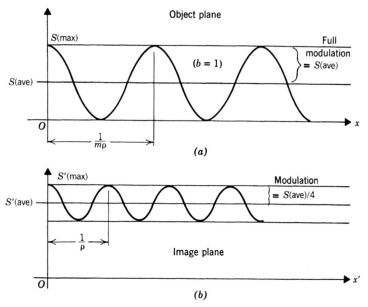

Fig. 8.24 Object flux density (*a*) and image flux density (*b*) when OTF = 1/4.

cannot devote much space here to the many details necessary for a good understanding. Some calculated results are shown in Figs. 8.25 and 8.27. A circular aperture of radius $\tilde{r}_0$ is assumed in all cases.

Figure 8.25 covers two cases where the aberration function does not depend on P'. The spread function is then a function only of $\sqrt{(x' - x'_0)^2 + (y' - y'_0)^2}$, and the OTF is then a function only of $\rho = \sqrt{u^2 + v^2}$ and is real. For ℓ wavelengths of focusing error the aberration function is

$$W = \ell \left(\frac{\tilde{r}}{\tilde{r}_0} \right)^2 \lambda \tag{8.186}$$

The resulting OTF curves are shown in Fig. 8.25*a* for $\ell\pi = 0, 1, 2, 3, 4$. For ℓ wavelengths of primary spherical aberration we have

$$W = \ell \left(\frac{\tilde{r}}{\tilde{r}_0} \right)^4 \lambda$$

This is for an observation plane at the paraxial focus. If we move the observation plane along the axis, we must add a term like Eq. (8.185). This gives

$$W = \ell\lambda \left[\left(\frac{\tilde{r}}{\tilde{r}_0} \right)^4 - 2\mu \left(\frac{\tilde{r}}{\tilde{r}_0} \right)^2 \right] \tag{8.187}$$

The parameter μ is zero at the paraxial focus and unity at the marginal focus. Curves for the OTF for $\ell = 1$ are shown in Fig. 8.25*b* for $\mu = 0, 1/2, 1$. When ℓ is

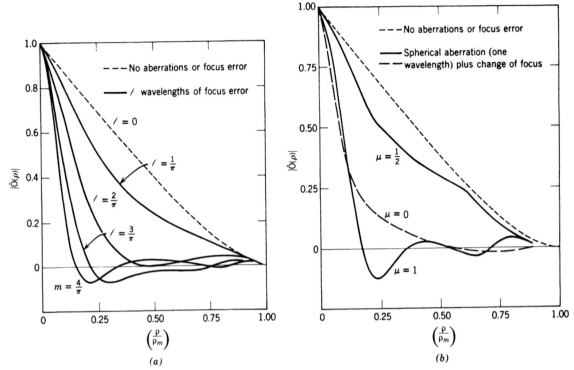

Fig. 8.25 (a) Modulation transfer function for a system with focus error. [After H. H. Hopkins, *Proc. Roy. Soc.* (London) *A231*, 91 (1955). Used with permission of The Royal Society.] (b) Modulation transfer function for one wavelength of spherical aberration. $\mu = 0$ corresponds to the paraxial focus, $\mu = 1$ to the marginal focus. [After G. Black and E. H. Linfoot, *Proc. Roy. Soc.* (London) *A239*, 522 (1957). Used with permission of The Royal Society.]

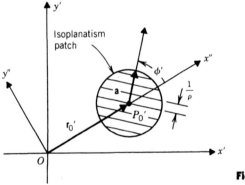

Fig. 8.26

small, the best performance is obtained for $\mu = 1/2$, halfway between paraxial and marginal foci. Note that some of the curves in Fig. 8.25 have the OTF negative. This can be described by a sudden jump in the phase ψ from 0 to π. It corresponds to a reversal of contrast (dark regions in the object appear light) and leads to a spurious resolution, because for objects containing spatial frequencies where the OTF is negative we have *apparent* resolution, whereas for objects containing lower spatial frequencies near where the OTF is passing through zero there is little or no resolution.

Off-axis aberrations will introduce a dependence on the off-axis distance $O'P_0'$ into the spread function and the OTF. The spread function will no longer have rotational symmetry about P_0', but it will be symmetric about the line $O'P_0'$. Because of this, it is convenient to define a rotated coordinate system with the x'' axis along this line. The OTF will then depend both on the magnitude ρ of the spatial frequency and on the angle ϕ' between the normal to the wave crests in the sinusoidal pattern and the x'' axis. This is shown in Fig. 8.26. When $\phi' = \pi/2$, the crests in the sinusoidal pattern are parallel to the symmetry axis. There can be no phase shift apart from a sudden reversal of contrast. For other values of ϕ', ψ will generally be nonzero.

Actually for primary astigmatism, the phase ψ is either zero or π as with the other cases discussed so far. The reason is that the point-spread function, while no longer possessing rotational symmetry about P_0', still has inversion symmetry about P_0', that is, the same at $\mathbf{r}_0' + \mathbf{a}$ and at $\mathbf{r}_0' - \mathbf{a}$ in Fig. 8.26. Such a function will have no sine Fourier transform, only a cosine Fourier transform and an exponential transform that is purely real. Halfway between tangential and sagittal image planes the aberration function for ℓ wavelengths of primary astigmatism is given by

$$W = \ell\lambda(\cos^2 \tilde{\phi} - \tfrac{1}{2})\left(\frac{\tilde{r}}{\tilde{r}_0}\right)^2 \tag{8.188}$$

This is where geometrical optics gives the circle of least confusion. The point-

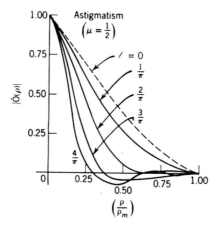

Fig. 8.27 Modulation transfer function at the circle of least confusion for ℓ wavelengths of primary astigmatism. (After M. De, *Proc. Roy. Soc.* (London) *A233*, 91 (1955). Used with permission of The Royal Society.]

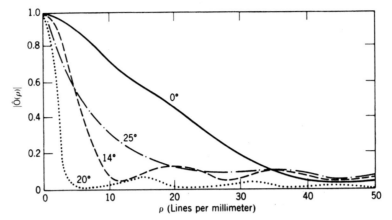

Fig. 8.28 The modulation transfer function for a Cooke triplet camera lens of 100-mm focal length. (After Kingslake in R. Kingslake, ed., *Applied Optics and Optical Engineering*, Vol. 3, p. 19. Academic Press, New York, 1965.)

spread function has fourfold symmetry about P_0'. This means that the OTF for angle ϕ' is the same as that for angle $\pm(\pi/2) \pm \phi'$. Some OTF curves are shown in Fig. 8.27 for $\phi' = \pi/4$.

For real lenses the small-aberration results just discussed are often inadequate, and other methods based on ray tracing or other digital computational techniques must be employed.

As an example, we show the MTF for a Cooke triplet camera lens in Fig. 8.28

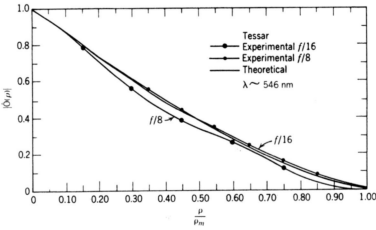

Fig. 8.29 The modulation transfer function for a Tessar lens at two different f numbers. Note that ρ_m is different in the two cases. (After R. R. Shannon in R. Kingslake, ed., *Applied Optics and Optical Engineering*, Vol. 3, p. 220. Academic Press, New York, 1965.)

for various values of the off-axis angle. As another example, plots are given in Fig. 8.29 of the on-axis MTF for a Tessar lens at two different f numbers. This shows that the theoretical diffraction-limited behavior is closely approached for well-designed lens used at small apertures.

REFERENCES

Arfken, G. *Mathematical Methods for Physicists.* Academic Press, New York, 1970.

Barnes, K. R. *The Optical Transfer Function.* Elsevier, New York, 1971.

Beran, M. J., and G. B. Parrent, Jr. *Theory of Partial Coherence.* Prentice-Hall, Englewood Cliffs, N.J., 1964.

Born, Max, and Emil Wolf. *Principles of Optics.* Pergamon Press, Oxford, 1980.

Breene, R. G. *The Shift and Shape of Spectral Lines.* Pergamon Press, New York, 1961.

Cummins, Herman Z., and Edward Roy Pike, ed. *Photon Correlation and Light Beating Spectroscopy;* (*Proceedings of NATO Advanced Study Institute, Capri, 1973*). Plenum Press, New York, 1974.

Dainty, J. C., ed. *Laser Speckle and Related Phenomena.* Springer-Verlag, Berlin, 1984.

Duffieux, P. M. *The Fourier Transform and Its Application to Optics.* Wiley, New York, 1983.

Françon, Maurice. *Diffraction: Coherence in Optics.* Pergamon Press, Oxford, 1966.

Françon, Maurice. *Optical Image Formation and Processing.* Academic Press, New York, 1979.

Garbuny, Max. *Optical Physics.* Academic Press, New York, 1965.

Goodman, Joseph W. *Introduction to Fourier Optics.* McGraw-Hill, New York, 1968.

Goodman, Joseph W. *Statistical Optics.* Wiley, New York, 1984.

Kingslake, Rudolf. *Lens Design Fundamentals.* Academic Press, New York, 1978.

Klauder, John R., and E. C. D. Sudarshan. *Fundamentals of Quantum Optics.* Benjamin, New York, 1968.

Lengyel, Bela A. *Introduction to Laser Physics.* Wiley, New York, 1966.

Levi, Leo. *Applied Optics.* Vol. II Wiley, New York, 1980.

Linfoot, E. H. *Fourier Methods in Optical Image Evaluation.* Focal Press, London, 1964.

Loudon, R. *The Quantum Theory of Light.* Oxford Clarendon Press, London, 1973.

Marathay, Arvind S. *Elements of Optical Coherence Theory.* Wiley, New York, 1982.

Martin, L. C. *The Theory of the Microscope.* Elsevier, New York, 1965.

Mertz, Lawrence. *Transformations in Optics.* Wiley, New York, 1965.

O'Neill, E. L. *Introduction to Statistical Optics.* Addison-Wesley, Reading, Mass., 1963.

Smith, William V., and Peter P. Sorokin. *The Laser.* McGraw-Hill, New York, 1966.

Steel, W. H. *Interferometry.* Cambridge University Press, Cambridge, 1967.

Troup, C. J. *Optical Coherence Theory—Recent Developments.* Metheun, London, 1967.

PROBLEMS

Section 8.1 Temporal Coherence

1. Show that the minimum visibility that will be observed with a source emitting two infinitely narrow spectral lines at frequencies v_i and v_j with associated flux densities S_i and $S_j < S_i$ is

$$\frac{S_i - S_j}{S_i + S_j}$$

Is it possible to determine which frequency is larger from a study of the visibility curve? Explain.

2. Prove that $P(v)$ and $\gamma(\tau)$ are Fourier transforms of each other. That is, demonstrate that if

$$\gamma(\tau) = \int_0^\infty P(v) \cos(2\pi v\tau)\, d\tau$$

then

$$P(v) = 4 \int_0^\infty \gamma(\tau) \cos(2\pi v\tau)\, d\tau$$

3. Demonstrate that for a quasimonochromatic source whose frequency is concentrated about v_0 the superposition integral Eq. (8.3a) is equivalent to Eq. (8.3b). That is,

$$A(t)\cos[2\pi v_0 t + \varphi(t)] = \int_0^\infty A(v)\cos[2\pi vt + \varphi(v)]\, dv$$

Determine the functions $A(t)$ and $\varphi(t)$.

4. Discuss quantitatively the interference pattern that would be observed as a function of τ if the spectral distribution function is flat from v_0 to $2v_0$ but otherwise zero, that is, for

$$S(v) = S_0, \quad \text{constant for } v_0 < v \le 2v_0$$
$$S(v) = 0, \quad \text{otherwise}$$

5. Prove Eqs. (8.40) through (8.42), which define $\gamma(\tau)$, $U(\tau)$ and $\Phi(\tau)$ for a two-frequency source.

6. Consider a source of light consisting of two quasi-monochromatic lines of the same shape centered at frequencies v_i and v_j. The spectral distribution function will be

$$P(v) = \tfrac{1}{3}D(v - v_i) + \tfrac{2}{3}D(v - v_j)$$

Further suppose that $D(\mu)$ is symmetric about $\mu = 0$ and that

$$\int_{-\infty}^\infty D(\mu)\cos 2\pi\mu\tau\, d\mu = \exp\left(-\frac{\tau^2}{\tau_0^2}\right)$$

Let

$$v_i > v_j, \quad \frac{1}{\tau_0} \ll (v_i - v_j) \ll v_j.$$

Calculate the normalized oscillatory function $\gamma(\tau)$ and sketch it. How does it compare with Fig. 8.4b?

7. In terms of the mirror separation d, find the period of the beat pattern in the fringe system and the separation between fringes for interference in a Michelson interferometer. Use light from the sodium doublet. Assume

both components have the same strength and have wavelengths 589.593 nm and 588.996 nm. Assume that the line widths are much smaller than the separation between components.

8. Determine the normalized oscillatory term in the interference pattern of a Michelson interferometer exposed to quasimonochromatic radiation defined by

$$D(\mu) = D_0\left[1 + \cos\left(\frac{2\pi\mu}{\Delta v}\right)\right], \quad |\mu| \le \frac{\Delta v}{2}$$
$$= 0, \quad |\mu| > \frac{\Delta v}{2}$$

9. This problem is concerned with the quantitative determination of the resolving power of an interference spectrometer. Let the detector signal be proportional to $S = S_0[1 + \gamma(\tau)]$ and use

$$\gamma(\tau) = \int_0^\infty P(v)\cos(2\pi v\tau)\, dv$$

Because the retardation time is limited by the travel of the mirror, τ must be less than

$$\tau_{max} = \frac{2d_{max}}{c}$$

What is actually measured is a truncated autocorrelation function

$$\gamma^T(\tau) = \gamma(\tau) \quad \text{if } |\tau| \le \tau_{max}, \quad = 0 \quad \text{otherwise}$$

(a) Show that if $\gamma^T(\tau)$ is used in the inverse transform equation

$$P(\mu) = 4\int_0^\infty \gamma(\tau)\cos(2\pi\mu\tau)\, d\tau$$

a measured frequency distribution will be obtained that is given by

$$P_m(v) = \int_0^\infty P(v')\left[\frac{\sin 2\pi(v' - v)\tau_{max}}{\pi(v' - v)}\right.$$
$$\left. + \frac{\sin 2\pi(v' + v)\tau_{max}}{\pi(v + v')}\right] dv'$$

Now consider $P(v)$ of the form shown in Fig. 8.30. Here $v_0\tau_{max} \gg 1$. Then only the first term in the measured normalized spectral distribution function integral just presented is important for positive frequency. Sketch the behavior of $P_m(v)$ in the following limiting cases.

(b) When $\Delta v \gg 1/\tau_{max}$.

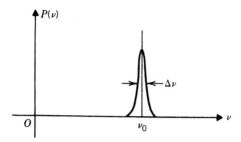

Fig. 8.30

(c) When $\Delta v \ll 1/\tau_{max}$. You will not require an explicit functional form for $P(v)$.

(d) What does the result of part (c) have to say about the resolving power?

10. Suppose that a very narrow quasimonochromatic line of frequency v_0 is being measured on an interference spectrometer. The line width is assumed to obey $\Delta v \ll 1/\tau_{max}$. Instead of the abrupt truncation obtained by direct use of the experimental $\gamma^T(\tau)$ discussed in Problem 4, suppose that we smooth γ linearly to obtain a new function

$$\gamma^S(\tau) = \gamma_0(\tau)\left(1 - \frac{|\tau|}{\tau_{max}}\right), \qquad |\tau| \leq \tau_{max}$$

$$= 0, \qquad\qquad\qquad \text{otherwise}$$

where $\gamma_0(\tau) \approx \cos(2\pi v_0 \tau)$. Discuss with equations and a sketch the frequency spectrum obtained if $\gamma^S(\tau)$ is used in Eq. (8.22), and compare this with the result if $\gamma^T(\tau)$ of the preceding problem is used with $\gamma(\tau) = \gamma_0(\tau) \approx \cos(2\pi v_0 \tau)$ (see Fig. 8.8.) Which method do you think gives the greater resolving power? Which method gives fewer spurious wiggles in the computed spectrum? (The operation described is called *apodization*, which means "removing the feet.")

Section 8.2 Statistical Optics

11. Derive Eq. (8.82). This is one form of Parseval's theorem.

12. Derive the results of Eqs. (8.86). Find either Δv or τ_c.

13. A pulsed laser operates at a wavelength of 532 nm with a pulse width of 70 nsec. The lineshapes of the allowed frequencies in a laser are defined by the resonant characteristics of the cavity. However, assuming

that truncation broadening influences the lineshapes as well, determine the spectral distribution function and the normalized autocorrelation function for a single line resulting from the finite pulse duration. Over what range of optical path differences would a Michelson interferometer need to operate so that this lineshape is resolved?

14. An instructive application of Eqs. (8.63) is to a steady source ($A = $ constant) where the phase difference

$$\Delta\phi(\tau) \equiv \phi(t) - \phi(t - \tau)$$

is a Gaussian random variable. Using the theory of probability we can show that $\langle [\Delta\phi(\tau)]^n \rangle = 0$ for all odd powers n. For even powers the averages can be related to the mean square value of the phase difference as follows. If we define $\langle [\Delta\phi(\tau)]^2 \rangle \equiv \sigma^2(\tau)$, then $\langle [\Delta\phi(\tau)]^4 \rangle \equiv 3\sigma^4$, $\langle [\Delta\phi(\tau)]^6 \rangle \equiv 3 \cdot 5\sigma^6$, $\langle [\Delta\phi(\tau)]^8 \rangle \equiv 3 \cdot 5 \cdot 7\sigma^8$, and so on. Using these results, show that

$$D_c(\tau) = \langle \cos \Delta\phi(\tau) \rangle = \exp\left(-\frac{\sigma^2(\tau)}{2}\right)$$

and

$$D_s(\tau) = \langle \sin \Delta\phi(\tau) \rangle = 0$$

(We can further show that $\sigma^2(\tau)$ can be written in the form $\sigma^2(\tau) = |\tau|/\tau_1$. τ_1 then represents the retardation time for which the mean square phase change is 1 rad. The autocorrelation function is then

$$\gamma(\tau) = \exp\left(-\frac{|\tau|}{2\tau_1}\right)\cos(2\pi v_0 \tau)$$

15. Show that $D_c(\tau)$ as defined by Eq. (8.30a) is even, $D_c(\tau) = D_c(-\tau)$, and that $D_s(\tau)$ as defined by Eq. (8.30b) is odd, $D_s(\tau) = -D_s(-\tau)$. Show further that Eqs. (8.30a and b) can be combined into a single complex Fourier integral, which when inverted gives

$$D(\mu) = \int_{-\infty}^{\infty} U(\tau)e^{i\Phi(\tau)} e^{-i2\pi\mu\tau} d\tau$$

$$= \int_{-\infty}^{\infty} [D_c(\tau) - iD_s(\tau)]e^{-i2\pi\mu\tau} d\tau$$

Now split $D(\mu)$ into even and odd parts:

$$D(\mu) = D_e(\mu) + D_0(\mu)$$

with

$$D_e(\mu) = \tfrac{1}{2}[D(\mu) + D(-\mu)]$$

$$D_0(\mu) = \tfrac{1}{2}[D(\mu) - D(-\mu)]$$

Then derive the results

$$D_e(\mu) = \int_{-\infty}^{\infty} D_c(\tau) \cos(2\pi\mu\tau)\, d\tau$$

$$= 2 \int_0^{\infty} D_c(\tau) \cos(2\pi\mu\tau)\, d\tau$$

$$-D_0(\mu) = \int_{-\infty}^{\infty} D_s(\tau) \sin(2\pi\mu\tau)\, d\tau$$

$$= 2 \int_0^{\infty} D_s(\tau) \sin(2\pi\mu\tau)\, d\tau$$

16. Consider the superposition of two cosine waves, each having the same slowly varying envelope:

$$E(t) = A(t) \cos(2\pi\nu_i t) + A(t) \cos(2\pi\nu_j t)$$

$$= 2A(t) \cos(2\pi\nu_0 t) \cos(2\pi f t)$$

where

$$\nu_0 = \tfrac{1}{2}(\nu_i + \nu_j), \quad f = \tfrac{1}{2}(\nu_i - \nu_j) > 0$$

Assume that $f \ll \nu_0$ and that $A(t)$ is slowly varying with respect to $\cos(2\pi f t)$.

(a) Find an approximate expression for the normalized autocorrelation function for $E(t)$ and sketch the result.

(b) Using the result of **(a)**, find an approximate expression for the spectral distribution function for this source by applying Eq. (8.70). The results should look like Fig. 8.31.

17. Consider a steady source emitting two steady quasimonochromatic lines so that the electric field is of the form

$$E(t) = A_i \cos[2\pi\nu_i t + \varphi_i(t)]$$
$$+ A_j \cos[2\pi\nu_j t + \varphi_j(t)], \quad A_i, A_j \text{ constants}$$

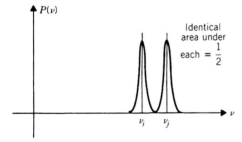

Fig. 8.31

where

$$\langle \cos[\varphi_i(t) - \varphi_i(t - \tau)] \rangle = \exp(-\tau^2/t_i^2)$$
$$\langle \cos[\varphi_j(t) - \varphi_j(t - \tau)] \rangle = \exp(-\tau/t_j^2)$$

and where the averages of the sines are zero, with

$$\nu_i t_i \gg 1, \quad \nu_j t_j \gg 1, \quad t_i \sim t_j, \quad \nu_i \gg \nu_i - \nu_j > 0$$

Find expressions for $\gamma(\tau)$ and for $P(\nu)$.

18. Light from a quasimonochromatic source at 656.3 nm is under analysis with a Michelson interferometer. The interference conditions can be maintained over a path length difference of 20 cm. What is the line width assuming collision broadening?

19. Over what range of optical path length difference would an interference spectrometer need to operate so that it would be able to resolve the "natural line width" of an atomic transition defined by lifetime broadening? A typical value of t_ℓ is 10^{-8} sec.

Section 8.3 Spatial Coherence

20. Show explicitly that the zero-order fringe moves up an angle $\theta' = \theta$ when the source moves down an angle θ. Include a detailed derivation of Eq. (8.88).

21. Show that Eq. (8.90) for two equal strength point sources can be written in the form $S = S_0(1 + \gamma_{12})$, where $S_0 = S_I + S_{II}$ and determine γ_{12} if the sources are located at $\theta = \theta_I$ and $\theta = \theta_{II}$, respectively. Assume the sources have the same, nearly monochromatic frequency. Derive an equation for the visibility $V(a)$.

22. Derive the equivalent of Eq. (8.93) for three point sources.

23. A screen is placed in front of a large diffuse incoherent source of quasimonochromatic light of wavelength λ. It has two openings in the form of long, narrow slits a distance a_1 apart. At a distance $D_1 \gg a_1$ from the first screen is a second screen with a pair of similar long, narrow slits with separation a_2 oriented parallel to the first pair of slits. The interference fringes are observed on a third screen a distance $D_2 \gg a_2$ from the second screen. For fixed values of D_2, a_1, and a_2, at what values of D_1 will the fringes disappear? (See Fig. 8.32.)

24. Show that the equal displacement of all sources in a Young's double-slit experiment by the angle θ_0 results

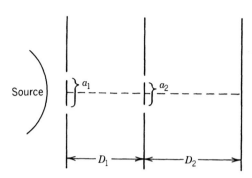

Fig. 8.32

in a change in phase of the complex degree of coherence given by

$$2\pi\left(\frac{\theta_0 a}{\lambda_0}\right)$$

25. Consider a double star. The components have equal intensities. Their separation is so small that the visibility of the fringes at the maximum-slit separation a_{max} is still close to unity. What would this visibility be if a_{max} is one tenth of the value needed to give the first zero in visibility?

26. What is the smallest angular diameter of stars that can be measured by the Michelson stellar interferometer using light of wavelength near 550 nm and a_{max} equal to 6 m?

27. Find the maximum size of a spot illuminated by the sun over which the illumination is nearly coherent. The sun's disk subtends an angle of 32 seconds of arc and the solar radiation is maximized around 550 nm.

28. A zirconium lamp has an arc 10^{-3} in. in diameter. It is used with a filter that limits the wavelength to the vicinity of 550 nm. For use as a source in a spatial filter we would require almost complete transverse coherence of the incident light across the object. Suppose that the focal length f of the first lens in Fig. 8.33 is 20 cm and its diameter is 3 cm. Will the source have the necessary coherence?

29. Quantitatively describe how the single-slit diffraction pattern is modified when the point source used in Chapter 6 is replaced by a line source extending to $\Delta\theta$ on either side of the optical axis (which is perpendicular to the diffracting slit panel). Use the far-field approximation.

30. We can define the angular spread of the source by

$$\Delta\theta \equiv \frac{1}{\int_{-\infty}^{\infty} i^2(\theta)\, d\theta}$$

and the transverse coherence length by

$$\ell_t = 2\int_0^\infty \left| I\left(\frac{v_0 a}{c}\right)\right|^2 da = 2\int_0^\infty V^2(a)\, da$$

(a) Show that $\ell_t = \lambda/\Delta\theta$.
(b) Show that for a uniform line source extending from θ_I to θ_{II}, $\Delta\theta$ as defined here equals $\theta_{II} - \theta_I$.

31. In the text we showed that in principle one may determine the amplitude and phase of the complex valued function $I(w) = \mathcal{F}[i(\theta)]$

$$w = \frac{v_0 a}{c} = \frac{a}{\lambda}$$

where $i(\theta)$ is the angular distribution function for the source distribution as projected onto the plane of the

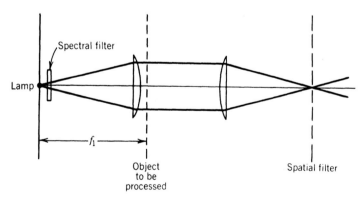

Fig. 8.33

drawing in Fig. 8.10b. From the known function $I(w)$, we can compute $i(\theta)$ by calculating the inverse Fourier transform

$$i(\theta) = \int_{-\infty}^{\infty} I(w)e^{i2\pi w\theta}\,du$$

Because a is positive, so is w. The integral can be converted to a sum of integrals involving $\text{Re}\{I(w)\}$ and $\text{Im}\{I(w)\}$ for positive w only. Do this.

32. A uniform disk subtends an angle of 0.01 rad at the slits of a Young's interferometer. The light from the disk demonstrates a Gaussian lineshape of width 100 nm about 600 nm. Derive an expression for $S(\theta')/S_0$ if the slits of the interferometer are set at 6 mm. Repeat the derivation for a slit spacing that is 0.1% larger.

33. Use the statistical expression for the correlation function Eq. (8.129) to calculate S, the flux density in the interference pattern of a Young's interferomter, as a function of a and θ' where the source consists of two points at θ_I and θ_{II} that start out with equal strength at $t = 0$ but that decay with different rates. Assume

$$E_I = A\,e^{-t/\tau_I}\cos(2\pi\nu_0 t)$$

$$E_{II} = A\,e^{-t/\tau_{II}}\cos(2\pi\nu_0 t)$$

Section 8.4 Fluctuations

34. Calculate the degeneracy parameters for light from the following sources, assumed to be blackbodies.

(a) A star at 10,000 K for λ near 500 nm.

(b) A gas discharge at 3000 K for λ at 546 nm.

35. A flux-density fluctuation interferometer is used to measure stellar diameters, in particular the diameter of the star in the previous problem. The detectors each have an area that is 0.001 times the area of coherence. Their quantum yield is 0.10, and their resolving time is 15 nsec. What is the minimum value of V that can be measured in a time of 3000 sec with a signal-to-noise ratio of unity? Assume the slow-detector limit.

36. The derivations in this section assumed that the light was in a definite state of polarization. Unpolarized light is an equal mixture of two mutually incoherent components in two independent states of polarization. For instance, these two states may be taken to be linearly-polarized along two mutually perpendicular directions. Let ρ_1 (unpol) be the average rate of emission of photoelectrons from each phototube when

unpolarized light is used and ρ_1(pol) the rate when a polarizer is inserted in the beam. Then ρ_1(unpol) = $2\rho_1$(pol) and η(unpol) = $\frac{1}{2}\eta$(pol).

(a) Explain why the second equation should hold on the basis of the statement at the beginning of this problem.

(b) Find the expression corresponding to Eq. (8.156) for the SNR for unpolarized light.

Section 8.5 Image Formation: Incoherent Objects

37. Equation (8.166), which defines the point spread function, shows that, except for constants, the OTF is the Fourier transform of $T_L T_L^*$. Through the convolution theorem [Eq. (6.97)] the Fourier transform of a product appears as the convolution of the Fourier transforms. Use this theorem starting with Eq. (8.166) and, including all the constants, prove Eq. (8.181).

38. (a) Prove that for any simple aperture Eq. (8.170) is true:

$$O_0(0, 0) = \frac{\sigma_L}{(\lambda S')^2}$$

(b) For a nonsimple aperture, show that the result in (a) generalizes to

$$O_0(0, 0) = \frac{1}{(\lambda S')^2}\int\!\!\!\int_{-\infty}^{\infty} |\tilde{\tau}_L(\tilde{x}, \tilde{y})|^2\,d\tilde{x}\,d\tilde{y}$$

39. Use the moment expansion technique to derive Eq. (8.171) for the point-spread function in the case that the aberration function is small.

40. Because of Eq. (8.170) we may rewrite Eq. (8.171) to give

$$O(0, 0; P') = O_0(0, 0)[1 - D]$$

$$D = \frac{4\pi^2}{\lambda^2}(\langle W^2 \rangle - \langle W \rangle^2)$$

where D represents the fractional decrease in $O(0, 0; P')$ because of a small amount of aberration. Calculate D for the following aberrations for a circular exit pupil of radius $\tilde{r}_0$ ($\tilde{\rho} = \tilde{r}/\tilde{r}_0$).

(a) m wavelengths of focus error (curvature of field):

$$W = m\lambda\tilde{\rho}^2$$

$Answer$: $D = \dfrac{\pi^2}{3}\,m^2$.

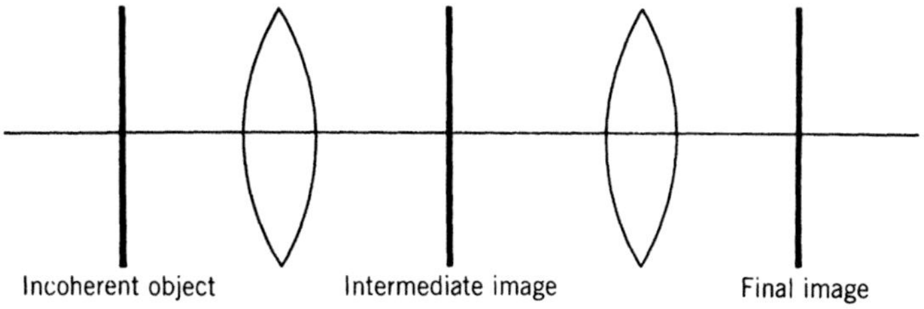

Incoherent object   Intermediate image   Final image

Fig. 8.34

(b) m wavelengths of primary spherical aberration:

$$W = m\lambda\tilde{\rho}^4$$

Answer: $D = \dfrac{16\pi^2}{45} m^2$.

(c) m wavelengths of spherical aberration plus change in focal plane:

$$W = m\lambda\tilde{\rho}^2(\tilde{\rho}^2 - 2\mu)$$

Show that D is minimized when $\mu = 1/2$. (Halfway between paraxial and marginal foci.)

(d) m wavelengths of astigmatism and curvature of field:

$$W = m\lambda\tilde{\rho}^2(\cos^2\tilde{\phi} - \mu)$$

Show that D is minimized when $\mu = 1/2$. (Halfway between paraxial and sagittal foci, at the circle of least confusion.)

(e) m wavelengths of coma:

$$W = m\lambda\tilde{\rho}^3 \cos\tilde{\phi}$$

Answer: $D = \dfrac{\pi^2}{2} m^2$.

41. Despite the presence of aberrations, the initial slope of the MTF is the same as that for a diffraction-limited system. Show that for a circular exit pupil of radius $\tilde{r}_0$ the modulation transfer function obeys

$$\frac{d}{d\rho}|\hat{O}(\rho)|_{\rho=0} = \frac{d}{d\rho}\hat{O}_0(\rho)\bigg|_{\rho=0} = \frac{-4}{\pi\rho_m}, \quad \rho_m = \frac{2\tilde{r}_0}{\lambda S'}$$

42. Show that for a rectangular aperture $2\tilde{x}_0$ by $2\tilde{y}_0$ the diffraction-limited OTF is

$$\hat{O}_0(u, v) = \left(1 - \left|\frac{u}{u_m}\right|\right)\left(1 - \left|\frac{v}{v_m}\right|\right)$$

$$\text{for } |u| \le u_m, |v| \le v_m$$

where

$$u_m = \frac{2\tilde{x}_0}{\lambda S'}, \quad v_m = \frac{2\tilde{y}_0}{\lambda S'}$$

43. Show that with a focusing error where $W = [n(\tilde{x}/\tilde{x}_0)^2 + m(\tilde{y}/\tilde{y}_0)^2]$ the optical transfer function for a $2x_0$ by $2y_0$ rectangular pupil is given by multiplying the result of the previous problem by $(\text{sinc } t_x \cdot \text{sinc } t_y)$, where

$$t_x = 8\pi n \frac{u}{u_m}\left|1 - \frac{u}{u_m}\right|, \quad t_y = 8\pi m \frac{v}{v_m}\left|1 - \frac{v}{v_m}\right|$$

44. It is not true that the OTF for the combined two-lens system shown in Fig. 8.34 is equal to the product of the individual OTF's for the two lenses. Prove this by assuming it to be true and then deriving a property of the point-spread function that can be shown to be false. (*Hint:* See Problem 37. If the light at the intermediate image were completely incoherent, then the OTF's would multiply. But the light is partially coherent. Why?)

9 Polarization

In Chapters 1 and 2 we introduced the electromagnetic theory of light that is based on Maxwell's equations. We found that the theory leads to wave equations for the components of the electric field vector $\mathbf{E} = (E_x, E_y, E_z)$ and the magnetic induction vector $\mathbf{B} = (B_x, B_y, B_z)$. In the development of most of geometrical optics and for a description of interference and diffraction in homogeneous media, the vector character of the light waves can be ignored. In this chapter we discuss important optical phenomena that can be understood only in terms of the transverse vector-wave character of the more complete model.

9.1 Polarized Light

Because the field vectors are perpendicular to the propagation direction, there is a degree of freedom involving their orientation not available in the case of scalar waves. The polarization properties of light are manifestations of this degree of freedom. We restrict our discussion to consideration of monochromatic radiation of the plane-wave type. The time and space dependence of these waves has the form

$$\cos(\omega t - \mathbf{k} \cdot \mathbf{r} + \phi)$$

with

$$\mathbf{k} = \frac{\omega n \hat{\mathbf{s}}}{c}$$

Care must be exercised when dealing with anisotropic media in the interpretation of the index of refraction n. However, the phase dependence will retain this form. It is sufficient to specify the behavior of the electric field $\mathbf{E}(\mathbf{r}, t)$ for then the magnetic induction $\mathbf{B}(\mathbf{r}, t)$ can be determined from $\mathbf{E}$ via Maxwell's equations.

A. Types of Polarized Light

Because the electric field is a vector quantity, we must specify both its magnitude and its direction. Consider a plane wave with wave vector **k**. The electric field **E** may point along any direction in the plane perpendicular to **k**, and **B** is along the vector **k** × **E**. Let **k** be along the z-axis and have magnitude $k = \omega/v$, where $v = c/n$ is the phase velocity. The monochromatic wave can then be written in the form

$$\mathbf{E}(x, y, z, t) = E_x \hat{\mathbf{x}} + E_y \hat{\mathbf{y}} \tag{9.1}$$

where

$$E_x(x, y, z, t) = A_x \cos(\omega t - kz + \phi_x) \tag{9.2}$$

$$E_y(x, y, z, t) = A_y \cos(\omega t - kz + \phi_y) \tag{9.3}$$

with positive real amplitudes A_x, A_y and phase angles ϕ_x and ϕ_y.

The classification of the types of polarized light will depend on the relative phase

$$\phi = \phi_y - \phi_x \tag{9.4}$$

and on the relative sizes of A_x and A_y.

In our previous discussions we have established a coordinate system with the x-axis vertical, the y-axis horizontal, and the z-axis along the general direction of propagation. This will also be used here. However, consistent with historical convention we display the system as viewed in the negative z-direction, that is, opposite to the direction of propagation. The diagrams thus appear with the x-axis up and the y-axis to the left.

1. Linearly Polarized Light. Sometimes called plane polarized light, *linearly polarized light* results when

$$\phi = 0 \quad \text{or} \quad \phi = \pi$$

for then E_y is proportional to E_x:

$$E_y = \frac{A_y}{A_x} E_x \quad \text{for} \quad \phi = 0 \tag{9.5}$$

$$E_y = -\frac{A_y}{A_x} E_x \quad \text{for} \quad \phi = \pi \tag{9.6}$$

using Eq. (9.2) and Eq. (9.3). At a fixed position z, the electric field undergoes simple harmonic motion along a *line* (Fig. 9.1). At a given value of t the electric field has a sinusoidal z-dependence and is found in the *plane* containing the z-axis and the line $y = +(A_y/A_x)x$ for $\phi = 0$, or $y = -(A_y/A_x)x$ for $\phi = \pi$ (Fig. 9.2).

2. Circularly Polarized Light. When

$$A_x = A_y = A \quad \text{and} \quad \phi = \pm \frac{\pi}{2}$$

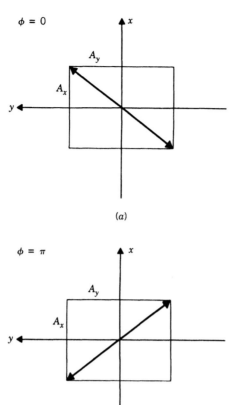

$\phi = 0$

A_y

A_x

(a)

$\phi = \pi$

A_y

A_x

(b)

Fig. 9.1 Linearly polarized light.

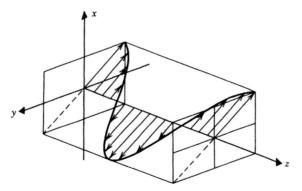

Fig. 9.2 Representation of the electric field vector in space at fixed time for linearly polarized light. Here $\phi = \pi$.

circularly polarized light results. Then

$$E_x = A \cos(\omega t - kz + \phi_x) \tag{9.7}$$

$$E_y = A \cos\left(\omega t - kz + \phi_x \pm \frac{\pi}{2}\right)$$

$$= \mp A \sin(\omega t - kz + \phi_x) \tag{9.8}$$

When $\phi = +\pi/2$, E_y leads E_x by $\pi/2$ rad; that is, E_y reaches its maximum value a quarter of a cycle before E_x does. As a function of time the electric field describes a *clockwise circle* in the *x-y* plane as seen head-on. Using the language of modern particle physics, we would say that this light has "negative helicity." It is conventional in optics, however, to call this case where $\phi = +\pi/2$, *right circularly polarized* (RCP).

The other case of *left circularly polarized* (LCP) light or light of positive helicity occurs when $\phi = -\pi/2$. Then E_y lags E_x by a quarter of a cycle, and the electric field describes a counterclockwise circle in the *x-y* plane (Fig. 9.3*a*).

We also obtain circular polarization when $\phi = \pm 3\pi/2$, for then Eq. (9.8) becomes

$$E_y = \pm A \sin(\omega t - kz + \phi_x)$$

In this case the minus sign gives clockwise rotation and right circular polarization, the plus sign counterclockwise rotation and left circular polarization. We summarize in Table 9.1.

For a fixed value of *t*, the electric field describes a spiral on the surface of a circular cylinder of radius A with its axis along z (Fig. 9.3*b*). Because an increase in z amounts to a *retardation* in phase, RCP light will spiral counterclockwise along the *z*-axis for fixed *t* and hence look like a *right-handed* screw. This explains the conventional optical terminology. As *t* increases, this screw rotates clockwise, that is, backward.

3. Elliptically Polarized Light. When A_x no longer equals A_y, but when we still have $\phi = \pm \pi/2$, then the curve traced out by the **E** vector in the *x-y* plane becomes an ellipse:

$$E_x = A_x \cos(\omega t - kz + \phi_x)$$

$$E_y = \mp A_y \sin(\omega t - kz + \phi_x) \tag{9.9}$$

We can transform this into a familiar form:

$$\frac{E_x^2}{A_x^2} + \frac{E_y^2}{A_y^2} = \cos^2(\omega t - kz + \phi_x) + \sin^2(\omega t - kz + \phi_x) = 1 \tag{9.10}$$

Thus the semiaxes of the ellipse are given by A_x and A_y, and it is oriented along the *x*- and *y*-axes. In this case the electric field traverses an elliptical path in the *xy* plane, the sense of traversal being clockwise (or right-handed) for $\phi = +\pi/2$ and

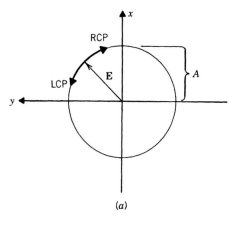

(a)

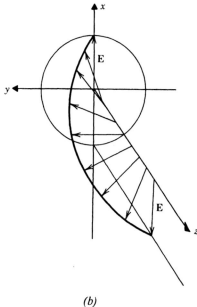

(b)

Fig. 9.3 (*a*) Circularly polarized light as seen in a plane perpendicular to the direction of propagation. The electric-field vector sweeps out a circle as time evolves. The convention for direction of rotation is shown. The view is with the light coming *toward* the observer. (*b*) Representation of the electric field vector in space at fixed time for circularly polarized light. This is RCP light. The direction of rotation with z is opposite to the direction of rotation with t.

Table 9.1 Circularly Polarized Light

ϕ	$+\pi/2$ or $-3\pi/2$	$-\pi/2$ or $+3\pi/2$
Rotation with t	Clockwise	Counterclockwise
Rotation with z	Counterclockwise	Clockwise
Sense of polarization	RCP	LCP
Helicity	−	+

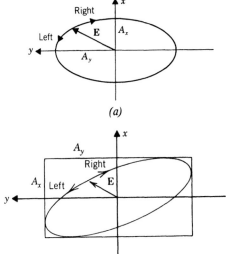

Fig. 9.4 (a) Elliptically polarized light. Same convention as in Fig. 9.3a. Here $\phi = \pm \pi/2$. (b) Same as in (a), except $\phi \neq \pm \pi/2$.

counterclockwise (or left-handed) for $\phi = -\pi/2$ (Fig. 9.4a). This is a special case of *elliptically polarized light*.

The general case occurs when $\phi \neq \pm \pi/2, \pm \pi$. We put $\phi_y = \phi_x + \phi$ and set

$$\phi_0 = (\omega t - kz) \tag{9.11}$$

to obtain

$$\begin{aligned} E_x &= A_x \cos(\phi_0 + \phi_x) \\ E_y &= A_y \cos(\phi_0 + \phi_y) \end{aligned} \tag{9.12}$$

The ellipse described by Eq. (9.12) is inscribed inside a rectangle having dimensions $2A_x$ by $2A_y$, as shown in Fig. 9.4b. The eccentricity of the ellipse depends on the relative phase difference $\phi = \phi_y - \phi_x$, reaching a minimum when $\phi = \pm \pi/2$ (Fig. 9.3) and a maximum when $\phi = \pm \pi$ (Fig. 9.1). To see this more clearly, we need to introduce some new representations for polarized light.

B. Representations for Elliptically Polarized Light

We will concentrate on elliptical polarization because it covers linear and circular polarization as special cases.

1. Linearly Polarized Basis. If we regard Eqs. (9.12) as specifying linearly polarized components of the total field, then we can identify a new complex notation that will help us understand the behavior of the electric-field vector in physical space. To accomplish this, we associate the x-axis in physical space with the real axis of a complex plane and the y-axis in physical space with the imaginary axis. Contrary to common convention, the orientation of our complex plane is rotated by 90°

about the origin to correspond to the orientation of our standard coordinate system. The optical electric field acts like a complex phasor in this plane,

$$\tilde{E} = E_x + iE_y \tag{9.13}$$

The "basis" in this case is specified in terms of linearly polarized components along x and y, where E_x and E_y are given by Eqs. (9.12).

It will be useful to reexpress Eqs. (9.12) for the component fields entirely in terms of exponential functions. This will lead to another valuable representation for polarized light. We proceed as follows:

$$E_x = A_x \cos(\phi_0 + \phi_x) = \tfrac{1}{2}A_x e^{i\phi_x}e^{i\phi_0} + \tfrac{1}{2}A_x e^{-i\phi_x}e^{-i\phi_0}$$
$$= \tfrac{1}{2}a_x e^{i\phi_0} + \tfrac{1}{2}a_x^* e^{-i\phi_0} \tag{9.14}$$

$$E_y = A_y \cos(\phi_0 + \phi_y) = \tfrac{1}{2}A_y e^{i\phi_y}e^{i\phi_0} + \tfrac{1}{2}A_y e^{-i\phi_y}e^{-i\phi_0}$$
$$= \tfrac{1}{2}a_y e^{i\phi_0} + \tfrac{1}{2}a_y^* e^{-i\phi_0} \tag{9.15}$$

with

$$a_x = A_x e^{i\phi_x} \tag{9.16a}$$

$$a_y = A_y e^{i\phi_y} \tag{9.16b}$$

Then we can write

$$\tilde{E} = E_x + iE_y = \tfrac{1}{2}(a_x + ia_y)e^{i\phi_0} + \tfrac{1}{2}(a_x^* + ia_y^*)e^{-i\phi_0} \tag{9.17}$$

2. Circularly Polarized Basis. The form of Eq. (9.17) suggests that we define

$$a_L = \tfrac{1}{2}(a_x + ia_y) = A_L e^{i\phi_L} \tag{9.18a}$$

$$a_R = \tfrac{1}{2}(a_x^* + ia_y^*) = A_R e^{i\phi_R} \tag{9.18b}$$

where A_L, A_R, ϕ_L, and ϕ_R are positive real constants. The phasor representation of the field then may be written

$$\tilde{E} = a_L e^{i\phi_0} + a_R e^{-i\phi_0} \tag{9.19}$$

This is the sum of two phasors that rotate in opposite directions with constant amplitude. We have created a basis in terms of circularly polarized components. When $a_R = 0$, the light is left circularly polarized. At $\phi_0 = 0$ the phasor assumes some initial orientation defined by ϕ_L. As time evolves, ϕ_0 increases and the phasor rotates counterclockwise, thus describing a circle of radius A_L. This is shown in Fig. 9.5a. When $a_L = 0$ we have right circular polarization as demonstrated in Fig. 9.5b. This rotates clockwise as time advances.

We can recover the components of our linearly polarized basis from Eq. (9.18b) by taking its complex conjugate and combining this form with Eq. (9.18a) to yield

$$a_x = a_L + a_R^* \tag{9.20}$$

$$a_y = -i(a_L - a_R^*) \tag{9.21}$$

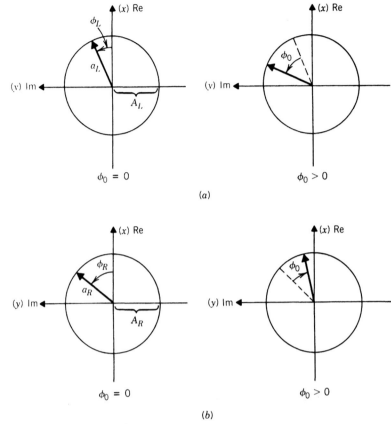

Fig. 9.5 Circular basis for the representation of polarized light. These take the form of phasors in the complex plane: (*a*) LCP phasor; (*b*) RCP phasor.

3. Interpretation. The advantage of using Eq. (9.19) rather than Eq. (9.13) lies in the greater ease of interpretation when circularly polarized components are used as a basis. We introduce the sum and difference

$$\beta \equiv \phi_L + \phi_R, \quad \alpha \equiv \phi_R - \phi_L \tag{9.22}$$

Then we may write Eq. (9.19) as

$$\tilde{E} = E_x + iE_y = e^{i\beta/2}[A_L e^{i(\phi_0 - \alpha/2)} + A_R e^{-i(\phi_0 - \alpha/2)}] \tag{9.23}$$

The behavior of the two terms in square brackets in Eq. (9.23) is indicated in Fig. 9.6. The phasors are shown when $\phi_0 = \alpha/2$, at which time they are parallel. Then we have

$$\tilde{E} = e^{i\beta/2}(A_L + A_R)$$

This corresponds to point 1 in Fig. 9.7. When $\phi_0 = \alpha/2 + \pi/2$ the phasor of Fig. 9.6*a* has rotated counterclockwise $\pi/2$ rad and that of Fig. 9.6*b* has rotated

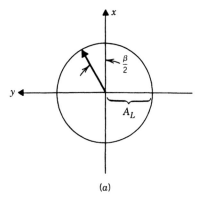

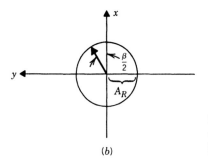

(b)

Fig. 9.6 Form of the basis for one example of elliptically polarized light.

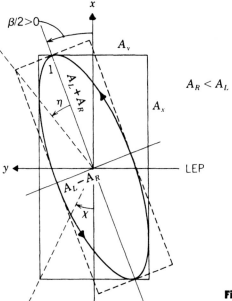

Fig. 9.7 Polarization ellipse for the basis in Fig. 9.6

clockwise $\pi/2$ rad. If A_R is less than A_L and if $(\pi/2) > (\beta/2) > 0$, the resulting phasor for $\tilde{E}$ will be in the second quadrant (in our convention) at the point 2 in Fig. 9.7. We see that the $\tilde{E}$ phasor traces an ellipse with semimajor axis of length $(A_L + A_R)$ at inclination $\beta/2$ with respect to the x-axis and semiminor axis of length $(A_L - A_R)$ at an inclination of $\beta/2 + \pi/2$. The light in this case is left elliptically polarized (LEP).

Other possibilities are illustrated in Fig. 9.8. When $A_L = A_R$ we obtain linearly polarized light with **E** along the line at $\beta/2$.

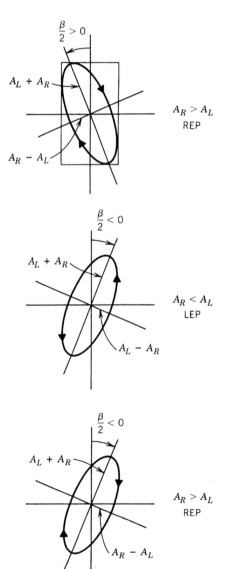

Fig. 9.8 Examples of the polarization ellipse.

With reference to Fig. 9.7, note that the elliptically polarized light is completely characterized by specifying the sense of rotation, the angle $\beta/2$, and the angle η, which gives the axial ratio of the ellipse via the equation

$$\tan \eta \equiv \frac{A_L - A_R}{A_L + A_R} \tag{9.24}$$

When the light is described in terms of its x and y components, as in Eq. (9.12), the appropriate parameters are the phase difference $\phi = \phi_y - \phi_x$ and the ratio A_y/A_x or χ where

$$\tan \chi \equiv \frac{|A_y|}{|A_x|} \tag{9.25}$$

The two sets of parameters are related by the following equations, which we state here without proof.

$$\tan \theta = \pm \frac{\tan 2\eta}{\sin \beta} \begin{Bmatrix} +\text{for REP} \\ -\text{for LEP} \end{Bmatrix} \tag{9.26}$$

$$\cos 2\chi = \cos \beta \cos 2\eta \tag{9.27}$$

C. Unpolarized Light

Unpolarized light is not an elementary state of polarization and is in a sense more complicated than the examples we have already discussed. In general an individual microscopic source of light (such as an atom or molecule) emits light that in a given direction has a well-defined polarization state. It is the superposition of many individual contributions to form the total macroscopic electric field that can produce unpolarized light. In the problems, you will be asked to show that the superposition of two *coherent* beams of elliptically polarized light will give another beam of elliptically polarized light, generally with a different ellipse. The nature of the resulting ellipse depends on the phase difference between the light in the two primary beams. If this phase difference varies, this ellipse may alter in orientation and shape. If the two beams have a finite coherence time, the phase difference between them will change randomly, with a root mean square change of about 1 rad in a time equal to the coherence time of the light. Thus, for two *incoherent* beams, where the measurement time is longer than the coherence time, the ellipse will change in a rapid, random way, and only its average properties will be measurable.

When many such incoherent beams are present, we can obtain unpolarized light if two conditions are satisfied. For the linearly polarized basis of Eq. (9.12), we require E_x and E_y to be incoherent and the amplitudes to have equal time averages: $\langle A_x^2 \rangle = \langle A_y^2 \rangle$. Because the total flux density is proportional to $\langle A^2 \rangle = \langle A_x^2 \rangle + \langle A_y^2 \rangle$, each component carries half the total energy. Because the coordinate system is arbitrary, a polarizer will always pass half the flux density independently of its orientation.

In the circularly polarized basis of Eq. (9.17), a similar statement can be made, namely, that

$$E_L = a_L e^{i\phi_0} \quad \text{and} \quad E_R = a_R e^{-i\phi_0}$$

must be incoherent and

$$\langle |a_L|^2 \rangle = \langle |a_R|^2 \rangle$$

A more powerful method of discussing unpolarized light and mixtures of unpolarized and polarized light (partially polarized light) will be developed in section 9.3.

One further point should be emphasized here, however. If we form a superposition of several incoherent beams, each in the same definite fully polarized state, the resulting beam will also be fully polarized in that state. The proof is quickly given. We use the linearly polarized basis. Suppose that each component beam has the same value of $\phi = \phi_y - \phi_x$ and the same value of the ratio $f = A_y/A_x$. Then we have for the jth beam

$$E_{xj} = A_{xj} e^{i\phi_{xj}} e^{i\phi_0}, \quad E_{yj} = f e^{i\phi} E_{xj}$$

The result is that $\mathbf{E}_{tot} = (E_{x\,tot}, E_{y\,tot})$ has components

$$E_{x\,tot} = \sum_j A_{xj} e^{i\phi_{xj}} e^{i\phi_0}, \quad E_{y\,tot} = \sum_j E_{yj} = f e^{i\phi} \sum_j E_{xj} = f e^{i\phi} E_{x\,tot} \qquad (9.28)$$

Even if the individual phases ϕ_{xj} and amplitudes A_{xj} fluctuate, the light will be in a pure polarization state described by f and ϕ.

9.2 Polarization-Sensitive Optical Elements

A. Producing Polarized Light

We now discuss briefly some of the methods commonly employed to produce polarized light. Most, but not all, of the simple methods yield plane polarized light.

1. Polarized Sources. One such source is the continuous *gas laser*. Such a device emits a very monochromatic narrow light beam that is often plane polarized. The laser emits light in such a polarized "mode" because of the use of windows set at *Brewster's angle* (see below).

Another source of polarized light, at least in principle, is provided when certain sources of narrow spectral lines are placed in a magnetic field. A splitting of the lines known as the *Zeeman effect* then occurs; each component is either linearly polarized or circularly polarized. This effect is difficult to produce and is not presently of practical importance as a source of polarized light.

Synchrotron radiation is produced when relativistic electrons are exposed to a magnetic field. They are thereby caused to travel in a circular path. The resulting acceleration to which the electrons are subjected leads to light emission tangent to the orbit and strongly concentrated in the forward direction. The radiation is highly polarized in the plane of the orbit. In addition to production of synchrotron radiation in astronomical phenomena, terrestrial artificial sources now exist. These are associated with high-energy particle accelerators and are now being built as dedicated light sources for fundamental studies of matter. The wavelength range of the emitted light depends on the characteristics of the orbit but is usually maximized in the x-ray range.

2. Reflection and Transmission Light reflected from a smooth interface becomes partially polarized if the original angle of incidence is nonzero. This effect becomes particularly strong and hence useful at a dielectric interface when the angle of incidence is at or near the so-called *Brewster's angle*. This was discussed in Chapter 2.

In Fig. 9.9 such an interface is shown in the x–z plane. When the reflected and transmitted rays meet at an angle of 90°, it is impossible for light to be reflected if its electric field lies in the plane of incidence (π polarized light). This light will be transmitted, however. Light with its electric field vibrating perpendicular to the plane of incidence (σ polarized light) may be reflected.

At Brewster's angle of incidence we must have

$$\theta_B + \theta_B' = \frac{\pi}{2} \tag{9.29}$$

Because Snell's law says that

$$\frac{n'}{n} = \frac{\sin \theta_B}{\sin \theta_B'}$$

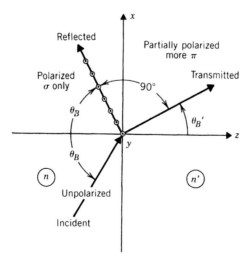

Fig. 9.9 Polarization by reflection and transmission through an interface.

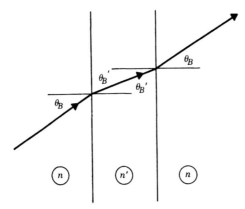

Fig. 9.10

and because $\sin \theta'_B = \cos \theta_B$ when $\theta_B + \theta'_B = \pi/2$, we must have

$$\tan \theta_B = \frac{n'}{n} \qquad (9.30)$$

at Brewster's angle, the same result as Eq. (2.78).

If the light incident at Brewster's angle is unpolarized, the reflected light has pure σ polarization, and the transmitted light is partially polarized with more π than σ polarization present.

This effect takes place on internal reflection also, as we have already discussed in Chapter 2. Thus, if the light is incident at Brewster's angle on a dielectric slab with parallel sides, as shown in Fig. 9.10, polarization selection will occur at both interfaces. This process may be repeated.

A pile of such parallel plates or slabs will then be an efficient polarizer for collimated light. The reflected beam will be pure σ-polarized; the transmitted beam can be made virtually pure π-polarized by using enough slabs.

If the angle of incidence is not exactly equal to θ_B, the reflected light will have a small component of π polarization, and the transmitted light will have correspondingly less π light.

Such a "pile of plates" can be used as a polarizer, either in reflection or in transmission, especially with well-collimated beams and where other more convenient polarizers are not available. Indeed, this type of polarizer was the first to be used in the 19th century.

3. Scattering. Light scattering can occur whenever there is an irregular distribution of small scattering centers. Examples are provided by dust particles in the air, small compared with the wavelength; the particles in a colloidal suspension; or even the gas molecules in the air. The basic mechanism for the production of polarized light from unpolarized light is similar to that just discussed. If the particles are small enough and simple enough, oscillations will be induced in them by the electric field of the incident light. There will be reradiation of light in all directions save that of the original electric field.

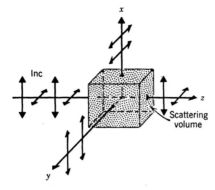

Fig. 9.11 Polarization by scattering.

This is shown in Fig. 9.11. The light scattered at a direction 90° to the incident propagation direction will be linearly polarized with the electric field in a direction perpendicular to the scattering plane, that is, the plane determined by the incident and scattered propagation directions. This mechanism of polarization by scattering is active in the earth's atmosphere. The blue color of the sky is caused by preferential scattering of short wavelength blue light over long wavelength red light. The blue light from the sky is partially polarized with the electric-field vector vibrating at right angles to the plane containing the incident and scattered light rays. At 90° viewing, as shown in Fig. 9.11, the polarization is complete.

4. Dichroism. There is a class of anisotropic optical media that exhibit a phenomenon known as *dichroism*, which means that light propagating in a given direction will be *absorbed* differently depending on the orientation of the electric field. The *absorption constant* coefficient [Eq. (2.44)] thus depends on the direction of vibration. In the extreme case, light vibrating in a particular direction will be essentially completely transmitted. The latter direction is called the *direction of easy passage*, or *pass direction*. A *tourmaline* crystal is an example of a dichroic polarizer, but the best-known example is a "polaroid" sheet, which consists of oriented dichroic molecules embedded in plastic.

In calculating the effect of a polarizer on an incident light beam, we should resolve the incident electric field along axes parallel and perpendicular to the pass direction. If ψ is the angle between the E field and the pass direction, the component passed is $E' = E \cos \psi$, and the transmitted flux density S' is given in terms of the incident flux density S by the *law of Malus*, (Fig. 9.12).

$$S' = S \cos^2 \psi \tag{9.31}$$

5. Double Refraction. The phenomenon of *double refraction* occurs in optically anisotropic media, that is, media having optical properties that depend on direction. The phenomenon is also referred to as *birefringence*. A detailed discussion of this subject will be given in section 9.4. Here we introduce the elementary concepts so that polarizing optics can be discussed. The "double" or "bi" in the

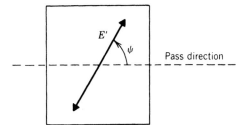

Fig. 9.12 Polarization by dichroism.

name of this effect refers to the *two* different directions of propagation that a given incident ray can take in such media, depending on the direction of polarization.

We discuss here the simplest kind of birefringent medium, a *uniaxial medium*. If diffraction effects are neglected, propagation of waves in such media can be described by a generalization of Huygens' principle. The spherical secondary wavelets must be replaced by wave surfaces of greater complexity, consisting of a spherical wave plus a wave that is an ellipsoid of revolution. The ellipsoid is tangent to the sphere at two points that lie on a line through the center of the elliptical waves. This line is called the *optic axis* (O.A.). There is one such axis, hence the term "uniaxial." It forms the axis for the ellipsoid of revolution. See Fig. 9.13. A plane is defined by the direction of propagation and the optic axis. This plane is called the *principal section*. The spherical wave will be linearly polarized with the electric field vibrating perpendicular to the principal section, that is, perpendicular to both the direction of propagation and the optical axis. The ellipsoidal wave will be linearly polarized with the electric vector in the principal section.

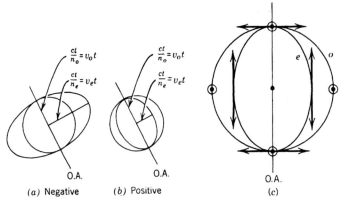

Fig. 9.13 Huygens' wavelets in uniaxial media. Double refraction: (*a*) negative; (*b*) positive; (*c*) polarization orientation.

If we pick a coordinate system having its z-axis along this optic axis, the equation for this ellipsoid may be written

$$\frac{x^2 + y^2}{v_e^2} + \frac{z^2}{v_o^2} = t^2 \tag{9.32}$$

The spherical wavelet spreads with a velocity $v_o = c/n_o$, *and so at time t its radius will be* $(c/n_o)t$. The ellipsoidal wavelet spreads with a velocity ranging between c/n_o *along the optic axis to a velocity* $v_e = c/n_e$ in a plane perpendicular to the O.A. For "positive" crystals, n_e is greater than n_o and for "negative" crystals, n_o *is greater. Examples are furnished by calcite* (negative) and *crystalline quartz* (positive). Quartz also exhibits a phenomenon known as *optical activity*, which we ignore for the moment.

The spherical wave is referred to as *ordinary* (*o* wave), and its refractive index n_o *as the ordinary index*. The elliptical wave is the *extraordinary wave* (*e* wave), and the extreme value of its refractive index n_e is called the *extraordinary index*.

The notion of a light ray must be redefined for the extraordinary wave. It represents the direction of energy flow and is given by a line from the origin to the point in question on the ellipsoid, such as OP in Fig. 9.14. The point P will move outward with the *ray velocity* v_r, intermediate between v_o *and* v_e, which depends on the angle γ_e of OP with respect to the O.A. Using $\overline{OP} = v_r t$, we can readily derive

$$\frac{1}{v_r^2} = \frac{\sin^2 \gamma_e}{v_e^2} + \frac{\cos^2 \gamma_e}{v_o^2} \tag{9.33}$$

The surfaces of constant phase (wavefronts) for the *e* wave are tangent to the *e* wave surface at P. The wave normal ŝ is not parallel to the ray direction OP. Because the distance $\overline{OP}$ is proportional to the ray velocity, these wave surfaces are also called *ray velocity surfaces*. These surfaces are useful for qualitative discussions of double refraction, but for quantitative work, other surfaces, the *index surfaces*, to be introduced in section 9.4, are more convenient. A novel aspect of the extraordinary wave is that except when $\gamma_e = 0°$ or $90°$ the electric displacement vector **D** is *not* perpendicular to the propagation direction OP. The electric field **E** is perpendicular to OP, however, as are the **B** and **H** fields. These matters will be discussed in section 9.4. The ordinary spherical wave is always linearly polarized perpendicular to the principal section.

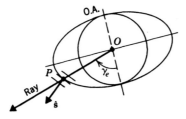

Fig. 9.14 Wave direction ŝ and ray direction in anisotropic medium.

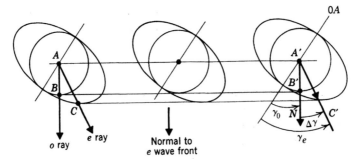

Fig. 9.15

A disturbance of *arbitrary* pure state of polarization will, in general, propagate in a complicated way in a birefringent medium. It can be resolved, however, into linearly polarized components, which become the e and o rays just discussed. The state of polarization of these components does not change, although that of the original disturbance can change quite markedly after passing through the birefringent medium. For example, it can change from linear to circular polarization.

Wavefronts associated with extraordinary rays will *not* be normal to the rays. They can be determined by Huygens' principle. An example is shown in Fig. 9.15. Suppose that the line AA' represents the location of the primary wave surface. Each point then acts as a source of secondary wavelets. The envelope of the spherical wavelets BB' then locates the ordinary wave surface at a later time. The envelope CC' locates the extraordinary wave surface, which is moving in the direction AC and making an angle $\Delta\gamma$ with the normal to the wavefront $A'N$. The *wave velocity* v_w is the velocity in the normal direction $A'N$ and is given by

$$v_w = v_r \cos \Delta\gamma \tag{9.34}$$

where v_r is given by Eq. (9.33).

The angular separation between the e ray and the o ray can be used to isolate one or the other of them and hence produce plane polarized light. One of the

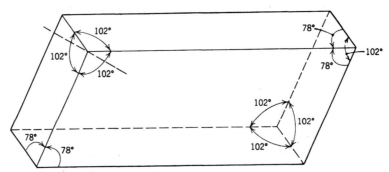

Fig. 9.16 A calcite rhomb showing natural cleavage faces.

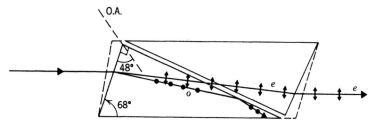

Fig. 9.17 Nicol prism.

common ways of doing this is employed in the *Nicol prism,* which is made from *calcite.* This prism is often employed as a high-precision polarizer.

This crystal cleaves naturally into a rhomb with parallel sides and edges. The direction of the optic axis makes equal angles with the edges at the most obtuse corner, where all the edges meet at an angle of 102°, as shown in Fig. 9.16.

The Nicol prism is cut from a calcite crystal as shown in Fig. 9.17. We start with a principal section of a cleaved rhomb containing the optic axis, as shown dashed in the figure. The ends are then cut to a more acute angle of 68°. A long diagonal cut is then made perpendicular to these new ends. The two halves of the cut crystal are then cemented together with *Canadian balsam.* The index of the balsam is 1.55, which lies between the index for the ordinary ray of 1.658 and that of the extraordinary ray propagating along the axis of the prism of 1.486. The angles have been chosen so that the *o* ray is totally internally reflected at the first calcite–balsam interface. This ray is then absorbed by a black coating on the sides of the prism.

The Nicol prism has several disadvantages, among which are the parallel displacement suffered by the transmitted beam and the relatively small angular spread of the beam that may be transmitted. The Glan-Thompson prism shown in Fig. 9.18 has neither of these disadvantages, although it makes less efficient use of calcite. It consists of two sections of calcite, both having ends cut perpendicular to the light beam with parallel optic axes as shown. The index of the cement is such that at the angle of the cut the ordinary wave suffers total internal reflection, whereas the extraordinary wave is transmitted with little loss.

In Fig. 9.19 two types of beam-splitting polarizing prism are shown. With unpolarized incident light they produce two transmitted beams with mutually perpendicular polarization directions and different directions of propagation.

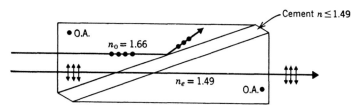

Fig. 9.18 Glan-Thompson polarizer.

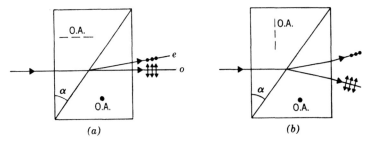

Fig. 9.19 (*a*) Rochon prism and (*b*) Wollaston prism. The optic axis for the right-hand parts of both is perpendicular to the plane of the drawing.

These prisms are usually made of either quartz or calcite and have their optic axes oriented as shown. In the first part of the Rochon prism, both waves have the ordinary index n_o. In the second part, the *o* wave vibrating in the plane of the diagram continues to have index n_o and is not deviated. The *e* wave vibrating perpendicular to the plane of the diagram has index n_e and is deviated up or down according to whether $n_e < n_o$ or $n_e > n_o$.

In the Wollaston prism the *o* wave in the first part vibrating perpendicular to the plane of the diagram becomes the *e* wave in the second part, and the *e* wave in the first part, vibrating in the plane of the diagram, becomes the *o* wave in the second part. If $n_e < n_o$, the rays will be both deviated as shown. If $n_e > n_o$, the polarizations of the two emerging beams will be reversed.

B. Shifting Phase

The characteristics of optically anisotropic media can be exploited to manipulate the phase of a light beam so as to produce, for instance, circularly polarized light from an incident linearly polarized beam.

A simple method of optical phase shifting employs a parallel slab of birefringent material. If the optic axis lies parallel to the surface and if light falls on the slab at normal incidence, as shown in Fig. 9.20, both *o* and *e* rays will continue to propagate in the same direction, but with different velocities. The phase change produced by propagation will then be different for the two rays.

The phase change in a distance *d* for the *o* ray will be given by

$$\Delta\phi_o = \frac{-2\pi}{\lambda} n_o d$$

and for the *e* ray

$$\Delta\phi_e = \frac{-2\pi}{\lambda} n_e d$$

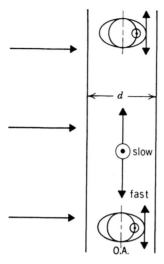

Fig. 9.20 Orientation of the optic axis in a compensator.

giving a phase difference

$$\Delta\phi = \Delta\phi_e - \Delta\phi_0 = \frac{-2\pi}{\lambda}(n_e - n_o)d \tag{9.35}$$

Such a phase shifter is often referred to as a *compensator*. If $n_e > n_o$ then $\Delta\phi < 0$ and the e ray lags the o ray in phase. If $n_e < n_o$, then the opposite is true.

1. Quarter-Wave Plate. A very useful device called a *quarter-wave plate* is produced when $|\Delta\phi| = \pi/2$, that is, when

$$|n_e - n_o|d = \frac{\lambda}{4} \tag{9.36}$$

Then the e ray is retarded or advanced by a quarter of a cycle with respect to the o ray; retardation occurs if $n_e > n_o$ (positive uniaxial crystal like quartz). The direction of polarization that is advanced in phase is along the *fast axis*; the retarded direction is the *slow axis*. The e ray is polarized along the slow axis for the example of quartz just presented. For negative calcite, the e ray is along the fast axis. Note that $\Delta\phi$ depends on λ. This limits the practical wavelength range for precision work.

A device called a *Fresnel rhomb* produces a $\pi/2$ relative shift by means of total internal reflection and is much less wavelength-dependent. The light beam undergoes a parallel displacement, however (see Problem 9.15).

A quarter-wave plate will convert linearly polarized light to elliptically polarized light, as shown in Fig. 9.21. Here the plane of polarization of the incident light makes an angle of 45° with the fast axis. In this special case, the $\lambda/4$ plate

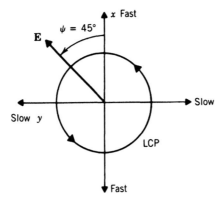

Fig. 9.21 Illustration of the polarization transformation from linear to circular that occurs on passage through a quarter-wave plate if the incident light is polarized at an angle of 45° with respect to the axes of the quarter-wave plate.

produces left circularly polarized light. In the general case where $\psi \neq 45°$ we would have

$$\text{Incident light} \begin{cases} E_x = A \cos \psi \cos \omega t \\ E_y = A \sin \psi \cos \omega t \end{cases} \tag{9.37}$$

After transmission, there will be a common phase shift $-\phi'$ plus a relative phase shift of $-\pi/2$:

$$\text{Transmitted light} \begin{cases} E_x = A \cos \psi \cos(\omega t - \phi') \\ E_y = A \sin \psi \cos\left(\omega t - \phi' - \dfrac{\pi}{2}\right) \\ \qquad = A \sin \psi \sin(\omega t - \phi') \end{cases} \tag{9.38}$$

This will give LEP light for $\psi > 0$ and REP light for $\psi < 0$. The axes of the ellipse coincide with the x and y axes and have a length ratio given by $\tan \psi$.

Elliptically polarized light will be converted to linearly polarized light when run through a quarter-wave plate, provided the axes of the quarter-wave plate and the ellipse coincide. The sequence from Eq. (9.38) to Eq. (9.37) is then essentially inverted.

Consider the elliptically polarized light appropriate to Fig. 9.22. Suppose that the light is REP. If the fast axis of the $\lambda/4$ plate makes an angle $\beta/2$ with the x axis, that is, if it coincides with the major axis of the ellipse ($\bar{x}$ axis), it will speed up $\bar{E}_x$, the component of E along the $\bar{x}$ axis, by a quarter of a cycle with respect to $\bar{E}_y$. In the original REP light $\bar{E}_x$ was lagging $\bar{E}_y$ by a quarter of a cycle. Thus it is now in phase, and the result is linearly polarized light vibrating at an angle η to the $\bar{x}$ axis, that is, at an angle $\beta/2 + \eta$ to the x axis. This polarized light and its orientation can be readily detected by use of a polarizer. The polarizer gives the angle $\beta/2 + \eta$ with respect to the x axis. The fast axis of the $\lambda/4$ plate gives the angle $\beta/2$. Subtraction gives η, which yields the axial ratio of the ellipse:

$$\frac{A_R - A_L}{A_L + A_R} = \tan \eta \tag{9.39}$$

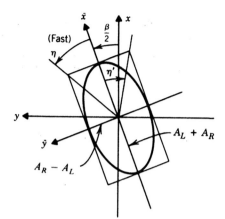

Fig. 9.22

If the light were LEP instead of REP, the resulting linearly polarized light would vibrate at the angle $\eta' = -\eta < 0$ with respect to the $\bar{x}$ axis and hence at the angle $\beta/2 + \eta' = \beta/2 - \eta$ with respect to the x axis. Equation (9.39) will now hold with η replaced by η' and will yield

$$\frac{A_R - A_L}{A_L + A_R} = \tan \eta' = -\tan \eta < 0$$

Hence we have $A_L > A_R$, as is necessary for LEP light. This difference can be used to decide between REP and LEP light.

2. Babinet Compensator. The *Babinet compensator* shown in Fig. 9.23 is similar to a Wollaston prism, but the angle $\alpha \approx d/L$ is sufficiently small that the angular deviation suffered by the two emerging beams can be neglected. The main effect

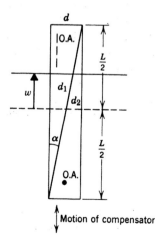

Fig. 9.23 Babinet compensator.

then is the production of a relative phase difference, which depends on the height of passage of the beam through the device. We assume here that the material is calcite, so that the fast axis is in the plane of the figure. In the first part of the compensator the relative phase change for the e wave with respect to the o wave is

$$\Delta\phi_1 = \frac{2\pi}{\lambda}(n_o - n_e)d_1$$

In the second part of the compensator the roles are reversed, and the relative phase change is

$$\Delta\phi_2 = \frac{-2\pi}{\lambda}(n_o - n_e)d_2$$

The total phase change is then

$$\Delta\phi = \Delta\phi_1 + \Delta\phi_2 = \frac{2\pi}{\lambda}(n_o - n_e)(d_1 - d_2)$$

Because

$$d_1 - d_2 = 2d_1 - d = \frac{2wd}{L}$$

this gives

$$\Delta\phi = \frac{4\pi}{\lambda L}(n_o - n_e)wd \tag{9.40}$$

as the phase of the e wave minus the phase of the o wave. When w is positive the e wave, which is polarized in the plane of the drawing, leads the o wave. When w is negative, the opposite is true. For proper operation the incident light must be in a *narrow* beam so that w is well-defined.

3. Optical Activity. Optical activity can be found in crystalline and monocrystalline solids, liquids, and gases. In crystals it is often accompanied by double refraction, and the resulting phenomena can be quite complex. Some cubic crystals—for example, sodium chlorate and sodium bromate—show optical activity uncomplicated by double refraction.

Optical activity originates in the inherent helical or screwlike character of the molecules of the material that produces it (Fig. 9.24). In its pure form, that is, without double refraction, it results in different refractive indices for left and right

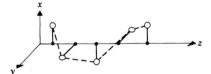

Fig. 9.24 Part of a helical molecule capable of producing optical activity.

circularly polarized light beams. Only such beams will propagate through the medium without change.

A linearly polarized beam will be modified; in fact its direction of vibration will rotate as it propagates through the medium. This is one of the clearest manifestations of optical activity. A brief description will now be given.

Let n_L and n_R be refractive indices for LCP and RCP, respectively. Let $\bar{n} = 1/2(n_L + n_R)$ be the mean index and $\bar{k} = 2\pi\bar{n}/\lambda$ the mean magnitude of the wave vector. The *specific rotary power* is defined to be

$$\rho \equiv \frac{\pi(n_L - n_R)}{\lambda} \tag{9.41}$$

Then the wave vectors for left and right circularly polarized light have magnitudes

$$k_L = \bar{k} + \rho$$
$$k_R = \bar{k} - \rho \tag{9.42}$$

It is convenient to use the complex notation of Eqs. (9.23) and Fig. 9.6. Let z be the direction of propagation and let $\tilde{E} = E_x + iE_y$. Suppose that in the plane $z = 0$ there is polarized light with amplitude A vibrating in the direction $\beta/2$, as shown in Fig. 9.25. Then

$$\tilde{E} = e^{i\beta/2} A \cos \omega t$$

or

$$E_x = A \cos \omega t \cos \frac{\beta}{2}$$

$$E_y = A \cos \omega t \sin \frac{\beta}{2}$$

This can also be written as a superposition of circularly polarized components

$$\tilde{E}_L = e^{i\beta/2} \tfrac{1}{2} A e^{i\omega t}$$
$$\tilde{E}_R = e^{i\beta/2} \tfrac{1}{2} A e^{-i\omega t}$$

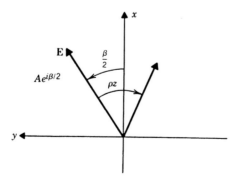

Fig. 9.25

For finite z we must make the replacement

$$\omega t \rightarrow \omega t - k_R z = \bar{\phi}_0 \mp \rho z$$

for the two components, where $\bar{\phi}_0 = \omega t - \bar{k}z$. This gives

$$\tilde{E}_L = e^{i\beta/2} \tfrac{1}{2} A e^{i(\bar{\phi}_0 - \rho z)}$$

$$\tilde{E}_R = e^{i\beta/2} \tfrac{1}{2} A e^{-i(\bar{\phi}_0 + \rho z)}$$

or

$$\tilde{E} = E_x + i E_y = e^{i\beta/2} \tfrac{1}{2} A [e^{i(\bar{\phi}_0 - \rho z)} + e^{-i(\bar{\phi}_0 + \rho z)}]$$

$$= e^{i(\beta/2 - \rho z)} \tfrac{1}{2} A (e^{i\bar{\phi}_0} + e^{-i\bar{\phi}_0}) = e^{i(\beta/2 - \rho z)} A \cos \bar{\phi}_0$$

Hence

$$\bar{\phi}_0 = \omega t - \bar{k}z \tag{9.43}$$

$$E_x = A \cos \bar{\phi}_0 \cos\left(\frac{\beta}{2} - \rho z\right)$$

$$E_y = A \cos \bar{\phi}_0 \sin\left(\frac{\beta}{2} - \rho z\right) \tag{9.44}$$

These equations describe linearly polarized light vibrating at angle $\beta/2 - \rho z$. Thus the parameter ρ gives the angle of rotation per unit distance of propagation.

Crystalline quartz is an example of a doubly refracting, optically active crystal. If it were merely doubly refracting, because it is uniaxial, light propagating along the optic axis would be expressed as a superposition of linearly polarized components with the same refractive index, the ordinary index. Instead, the two components that propagate unaltered are circularly polarized with indices n_L and $n_R \neq n_L$. The specific rotation at $\lambda = 589.3$ nm is $\rho = \pm 3.8$ rad/cm. Two types of quartz exist: positive or right-handed quartz for which ρ is positive, and negative or left-handed quartz for which ρ is negative.

For propagation in quartz not along the optic axis, the two states of polarization that propagate unchanged correspond to elliptically polarized light (Fig. 9.26). The ellipses for the two components are oriented at right angles. They become circles for propagation along the optic axis and degenerate into straight lines for propagation perpendicular to the optic axis.

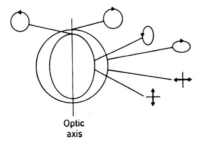

Optic
axis

Fig. 9.26 Wavefronts and polarization ellipses for the waves that propagate without change in crystalline quartz. This diagram is meant to be schematic and is not quantitatively accurate.

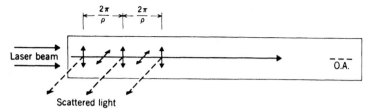

Fig. 9.27 Demonstration of optical activity in quartz.

A demonstration of optical activity in quartz is easily given using the polarized light beam from a powerful continuous gas laser. Figure 9.27 shows schematically the rotation of the plane of polarization of light propagating along the optic axis. This can be seen directly by means of scattering of the intense beam by imperfections in the crystal. These act like dust particles in a gas. As discussed in connection with Fig. 9.11, the scattering will be most intense in directions perpendicular to the E vector of the beam and zero in directions parallel to the E vector. When we look at right angles to the laser beam, the intensity of the scattered light varies periodically with maxima when E is perpendicular to the direction of observation. These occur at equally spaced distances of $2\pi/\rho = 5.5$ mm for $\lambda = 488$ nm.

9.3   Partially Polarized Light

A partially polarized light beam is a mixture of a beam in a pure polarization state and an unpolarized beam. This decomposition is unique. It depends on the mutual coherence between the electric fields vibrating along two perpendicular directions and can be determined experimentally. In this section we present the coherency matrix formalism with which partially polarized light may be characterized.

Consider a polarizer having its direction of easy passage, or pass direction, inclined at angle ψ to the x-axis (Fig. 9.28). If E_x and E_y represent the real rectangular components of the incident linearly polarized electric field, the projection at angle ψ is given by

$$E(\psi) = E_x \cos \psi + E_y \sin \psi$$

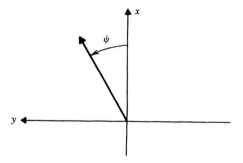

Fig. 9.28

The transmitted flux density will be proportional to the time average

$$\langle E(\psi)^2 \rangle = \langle E_x^2 \rangle \cos^2 \psi + \langle E_y^2 \rangle \sin^2 \psi + 2\langle E_x E_y \rangle \sin \psi \cos \psi \quad (9.45)$$

or, if the fields are represented by complex quantities as in Eqs. (9.14) and (9.15)

$$\langle E(\psi)^2 \rangle = \tfrac{1}{2}\langle EE^* \rangle = \tfrac{1}{2}\langle A_x^2 \rangle \cos^2 \psi + \frac{1}{2}\langle A_y^2 \rangle \sin^2 \psi$$

$$+ \tfrac{1}{2}[\langle E_x E_y^* \rangle + \langle E_x^* E_y \rangle] \sin \psi \cos \psi$$

$$\propto S(\psi) \quad (9.46)$$

If the correlation function $\langle E_x E_y \rangle$ is zero, the flux densities associated with the x and y components add. This can occur through either of two ways, or a combination of both: (1) E_x and E_y can have no mutual coherence, or (2) E_x and E_y can be correlated but have a relative phase difference of $\pi/2$ rad. If $\langle E_x, E_y \rangle \neq 0$, there must be mutual coherence between the components, but Eq. (9.45) does not suffice to determine it completely. Additional phase shifts must be introduced artificially, and then the light passed through a rotatable polarizer, for the light to be uniquely characterized.

Suppose that a compensator is inserted before the polarizer so that the phase of the y component is shifted from ϕ_y to $\phi_y + \phi' + \Delta\phi$ and the phase of the x component is shifted to $\phi_x + \phi'$. Then E_x is replaced by

$$A_x \, e^{i(\omega t - kz + \phi_x)} \, e^{i\phi'} = E_x e^{i\phi'}$$

and E_y is replaced by

$$A_y \, e^{i(\omega t - kz + \phi_y)} \, e^{i(\phi' + \Delta\phi)} = E_y \, e^{i(\phi' + \Delta\phi)}$$

and the field passed by the polarizer set at angle ψ becomes

$$E(\psi, \Delta\phi) = (E_x \cos \psi + e^{i\Delta\phi} E_y \sin \psi)e^{i\phi'}$$

with a flux density proportional to

$$\tfrac{1}{2}\langle EE^* \rangle = \tfrac{1}{2}\{\langle A_x^2 \rangle \cos^2 \psi + \langle A_y^2 \rangle \sin^2 \psi$$

$$+ [\langle E_x E_y^* \rangle e^{-i\Delta\phi} + \langle E_x^* E_y \rangle e^{i\Delta\phi}]\sin \psi \cos \psi \}$$

$$\propto S(\psi, \Delta\phi) \quad (9.47)$$

It is clear from Eq. (9.47) that $S(\psi, \Delta\phi)$ is sensitive only to the relative phase shift $(\Delta\phi)$ between E_y and E_x produced by the compensator, and insensitive to any common phase shift ϕ'.

A. The Coherency Matrix

1. Definition. There are four averages appearing in Eq. (9.47) that determine the function $S(\psi, \Delta\phi)$. They are:

$$\left.\begin{aligned}
J_{xx} &\equiv \langle E_x E_x^* \rangle = \langle A_x^2 \rangle \\
J_{yy} &\equiv \langle E_y E_y^* \rangle = \langle A_y^2 \rangle \\
J_{xy} &\equiv \langle E_x E_y^* \rangle \\
J_{yx} &\equiv \langle E_x^* E_y \rangle = J_{xy}^*
\end{aligned}\right\} \tag{9.48}$$

and they form the elements of a two-by-two matrix called the coherency matrix:

$$\mathbf{J} = \begin{pmatrix} J_{xx} & J_{xy} \\ J_{yx} & J_{yy} \end{pmatrix} \tag{9.49}$$

This matrix is *Hermitian*. For such a matrix the elements satisfy

$$J_{xx}^* = J_{xx}, \quad J_{xy}^* = J_{yx}, \quad J_{yy}^* = J_{yy} \tag{9.50}$$

The flux density (ignoring constants) can then be written

$$S(\psi, \Delta\phi) = J_{xx} \cos^2 \psi + J_{yy} \sin^2 \psi + \sin \psi \cos \psi (J_{xy} e^{-i\Delta\phi} + J_{yx} e^{i\Delta\phi}) \tag{9.51a}$$

Writing $J_{xy} = |J_{xy}| e^{-i\phi}$, $J_{yx} = |J_{xy}| e^{i\phi}$, [where $\phi = \phi_y - \phi_x$ as in Eq. (9.4)] we obtain an alternative form of Eq. (9.51a)

$$S(\psi, \Delta\phi) = J_{xx} \cos^2 \psi + J_{yy} \sin^2 \psi + 2|J_{xy}| \sin \psi \cos \psi \cos(\phi + \Delta\phi) \tag{9.51b}$$

If we average over ψ we find

$$\langle S(\psi, \Delta\phi) \rangle_\psi = \tfrac{1}{2} S_0 = \tfrac{1}{2}(J_{xx} + J_{yy})$$

Here S_0 represents the sum of flux densities measured in two perpendicular pass directions, at angle ψ and at angle $\psi \pm \pi/2$,

$$S_0 = J_{xx} + J_{yy} = S(\psi, \Delta\phi) + S\left(\psi \pm \frac{\pi}{2}, \Delta\phi\right) \tag{9.52}$$

and is the total flux density in the beam. Note that $J_{xx} = S(0, \Delta\phi)$ and $J_{yy} = S(\pi/2, \Delta\phi)$. The sum of the diagonal elements of a matrix is the trace of the matrix, thus

$$S_0 = \text{Tr } \mathbf{J} \tag{9.53}$$

The nature of the polarized light in the beam is independent of S_0 and depends on the ratios of the matrix elements J_{ij}. For this reason we define a *reduced* coherency matrix

$$\mathbf{j} = \frac{1}{S_0} \mathbf{J} \tag{9.54}$$

having elements $j_{ij} = J_{ij}/S_0$. Then

$$\text{Tr } \mathbf{j} = 1 \tag{9.55}$$

2. Determination of the Coherency Matrix Elements. Because we can write

$$J_{xy} = J_{yx}^* = |J_{xy}| e^{-i\phi}$$

there are four independent real quantities that determine $\mathbf{J}$: J_{xx}, J_{yy}, $|J_{xy}|$, and ϕ. They can be determined from four flux density measurements using a polarizer and a compensator. There is no unique way of doing this, but we give here an example of a set of four such measurements.

1. Polarizer along the x axis. Then we measure $S(0,0) = J_{xx}$.
2. Polarizer along the y axis. Then we measure $S(\pi/2, 0) = J_{yy}$.
3. Polarizer set at $45°$ to the axis. This gives

$$S\left(\frac{\pi}{4}, 0\right) = \frac{1}{2}[J_{xx} + J_{yy} + J_{xy} + J_{yx}] = \frac{1}{2}(J_{xx} + J_{yy}) + \mathrm{Re}\, J_{xy}$$

4. Quarter-wave plate with fast axis along x $(\Delta\phi = -\pi/2)$, followed by a polarizer at $45°$ to the x axis. This gives

$$S\left(\frac{\pi}{4}, \frac{-\pi}{2}\right) = \frac{1}{2}[(J_{xx} + J_{yy} + i(J_{xy} - J_{yx})] = \frac{1}{2}(J_{xx} + J_{yy}) - \mathrm{Im}\, J_{xy}$$

These four equations may be solved for J_{xx}, J_{yy}, $\mathrm{Re}\, J_{xy} = |J_{xy}| \cos\phi$, and $\mathrm{Im}\, J_{xy} = |J_{xy}|\sin\phi$.

B. Examples of Polarization

1. Linearly Polarized Light. Suppose that the plane of vibration of the light makes an angle χ with the x axis (Fig. 9.29) where

$$\tan\chi = A_y/A_x$$

When the pass direction ψ of the polarizer equals χ we observe a maximum transmitted flux density given by

$$S_0 = \langle A_x^2 \rangle + \langle A_y^2 \rangle$$

where S_0 is the total flux density in the beam. When ψ equals $\pm\chi \pm \pi/2$, S takes its minimum value of zero.

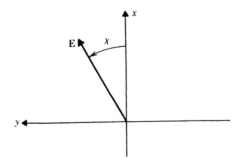

Fig. 9.29

For a general value of ψ, Eq. (9.51) holds with $\phi = 0$, and

$$J_{xx} = S_0 \cos^2 \chi, \quad J_{yy} = S_0 \sin^2 \chi$$
$$J_{xy} = J_{yx} = S_0 \sin \chi \cos \chi \tag{9.56}$$

The coherency matrix thus takes the form

$$\mathbf{J} = S_0 \begin{pmatrix} \cos^2 \chi & \sin \chi \cos \chi \\ \sin \chi \cos \chi & \sin^2 \chi \end{pmatrix} \tag{9.57}$$

For zero-compensating phase shift $\Delta\phi$, Eq. (9.51) can then be rearranged to give

$$S(\psi, 0) = S_0 \cos^2(\psi - \chi) \tag{9.58}$$

2. Unpolarized Light. Unpolarized light is characterized by a constant transmitted flux density independent of the polarizer angle ψ or the phase shift $\Delta\phi$. Thus

$$S(\psi, \Delta\phi) = \text{const} = S_0/2$$

This can only be true if

$$J_{xx} = J_{yy} = S_0/2, \quad J_{xy} = J_{yx} = 0$$

Here S_0 represents the total flux density measured when there is no polarizer in the beam. The coherency matrix then takes the form

$$\mathbf{J} = \frac{S_0}{2} \begin{pmatrix} 1 & 0 \\ 0 & 1 \end{pmatrix} \tag{9.59}$$

that is, it is a multiple of the unit matrix

$$\mathbf{I} \equiv \begin{pmatrix} 1 & 0 \\ 0 & 1 \end{pmatrix}$$

3. Circularly Polarized Light. When the light is circularly polarized we have $A_x = A_y = A$ and

$$\phi = \phi_y - \phi_x = -\frac{\pi}{2} \quad \text{for LCP}$$

$$\phi = +\frac{\pi}{2} \quad \text{for RCP}$$

Then

$$J_{xx} = J_{yy} = \langle A^2 \rangle$$
$$J_{xy} = \langle A^2 \rangle e^{i\phi} = \langle A^2 \rangle e^{-i\pi/2} = -i\langle A^2 \rangle = -J_{yx} \quad \text{for LCP}$$
$$J_{xy} = +i\langle A^2 \rangle = -J_{yx} \quad \text{for RCP}$$

When circularly polarized light passes through a quarter-wave plate with fast axis along the x or y axes, we obtain linearly polarized light at angle $\chi = \pm\pi/4$ and having a coherency matrix (for $\chi = +\pi/4$)

$$\mathbf{J} = \langle A^2 \rangle \begin{pmatrix} 1 & 1 \\ 1 & 1 \end{pmatrix}$$

The flux density is maximized at $\psi = \chi$ and equals $S_0 = 2J_{xy}$. Thus we may write

$$J_{xy} = \langle A^2 \rangle = \tfrac{1}{2}S_0$$

and the coherency matrix of the original light takes the form

$$\mathbf{J} = \frac{1}{2}S_0 \begin{pmatrix} 1 & -i \\ +i & 1 \end{pmatrix} \quad \text{for LCP} \tag{9.60a}$$

$$\mathbf{J} = \frac{1}{2}S_0 \begin{pmatrix} 1 & +i \\ -i & 1 \end{pmatrix} \quad \text{for RCP} \tag{9.60b}$$

4. Elliptically Polarized Light. For general elliptically polarized light we have (for $z = 0$)

$$E_x = A_x\, e^{i(\omega t + \phi_x)}$$
$$E_y = A_y\, e^{i(\omega t + \phi_y)}$$

where the ratio A_y/A_x and the phase difference $\phi = \phi_y - \phi_x$ must be constant. We use the shorthand forms

$$\chi = \tan^{-1}(A_y/A_x)$$

and

$$S_0 = J_{xx} + J_{yy} = \langle A_x^2 \rangle + \langle A_y^2 \rangle$$

Then the coherency matrix may be written as

$$\mathbf{J} = S_0 \begin{pmatrix} \cos^2 \chi & e^{-i\phi} \sin \chi \cos \chi \\ e^{i\phi} \sin \chi \cos \chi & \sin^2 \chi \end{pmatrix} \tag{9.61}$$

This includes Eqs. (9.57) and (9.60) as special cases.

Note that for all pure states of polarization

$$\det \mathbf{J} = J_{xx}J_{yy} - J_{xy}J_{yx} = 0 \tag{9.62}$$

If light described by Eq. (9.61) passes through a compensator so that the phase of E_y is retarded by an angle $-\phi$ with respect to that of E_x, then the coherency matrix becomes

$$\mathbf{J}' = S_0 \begin{pmatrix} \cos^2 \chi_0 & \sin \chi_0 \cos \chi_0 \\ \sin \chi_0 \cos \chi_0 & \sin^2 \chi_0 \end{pmatrix}$$

This has the same form as Eq. (9.56) and thus represents linearly polarized light in direction χ_0.

C. Combination of Light Beams

The matrix formalism allows us to document the characteristics of light-beam combinations with a compact notation.

1. Mutually Incoherent Beams. Two light beams propagating in the same direction will be mutually incoherent if their fields are uncorrelated, that is if

$$\langle E_x^{(1)} E_x^{(2)*} \rangle = \langle E_y^{(1)} E_y^{(2)*} \rangle = \langle E_x^{(1)} E_y^{(2)*} \rangle = \langle E_y^{(1)} E_x^{(2)*} \rangle = 0$$

The coherency matrix **J** for the combined light is then simply the sum of the individual coherency matrices,

$$\begin{pmatrix} \langle E_x E_x^* \rangle & \langle E_x E_y^* \rangle \\ \langle E_x^* E_y \rangle & \langle E_y E_y^* \rangle \end{pmatrix} = \begin{pmatrix} \langle E_x^{(1)} E_x^{(1)*} \rangle & \langle E_x^{(1)} E_y^{(1)*} \rangle \\ \langle E_x^{(1)*} E_y^{(1)} \rangle & \langle E_y^{(1)} E_y^{(1)*} \rangle \end{pmatrix}$$
$$+ \begin{pmatrix} \langle E_x^{(2)} E_x^{(2)*} \rangle & \langle E_x^{(2)} E_y^{(2)*} \rangle \\ \langle E_x^{(2)*} E_y^{(2)} \rangle & \langle E_y^{(2)} E_y^{(2)*} \rangle \end{pmatrix}$$

or

$$\mathbf{J} = \mathbf{J}^{(1)} + \mathbf{J}^{(2)}$$

If both beams have the same state of polarization, they will have the same reduced coherency matrix. Hence,

$$\mathbf{J} = S^{(1)} \mathbf{j} + S^{(2)} \mathbf{j}$$
$$= (S^{(1)} + S^{(2)}) \mathbf{j}$$

This says that the combination of two incoherent beams in the same state of polarization yields a beam having the same polarization as its components.

The notion of superposition of incoherent beams is useful for describing unpolarized light. For instance, unpolarized light can be described as a superposition of two equal incoherent plane-polarized components with mutually perpendicular planes of vibration. Let one plane have an inclination angle χ. Then one component coherency matrix is

$$\mathbf{J}^{(1)} = \mathbf{J}(\chi) = \frac{S_0}{2} \begin{pmatrix} \cos^2 \chi & \sin \chi \cos \chi \\ \sin \chi \cos \chi & \sin^2 \chi \end{pmatrix} \tag{9.63a}$$

The other component must then have inclination angle $\chi \pm \pi/2$ and coherency matrix

$$\mathbf{J}^{(2)} = \mathbf{J}\left(\chi \pm \frac{\pi}{2}\right) = \frac{S_0}{2} \begin{pmatrix} \sin^2 \chi & -\sin \chi \cos \chi \\ -\sin \chi \cos \chi & \cos^2 \chi \end{pmatrix} \tag{9.63b}$$

Then

$$\mathbf{J} = \mathbf{J}^{(1)} + \mathbf{J}^{(2)} = \frac{S_0}{2}\begin{pmatrix} 1 & 0 \\ 0 & 1 \end{pmatrix} = \frac{S_0}{2}\mathbf{I}$$

Unpolarized light may also be described as a superposition of equal incoherent right and left circularly polarized beams:

$$\mathbf{J}^{(1)} = \mathbf{J}(\text{RCP}) = \frac{1}{4}S_0\begin{pmatrix} 1 & i \\ -i & 1 \end{pmatrix} \tag{9.64a}$$

$$\mathbf{J}^{(2)} = \mathbf{J}(\text{LCP}) = \frac{1}{4}S_0\begin{pmatrix} 1 & -i \\ i & 1 \end{pmatrix} \tag{9.64b}$$

$$\mathbf{J} = \mathbf{J}^{(1)} + \mathbf{J}^{(2)} = \frac{1}{2}S_0\begin{pmatrix} 1 & 0 \\ 0 & 1 \end{pmatrix} \tag{9.65}$$

Two incoherent beams, each in a pure elliptically polarized state, may be combined to yield unpolarized light. The notation of Eq. (9.61) is convenient for this purpose. Let the coherency matrices for the two beams be

$$\mathbf{J}^{(1)} = S^{(1)}\begin{pmatrix} \cos^2 \chi_1 & e^{i\phi_1}\sin\chi_1\cos\chi_1 \\ e^{i\phi_1}\sin\chi_1\cos\chi_1 & \sin^2\chi_1 \end{pmatrix}$$

$$\mathbf{J}^{(2)} = S^{(2)}\begin{pmatrix} \cos^2 \chi_2 & e^{-i\phi_2}\sin\chi_2\cos\chi_2 \\ e^{i\phi_2}\sin\chi_2\cos\chi_2 & \sin^2\chi_2 \end{pmatrix} \tag{9.66}$$

Then let $\chi_2 = (\pi/2) - \chi_1$, $\phi_2 = \pi + \phi_1$, $S^{(2)} = S^{(1)} = S_0/2$ to obtain

$$\mathbf{J} = \mathbf{J}^{(1)} + \mathbf{J}^{(2)} = \frac{S_0}{2}\mathbf{I}$$

This last example includes the first two as special cases— when $\phi_1 = 0$, $\chi_1 = \chi$, the first example is recovered, and when $\phi_1 = -\pi/2$, $\chi_1 = \pi/4$, the second is recovered. The two different components in the three cases are shown in Fig. 9.30. Note that the major axes of the two ellipses are 90° apart and that the ellipses have the same size and shape with opposite senses of rotation.

2. Partially Polarized Light. We now show that partially polarized light can be expressed as a superposition of unpolarized light and of pure polarized light (elliptically polarized, in general), and that this decomposition is unique. To do this we show that an arbitrary coherency matrix $\mathbf{J}$ may be written in the form

$$\mathbf{J} = \mathbf{J}^{(1)} + \mathbf{J}^{(2)}$$

or

$$\begin{pmatrix} J_{xx} & J_{xy} \\ J_{xy}^* & J_{yy} \end{pmatrix} = \begin{pmatrix} A & 0 \\ 0 & A \end{pmatrix} + \begin{pmatrix} B & D \\ D^* & C \end{pmatrix} \tag{9.67}$$

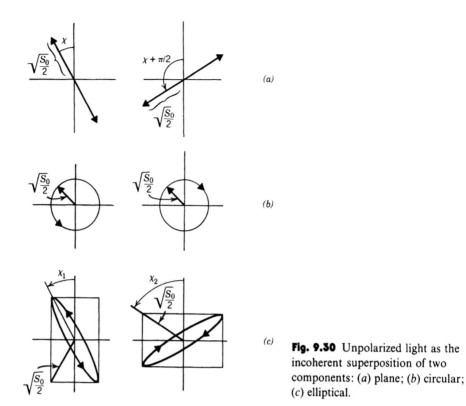

Fig. 9.30 Unpolarized light as the incoherent superposition of two components: (a) plane; (b) circular; (c) elliptical.

The second matrix $\mathbf{J}^{(2)}$ should represent a pure state of polarization. Hence by Eq. (9.62) its determinant must vanish:

$$BC - DD^* = 0 \tag{9.68}$$

Another requirement on $\mathbf{J}^{(2)}$ is that it gives nonnegative values of flux density when used in Eq. (9.46) or (9.47). In particular, the requirement $S(0, 0) \geq 0$ gives

$$B \geq 0 \tag{9.69}$$

and the requirement $S(\pi/2, 0) \geq 0$ gives

$$C \geq 0 \tag{9.70}$$

The matrix Eq. (9.67) gives four ordinary equations

$$J_{xx} = A + B$$
$$J_{xy} = D$$
$$J_{xy}^* = D^*$$
$$J_{yy} = A + C$$

Thus $B = J_{xx} - A$ and $C = J_{yy} - A$, and Eq. (9.68) becomes

$$(J_{xx} - A)(J_{yy} - A) - J_{xy}J_{yx} = 0 \tag{9.71}$$

or

$$A^2 - A(J_{xx} + J_{yy}) + J_{xx}J_{yy} - J_{xy}J_{yx} = 0$$

or

$$A^2 - A \, \text{Tr} \, \mathbf{J} + \det \mathbf{J} = 0 \tag{9.72}$$

with

$$\text{Tr} \, \mathbf{J} = J_{xx} + J_{yy}; \quad \det \mathbf{J} = J_{xx}J_{yy} - J_{xy}J_{yx}$$

Equation (9.71) also represents the so-called *characteristic equation* for the matrix $\mathbf{J}$. Such an equation is usually obtained in the form

$$\det(\mathbf{J} - A\mathbf{I}) = 0$$

where $\mathbf{I}$ is the unit matrix

$$\mathbf{I} = \begin{pmatrix} 1 & 0 \\ 0 & 1 \end{pmatrix}$$

or

$$\det \begin{pmatrix} J_{xx} - A & J_{xy} \\ J_{yx} & J_{yy} - A \end{pmatrix} = 0$$

There are two solutions of Eq. (9.72) given by

$$A_{1,2} = \frac{\text{Tr} \, \mathbf{J} \pm \sqrt{(\text{Tr} \, \mathbf{J})^2 - 4 \det \mathbf{J}}}{2} \tag{9.73}$$

The argument of the radical is

$$(J_{xx} + J_{yy})^2 - 4(J_{xx}J_{yy} - |J_{xy}|^2) = J_{xx}^2 + J_{yy}^2 - 2J_{xx}J_{yy} + 4|J_{xy}|^2$$
$$= (J_{xx} - J_{yy})^2 + 4|J_{xy}|^2 \geq 0 \tag{9.74}$$

so that the radical is real.

Both solutions $A_{1,2}$ are nonnegative. This is obvious for

$$A_1 = \frac{\text{Tr} \, \mathbf{J} + \sqrt{(\text{Tr} \, \mathbf{J})^2 - 4 \det \mathbf{J}}}{2}$$

To see it for

$$A_2 = \frac{\text{Tr} \, \mathbf{J} - \sqrt{(\text{Tr} \, \mathbf{J})^2 - 4 \det \mathbf{J}}}{2}$$

we need only note that

$$\sqrt{(\text{Tr} \, \mathbf{J})^2 - 4 \det \mathbf{J}} \leq \text{Tr} \, \mathbf{J}$$

so that

$$-\sqrt{(\text{Tr }\mathbf{J})^2 - 4\det \mathbf{J}} \geq -\text{Tr }\mathbf{J}$$

and that

$$A_2 = \frac{\text{Tr }\mathbf{J} - \sqrt{(\text{Tr }\mathbf{J})^2 - 4\det \mathbf{J}}}{2} \geq \frac{\text{Tr }\mathbf{J} - \text{Tr }\mathbf{J}}{2} = 0$$

We also have from Eq. (9.74)

$$\sqrt{(\text{Tr }\mathbf{J})^2 - 4\det \mathbf{J}} \geq |J_{xx} - J_{yy}|$$

so that

$$A_1 \geq J_{xx} \quad \text{if } J_{xx} \geq J_{yy}$$

or

$$A_1 \geq J_{yy} \quad \text{if } J_{yy} \geq J_{xx} \tag{9.75}$$

This violates one or the other of the inequalities (9.69) or (9.70):

$$B = J_{xx} - A \geq 0$$

$$C = J_{yy} - A \geq 0$$

Thus our physically meaningful solution must be

$$A = A_2 = \frac{\text{Tr }\mathbf{J} - \sqrt{(\text{Tr }\mathbf{J})^2 - 4\det \mathbf{J}}}{2} \tag{9.76}$$

Then

$$B = J_{xx} - A_2 = \frac{J_{xx} - J_{yy} + \sqrt{(\text{Tr }\mathbf{J})^2 - 4\det \mathbf{J}}}{2} \tag{9.77}$$

and

$$C = J_{yy} - A_2 = \frac{J_{yy} - J_{xx} + \sqrt{(\text{Tr }\mathbf{J})^2 - 4\det \mathbf{J}}}{2} \tag{9.78}$$

The total flux density is given by Eq. (9.53) as

$$S_0 = \text{Tr }\mathbf{J} = A_1 + A_2 \tag{9.79}$$

The flux density of the polarized part is

$$S_{\text{pol}} = \text{Tr }\mathbf{J}^{(2)} = B + C = (A_1 - A_2) \tag{9.80}$$

and the flux density of the unpolarized part is

$$S_{\text{un}} = S_0 - S_{\text{pol}} = 2A_2 \tag{9.81}$$

D. Polarization Specification

1. Degree of Polarization. The ratio of flux density of the polarized part to that of the total of a partially polarized light beam is defined to be the *degree of polarization P*:

$$P = \frac{S_{\text{pol}}}{S_0} = \left[1 - \frac{4 \det \mathbf{J}}{(\operatorname{Tr} \mathbf{J})^2}\right]^{1/2} \tag{9.82}$$

Here *P* can also be written

$$P = \frac{A_1 - A_2}{A_1 + A_2} \tag{9.83}$$

The matrix $\mathbf{J}^{(1)}$ represents an unpolarized beam:

$$\mathbf{J}^{(1)} = \begin{pmatrix} A_2 & 0 \\ 0 & A_2 \end{pmatrix} = \frac{1}{2} S_{\text{un}} \begin{pmatrix} 1 & 0 \\ 0 & 1 \end{pmatrix} \tag{9.84}$$

The matrix $\mathbf{J}^{(2)}$ represents a pure elliptically polarized beam and may be written

$$\begin{aligned} \mathbf{J}^{(2)} &= \begin{pmatrix} B & J_{xy} \\ J_{xy}^* & C \end{pmatrix} = \begin{pmatrix} B & |J_{xy}|e^{-i\phi} \\ |J_{xy}|e^{i\phi} & C \end{pmatrix} \\ &= S_{\text{pol}} \begin{pmatrix} \cos^2 \chi & e^{-i\phi} \sin \chi \cos \chi \\ e^{i\phi} \sin \chi \cos \chi & \sin^2 \chi \end{pmatrix} \end{aligned} \tag{9.85}$$

with

$$\chi = \tan^{-1}\left(\frac{C}{B}\right)^{1/2}$$

If the beam is passed through a compensator with phase shift $\Delta\phi = -\phi$, $\mathbf{J}^{(2)}$ is changed into

$$\mathbf{J}^{(2)'} = \begin{pmatrix} B & |J_{xy}| \\ |J_{xy}| & C \end{pmatrix} = S_{\text{pol}} \begin{pmatrix} \cos^2 \chi & \sin \chi \cos \chi \\ \sin \chi \cos \chi & \sin^2 \chi \end{pmatrix}$$

Here $\mathbf{J}^{(1)}$ is left unchanged. Then if the beam is passed through a polarizer with pass angle ψ, the measured intensity $S(\psi, -\phi)$ in our original notation [Eq. (9.51)], will be given by

$$S(\psi, -\phi) = \tfrac{1}{2}S_{\text{un}} + S_{\text{pol}} \cos^2(\psi - \chi) = S_0\{\tfrac{1}{2} + P[\cos^2(\psi - \chi) - \tfrac{1}{2}]\} \tag{9.86}$$

This has a maximum value of

$$S_{\text{max}} = \tfrac{1}{2}S_0(1 + P) = A_1 \tag{9.87a}$$

and a minimum value of

$$S_{\text{min}} = \tfrac{1}{2}S_0(1 - P) = A_2 \tag{9.87b}$$

Thus the degree of polarization can be written

$$P = \frac{S_{max} - S_{min}}{S_{max} + S_{min}} \qquad (9.88)$$

When $\Delta\phi$ is not equal to $-\phi$, the maximum and minimum values of $S(\psi, \Delta\phi)$ will be respectively smaller and larger than the values just given.

2. Experimental Characterization. A general partially polarized light beam is a mixture of elliptically polarized light and unpolarized light. It will be uniquely specified by five parameters, the total flux density S_0, the degree of polarization P, the orientation angle $\beta/2$ of the major axis of the ellipse, the ratio $\tan(\eta)$ of the lengths of minor to major axes, and the sense of rotation or helicity of the light (see Fig. 9.22).

We need a polarizer and a quarter-wave plate to determine these parameters. The polarizer is first placed alone in the beam and rotated while the transmitted flux density is observed. The two pass directions giving maximum flux density correspond to **E** vibrating along the major axis of the ellipse and hence correspond to the angles $\beta/2$ and $\beta/2 + \pi$. The pass directions giving minimum flux density are at $\beta/2 \pm \pi/2$. So far the light could lie between the following two extremes: (1) pure elliptically polarized light with $\tan\eta = \sqrt{S_{min}/S_{max}}$, and (2) a mixture of unpolarized light and plane polarized light vibrating along the direction $\beta/2$.

It is convenient to consider a new set of axes along $\bar{x}$ and $\bar{y}$ as shown in Fig. 9.22. Place a quarter-wave plate in the incident beam so that its fast axis is along the $\bar{x}$ axis. Any elliptically polarized light with major axis along $\bar{x}$ will then be converted to plane polarized light vibrating at an angle of $\pm\eta$ with respect to the $\bar{x}$ axis, where the plus sign refers to right elliptically polarized light and the minus sign to left elliptically polarized light. If the original light is a mixture of unpolarized light and plane polarized light vibrating along $\bar{x}$, the quarter-wave plate will have no effect.

For the mixed case, there will be an unpolarized component present after the light has passed through the quarter-wave plate. The angle η can be determined by using a polarizer after the quarter-wave plate. With the pass direction of the polarizer along η or $\eta + \pi$ (with respect to the $\bar{x}$ axis), the flux density will have a maximum value of

$$S_{max} = S_{pol} + \tfrac{1}{2}S_{un}$$

With the pass direction along $\eta + \pi/2$, the flux density will have a minimum value of

$$S_{min} = \tfrac{1}{2}S_{un}$$

The degree of polarization is then

$$P = \frac{S_{max} - S_{min}}{S_{max} + S_{min}}$$

and the total flux density is

$$S_0 = S_{max} + S_{min}$$

9.4 Crystal Optics

Many crystals have symmetry sufficiently low that their optical properties are anisotropic (depend on the direction of propagation and polarization). The result is double refraction. We have already introduced this subject in section 9.2. Here we wish to elaborate on the phenomenon with a presentation based on Maxwell's equations. We will not discuss the related effects such as electrooptics, magnetooptics, and nonlinear optical properties but will concentrate on the fundamentals of light propagation in anisotropic media.

A. Electromagnetic Waves in Anisotropic Dielectrics

The four Maxwell equations in the absence of free charges or currents are

$$\text{I} \quad \mathbf{V} \cdot \mathbf{D} = 0$$

$$\text{II} \quad \mathbf{V} \cdot \mathbf{B} = 0$$

$$\text{III} \quad \mathbf{V} \times \mathbf{E} = -\frac{\partial \mathbf{B}}{\partial t}$$

$$\text{IV} \quad \mathbf{V} \times \mathbf{H} = \frac{\partial D}{\partial t}$$

The medium will be assumed nonmagnetic so that $\mu = \mu_0$ and $\mathbf{B} = \mu_0 \mathbf{H}$. We seek solutions of the plane wave form with a phase dependence $\exp[i(\omega t - \omega n \hat{s} \cdot \mathbf{r}/c)]$ as suggested in the introduction to this chapter. This enables the application of the operator shortcut of Eq. (1.28) to Maxwell equations III and IV:

$$\mathbf{V} \times \mathbf{E} = -i\omega\mu_0 \mathbf{H}$$

$$\mathbf{V} \times \mathbf{H} = i\omega \mathbf{D}$$

or

$$\mathbf{V} \times (\mathbf{V} \times \mathbf{E}) = \mu_0 \omega^2 \mathbf{D} \tag{9.89}$$

1. Relationship Between D and E. The remaining spatial dependence of the phase contains the unit vector $\hat{s}$, which is normal to the plane surfaces of constant phase, and n, the index of refraction. The magnitude of the *phase velocity* c/n will then be the speed at which the surfaces of constant phase propagate. This will also be denoted as the *wave velocity*

$$\mathbf{v}_w = \frac{\hat{s}c}{n} \tag{9.90}$$

Applying the second operator shortcut [Eq. (1.29)] we find

$$\mathbf{V} \times (\mathbf{V} \times \mathbf{E}) = -\frac{\omega^2 n^2}{c^2} \hat{s} \times (\hat{s} \times \mathbf{E}) = -\frac{\omega^2 n^2}{c^2} [\hat{s}(\hat{s} \cdot \mathbf{E}) - \mathbf{E}]$$

because $\hat{s} \cdot \hat{s} = 1$. (We have used a vector identity also.) This leads to the following expressions for $\mathbf{D}$ in terms of $\mathbf{E}$:

$$\mathbf{D} = \varepsilon_0 n^2 [\mathbf{E} - \hat{s}(\hat{s} \cdot \mathbf{E})] \tag{9.91}$$

where $\mathbf{D}$ is perpendicular to $\hat{s}$.

Equation (9.91) has been derived from Maxwell's equations using only the assumption of the plane-wave space–time dependence. The properties of the dielectric medium are introduced by the additional equations

$$D_x = \varepsilon_{xx} E_x + \varepsilon_{xy} E_y + \varepsilon_{xz} E_z$$

$$D_y = \varepsilon_{yx} E_x + \varepsilon_{yy} E_y + \varepsilon_{yz} E_z$$

$$D_z = \varepsilon_{zx} E_x + \varepsilon_{zy} E_y + \varepsilon_{zz} E_z$$

or

$$D_i = \sum_{j=1}^{3} \varepsilon_{ij} E_j \tag{9.92}$$

where $i, j = 1$ for x, 2 for y, and 3 for z. The form of these equations suggests that we express the relationship in terms of matrices

$$\mathbf{D} = \varepsilon \mathbf{E} \tag{9.93}$$

where

$$\mathbf{D} = \begin{pmatrix} D_x \\ D_y \\ D_z \end{pmatrix}, \quad \mathbf{E} = \begin{pmatrix} E_x \\ E_y \\ E_z \end{pmatrix} \tag{9.94}$$

and

$$\varepsilon = \begin{pmatrix} \varepsilon_{xx} & \varepsilon_{xy} & \varepsilon_{xz} \\ \varepsilon_{yx} & \varepsilon_{yy} & \varepsilon_{yz} \\ \varepsilon_{zx} & \varepsilon_{zy} & \varepsilon_{zz} \end{pmatrix} \tag{9.95}$$

The 3×3 matrix in Eq. (9.95) is a special case of a *tensor*. Tensors can be defined to describe a variety of direction-dependent properties of the medium. Frequently higher dimensionality is required. Rules exist for the algebraic operations between tensors. In three dimensions these are the same as those for matrix algebra. We will not discuss tensors of higher dimensionality in this book. However, we identify ε as the *dielectric tensor*.

In nonabsorbing media with no optical activity, it can be shown that ε is real and symmetric. Thus

$$\varepsilon_{ij} = \varepsilon_{ij}^* \tag{9.96}$$

and

$$\varepsilon_{ij} = \varepsilon_{ji} \tag{9.97}$$

For such a tensor we can always find a rectangular coordinate system in terms of which the matrix of Eq. (9.95) is *diagonal*. There are then only three independent elements. The coordinate axes in this special system are called *principal axes*. With respect to them ε takes the form

$$\varepsilon = \begin{pmatrix} \varepsilon_x & 0 & 0 \\ 0 & \varepsilon_y & 0 \\ 0 & 0 & \varepsilon_z \end{pmatrix} \tag{9.98}$$

The quantities ε_x, ε_y, and ε_z are called *principal values* of ε. They are all nonnegative and are functions of the frequency ω.

When we combine Eqs. (9.91) and (9.92), we find that for principal axes

$$\varepsilon_0 n^2 \left[E_i - \hat{s}_i \sum_j \hat{s}_j E_j \right] - \varepsilon_i E_i = 0 \tag{9.99}$$

To better understand this equation, we can write it in two alternative forms. The first version is

$$\mathbf{ME} = 0 \tag{9.100}$$

with

$$\mathbf{M} = \begin{vmatrix} n^2(1 - \hat{s}_x^2) - \dfrac{\varepsilon_x}{\varepsilon_0} & -n^2 \hat{s}_x \hat{s}_y & -n^2 \hat{s}_x \hat{s}_z \\[2mm] -n^2 \hat{s}_x \hat{s}_y & n^2(1 - \hat{s}_y^2) - \dfrac{\varepsilon_y}{\varepsilon_0} & -n^2 \hat{s}_y \hat{s}_z \\[2mm] -n^2 \hat{s}_x \hat{s}_z & -n^2 \hat{s}_y \hat{s}_z & n^2(1 - \hat{s}_z^2) - \varepsilon_z/\varepsilon_0 \end{vmatrix} \tag{9.101}$$

Provided that the elements of **M** are known, we can use Eq. (9.100) to find the elements of **E**. Presumably we would have access to the elements of the dielectric tensor ε and would be able to specify the propagation direction by unit vector $\hat{s} = (\hat{s}_x, \hat{s}_y, \hat{s}_z)$. We must determine n^2. This can be accomplished by recognizing that Eqs. (9.100) have a solution if and only if the *characteristic equation* is satisfied

$$\det \mathbf{M} = 0 \tag{9.102}$$

An equivalent version may be written as three separate equations of the form

$$\left(n^2 - \frac{\varepsilon_x}{\varepsilon_0} \right) E_x = n^2 \hat{s}_x (\hat{s} \cdot \mathbf{E})$$

or

$$E_x = \frac{n^2}{n^2 - \dfrac{\varepsilon_x}{\varepsilon_0}} \hat{s}_x (\hat{s} \cdot \mathbf{E})$$

and so on for E_y and E_z. Or

$$\hat{s}_x \hat{E}_x = \frac{n^2 \hat{s}_x^2}{n^2 - n_x^2} (\hat{s} \cdot E) \qquad (9.103)$$

and so on, where

$$\left. \begin{aligned} n_x^2 &\equiv \frac{\varepsilon_x}{\varepsilon_0} \\[2ex] n_y^2 &\equiv \frac{\varepsilon_y}{\varepsilon_0} \\[2ex] n_z^2 &\equiv \frac{\varepsilon_z}{\varepsilon_0} \end{aligned} \right\} \qquad (9.104)$$

When the three equations of the form of Eq. (9.103) are summed we find

$$(\hat{s} \cdot E) = \left[\frac{n^2 \hat{s}_x^2}{n^2 - n_x^2} + \frac{n^2 \hat{s}_y^2}{n^2 - n_y^2} + \frac{n^2 \hat{s}_z^2}{n^2 - n_z^2} \right] (\hat{s} \cdot E)$$

When $(\hat{s} \cdot E) \neq 0$, it may be canceled from both sides, leaving *Fresnel's equation*

$$\frac{1}{n^2} = \frac{\hat{s}_x^2}{n^2 - n_x^2} + \frac{\hat{s}_y^2}{n^2 - n_y^2} + \frac{\hat{s}_z^2}{n^2 - n_z^2} \qquad (9.105)$$

Equations (9.102) and (9.105) both appear to be cubic in n^2, but in both cases when they are written in terms of a polynomial in n^2, the coefficients of n^6 vanish identically, leaving a quadratic equation in n^2, which can be shown to have two positive roots, n_1^2 and n_2^2.

Thus, for a given propagation direction $\hat{s}$, there are in general two values of the refractive index, n_1 and n_2. If these are substituted one at a time into the matrix **M**, Eq. (9.100) may be solved for the components of **E**, to within an undetermined multiplicative constant. Let the two solutions for **E**, corresponding to n_1 and n_2 be denoted by E_1 and E_2, respectively. The displacement vector is then given by Eq. (9.91). If **E** is resolved into components parallel and perpendicular to $\hat{s}$ as shown in Fig. 9.31

$$E = E_{//} + E_\perp$$

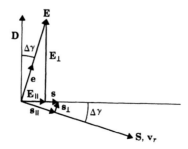

Fig. 9.31

Then $E_{//} = (\mathbf{E} \cdot \mathbf{\hat{s}})\mathbf{\hat{s}}$ and $\mathbf{E}_\perp = \mathbf{E} - (\mathbf{E} \cdot \mathbf{\hat{s}})\mathbf{\hat{s}}$, and $\mathbf{D}$ is given by the simple relation

$$\mathbf{D} = \varepsilon_0 n^2 \mathbf{E}_\perp \tag{9.106}$$

Here $\mathbf{D}$, $\mathbf{E}$, and $\mathbf{\hat{s}}$ are seen to be coplanar.

Actually there are two equations of the form of Eq. (9.106), one for each solution of Eq. (9.99):

$$\mathbf{D}_1 = \varepsilon_0 n_1^2 \mathbf{E}_{1\perp} \tag{9.107a}$$

$$\mathbf{D}_2 = \varepsilon_0 n_2^2 \mathbf{E}_{2\perp} \tag{9.107b}$$

These two allowed values of $\mathbf{D}$ are perpendicular. To demonstrate this we need to prove

$$(\mathbf{D}_1 \cdot \mathbf{D}_2) = 0 \tag{9.108a}$$

or equivalently

$$(\mathbf{E}_{1\perp} \cdot \mathbf{E}_{2\perp}) = 0 \tag{9.108b}$$

Consider the scalar product

$$(\mathbf{E}_2 \cdot \mathbf{D}_1) = \sum_i E_{2i} D_i = \sum_{ij} E_{2i} \varepsilon_{ij} E_{1j} = \sum_{ij} E_{1j} \varepsilon_{ji} E_{2i}$$

$$= \sum_j E_{1j} D_{2j} = (\mathbf{E}_1 \cdot \mathbf{D}_2)$$

The third equality made use of the symmetry of the dielectric tensor. Use of Eqs. (9.107) gives

$$n_1^2 (\mathbf{E}_2 \cdot \mathbf{E}_{1\perp}) = n_2^2 (\mathbf{E}_1 \cdot \mathbf{E}_{2\perp})$$

Set $\mathbf{E}_2 = \mathbf{E}_{2\perp} + \mathbf{E}_{2\parallel}$, where $\mathbf{E}_{2\parallel}$ is parallel to $\mathbf{\hat{s}}$ and hence perpendicular to $\mathbf{E}_{1\perp}$. Then $(\mathbf{E}_2 \cdot \mathbf{E}_{1\perp}) = (\mathbf{E}_{2\perp} \cdot \mathbf{E}_{1\perp})$. Similarly $(\mathbf{E}_1 \cdot \mathbf{E}_{2\perp}) = (\mathbf{E}_{1\perp} \cdot \mathbf{E}_{2\perp})$. Thus

$$(n_1^2 - n_2^2)(\mathbf{E}_{2\perp} \cdot \mathbf{E}_{1\perp}) = 0$$

When $n_1^2 \neq n_2^2$, this gives Eq. (9.108b). The case $n_1 = n_2$ occurs in isotropic media and along certain directions (wave-optic axes) in anisotropic media. The directions of $\mathbf{D}_1$ and $\mathbf{D}_2$ are then arbitrary as long as they are perpendicular to $\mathbf{\hat{s}}$, and it is convenient to choose them to be mutually perpendicular in this case also.

2. The Poynting Vector and Ray Velocity. The Poynting vector $\mathbf{S}$ gives the direction and rate of energy flow. From Eq. (2.20) this is

$$\mathbf{S} = \mathbf{E} \times \mathbf{H} \tag{9.109}$$

The magnetic field $\mathbf{H}$ is related to $\mathbf{E}$ through Eq. (2.35)

$$\mathbf{B} = \frac{\mathbf{k} \times \mathbf{E}}{\omega} = \frac{n}{c}\mathbf{\hat{s}} \times \mathbf{E} = n\varepsilon_0 \mu_0 c\mathbf{\hat{s}} \times \mathbf{E} = \mu_0 \mathbf{H}$$

So

$$\mathbf{H} = \varepsilon_0 nc\mathbf{\hat{s}} \times \mathbf{E} \tag{9.110}$$

where $\mathbf{S}$ is perpendicular to both $\mathbf{E}$ and $\mathbf{H}$, and because $\mathbf{H}$ is normal to the plane of $\mathbf{D}$, $\mathbf{E}$, and $\hat{\mathbf{s}}$, $\mathbf{S}$ will lie in that plane. However, it will, in general, not be parallel to the wave normal $\hat{\mathbf{s}}$. In fact, it makes the same angle $\Delta\gamma$ with $\hat{\mathbf{s}}$ as $\mathbf{E}$ does with $\mathbf{D}$. This is illustrated in Fig. 9.31.

An explicit expression for $\mathbf{S}$ may be obtained by writing $\mathbf{E} = E\hat{\mathbf{e}}$ where $\hat{\mathbf{e}}$ is a unit vector in the direction of $\mathbf{E}$. Then $\mathbf{S}$ is given by

$$\mathbf{S} = n\varepsilon_0 cE^2[\hat{\mathbf{s}} - \hat{\mathbf{e}}(\hat{\mathbf{s}} \cdot \hat{\mathbf{e}})] \tag{9.111}$$

Now $\hat{\mathbf{e}}(\hat{\mathbf{s}} \cdot \hat{\mathbf{e}})$ is the component $\hat{\mathbf{s}}_\perp$ of $\hat{\mathbf{s}}$ along $\hat{\mathbf{e}}$, and $\hat{\mathbf{s}} - \hat{\mathbf{e}}(\hat{\mathbf{s}} \cdot \hat{\mathbf{e}})$ is the component $\hat{\mathbf{s}}_\parallel$ of $\hat{\mathbf{s}}$ along $\mathbf{S}$. Thus Eq. (9.111) may also be written

$$\mathbf{S} = n\varepsilon_0 cE^2\hat{\mathbf{s}}_\parallel \tag{9.112}$$

The Poynting vector $\mathbf{S}$ defines the ray direction. We have just seen that, in general, for anisotropic media this is not the direction of the wave normal $\hat{\mathbf{s}}$. It can be shown that when dispersion is present, that is, when the elements of ε depend on frequency, the *group velocity* vector for a *wave packet* will be along $\mathbf{S}$.

The wave velocity $\mathbf{v}_w$ is the velocity of propagation of the surfaces of constant phase $\omega t - \omega n(\hat{\mathbf{s}} \cdot \mathbf{r})/c = (\text{const})$ and is given by Eq. (9.90)

$$\mathbf{v}_w = \frac{\hat{\mathbf{s}}c}{n}$$

The ray velocity $\mathbf{v}_r$ is along $\mathbf{S}$ and has a magnitude according to Eq. (9.34)

$$v_r = \frac{v_w}{\cos \Delta\gamma} = \frac{c}{n \cos \Delta\gamma} \tag{9.113}$$

A unit vector along $\mathbf{S}$ may be constructed from the parallel vector

$$\hat{\mathbf{s}}_\parallel = \hat{\mathbf{s}} - \hat{\mathbf{e}}(\hat{\mathbf{s}} \cdot \hat{\mathbf{e}})$$

which has magnitude $\cos \Delta\gamma$. Thus the desired unit vector is $\hat{\mathbf{s}}_\parallel/\cos \Delta\gamma$, and $\mathbf{v}_r$ may be written

$$\mathbf{v}_r = \frac{c}{n \cos^2 \Delta\gamma} \hat{\mathbf{s}}_\parallel = \frac{c}{n \cos^2 \Delta\gamma} [\hat{\mathbf{s}} - \hat{\mathbf{e}}(\hat{\mathbf{s}} \cdot \mathbf{e})] \tag{9.114}$$

Another version is obtained by noting that

$$|\hat{\mathbf{s}}_\parallel|^2 = \cos^2 \Delta\gamma = 1 - (\hat{\mathbf{s}} \cdot \hat{\mathbf{e}})^2$$

Hence

$$\mathbf{v}_r = \frac{c[\hat{\mathbf{s}} - \hat{\mathbf{e}}(\hat{\mathbf{s}} \cdot \hat{\mathbf{e}})]}{n[1 - (\hat{\mathbf{s}} \cdot \hat{\mathbf{e}})^2]} \tag{9.115}$$

Yet another version may be easily derived:

$$\mathbf{v}_r = \frac{n\varepsilon_0 c[\hat{\mathbf{s}} - \hat{\mathbf{e}}(\hat{\mathbf{s}} \cdot \hat{\mathbf{e}})]}{\sum_{ij} e_i \varepsilon_{ij} e_j} = \frac{\mathbf{S}}{\mathbf{E}\varepsilon\mathbf{E}} \tag{9.116}$$

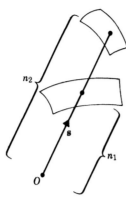

Fig. 9.32 Portions of index surface.

B. Index Surfaces

There are several constructions useful in solving problems of wave propagation in optically anisotropic crystals. One of the most useful is that of the *index surface*. A given index surface may be expressed as a function $n(\hat{s})$ and actually represents two surfaces for each value of $\hat{s}$. To construct the index surfaces we must first find the two indices n_1 and n_2 for each direction $\hat{s}$ using Fresnel's equation (9.105) or an equivalent formalism. Starting at a suitable origin O, we proceed along appropriate $\hat{s}$ directions, distances proportional to n_1 and n_2 thus locating two points. The loci of all such points form the two index surfaces (Fig. 9.32).

1. Ray Directions. The index surface $n(\hat{s})$ may be used to determine the directions of **E, H, D,** and **S** for the electromagnetic wave with normal $\hat{s}$. It can be shown that the direction of the Poynting vector **S**, that is the ray direction associated with the wave normal $\hat{s}$, is perpendicular to the index surface at the points identified by n_1 and n_2 along $\hat{s}$. The plane of $\hat{s}$ and **S** is along the plane of **D** and **E**. As illustrated in Fig. 9.31, it is then possible to locate in this plane the direction of **D** (perpendicular to $\hat{s}$) and **E** (perpendicular to $\hat{S}$). These concepts are clarified in Fig. 9.33 where $n_1 < n_2$. The index surfaces are shown in Fig. 9.33a with the ray and wave velocities and the field vectors **E** and **D**.

For comparison, the Huygens' wavelet construction for the same material is shown in Fig. 9.33b. Note the inverse relationship of the ellipsoids between the two diagrams. Care should be exercised not to confuse these two representations. In Fig. 9.33a the index surfaces are merely a mechanism for identifying the relevant vector quantities at a particular point in the crystal. The distances involved are not associated with propagation. In that sense, the vectors shown in Fig. 9.33a are really at the origin. In Fig. 9.33b the surfaces are actual wavefronts after a time has elapsed. The disturbance originates at the origin of that diagram at $t = 0$. We can construct Fig. 9.33b, a physical picture of the wavefronts, from the information obtained in Fig. 9.33a.

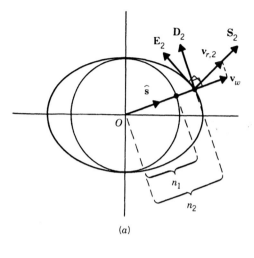

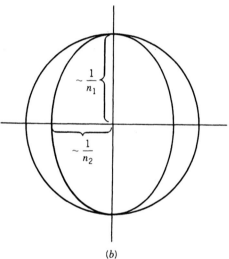

Fig. 9.33 Comparison of (a) the index surfaces and (b) Huygens' wavelets in the same crystal.

2. Reflection and Refraction. Consider an interface between two nonabsorbing, nonoptically active media. Both will be anisotropic, in general. Let a plane wave be incident in the unprimed medium with wave normal direction $\hat{s}$ and index n. Then there will generally be two transmitted waves in the second medium and two reflected waves in the incident medium. The rays in the incident medium will have, in general, three different indices n, n'_1, and n'_2, and three different directions of linear polarization (Fig. 9.34).

We will not attempt to solve the entire reflection-transmission problem, which is to determine the two transmitted and the two reflected fields, if the incident electric field is known. (This can be done by requiring that the tangential components of **E** and **H** be continuous across the interface.) Instead, we ask only

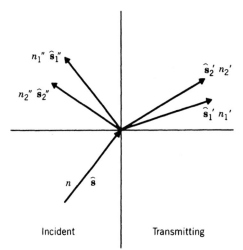

Fig. 9.34 Relevant wave normals in reflection and transmission at an interface between anisotropic media.

for the directions $\hat{\mathbf{s}}$ of the four wave normals and the corresponding values of the indices n. Let $\hat{\boldsymbol{\eta}}$ be the normal to the interface. With the origin O as shown in Fig. 9.35, the plane interface obeys the equation $\hat{\boldsymbol{\eta}} \cdot \mathbf{r} = 0$ where $\mathbf{r}$ is on the interface. The incident wave will be of the form

$$\mathbf{E} = \hat{\mathbf{e}} A\, e^{i\omega t}\, e^{-i\omega n \mathbf{s} \cdot \mathbf{r}/c}$$

with two reflected waves, one of which is

$$\mathbf{E}''_1 = \hat{\mathbf{e}}''_1 A''_1\, e^{i\omega t}\, e^{-i\omega n''_1 \mathbf{s}''_1 \cdot \mathbf{r}/c}$$

and two transmitted waves, one of which is

$$\mathbf{E}'_1 = \hat{\mathbf{e}}'_1 A'_1\, e^{i\omega t}\, e^{-i\omega n'_1 \mathbf{s}'_1 \cdot \mathbf{r}/c}$$

Imposition of the boundary conditions on $\mathbf{E}_{\text{tan}}$ and $\mathbf{H}_{\text{tan}}$ (here $\mathbf{B}_{\text{tan}}$) will lead, among

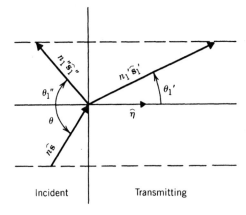

Fig. 9.35 Relationship among scaled vectors $\mathbf{n} \cdot \hat{\mathbf{s}}$ for one of the waves.

other things, to the condition that the phases in the exponents of each term must match at the interface. Thus for all $\mathbf{r}$ such that $\hat{\mathbf{\eta}} \cdot \mathbf{r} = 0$, we must have

$$n_1'(\hat{\mathbf{s}}_1' \cdot \mathbf{r}) = n_2'(\hat{\mathbf{s}}_2' \cdot \mathbf{r}) = n_1''(\hat{\mathbf{s}}_1'' \cdot \mathbf{r}) = n_2''(\hat{\mathbf{s}}_2'' \cdot \mathbf{r}) = n(\hat{\mathbf{s}} \cdot \mathbf{r}) \tag{9.117}$$

This leads to four equations of the form

$$n_1''\hat{\mathbf{s}}_{1,\sigma}'' = n\hat{\mathbf{s}}_\sigma \tag{9.118}$$

where the extra subscript σ means the component *in* the plane of the interface. Then the vectors

$$n_1''\hat{\mathbf{s}}_1'' - n\hat{\mathbf{s}}$$

(and so on) must be parallel to the normal $\hat{\mathbf{\eta}}$, because the components perpendicular to $\hat{\mathbf{\eta}}$ are zero by Eq. (9.118). This leads to four equations of the form

$$n_1''\hat{\mathbf{s}}_1'' = n\hat{\mathbf{s}} + (\text{const})\hat{\mathbf{\eta}} \tag{9.119}$$

Thus all the wave normal vectors $\hat{\mathbf{s}}_1''$, $\hat{\mathbf{s}}_2''$, $\hat{\mathbf{s}}_1'$, and $\hat{\mathbf{s}}_2'$ shown in Fig. 9.34 are in the plane determined by $\hat{\mathbf{s}}$ and $\hat{\mathbf{\eta}}$, that is, the *plane of incidence*. Because the components of the vectors $n\hat{\mathbf{s}}$ in the plane of the interface are the same, we obtain a generalization of Snell's law for anisotropic media:

$$n_1'' \sin \theta_1'' = n \sin \theta \tag{9.120a}$$

$$n_2'' \sin \theta_2'' = n \sin \theta \tag{9.120b}$$

$$n_1' \sin \theta_1' = n \sin \theta \tag{9.120c}$$

$$n_2' \sin \theta_2' = n \sin \theta \tag{9.120d}$$

The simple law of reflection $\theta_1'' = \theta$ and $\theta_2'' = \theta$ will not generally hold, because the n's are different for the reflected waves.

Total internal reflection can occur in two ways for light incident onto a doubly refracting medium. They are illustrated in Fig. 9.36. The medium on the right is anisotropic, and one of the index surfaces is drawn. For case (a) we increase the angle of incidence θ until $n \sin \theta = \overline{OP}$. The transmitted wave normal $\hat{\mathbf{s}}'$ is then along the interface. The transmitted ray velocity $\mathbf{v}_r$ for this case is not yet grazing the interface, but nevertheless for larger values of θ than that shown the wave is totally reflected, because Eq. (9.120c or d) can no longer be satisfied by a real angle θ'.

In case (b), as θ is increased, a point is reached for which $\mathbf{v}_r$ is parallel to the interface, whereas $\hat{\mathbf{s}}'$ is not yet grazing. The Poynting vector is then parallel to OP, and for any larger value of θ beyond that for which $n \sin \theta = \overline{OP}$ there can be no energy flow into the upper medium for this branch of the index surface. This situation also corresponds to total reflection.

3. Example of a Uniaxial Medium. In a uniaxial medium the dielectric tensor, when expressed in diagonal form, has two equal elements. If the axes are chosen so

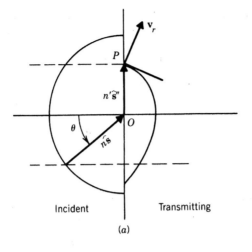

Incident   Transmitting

(a)

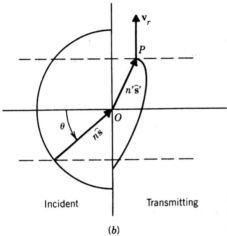

Incident   Transmitting

(b)

Fig. 9.36 Two types of total internal reflection at an interface between anisotropic media. $\mathbf{v}_r$ is the refracted ray velocity.

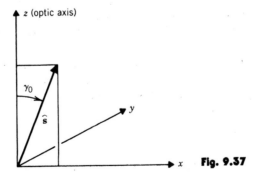

Fig. 9.37

that the third element is different, it takes the form

$$\varepsilon = \varepsilon_0 \begin{pmatrix} n_o^2 & 0 & 0 \\ 0 & n_o^2 & 0 \\ 0 & 0 & n_e^2 \end{pmatrix} \tag{9.121}$$

Thus $\varepsilon_x/\varepsilon_0 = \varepsilon_y/\varepsilon_0 = n_o^2$ and $\varepsilon_z/\varepsilon_0 = n_e^2$, where n_o and n_e are the ordinary and the extraordinary indices of refraction, respectively. The optical properties are now invariant with respect to rotation of the crystal about the z axis, which is the optic axis (Fig. 9.37). Without loss of generality we can assume that the wave normal $\hat{s}$ is in the x-z plane. Hence, $\hat{s} = (\sin \gamma_0, 0, \cos \rho_0)$, and Eqs. (9.99), when written out in detail, give

$$(n^2 - n_o^2)E_x - n^2 \sin \gamma_0 (\sin \gamma_0 E_x + \cos \gamma_0 E_z) = 0 \tag{9.121a}$$

$$(n^2 - n_o^2)E_y \qquad\qquad\qquad\qquad\qquad = 0 \tag{9.121b}$$

$$(n^2 - n_e^2)E_z - n^2 \cos \gamma_0 (\sin \gamma_0 E_x + \cos \gamma_0 E_z) = 0 \tag{9.121c}$$

One solution of Eqs. (9.121) corresponds to the ordinary wave $\mathbf{E}_o$ and is obtained by setting

$$E_{ox} = E_{oz} = 0, \quad E_{oy} \neq 0, \quad n^2 = n_o^2 \tag{9.122}$$

The surface for $n(\hat{s})$ is a sphere of radius n_o. Here $\mathbf{E}_o$ is perpendicular to both the wave normal $\hat{s}$ and to the optic axis $\hat{z}$. The corresponding $\mathbf{D}$ field has components

$$D_{ox} = D_{oz} = 0, \quad D_{oy} = \varepsilon_0 n_o^2 E_{oy} \tag{9.123a}$$

Hence

$$\mathbf{D}_o = \varepsilon_0 n_o^2 \mathbf{E}_o \tag{9.123b}$$

The wave and ray velocities are from Eqs. (9.90) and (9.115)

$$\mathbf{v}_w = \mathbf{v}_r = \hat{s}\,\frac{c}{n_o} \tag{9.124}$$

A second solution corresponds to the extraordinary wave $\mathbf{E}_e$ and is obtained by setting

$$E_{ey} = 0, \quad E_{ex} \neq 0, \quad E_{ez} \neq 0 \tag{9.125}$$

and solving the two remaining equations, which can be written

$$(n_o^2 - n^2 \cos^2 \gamma_0)E_{ex} + (n^2 \sin \gamma_0 \cos \gamma_0)E_{ez} = 0 \tag{9.126a}$$

$$(n^2 \sin \gamma_0 \cos \gamma_0)E_{ex} + (n_e^2 - n^2 \sin^2 \gamma_0)E_{ez} = 0 \tag{9.126b}$$

For a solution to exist, the determinant of the coefficients of E_{ex} and E_{ey} must equal zero:

$$n_o^2 n_e^2 + n^4 \sin^2 \gamma_0 \cos^2 \gamma_0 - n^4 \sin^2 \gamma_0 \cos^2 \gamma_0 - n^2(n_o^2 \sin^2 \gamma_0 + n_e^2 \cos^2 \gamma_0) = 0$$

The second and third terms cancel, giving a first-order equation for n^2, which can be written

$$\frac{1}{n^2} = \frac{\sin^2 \gamma_0}{n_e^2} + \frac{\cos^2 \gamma_0}{n_o^2} \tag{9.127}$$

This then represents the equation for the e-wave index surface. For propagation along the optic axis ($\gamma_0 = 0°$, $180°$), the index n is n_o; for propagation in the x-y plane ($\gamma_0 = 90°$), the index n is n_e. The index surface is an ellipsoid of revolution that is tangent to the sphere defined by Eq. (9.122). The index surfaces for two examples are shown in Fig. 9.38.

The ray velocity for the extraordinary wave is provided through an application of Eq. (9.116). It is useful to manipulate this so as to arrive at the form of Eq. (9.33), which was stated without proof. With this in mind we form

$$\frac{1}{\mathbf{v}_r \cdot \mathbf{v}_r} = \frac{1}{v_r^2} = \frac{\mathbf{E} \varepsilon \mathbf{E}}{\mathbf{S} \cdot \mathbf{v}_r} \tag{9.128}$$

In the uniaxial case

$$\mathbf{E} \varepsilon \mathbf{E} = \varepsilon_0 n_o^2 E_{ex}^2 + \varepsilon_0 n_e^2 E_{ez}^2 \tag{9.129}$$

and from Eq. (9.112) we have

$$\mathbf{S} \cdot \mathbf{v}_r = n\varepsilon_0 c E_e^2 \hat{\mathbf{s}}_{\parallel} \cdot \mathbf{v}_r = n\varepsilon_0 c E_e^2 v_r \cos \Delta\gamma = \varepsilon_0 c^2 E_e^2 \tag{9.130}$$

where use has been made of Eq. (9.113) in the last step. Together these relations lead to

$$\frac{1}{v_r^2} = \frac{n_o^2}{c^2} \frac{E_{ex}^2}{E_e^2} + \frac{n_e^2}{c^2} \frac{E_{ez}^2}{E_e^2} \tag{9.131}$$

This can be simplified by recognizing that the ray velocity $\mathbf{v}_r$ is perpendicular to $\mathbf{E}$, thus

$$\frac{v_{rx}}{v_r} = \sin \gamma_e = \frac{E_{ez}}{E_e} \tag{9.132a}$$

and

$$\frac{v_{rz}}{v_r} = \cos \gamma_e = \frac{E_{ex}}{E_e} \tag{9.132b}$$

Using Eqs. (9.132) in Eq. (9.131) provides us with the final result

$$\frac{1}{v_r^2} = \frac{\cos^2 \gamma_e}{c^2/n_o^2} + \frac{\sin^2 \gamma_e}{c^2/n_e^2} \tag{9.133}$$

which has the same form as Eq. (9.128).

The ordinary ray velocity defines a surface that is a sphere; the extraordinary ray velocity identifies an ellipsoid of revolution. As mentioned before, the shape of

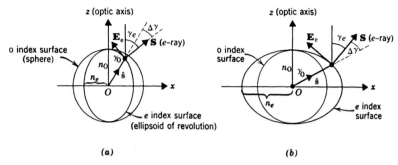

Fig. 9.38 Index surfaces for a uniaxial crystal: (*a*) negative uniaxial crystal ($n_e < n_0$); (*b*) positive uniaxial crystal ($n_e > n_0$).

the ellipsoid for the wave surface will be the inverse of the ellipsoid pertaining to the index surface. In other words, if the index surface ellipsoid is oblate with respect to the optic axis, then the ray velocity surface, which is the same as the Huygens construction wavefront, will be prolate. Examples of this are illustrated in Fig. 9.39 where parts (*a*) and (*b*) go with (*a*) and (*b*) in Fig. 9.38.

C. Biaxial Crystals

For biaxial anisotropic crystals, all three principal values of the dielectric tensor are different. This leaves us with the general form of Eq. (9.98) for ε. Equations (9.100) and (9.101) must be dealt with without further simplication in the general case in order to determine $n(\hat{s})$ and the fields **E**.

The treatment here will be limited to the case where the wave normal $\hat{s}$ lies in a plane perpendicular to one of the principal axes of the dielectric tensor. Assume for

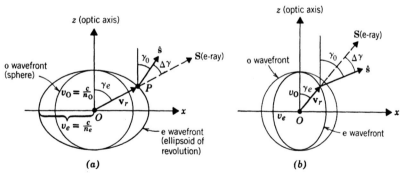

Fig. 9.39 Ray velocity surfaces corresponding to Fig. 9.38. (Wave surfaces or Huygens' wavelets at unit time after propagation from the origin.) (*a*) Negative uniaxial crystal ($v_0 < v_e$). (*b*) Positive uniaxial crystal ($v_0 > v_e$).

convenience that $n_x < n_y < n_z$, and suppose that ŝ lies in the y-z plane. Then ŝ = $(0, s_y, s_z)$, and Eqs. (9.100) give

$$(n^2 - n_x^2)E_x \qquad\qquad\qquad = 0 \qquad\qquad (9.134a)$$

$$(n_y^2 - n^2 s_z^2)E_y - n^2 s_y s_z E_z \quad = 0 \qquad\qquad (9.134b)$$

$$-n^2 s_y s_z E_y + (n_z^2 - n^2 s_y^2)E_z = 0 \qquad\qquad (9.134c)$$

One solution is simply

$$E_y = 0, \quad E_z = 0, \quad E_x \neq 0; \quad n = n_x \qquad\qquad (9.135)$$

The index surface in this case is a circle in the y-z plane. The **D** field has components

$$D_y = D_z = 0, \quad D_x = \varepsilon_0 n_x^2 E_x \qquad\qquad (9.136a)$$

therefore

$$\mathbf{D} = \varepsilon_0 n_x^2 \mathbf{E} \qquad\qquad (9.136b)$$

The ray velocity and the wave velocity are both along ŝ for this solution. The **D** and **E** directions are normal to the y-z plane, as shown in Fig. 9.40.

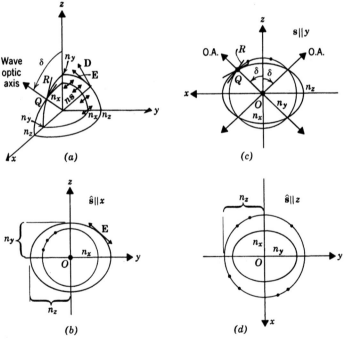

Fig. 9.40 Sections of the index surfaces for a biaxial crystal with $n_x < n_y < n_z$. The directions of **D** and **E** for the circles are out of the plane of the drawing as indicated by the dots.

Another solution of Eqs. (9.134) will have

$$E_x = 0, \quad E_y \neq 0, \quad E_z \neq 0 \tag{9.137}$$

Equations (9.134b, c) have the same form as Eqs. (9.126), and the solutions for $n(\hat{s})$ will have the same form as Eq. (9.127), namely,

$$\frac{1}{n^2} = \frac{s_y^2}{n_z^2} + \frac{s_z^2}{n_y^2} \tag{9.138}$$

Thus, for this solution when $\hat{s}$ is along the y axis, $n = n_z$, and when $\hat{s}$ is along the z axis, $n = n_y$. The **D** vectors will be in the y-z plane tangent to the ellipse formed by the intersection of the wave surface for this solution with that plane. The ellipse has semimajor axis n_z and semiminor axis n_y.

Similar results are obtained for $\hat{s}$ perpendicular to the y and x axes. In the former case, one of the index surfaces intersects the x-y plane in a circle of radius n_y. The other index surface intersects the plane in the ellipse.

$$\frac{1}{n^2} = \frac{s_z^2}{n_x^2} + \frac{s_x^2}{n_z^2} \tag{9.139}$$

Because $n_x < n_y < n_z$, this ellipse will intersect the circle of radius n_y at a point Q. For $\hat{s}$ along OQ, the two indices of refraction are both equal to n_y.

The direction OQ is one of the two *wave optic axes*. It makes an angle δ with the $+z$ axis that can be determined as follows. Let a bar over $\hat{s}$ indicate that it is along an optic axis. Then

$$\bar{s}_z = \cos \delta, \quad \bar{s}_x = \sin \delta, \quad \text{and} \quad n = n_y$$

Equation (9.139) then gives

$$\frac{1}{n_y^2} = \frac{\cos^2 \delta}{n_x^2} + \frac{\sin^2 \delta}{n_z^2} = \cos^2 \delta \left(\frac{1}{n_x^2} - \frac{1}{n_z^2} \right) + \frac{1}{n_z^2}$$

or.

$$\cos^2 \delta = \frac{n_y^{-2} - n_z^{-2}}{n_x^{-2} - n_z^{-2}} \tag{9.140}$$

From this result we readily obtain

$$\sin^2 \delta = \frac{n_x^{-2} - n_y^{-2}}{n_x^{-2} - n_z^{-2}} \tag{9.141}$$

$$\tan^2 \delta = \frac{n_x^{-2} - n_y^{-2}}{n_y^{-2} - n_z^{-2}} \tag{9.142}$$

If δ is less than 45°, the crystal is said to be a *positive biaxial crystal*. If δ is greater than 45°, the crystal is said to be *negative*. The other wave optic axis is also in the x-z plane. It makes an angle δ with the z axis. When δ is zero or 90°, the two optic axes coincide, and the crystal becomes a positive or negative uniaxial crystal, respectively.

Figure 9.40 enables us to visualize one octant of the index surfaces for a biaxial medium. The full surfaces may be obtained by reflection in the x-y, y-z, and z-x planes. Note that there are two interpenetrating sheets that meet at the optic axes. Figures 9.40b, c, and d show the intersections of the full surfaces with each coordinate plane. Note that these intersections consist of a circle and an ellipse. The circle corresponds to light with $\mathbf{E}$ and $\mathbf{D}$ perpendicular to the plane, the ellipse to light with $\mathbf{E}$ and $\mathbf{D}$ in the plane.

REFERENCES

Arfken, G. *Mathematical Methods for Physicists.* Academic Press, New York, 1970.

Azzam, R. M. A., and N. M. Bashara. *Ellipsometry and Polarized Light.* North Holland, New York, 1977.

Beran, M. J., and G. B. Parrent, Jr. *Theory of Partial Coherence.* Prentice-Hall, Englewood Cliffs, N. J., 1964.

Born, Max, and Emil Wolf. *Principles of Optics.* Pergamon Press, Oxford, 1980.

Cagnet, Michel, Maurice Françon, and Jean-Claude Thrierr. *Atlas of Optical Phenomena.* Springer-Verlag, Berlin, 1962.

Clarke, David. *Polarized Light and Optical Measurement.* Pergamon Press, Oxford, 1971.

Driscoll, Walter G., and William Vaughan. *Handbook of Optics.* McGraw-Hill, New York, 1978.

Rossi, Bruno. *Optics.* Addison-Wesley, Reading, Mass., 1957.

Shurecliff, W. A., and S. S. Balard. *Polarized Light.* Van Nostrand, Princeton, N.J., 1964.

Wood, Elizabeth A. *Crystals and Light: An Introduction to Optical Crystallography.* Van Nostrand, Princeton, N.J., 1964.

PROBLEMS

Section 9.1 Polarized Light

1. Describe semiquantitatively with a sketch the state of polarization of light waves obeying the following equations:

(a) $E_x = A \cos(\omega t - kz)$

$E_y = 2A \cos(\omega t - kz)$

(b) $E_x = A \cos\left(\omega t - kz + \dfrac{\pi}{4}\right)$

$E_y = A \cos(\omega t - kz)$

(c) $E_x = A \cos\left(\omega t - kz - \dfrac{\pi}{4}\right)$

$E_y = \left(\dfrac{A}{2}\right)\cos(\omega t - kz)$

2. Show that $A_L^2 + A_R^2 = \tfrac{1}{2}(A_x^2 + A_y^2)$

3. Show that, in general, the superposition of two mutually coherent quasimonochromatic beams of elliptically polarized light will yield a beam of elliptically polarized light.

4. **(a)** Show that circularly polarized light can be represented as the coherent superposition of two linearly polarized waves vibrating at right angles.

(b) Show that linearly polarized light can be represented as the coherent superposition of right circularly polarized light and left circularly polarized light.

(c) Show further that changing the sum of the phases of the RCP and LCP light in **(b)** changes the plane of vibration of the resultant linearly polarized light.

5. Find numerical values for A_L, A_R, ϕ_L, and ϕ_R, sketch the ellipse, and indicate its sense of rotation for light described by

$$E_x = A_x \cos \phi_0$$

$$E_y = A_y \cos(\phi_0 + \phi)$$

with

(a) $\phi = \dfrac{\pi}{2}$ $A_y = \dfrac{1}{2} A_x$

(b) $\phi = \dfrac{\pi}{4}$ $A_y = A_x$

(c) $\phi = \dfrac{\pi}{4}$ $A_y = \dfrac{1}{2} A_x$

(d) $\phi = \dfrac{5\pi}{4}$ $A_y = A_x$

(e) $\phi = \dfrac{-\pi}{6}$ $A_y = A_x$

6. In the general case of elliptically polarized light specified by $E_x = A_x \cos \phi_0$, $E_y = A_y \cos (\phi_0 + \phi)$ show that the polarization ellipse obeys the following formula

$$\left(\frac{E_x}{A_x}\right)^2 + \left(\frac{E_y}{A_y}\right)^2 - 2 \frac{E_x E_y}{A_x A_y} \cos \phi = \sin^2 \phi$$

7. In Fig. 9.7 the general polarization ellipse is rotated by $\beta/2$ with respect to the coordinate axes. Define new rectangular coordinate axes which are rotated such that in the new system the field components are $E_{\bar{x}}$ and $E_{\bar{y}}$ where

$$E_{\bar{x}} = E_x \cos\left(\frac{\beta}{2}\right) + E_y \sin\left(\frac{\beta}{2}\right)$$

$$E_{\bar{y}} = -E_x \sin\left(\frac{\beta}{2}\right) + E_y \cos\left(\frac{\beta}{2}\right)$$

Show that in the rotated coordinate system the polarization ellipse takes the form

$$\left(\frac{E_{\bar{x}}}{A_{\bar{x}}}\right)^2 + \left(\frac{E_{\bar{y}}}{A_{\bar{y}}}\right)^2 = 1$$

8. Prove that the orientation of the general polarization ellipse with respect to the coordinate axes x and y is given by $\beta/2$ with

$$\tan(\beta) = \tan(2\chi) \cos \phi$$

where $\tan \chi = A_y/A_x$ and $\phi = \phi_y - \phi_x$.

9. Using the general equation for elliptically polarized light, which is given in Problem 6, verify that linear polarization and that both right and left circular polarization are special cases. Determine, in the general case, the points where the polarization ellipse touches the rectangle that is defined by $E_x = \pm A_x$, $E_y = \pm A_y$.

10. Derive Eq. (9.23) for the circular representation of the field phasor in terms of the sum and difference angle α and β.

11. Linearly polarized light is incident at an angle of $65°$ on a crystal of gallium phosphide. The wavelength is 288 nm, which corresponds to a complex index of refraction of $(3.862) - i(1.481)$. Specify the orientation and eccentricity of the polarization ellipse after reflection if the incident light is polarized $45°$ with respect to the plane of incidence.

Section 9.2 Polarization-Sensitive Optical Elements

12. Prove that for a plane parallel slab if Brewster's condition is satisfied at the top interface it will be satisfied also for internal incidence onto the second interface.

13. A pile of six parallel plates is used at Brewster's angle. If $n = 1.62$, find the fraction of the transmitted light that is polarized in the plane of incidence and perpendicular to the plane of incidence. Assume that unpolarized radiation is incident. Remember that multiple reflection will play a role; however, you may assume that the multiple reflections are incoherent.

14. Consider total internal reflection of an incident light beam linearly polarized at $45°$ to the plane of incidence of a dielectric interface. Assume that what we have here called the x-axis is associated with light polarized perpendicular to the plane of incidence. Using the formalism of Chapter 2, show that the relative phase shift $\phi = \phi_y - \phi_x$ obeys the equation

$$\tan \frac{\phi}{2} = \gamma \frac{\cos^2 \theta}{\sin^2 \theta}, \quad \gamma = \frac{\sqrt{n^2 \sin^2 \theta - n'^2}}{n \cos \theta}$$

Show that the relative phase shift is a maximum when

$$\cos \theta = \sqrt{\frac{n^2 - n'^2}{n^2 + n'^2}}$$

15. With reference to the previous problem, consider the prism shown in Fig. 9.41, for which two internal

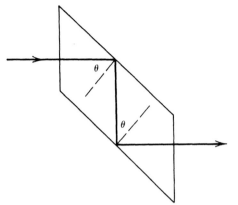

Fig. 9.41

reflections occur. This doubles the relative phase shift, producing a net shift of 2ϕ. When $2\phi = \pi/2$, the device is called a *Fresnel rhomb*. It then functions like a quarter-wave plate. With $n = 1$ and n' as a parameter, find the minimum value of n' and the corresponding angle of incidence that can be used in such a rhomb.

16. Derive Eq. (9.33) from Eq. (9.32).

17. The indices of refraction for calcite and quartz for the sodium line are

$$\text{Calcite:} \quad n_o = 1.654, \quad n_e = 1.486$$

$$\text{Quartz:} \quad n_o = 1.544, \quad n_e = 1.553$$

Calculate the thickness of a quarter-wave plate made from these materials and say whether the optic axis is a fast or slow axis of the plate.

18. A point source of yellow light is place within a calcite crystal. Determine the shape and find the size of the spherical and the ellipsoidal wavefronts that exist 10^{-11} sec later. Use the data from Problem 17.

19. Discuss the action of a "half-wave" plate (for which $|n_e - n_o|d = \lambda/2$) on various types of polarized light.

20. Let the incident beam have polarization parameters $A_L/A_R \neq 1$ and $\alpha = \phi_L - \phi_R$, as in Fig. 9.42. Show that there are four values of ψ_1 for which extinction can be obtained with the polarizer, and find the resulting value of ψ_1 and ψ_2.

21. A plane-parallel plate is cut from a calcite crystal to be 3-cm thick with the optic axis parallel to the faces of the plate. Light that is unpolarized at a wavelength of 589 nm (sodium light) is incident on the face of the crystal at an angle of 45°. Compute the separation of the ordinary and the extraordinary rays as they emerge from the plate on the opposite side.

22. Linearly polarized white light is normally incident on a quartz plate that is 0.865-mm thick. The optic axis is parallel to the faces of the plate and 45° with respect to the orientation of the incident polarization. Assuming that we can use the indices given in Problem 17 for quartz and that these are independent of wavelength (only an approximation), which wavelengths in the band from 600 to 700 nm will emerge circularly polarized? Which wavelengths emerge linearly polarized along the direction of the incident light?

23. Compute the angle through which linearly polarized light is rotated in traveling through a 3-mm thick quartz plate cut with faces perpendicular to the optic axis. The incident light has a wavelength of 397 nm, which corresponds to indices of 1.55821 and 1.55810 for left and right circular polarization, respectively.

24. Consider a quarter-wave plate designed for wavelength λ_0. Discuss its behavior for wavelengths λ near λ_0, that is, for small $\varepsilon = [(\lambda_0/\lambda) - 1]$. Imagine the use of this plate in the arrangement of Problem 9.20. Show

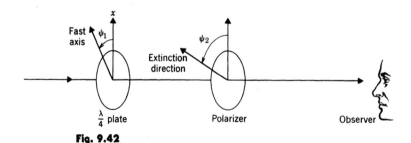

Fig. 9.42

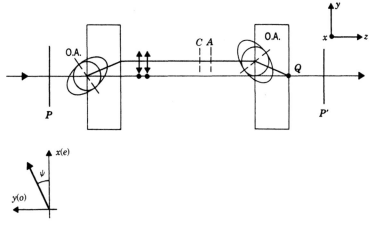

Fig. 9.43

that for $\varepsilon = \pm 0.1$ and for circularly polarized incident light the ratio of maximum to minimum transmitted flux density (as a function of the angle ψ_2) is 164.

Section 9.3   Partially Polarized Light

25. Partially polarized light can be expressed in a unique way as the incoherent superposition of two elliptically polarized beams. The ellipses have the same ellipticity, opposite rotation sense, and major axes oriented at right angles.

(a) Show that we can write

$$\mathbf{J} = A_1 \begin{pmatrix} \cos^2 \chi & e^{-i\phi} \sin \chi \cos \chi \\ e^{i\phi} \sin \chi \cos \chi & \sin^2 \chi \end{pmatrix}$$
$$+ A_2 \begin{pmatrix} \sin^2 \chi & -e^{-i\phi} \sin \chi \cos \chi \\ e^{i\phi} \sin \chi \cos \chi & \cos^2 \chi \end{pmatrix}$$

where the phase is defined by $J_{xy} = |J_{xy}|e^{-i\phi}$, where $A_{1,2}$ are defined in Eq. (9.73) and B and C in Eqs. (9.77) and (9.78). Then $\chi = \tan^{-1}(C/B)$.

(b) Then justify the statements made at the beginning of this problem.

26. Two calcite plates may be used to isolate spatially the e ray from the o ray as shown in Fig. 9.43. The thicknesses of the two plates are the same, and their optic axes have opposite tilt angles. The e ray has electric field vibrations in the plane of the drawing along the x direction, and the o ray has vibrations perpendicular to the plane of the drawing along the y direction. The second plate brings the two beams

together at Q. The symbol C is a transparent plate used as a phase shifter to make the total optical path lengths for o and e rays nearly the same, in particular to make the optical path difference less than the longitudinal coherence length for the light. Note that because of its larger wave surface and greater velocity, it is the e ray that must be retarded to match the optical path lengths. Let $\Delta\phi$ be the net resulting phase difference $\Delta\phi = \phi_e - \phi_o$.

(a) Crossed polarizers are placed at P and P' having pass angles $\psi = \pi/4$ and $\psi' = 3\pi/4$. Let S_m be the maximum transmitted flux density as a function of $\Delta\phi$. Then $\Delta\phi$ is adjusted to give zero-transmitted flux density. One of the two beams in the region between the plates is blocked out. Light of flux density $S_m/4$ is now transmitted! Explain.

(b) Let partially polarized light with a general coherence matrix

$$\mathbf{J} = \begin{pmatrix} J_{xx} & J_{xy} \\ J_{yx} & J_{yy} \end{pmatrix}$$

be incident from the left. In addition to the phase shifter C, an absorber A is now inserted in the e beam. It produces no phase shift but changes the amplitude by the factor $\tau_0 < 1$. What is the resulting coherency matrix $\mathbf{J}'$ at Q for general values of $\Delta\phi$ and τ_0?

27. Consider unpolarized laser light with the following properties.

(a) Coherence time τ long enough to be readily accessible to direct measurement, say 10^{-3} sec.

(b) Total intensity that is constant on a time scale much less than the coherence time (but much longer than the period of the light $2\pi/\omega$). Discuss qualitatively the nature of the flux-density fluctuations observed when this light is passed through a polarizer.

28. Determine the coherency matrix for partially polarized light having the parameters defined in section 9.D2. Write the components of the matrix first with respect to the $\bar{x}$-$\bar{y}$ coordinate system (Fig. 9.22). Then rotate coordinates and calculate the elements of the matrix with respect to the x-y system.

29. A diffraction grating is covered with two polaroid sheets with mutually perpendicular pass directions oriented at 45° to the grating rulings (Fig. 9.44). What does this do to the ultimate resolution of the grating for

    (a) Light polarized along one of the pass directions.

    (b) Light polarized parallel to the rulings.

    (c) Light polarized perpendicular to the rulings.

    (d) Unpolarized light.

Section 9.4 Crystal Optics

30. Prove that the values of n^2 that solve Eqs. (9.99) and (9.100) must be positive. (*Hint:* Eqs. (9.99) and (9.100) are equivalent to

$$\mathbf{D} = \varepsilon\mathbf{E} = n^2\mathbf{E}_\perp$$

the tensor ε is *positive definite*, which means that

$$\sum_{i,j} \varepsilon_{ij}E_iE_j > 0$$

unless $E_i = 0$ for all i.)

31. Solve Eq. (9.105) to find the two positive roots n_1^2 and n_2^2.

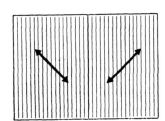

Fig. 9.44

32. Show that the two linearly polarized waves for a given wave normal $\hat{s}$ have electric fields $\mathbf{E}_1$ and $\mathbf{E}_2$ that are in general *not* perpendicular to each other.

33. Describe in detail how an index surface $n(\hat{s})$ allows us to find all quantities $\mathbf{D}$, $\mathbf{E}$, $\mathbf{H}$, and $\mathbf{S}$, given $\hat{s}$ and the magnitude of $\mathbf{E}$.

34. Prove Eq. (9.113), which gives the ray velocity in terms of the wave velocity.

35. Derive Eq. (9.138) from Fresnel's equation, Eq. (9.105), with $\hat{s}_x = 0$.

Problems 36 through 38 refer to anisotropic crystals cut into prisms to be used in a spectrometer such as that of Fig. 6.40 or a similar instrument with nearly parallel light. The plane of incidence is perpendicular to the apex AC in Fig. 9.45.

36. Show that when a uniaxial crystal is made into a prism with its apex AC parallel to the optic axis, both ordinary and extraordinary rays obey the ordinary law of refraction with indices n_o and n_e, respectively.

37. Consider a prism made from a uniaxial crystal with its optic axis parallel to AB. Show that light with $\mathbf{E}$ polarized perpendicular to the plane of incidence is refracted in the usual way with index n_o. Show that light with $\mathbf{E}$ vibrating in the plane of incidence behaves in a more complicated fashion unless the deviation θ_D is a minimum $\theta_{D,\min}$, at which point the simple law of refraction holds with index n_e given by Eq. (2.98)

$$n_e = \frac{\sin\left[\dfrac{(\alpha + \theta_{D,\min})}{2}\right]}{\sin(\alpha/2)}$$

38. If a biaxial crystal is made into a prism as shown in Fig. 9.45 with the apex AC parallel to the jth principal axis of the dielectric tensor, show that light polarized parallel to AC will obey the simple form of Snell's law

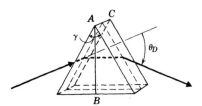

Fig. 9.45

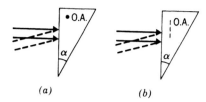

(a)   (b)

Fig. 9.46

with index n_j. If, in addition, another principal axis, say the kth, is oriented parallel to AB, then light polarized in the plane of incidence will have an angle θ_D of minimum deviation satisfying

$$n_k = \frac{\sin\left[\dfrac{(\alpha + \theta_{D,\min})}{2}\right]}{\sin(\alpha/2)}$$

Show this.

39. Show that in a uniaxial crystal the largest angle $\Delta\gamma = \gamma_e - \gamma_o$ between the ray direction and the direction of the wave normal $\hat{s}$ occurs when $\hat{s}$ makes an angle with the optic axis of

$$\gamma_o = \tan^{-1}\left(\frac{n_e}{n_o}\right)$$

and that

$$\tan \Delta\gamma_{\max} = \frac{n_o^2 - n_e^2}{2 n_e n_o}$$

40. Show that it is possible to make a quarter-wave plate to use near normal incidence from a thin slice of a biaxial crystal. Thin sheets of mica oriented almost exactly parallel to the y-z plane may be prepared easily by cleaving. What thickness should such a sheet have to

make a quarter-wave plate for light with wavelength 500 nm if the principal indices are $n_x = 1.5601$, $n_y = 1.5936$, and $n_z = 1.5877$?

41. Consider the calcite prisms in Fig. 9.46 used with unpolarized light at normal incidence as shown. Use $n_o = 1.668$ and $n_e = 1.491$ and calculate the range of α for which only one polarized component is critically reflected at the second surface. Which one is it? Are prisms (a) and (b) equivalent in this regard? Are they equivalent for the dashed rays?

42. A parallel unpolarized light beam (Fig. 9.47) is incident normally onto one face of a turquoise crystal, which is biaxial and has principal indices of $n_x = 1.520$, $n_y = 1.523$, and $n_z = 1.530$. With $\alpha = 30°$, calculate the angles of deviation for the following ways of cutting the prism: $AB//x$ and $BC//y$; with $AB//y$ and $BC//z$; and with $AB//z$ and $BC//x$. With each of these orientations, find the range of α for which only one of the two mutually perpendicularly polarized rays is critically reflected.

43. Consider Rochon and Wollaston prisms such as those of Fig. 9.19 made of quartz for which $n_o = 1.54467$ and $n_e = 1.55379$. With unpolarized light incident normally, calculate the angles made by both ordinary and extraordinary rays with the normal to the second surface after they have emerged from it ($\alpha = 35°$).

44. Consider an air-spaced polarizing prism made of calcite, Fig. 9.48. It works by totally reflecting the ordinary ray at the air gap, while passing the extraordinary ray. There is only a finite range of the angle of incidence θ for which both of these effects occur. The prism is designed so that if the o ray is incident below the normal at an angle greater than θ_o, it ceases to be

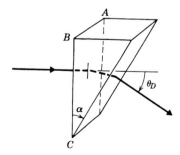

Fig. 9.47

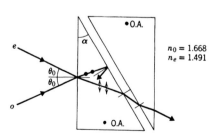

Fig. 9.48

totally reflected, whereas if the *e* ray is incident above the normal at an angle greater than θ_o, it ceases to be transmitted and is also totally reflected. Use these two criteria to determine the maximum angular spread $2\theta_o$ and the prism angle α.

45. When the air gap of the prism of Problem 44 is replaced by linseed oil ($n = 1.494$), the angular aperture $2\theta_o$ can be considerably increased provided that the angle α is made much larger (Fig. 9.49). This gives the

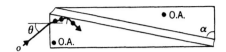

Fig. 9.49

Glan-Thompson polarizer of Fig. 9.19. With $\alpha = 76.4°$, find the maximum value of the angle of incidence for the *o* ray for which it will still be totally reflected.

Appendix:
The Fresnel–Kirchhoff
Integral

Here we will put the theory of diffraction on a mathematical foundation firmer than that provided by the intuitive approach of Chapter 6. The line of reasoning, which was developed by Kirchhoff, is based ultimately on the time-independent wave equation.

To deemphasize the time dependence of the optical field, let us recall that, through the superposition principle, we can add components having different frequencies to produce a polychromatic diffraction effect. Assume that we are then dealing with one monochromatic component at a time. This has the form

$$\text{Physical optical electric field} = \text{Re}(E e^{i\omega t})$$

Here E contains only the spatial dependence of the optical field (unlike in Chapter 6 where we used the same notation to represent the time dependence as well). If we use this form in the time-dependent three-dimensional wave equation, we arrive at

$$\nabla^2 E + k^2 E = 0 \qquad \qquad \text{(A.1)}$$

where $k = 2\pi/\lambda = \omega/v$. This is called the *Helmholtz equation*. Its solutions contain only the spatial dependence of the physical optical field. Because, in the end, we will time average the field to obtain the measured flux density, the solutions to the Helmholtz equation contain all of the important information for diffraction solutions at our level. In what follows we use $E(\mathbf{r})$ to describe the optical field at the source, $E(\bar{\mathbf{r}})$ to describe the field at a point $\tilde{P}$ on an integration surface in the

647

presence of the aperture, and $E(\mathbf{r}')$ to describe the field at the observation point P'. In this notation $R' = |\mathbf{r}' - \tilde{\mathbf{r}}|$ and $R = |\tilde{\mathbf{r}} - \mathbf{r}|$. We also will need ray direction unit vectors $\hat{\mathbf{n}}' = (\mathbf{r}' - \tilde{\mathbf{r}})/R'$ and $\hat{\mathbf{n}} = (\tilde{\mathbf{r}} - \mathbf{r})/R$.

A. Derivation of the Fresnel–Kirchhoff Integral from the Helmholtz–Kirchhoff Theorem

The basic integral of the Fresnel-Kirchhoff scalar theory of direction that is used in this book is obtained by combining Eqs. (6.7), (6.8), and (7.37) (in the new notation)

$$E(\mathbf{r}') = \frac{i}{\lambda} \iint_{\sigma_o} E_{\text{inc}}(\tilde{\mathbf{r}}) \frac{e^{-ikR'}}{R'} \frac{(\cos\theta_\eta - \cos\theta'_\eta)}{2} \, d\tilde{\sigma} \tag{A.2}$$

We now discuss the extent to which this equation can be obtained from first principles. To what more basic equation is it an approximation? It is an approximation to the following equation, known as the Helmholtz-Kirchhoff theorem:

$$E(\mathbf{r}') = \frac{1}{4\pi} \oiint_{\sigma} \left[\frac{e^{-ik|\mathbf{r}' - \tilde{\mathbf{r}}|}}{|\mathbf{r}' - \tilde{\mathbf{r}}|} \frac{\partial E(\tilde{\mathbf{r}})}{\partial n_\sigma} - E(\tilde{\mathbf{r}}) \frac{\partial}{\partial n_\sigma} \frac{e^{-ik|\mathbf{r}' - \tilde{\mathbf{r}}|}}{|\mathbf{r}' - \tilde{\mathbf{r}}|} \right] d\tilde{\sigma} \tag{A.3}$$

This is a rigorous mathematical expression following from the wave equation in the nonretarded limit.

Equation (A.3) gives the electric field at a point $\mathbf{r}'$ in terms of the field $E(\tilde{\mathbf{r}})$ and its normal derivative evaluated on a closed surface σ that either surrounds the point $P'(\mathbf{r}')$ (Fig. A.1b) or excludes P' (Fig. A.1a). The normal $\mathbf{n}_\sigma$ is assumed to point away from the region that contains P'. Furthermore, the region containing P' must be free of any sources of optical radiation. This includes secondary sources of reradiated or scattered light. Thus, strictly speaking, the region must be free of optical material of any kind. This means the sources must be on one side of σ and P' must be on the other side. We will derive Eq. (A.3) in the section that follows.

It is important to realize that $E(\tilde{\mathbf{r}})$ and $\partial E(\tilde{\mathbf{r}})/\partial n_\sigma$ represent the field and its derivative at the closed surface σ *in the presence of the diffracting aperture*, if any. To obtain Eq. (A.2) from Eq. (A.3), we must let σ include σ_o, the open part of the diffracting aperture. We also let it include σ_c, a surface just in back of the nontransmitting part of the diffracting aperture. Because σ must be a closed surface, we must add a mathematical surface σ_a or σ_b as shown in Fig. A.2. Thus, corresponding to Fig. A.1a, we have

$$\sigma = \sigma_o + \sigma_c + \sigma_a$$

and corresponding to Fig. A.1b we have

$$\sigma = \sigma_o + \sigma_c + \sigma_b$$

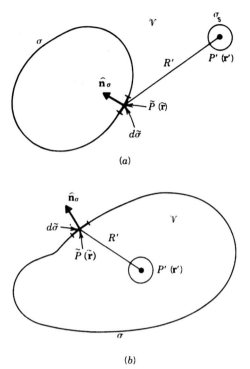

(a)

(b)   **Fig. A.1**

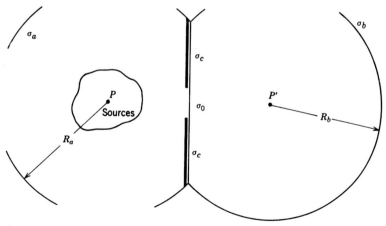

Fig. A.2

Here σ_a is useful if all the primary and secondary optical sources except the aperture are located in a finite region of space. It can be taken to be part of a large sphere of radius R_a about these sources; σ_b can be part of a similar sphere about P'. We can show that with the type of physical electric fields encountered in optics, the integrals in Eq. (A.3) over σ_a or σ_b will tend to zero as R_a and R_b tend to infinity.

It remains to integrate Eq. (A.3) over $\sigma_o + \sigma_c$. The actual field on these surfaces is almost impossible to determine except in the simplest cases. This is partly because the boundary conditions for the optical field and its derivatives are apt to be complicated for real physical apertures, which will partially absorb, partially reflect, and partially scatter the incident light. The first and key approximation that we make to go from Eq. (A.3) to Eq. (A.2) "cuts the Gordian knot" posed by the real boundary conditions by applying the so-called *Kirchhoff boundary conditions*. These replace the actual field in the presence of the aperture by the incident field wherever the aperture is open and by zero wherever the aperture is nontransmitting. Thus we assume

$$E(\tilde{r}) = 0 \tag{A.4a}$$

$$\frac{\partial E(\tilde{r})}{\partial n_\sigma} = 0 \tag{A.4b}$$

for $\tilde{r}$ on σ_c and

$$E(\tilde{r}) = E_{\text{inc}}(\tilde{r}) \tag{A.5a}$$

$$\frac{\partial E(\tilde{r})}{\partial n_\sigma} = \frac{\partial E_{\text{inc}}(\tilde{r})}{\partial n_\sigma} \tag{A.5b}$$

for $\tilde{r}$ on σ_o, the open part of σ where $E_{\text{inc}}(\tilde{r})$ is the field at that position in space in the absence of any apertures. Then Eq. (A.3) becomes

$$E(\mathbf{r}') = \frac{1}{4\pi} \iint_{\sigma_o} \left[\frac{e^{-ik|\mathbf{r}'-\mathbf{r}|}}{|\mathbf{r}'-\tilde{r}|} \frac{\partial E_{\text{inc}}(\tilde{r})}{\partial n_\sigma} - E_{\text{inc}}(\tilde{r}) \frac{\partial}{\partial n_\sigma} \frac{e^{-ik|\mathbf{r}'-\tilde{r}|}}{|\mathbf{r}'-\tilde{r}|} \right] d\tilde{\sigma} \tag{A.6}$$

Equation (A.6) is unnecessarily complicated for use with light. A simplification occurs if the incident light has a well-defined direction of propagation $\hat{n}$ at the point $\tilde{P}$ on the surface σ_o. This would be the case if the light were coming from a point source P (Fig. A.3). Then the gradient of E at $\tilde{P}$ is in the direction of $\hat{n}$, and we can write

$$\frac{\partial E_{\text{inc}}(\tilde{r})}{\partial n_\sigma} = \hat{n}_\sigma \cdot \nabla E_{\text{inc}}(\tilde{r}) = \frac{(\hat{n}_\sigma \cdot \hat{n}) \partial E_{\text{inc}}}{\partial n_\sigma} = -\cos\theta_\eta \frac{\partial E_{\text{inc}}}{\partial n_\sigma} \tag{A.7}$$

where θ is the angle between $(-\hat{n}_\sigma)$ or $\hat{\eta}$ and $\hat{n}$. The other normal derivative appearing in Eq. (A.3) can also be rewritten. Recall $R' = |\mathbf{r}' - \tilde{r}|$. Then we have

$$\frac{\partial}{\partial n_\sigma} \frac{e^{-ikR'}}{R'} = \cos\theta'_\eta \frac{\partial}{\partial R'} \left(\frac{e^{-ikR'}}{R'} \right) = -\cos\theta'_\eta \left(ik + \frac{1}{R'} \right) \frac{e^{-ikR'}}{R'} \tag{A.8}$$

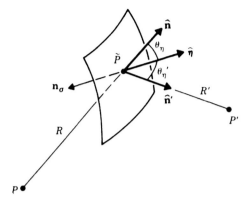

Fig. A.3

Here θ_η' is the angle between $(\mathbf{r}' - \tilde{\mathbf{r}})$ or $\hat{\mathbf{n}}'$ and $(-\hat{\mathbf{n}}_\sigma)$ or $\hat{\boldsymbol\eta}$. For light waves the distance R' from $\tilde{P}(\tilde{\mathbf{r}})$ to $P'(\mathbf{r}')$ is usually much larger than the wavelength $\lambda = 2\pi/k$. Thus

$$\frac{1}{\lambda} \gg \frac{2}{R'} \quad \text{and} \quad k = \frac{2\pi}{\lambda} \gg \frac{1}{R'}$$

and the second term in Eq. (A.8) is negligible compared with the first. This gives

$$\frac{\partial}{\partial n_\sigma} \frac{e^{-ikR'}}{R'} \approx -ik \cos\theta_\eta' \frac{e^{-ikR'}}{R'} \tag{A.9}$$

We can simplify Eq. (A.7) in a similar way. Light with a well-defined direction $\hat{\mathbf{n}}$ of propagation at $\tilde{P}$ will have the approximate local spatial dependence

$$E_{\text{inc}}(\tilde{\mathbf{r}}) \approx \text{const} \cdot e^{-ikR}$$

where R is a distance measured in the direction of $\hat{\mathbf{n}}$. Hence

$$\frac{\partial}{\partial n_\sigma} E_{\text{inc}}(\tilde{\mathbf{r}}) = \frac{\partial E_{\text{inc}}}{\partial R} \approx -ik E_{\text{inc}}(\tilde{\mathbf{r}}) \tag{A.10}$$

Equation (A.10) is valid provided that the source P is very many wavelengths from the surface σ_o.

We may use Eqs. (A.9) and (A.10) to write Eq. (A.6) in the form

$$E(\mathbf{r}) = \frac{i}{\lambda} \iint_{\sigma_o} \frac{e^{-ikR'}}{R'} E_{\text{inc}}(\tilde{\mathbf{r}}) \left(\frac{\cos\theta_\eta + \cos\theta_\eta'}{2} \right) d\tilde{\sigma}$$

which is Eq. (A.2). The angle θ_η' is well defined in all cases, but θ_η, the angle between the direction of propagation from the source and $\hat{\boldsymbol\eta}$, may not be well defined if a spatial distribution of sources is present instead of just one at P. If this distribution is present, the integral in Eq. (A.2) will have to be replaced by a sum of integrals, with a different source point and a different value of $\cos \theta'_n$ appearing in each one. Fortunately, this extra complexity is unnecessary when the range of values of θ'_n is small.

B.   Derivation of the Helmholtz–Kirchhoff Theorem from the Wave Equation

The derivation of Eq. (A.3) starts with Green's theorem, which can be written

$$\iiint_{\mathscr{V}} (\psi \nabla^2 \phi - \phi \nabla^2 \psi)\, d\mathscr{V} = \oiint_{\sigma_t} \left(\psi \frac{\partial \phi}{\partial n_\sigma} - \phi \frac{\partial \psi}{\partial n_\sigma} \right) d\sigma$$

Here the volume integral extends throughout a volume $\mathscr{V}$, which is bounded by the surface σ_t. The normal derivatives $\partial \phi / \partial n_\sigma$ and $\partial \psi / \partial n_\sigma$ are taken with respect to the outward normal of σ_t. Then ψ and ϕ may be arbitrary functions of position that are sufficiently well behaved. Green's theorem may be proved by applying Gauss's theorem.

$$\iiint_{\mathscr{V}} \nabla \cdot \mathbf{A}\, d\mathscr{V} = \oiint_{\sigma_1} \mathbf{A} \cdot d\boldsymbol{\sigma}$$

successively to $\mathbf{A} = \psi \nabla \phi$ and $\mathbf{A}' = \phi \nabla \psi$ and subtracting.

We choose $\phi(\tilde{\mathbf{r}}) = E(\tilde{\mathbf{r}})$ and choose the volume $\mathscr{V}$ to contain no primary or secondary sources of the electric field, so that the free-space time-independent wave equation

$$\nabla^2 \phi = -k^2 \phi$$

holds there. We also set

$$\psi(\tilde{\mathbf{r}}) = \frac{e^{-ik|\mathbf{r}' - \tilde{\mathbf{r}}|}}{|\mathbf{r}' - \tilde{\mathbf{r}}|}$$

From Chapter 1 we know that this is a spherically symmetric solution of our wave equation, so that

$$\nabla^2 \psi = -k^2 \psi$$

except when $\tilde{\mathbf{r}} = \mathbf{r}'$, and we must exclude this point from our integration volume $\mathscr{V}$. We do this by surrounding the point $P'(\mathbf{r}')$ by a small sphere of radius ε, and exclude this volume from $\mathscr{V}$. The surface σ_t is then given by the sum $\sigma + \sigma_s$, where σ has already been defined and where σ_s is the surface of the small sphere, as shown in Fig. A.1. This choice of $\mathscr{V}$ and of σ guarantees that both ϕ and ψ satisfy the Helmholtz equation everywhere in $\mathscr{V}$. Thus, in $\mathscr{V}$ we have

$$\psi \nabla^2 \phi - \phi \nabla^2 \psi = k^2 \psi \phi - k^2 \phi \psi = 0$$

Then Green's theorem gives

$$0 = \oiint_{\sigma_t} \left(\psi \frac{\partial \phi}{\partial n_\sigma} - \phi \frac{\partial \psi}{\partial n_\sigma} \right) d\tilde{\sigma} = \oiint_{\sigma} \left(\psi \frac{\partial \phi}{\partial n_\sigma} - \phi \frac{\partial \psi}{\partial n_\sigma} \right) d\tilde{\sigma}$$

$$+ \oiint_{\sigma_t} \left(\psi \frac{\partial \phi}{\partial n_\sigma} - \phi \frac{\partial \psi}{\partial n_\sigma} \right) d\tilde{\sigma}$$

We now explicitly evaluate the integral over the little sphere σ_s'. The sphere is outside the region $\mathscr{V}$; thus the outward normal to σ_s' is along its radius, directed inward toward $P'(\mathbf{r}')$. We can write

$$\oiint_{\sigma_s'} \left(\psi \frac{\partial \phi}{\partial n_\sigma} - \phi \frac{\partial \psi}{\partial n_\sigma} \right) d\tilde{\sigma} = \oiint_{\sigma_s'} \left[\frac{e^{-ik\varepsilon}}{\varepsilon} \frac{\partial E}{\partial n_\sigma} + E \left(-ik - \frac{1}{\varepsilon} \right) \frac{e^{-ik\varepsilon}}{\varepsilon} \right] d\tilde{\sigma}$$

$$= \oiint_{\sigma_s'} \frac{e^{-ik\varepsilon}}{\varepsilon} \frac{\partial E}{\partial n_\sigma} d\tilde{\sigma} + \oiint_{\sigma_s'} (-ik) E \frac{e^{-ik\varepsilon}}{\varepsilon} d\tilde{\sigma}$$

$$+ \oiint_{\sigma_s'} \left(-\frac{1}{\varepsilon^2} \right) E\, e^{-ik\varepsilon}\, d\tilde{\sigma}$$

The value of $\partial E/\partial n_\sigma$ will be bounded in the neighbourhood of $\mathbf{r}'$, so that the absolute value of the first term is less than

$$\left| \frac{\partial E}{\partial n_\sigma} \right|_{max} \oiint_{\sigma_s'} \frac{d\sigma}{\varepsilon} = \left| \frac{\partial E}{\partial n_\sigma} \right|_{max} \cdot 4\pi\varepsilon$$

This tends to zero as $\varepsilon \to 0$. The second term is less in absolute value than

$$k|E|_{max} 4\pi\varepsilon$$

which also tends to zero. For the third term, because $E(\tilde{\mathbf{r}})$ is regular at $\tilde{\mathbf{r}} = \mathbf{r}'$ (there is no source there), we may treat it as a constant for small ε and take it outside the integral. The last terms then give

$$-\frac{E(\mathbf{r}')}{\varepsilon^2} \oiint_{\sigma_s'} d\sigma = -4\pi E(\mathbf{r}')$$

as $\varepsilon \to 0$. Recalling that

$$\phi(\tilde{\mathbf{r}}) = E(\tilde{\mathbf{r}}) \quad \text{and} \quad \psi(\tilde{\mathbf{r}}) = e^{-ik|\mathbf{r}'-\tilde{\mathbf{r}}|}/|\mathbf{r}' - \tilde{\mathbf{r}}|$$

We have proved

$$\oiint_\sigma [\psi(\partial\phi/\partial n_\sigma) - \phi(\partial\psi/\partial n_\sigma)]\, d\tilde{\sigma} = 4\pi E(\mathbf{r}'),$$

which is Eq. (A.3).

Index

Printed in the United States
151738LV00005B/5/A

9 780471 872979